Gardens, Flowers, and Fruit
Proceedings of the Oxford Symposium on Food and Cookery 2024

Gardens, Flowers, and Fruit

Proceedings of the Oxford Symposium on Food and Cookery 2024

Edited by Mark McWilliams

SHEFFIELD UK   BRISTOL CT

The views expressed in this volume are those of the authors themselves and do not reflect the policy or positions of their affiliated institutions or of the Oxford Symposium on Food and Cookery.

Prospect Books is an imprint of Equinox Publishing Ltd.

UK: Office 415, The Workstation, 15 Paternoster Row, Sheffield, South Yorkshire S1 2BX

USA: ISD, 70 Enterprise Drive, Bristol, CT 06010

www.equinoxpub.com

First published 2025

© 2025 as a collection Equinox Publishing Ltd.
© 2025 in individual articles rests with the authors.

All rights reserved. No part of this publication may be reproduced or transmitted in any form or by any means, electronic or mechanical, including photocopying, recording or any information storage or retrieval system, without prior permission in writing from the publishers.

British Library Cataloguing-in-Publication Data

A catalogue record for this book is available from the British Library.

ISBN-13	978 1 80050 688 6	(paperback)
	978 1 80050 689 3	(ePDF)
	978 1 80050 715 9	(ePub)

Library of Congress Cataloging-in-Publication Data

Names: Oxford Symposium on Food & Cookery (2024 : Oxford, England) author | McWilliams, Mark editor
Title: Gardens, flowers, and fruit : proceedings of the Oxford Symposium on Food and Cookery 2024 / edited by Mark McWilliams.
Description: Bristol, CT : Prospect Books, 2025. | Includes bibliographical references. | Summary: "Gardens, Flowers, and Fruit includes selected papers from the 2024 Oxford Food Symposium. Grounded in a number of different disciplines, writers from around the globe consider the precariousness of our fragile lifestyles, celebrating the joys of the natural world while conscious of the significant stresses on our environment"-- Provided by publisher.
Identifiers: LCCN 2025006977 (print) | LCCN 2025006978 (ebook) | ISBN 9781800506886 paperback | ISBN 9781800506893 pdf | ISBN 9781800507159 epub
Subjects: LCSH: Gardens--Congresses | Flowers--Congresses | Fruit--Congresses | LCGFT: Conference papers and proceedings | Conference papers and proceedings
Classification: LCC TX406 .O94 2024 (print) | LCC TX406 (ebook) | DDC 635--dc23/eng/20250616
LC record available at https://lccn.loc.gov/2025006977
LC ebook record available at https://lccn.loc.gov/2025006978

Typeset by Scribe Inc.

Printed and bound by CPI Group (UK) Ltd, Croydon, CR0 4YY

Contents

Foreword   9
Mark McWilliams

1 Spoons in the Garden   11
Ken Albala

2 How Does Your Justice Grow: Gardens and Gardening in
Agatha Christie's Fiction   20
Tatiana Alekseeva

3 Floating Garden Geopoetics: More-than-Human Collaborations
in Xochimilca Foodways   30
Diego Astorga de Ita

4 The 'Apple Isle': Tasmanian Apples as a Weapon of Ecological
Colonization and Icon of Botanic Imperialism   43
Carla Baker

5 More Than Wine: Vineyards and Grapes in Medieval and Early
Modern Dutch Gardens   52
Mariëlla Beukers

6 Some Mediterranean Culinary Isoglosses: Perry's 'Olive Line' in the
Broader Context of the Umami-Enhancer Isogloss Bundle   61
Anthony F. Buccini

7 Cultivating Liberation: Gardens as Agents of Connection
and Empowerment in Havana, Cuba   70
Mallory Cerkleski

8 The Blight of #Cottagecore: Past and Present S/Place-Making
in the Garden and Home   79
Jolin Chan

9 Preserving Fruit at the Nexus of 'Lay' and 'Expert' Knowledge   89
Danille Elise Christensen

Contents

10 Syl Anagist Was a Garden: Food and Vitalism in *The Broken Earth*   102
Sara Clugage

11 Towards a Gastronomic Taxonomy for Fruits and Flowers: Beyond Linnaeus's Garden   113
Len Fisher and Anders Sandberg

12 Gardens and Ghosts: Arts of Living in a Contemporary Bulgarian Village   121
Lindsey Foltz

13 Roman Edens: Virgil's Tarentum and Pompeiian Garden Painting (Re)Considered   130
Christopher Grocock

14 Two Cherry Dishes from Medieval China: A Fruit of Ritual Reciprocity and Sensual Imagination   143
Zihan Guo

15 Growing Rice, Growing Taste   154
Chu Hao Pei

16 On Gardens, Flowers, and (Possibly) Fruit in Cheese   170
Ursula Heinzelmann

17 Postnatural Apple: Re-Defining Fruit and Re-Thinking Food Systems through Design   176
Leonie Hochstrasser and Katharina Mludek

18 Śiva's Flora   189
Soham Kacker and Deepa S. Reddy
with illustrations by Chippy Diac Vivekanandah

19 Ancestral Fruits: Strawberries, 'Strawberries' and Acorns in Roman Gardens and Texts   202
Erzsébet Kovács

20 The *Petit Fruit* that Could: How the *Camerise* Captivated Québec   210
Ivy Lerner-Frank

21 #mangowars: Why Are Some Varieties of Fruit More Popular than Others?   221
Priya Mani

Contents

22 Resilience of Indigenous Food Systems: A Study of Cacao Grown in *Chakras* in the Ecuadorian Amazon   233
Camila Marcías Álvarez

23 Doomsday Plots: The High-Stakes Gardens of America's Preppers   243
Rebecca D. Mazumdar

24 Orange Blossoms and the Holy Grail of Persian Jams   252
Nader Mehravari

25 If This Field Could Talk: Considering Soil as a Site of Hospitality and Diplomacy between Humans and Non-Humans   263
Jennie Moran

26 Common Gardens to Forbidden Forests: Food Procurement Pollution and Policy on Tokunoshima   270
Hanika Nakagawa

27 Seeding Resistance: Urban Food Forests and Indigenous Experiences   280
Lotta Ortheil

28 The Seven Sisters and South Downs: A Culinary Exploration of Southern England's Wild Edibles   291
Cordula C. Peters

29 The Fall of the Acadian Belliveau: Genetics, Genealogy, and Recovery of Pomological *Matrimoine*   301
Karen Pinchin and Simon Thibault

30 The Rebellious Heart: Food and Flowers in the Chinese Classical Garden   313
Wena Poon

31 Flowers and Fruit in Early Modern Mesoamerica: Chocolate as a Case Study   323
Kathryn E. Sampeck

32 Bringing the Kitchen Garden into the Kitchen: The Research and Writing of Three Inspirational Women   335
Barbara Segall

33 Welcome to Utopia! Hungry?   345
Laura Shapiro

Contents

34 Gardens, Markets, and Migrants 352
Jayeeta Sharma and Sarah Elton

35 *Le Potager du Roi*: The 'King's Kitchen Garden' at Versailles as Political Metaphor and Gastronomic Laboratory 363
Richard Warren Shepro

36 The Garden: Keynote Address 375
Carolyn Steel

37 Ayurvedic Renaissance: Exploring Fruits, Flowers, and Well-Being in Colonial Western India 399
Maithili Tagare

38 Opening the Garden Gate: Wild Gardens and Indigenous Culinary Knowledge 410
Rachel Thomas Tharmabalan and Jeremy Morell

39 English Commercial and Private Garden Production from the Sixteenth Century until the Coming of the Railways 420
Malcolm Thick

40 The Emperor, the Major, and the Jackfruit Tree Disaster 430
Marcia Zoladz

Foreword

What is a garden? A surprising amount of energy was devoted to that question at this year's Oxford Food Symposium, triggered by one symposiast's insistence that a pasture is a garden. We ended up agreeing, mostly, that any natural space shaped or tended by human hands could qualify, meaning that not just the royal gardens at Versailles and the raised bed out back count, but also urban gardens, forest gardens, probably that pasture, and certainly whole swaths of landscapes managed so skillfully by Indigenous peoples that their care could not be seen by European eyes.

Those Europeans, like people in Persia or China or any number of other places, were used to gardens with walls. Walls give gardens shape, borders, but they also exclude – indeed their intent is exclusion. That exclusion has ideology; it has economics and politics.[1] It allows the wealthy to exert control – think of the English Enclosure Acts – and then to measure themselves by that control. As Annalisa Marzano notes of Ancient Rome, these kinds of gardens become a kind of 'competitive display' of wealth and power.[2] That holds true for the rich in other times and places, but the control behind it is also sought by groups like modern-day preppers concerned only with their own survival.

That limited view will not do. We need gardens that are about inclusion, not exclusion. As Olivia Laing explains:

> We need gardens and the life they support established everywhere if we are to survive, and they must extend beyond the private realm, to form part of a cherished common wealth, while retaining their intimate and wayward qualities, where individual creativity can flourish.[3]

Now more than ever, in times of catastrophic climate, devastating conflicts, looming pandemics, and divisive politics, we need the shared spaces of gardens and the healthful practices of tending them.[4]

Gardens of all types – and especially the flowers and fruits grown in them – were the subject of this year's Oxford Food Symposium. Fittingly, our in-person gathering happened not in the familiar halls of Oxford's St Catherine's College, but in its gardens, in tents set up along the River Cherwell, and conversations began there continued over the two-week virtual conference. All in attendance shared Cristina Mazzoni's belief that 'a deepened understanding of what we see, hear, taste, touch, smell expands the intellectual, physical, and emotional appreciation'.[5] As the papers in this volume attest, those conversations were wide-ranging, rigorously informed, and generously nurturing.

While the events symposiasts share may be the life of our gatherings in-person and online, this book, alongside its predecessors, represents the real legacy of the Oxford Food Symposium. For the last several years, these volumes have been made possible by grants from Tokyo's RINRI Institute of Ethics, testifying to the Symposium's international community and its commitment to lifelong learning. That support matters: in print and online, the papers in this volume will reach legions more readers than those who first heard them presented. I thank those who helped see it to press, including Cathy Kaufman, Richard Shepro, David Matchett, Naomi Daguid, Carolyn Steel, Elisabeth Luard, Janet Beizer, and Jake Tilson, along with Equinox Publishing's Brendan King, Sarah Lee, and Janet Joyce. The real appreciation goes, of course, to the authors who, though cultivating their own gardens, have been generous enough to share the produce with all of us.

Mark McWilliams
Editor, Oxford Symposium on Food and Cookery

About the Editor

Mark McWilliams, Professor of English at the United States Naval Academy, has served as the Editor of the Oxford Food Symposium since 2011. The views expressed in this volume are those of the author(s) and do not reflect the official policy or position of the US Naval Academy, Department of the Navy, the Department of Defense, or the US Government.

Notes

1 This line of thought is inspired by Olivia Laing, who insists that, 'while the spell of the garden does lie in its suspension, its seeming separation, from the larger world, the idea that it exists outside of history or politics is not a possibility' (*The Garden Against Time: In Search of a Common Paradise* (Picador, 2024), p. 12).
2 Annalisa Marzano, 'Rome's Horticultural "Revolution": Ideology, Display, and the Economy', Oxford Food Symposium, 6 July 2024 <https://www.youtube.com/watch?v=eRt4Rxgyur0> [accessed 15 May 2025].
3 Laing, p. 284.
4 See e.g. Sue Stuart-Smith, *The Well-Gardened Mind: The Restorative Power of Nature* (Scribner, 2021).
5 Cristina Mazzoni, 'Love, Magic, and the Bittersweet Flavor of Oranges in Italian Culture', Oxford Food Symposium, 7 July 2024 <https://www.youtube.com/watch?v=LmwdmrdaM7o> [accessed 15 May 2025].

1
Spoons in the Garden

Ken Albala

If you have ever strolled through a garden with food-obsessed people, you will notice that they find it impossible not to think of the space in gastronomic terms. 'That rosemary would be so nice with lamb.' 'I could make such a great strudel with those apples.' I have always been such a person. My own garden in California is mostly edible. Olive trees bear fruit cured in brine, grapes festooning the fence are eaten or pressed into wine, bay leaves go in the stew pot, the bamboo is sliced and stir-fried, Meyer lemons get pickled or squeezed into cocktails. A few years ago, however, I began to see gardens in a completely different light, for their potential as utensils and vessels.

It began with a routine pruning of my olive trees (Scylla and Charybdis) in spring 2002 and a large branch that I thought might make a nice spoon. I had no idea how to do that at first, and it took quite some time to attain skill, build up callouses and physical strength, buy the right tools, and simply log enough hours carving to understand how wood works. I sourced everything locally from my own garden and from trimmings found in the nearby park, college campus, or curbside. This was the aesthetic equivalent of locavorism, and it seemed to have an environmental angle as well – just think if everyone used a handmade spoon they really like and never touched plastic utensils again. I didn't exactly know what compelled me to carve, but it became completely engrossing.

Figure 1. A hand holding an impossibly wide fork by Martin Kullik and Jouw Wijnsma, in a presentation at the Oxford Symposium in 2023.

Then, a moment of revelation occurred while listening to a plenary session at the 2023 Oxford Symposium on Food and Cookery given by two designers, Martin Kullik and Jouw Wijnsma. They detailed highly unusual cutlery they designed and used in formal dinners, including a spoon with holes in it, a fork with tines bent backwards, a rake-like fork with forty tines, so wide it could never fit in your mouth, and spoons whose handles were so long they were impossible to use. The obvious intention was to make people think about common objects taken for granted and carefully reconsider how we use them. In pictures of these meals, diners were clearly fooling around and having fun as they attempted to use the cutlery, but I don't think anyone would want to take these home to use on a regular basis. Hilariously, though, they said people sometimes did steal the cutlery – I think to play gags on their friends and family.

My question is not whether this impossible cutlery constituted art or even a serious gastronomic experience – I believe it did – but rather whether this actually made diners reconsider their forks and knives, reevaluate how and why we use cutlery as we do and whether they are ideally suited to enhance the entire gastronomic experience. Did the shock value extend beyond the short meal and perhaps the stories diners told about it?

It then occurred to me that I had been engaged in a comparable project whose intent is to inspire people to reexamine the common objects they use at the table, but from a completely opposite perspective from messing with people's heads. Rather than confuse and perhaps annoy diners, I want them to experience utensils as a seamless extension of the body, to use them every day rather than in one performative event. I want to make spoons and bowls to match specific foods, so they are perfectly suited to each other not just physically, but in terms of mood and the ideas they evoke. In other words, I want diners to be comforted by the utensils, so they care about them, and want to take them home to use. What better way than to source the wood from locales people know intimately and care about deeply, such as gardens which evoke specific emotions?

Moreover, I believe that physical objects can in a sense be alive, not only because they are made from materials occurring in nature, but because of the attachments people have to them and the meanings people invest in them. This makes them revered rather than tossed in a trash heap after use. It is a craft aesthetic that resonates with Japanese Mingei and the Arts and Crafts Movement – that everyday objects should be those you know to be useful and believe to be beautiful, to paraphrase William Morris.

So I carved, and carved, with the intent of gaining mastery through sheer persistence. I watched a lot of videos on YouTube, posted my spoons on social media, gave them away to friends, even gave talks at academic conferences. After about two years and over three hundred spoons, I was at a party and randomly showed someone I had just met a spoon that happened to be in my pocket. Within a few days I was asked to do an exhibit which ran in the spring of 2024.

My ultimate intention is to write a gastronomic manifesto about spoons, pottery (which I've been doing for decades), and cooking, in an effort to show how the three might be harmonized. I'm not entirely certain what that means yet, but I can share some of my general thoughts here.

Beauty and the Graceful Line
It sounds a little silly nowadays to be talking about Beauty with a capital B in universal terms. We are so used to setting everything in historical context, being careful to understand the culture that produced an object without our own blinkered prejudices. We do this for good reason: it's difficult to escape the way our own culture shapes our

judgement, since we are so embedded in our own time and place and so swayed by the forces of fashion. We are hesitant to judge at all after so many centuries of revering the great masters enshrined in the canons of taste. Swinging in the complete opposite direction, the Romantics insisted that only the direct imitation of nature, even if gruesome, can serve as an objective standard, that which evokes emotion because it strikes us at the core without the intervention of intellectual rationale, or taste – which is just a byword for status. This was the complaint the energetic Baroque had with the cool intellectualism of sixteenth-century Mannerism, and how nineteenth-century Romantics rejected the Enlightenment, by turning to nature.

Nature does indeed offer lessons for craft, if not precisely in the way a painter approaches a landscape or a human body. We find objects beautiful precisely because they are well fitted to carry out specific functions. In nature, a flower is beautiful because it offers colour, aroma, and an inviting form to the bee, who finds it beautiful and spreads pollen for the plant. We find it beautiful for the same reason, though of course with added cultural baggage.

So too is there beauty in objects. It is not so simple that form should follow function as architect Louis Sullivan insisted, but that certain materials dictate certain shapes, or they simply collapse or refuse to work. A teapot may be gorgeous but if it doesn't pour, it's not a teapot – I've made a few of those. It may be art, but that's an entirely different topic. Here I am concerned with the relationship between objective standards of beauty and the utility of everyday objects. If we simply say everything is beautiful, then there are no standards and we are subjected to hideousness and filth.

What then constitutes Beauty? Let's begin with the line. They come curved, straight, bent, twisted, and curled. Each exists in nature and for quite specific reasons. In my own backyard garden there are grape vines that stretch out and tightly grab anything with little curled tendrils that support the plant's desire to stretch outward. The leaves thus achieve maximum exposure to the sun, and the result is sensuous curvaceous vines. The bamboo has an entirely different strategy: it bursts straight upward as fast as it can, in a week or two growing fifty or sixty feet, and then pops out leaves so high up that nothing can shade them. The higher it goes, the thicker the stem needs to be, and some can be a foot in diameter. Each is beautiful in its own unique way.

Leonardo da Vinci appreciated this. He drew the swirls of the horsehair equisetum plant and wondered why the whorls were similar to the eddies in gushing rivers and the flow of blood coursing through our bodies. I think he was considering precisely what physical properties can be considered universally Beautiful. Skeptics of course will point out that appreciating such forms is a cultural value, that even nature itself is culturally constructed. What one people find beautiful is strange or meaningless to another. The Enlightenment *philosophe* Voltaire quipped about this in his *Philosophical Dictionary* when he asked what is the *To Kalon*, or ideal form of beauty. To a frog it's a green complexion and bulging eyes. He was calling for a relativistic approach, to each his own. But Voltaire entirely misses the point: frogs are eminently beautiful – their slippery skin well suited for ponds, their bulging legs perfectly designed for jumping, their eyes ideally designed for circumspection.

How does this relate to human-made objects? Let's turn to pottery. If you lift the walls of a clay vessel, either wheel thrown or hand built, straight upward, eventually the form will twist and buckle and then probably collapse. To prevent this you have to make the walls so thick that the vessel is blunt and too heavy to use. But if you give the pot a shoulder or hip (ceramic forms are often anthropomorphized) so it bulges toward the top or bottom, the pot gains structural integrity. But it also becomes beautiful – the line becomes sensuous, much like the bend of a spoon – and it becomes stronger, lighter and more functional. An acute angle in a pot wall or spoon compromises its use. A bent spoon will spill soup in your lap. A pot that flares out at the top too sharply will flop down through the force of gravity. The 'well wrought urn' is not curved because that's the way the Greeks did it, but the Greeks did it that way because they had to. So did the Chinese masters of the Song Dynasty and the potters of Seagrove, South Carolina. The vase turns out to be a great example of Beauty.

In many cases the material dictates the form, as in architecture we build square rooms not because we are square but because straight wooden beams hold the structure together. For similar reasons our furniture is square or rectangular, not just our book shelves but even the books to fit in them. Likewise a round cooking vessel fits over a round gas jet or electric hob both of which replicate the shape of a natural fire, but there is no inherent reason why we need to continue using these shapes. For the moment though, their functionality is directly connected to their aesthetic value.

I hear your objections. Aren't there awkward angles in pots, and forms intended to disturb or shock the viewer/user? Obviously in museum art, but I'm not so sure about utilitarian wares. Think of food: it must be palatable, meaning it can't contain shards of glass or metal, and it can't be laced with arsenic, or it's no longer food. But don't some cultures adore foods that others abhor? Yes, but anyone can learn to overcome those culturally learned preferences, just as they can learn to like coffee or whiskey, or unlearn our instinctive predilection for sweetness. I mean, a baby would happily eat a worm if offered by a parent. Does that mean that there are some foods that are universally Beautiful for all cultures and all times in history? Of course there are, just as there are dreadful foods people learn to love – I'll take a marshmallow peep or gummy candy any day.

I argue that certain flavours, combinations of ingredients, cooking and preserving methods are universally tasty. Maybe some don't like too much heat, or are genetically averse to cilantro, but think of that innate preference for sugar. Or think of how smoky flavours are hardwired into our limbic brain after millennia of cooking over fire. Or even take a food spread around the world – like pasta, which is universally Beautiful. If anyone tells you they don't like pasta, they are either on a low-carb diet or lying. I would say the same for an elegantly sloping bowl: just holding it in your hands gives instinctive pleasure, as does the graceful arc of a well carved spoon. These transcend all time, geography, and culture prejudice.

You could equally argue, and this is perhaps the best example of what I mean, that flowers are universally Beautiful, transcending all time and space – precisely because they are so configured, coloured and disposed to attract pollinators and people. And they get us to plant them in gardens everywhere, as Michael Pollan would put it.

Matching Food with Vessels and Utensils

What exactly does *matching* mean in the context of food, clay and wood? Is it like clothes – combining the same colour pants and jacket or complimentary patterns with accessories? It can be, but I think we risk oversimplification if we try to capture a particular style or copy a historical example from the past. I can see why that would be a fun exercise, like designing an art deco set for a movie or recreating a nineteenth-century Thai banquet. It would take enormous skill to master a single style, let alone many. That's really not what I'm after. I want you to make objects that fit your own

Figure 2. Handmade placesetting: shagbark elm spoon poised over a brown stoneware bowl

time and place, things you will want to use every day. They should be made of materials at hand, with your own equipment and the time you have to devote. Think of the recipes you make time for in the kitchen, here that's extended to the bowls and spoons. Your favourite dishes will be different from mine and so will your utensils and vessels, so I don't intend to prescribe exactly how your creativity might be expressed. For example, I make handles on mugs in a particular, immediately recognizable way because my hand is used to that particular curve. I am also learning what kind of spoons fit best in my mouth, but they probably won't be the same as yours. I am inordinately fond of green but can't stand light blue. Your preferences will probably be very different from mine.

So then what does matching mean? It is finding the colour, form, figure – as in curves and lines, and most importantly, the mood that works for a particular season, occasion, setting. So in winter we generally like warmer colours, rounder and more capacious vessels, and food that is filling and comforting. This calls for a heavier and perhaps darker spoon, and vegetables that grow in cold weather. We do this kind of matching instinctively. A bright floral pattern and an evanescent meal is disappointing in winter – though sometimes intentionally flouting rules is itself fun.

Even this is oversimplification, though. Sometimes you will just feel that the parts go together, based on your mood, your own past experiences, memories, and preferences. Sometimes a nice bright celadon-glazed bowl and a light beechwood spoon contrasts nicely with a heavy soup. And sometimes you do want a big heavy meal in the summer. Isn't barbecue just that? That's to say, there are no hard and fast rules. There is definitely some logic though. Just like a heavy coat in summer makes you uncomfortable, so too would certain foods and colours. Just think of food, pots and utensils, and being in season and then take it from there. If a dish is bright and cheery with high acid notes, you want the rest to go along. If your meal is serious, festive, luxurious, extravagant, try to get everything to match. You might not be able to make all the wares yourself. For example, I really love majolica, but it's such a different clay and technique than what

I normally do, that I have a few bowls and plates when I'm feeling particularly Mediterranean. I have formal white dishes, glass stemware, and even metal cutlery because sometimes they are the best way to express what I've cooked. All that is to say, you don't have to only use the things you make – that would be as silly as saying you can only eat the food you cook yourself. But adding your own art to the mix certainly makes it more lovely. And when you can sit down to a meal where all the parts come from your hands, it can be truly magical.

My Sycamore

A grand sycamore stood sentinel by the sidewalk at the front lawn of the colonial revival house where I grew up in central New Jersey. The entire Levitt-planned community was planted with this and a handful of other species, and they must have been fairly large in 1967 because I could easily climb them a decade later. But I would hang out in this one particular sycamore. It had low limbs, evenly spaced, and was very easy to hoist up and into. I went as high as I possibly could, hugged the main trunk and let it sway in the breeze as I twisted along with it. My neighbours claimed that I would sing opera up there, which may be true. Eventually I nailed a plank of wood into the main V-section about twenty feet up as a seat. It's still there, the wood grew around it, and forty-five years later is almost completely engulfed. The lowest limb is now maybe forty feet up; it's a massive tree today.

What I liked most about my tree was its resonance. You could rap any limb, with your ear up against the bark and it would offer distinct mellifluous tones. You couldn't get a conventional scale out of it, but it was definitely worth slapping and knocking. I like to imagine that the tree knew me, responded to my moods, maybe even hugged me, as I definitely did its limbs. I came to know that strange mottled bark so intimately and even saved pieces as they peeled off.

After college, when I moved home between degrees, I spent more time in the tree. I wrote up there. I even threw parties in it. At one point there were six or seven people in the tree, drinking cocktails hoisted up in Ziplock bags and eating snacks in little bowls. I started dating the woman who is now my wife right up in that tree. I went away of course, and eventually so did my parents, literally. I still visit the tree though, thousands of miles from the other coast where I live now. The last time I saw it was long before I started carving, and a limb fell off. I took a few tiny pieces. What I would give now for that whole limb. Sycamore, as it turns out, is utterly gorgeous wood, with a clear set of rings, but also tiny vertical

Figure 3. Toyon (Hollywood) spoon and fork resting in a multicoloured stoneware bowl.

lines cutting across them that create a gorgeous pattern in carved wood. Even the smell of that tree I think I could recognize blindfolded. Apparently you can boil down the sap into syrup too.

I say all this because I think people can have very intimate relationships with other living beings that don't react like pets, but in a different, primordial way. When I encounter a huge old copper beach or a valley oak where I live now, it's more awe inspiring and intense than seeing a large animal. Just knowing they were alive hundreds of years ago thrills me. In the case of the towering redwoods in the California foothills, they were around at time of Plato. They stand as living witnesses to the past – if you are willing to listen.

Obviously I have a deep empathy for wood. So it felt strange the first time I carved it. Some trees have deep pink striated flesh that resembles tuna or rare beef. Sometimes wood will be so fresh and wet that it seems to bleed, or practically splashes with the fall of the axe. This isn't a bad feeling. It's no different than butchering an animal. I've never killed a large mammal but taking one apart I truly enjoy, cutting around the bones, following the sinews connecting to muscle tissue. The same is true of wood and its own internal structure. When a limb falls down in a storm, I'm happy to make use of it. Or when a tree needs pruning.

All this is to say that a tree you know well, have harvested fruit from, see every day perhaps, or best of all planted with your own hands, is one whose wood you will respect and cut with care. Even those in my neighbourhood, I treat with a certain reverence, knowing full well that a branch will end up chopped and used as compost. It gives me a certain satisfaction knowing something else useful might be done with it. At times I've been tempted to knock on people's doors, not to ask if I could take some wood, but to say I already had, and please accept this spoon, thanks so much. (Whenever anyone gives me wood, I do actually make a spoon for them.)

Figure 4. Sycamore with a seat inserted between the limbs in Manalapan, NJ, USA, circa 1975, where the author was known to sing opera as a youth.

Confidence

It is very easy to insist that you need confidence to do a task, whether it be cooking, throwing, carving, or basically anything else in life. The self-doubt, hesitancy, rethinking actually takes far more energy than just doing it. It is self-defeating. If only there were a simple way to switch off your superego and let the id fly – but there are a few tactics to achieve exactly that.

First, you must be willing to make mistakes boldly. People often say never try out a new dish on company – but that's exactly when I like to do it. It might not be perfect, but you probably won't mess up so badly that it's inedible. Having company means you're upping the stakes, you have to get down to business, the doorbell will ring, so there's no looking back. It forces you to act quickly. Urgency is a wonderful way to shed inhibition.

I was once being filmed for a video series on historic cooking and we had ten days to shoot twenty-eight episodes ranging from ancient times to the twentieth century. The plan was to do three episodes per day, with three recipes in each episode. Some were seriously difficult. The cameraman thought this was insane. And I will readily admit it was among the most physically demanding things I have ever done in this life. The only reason it worked was because I didn't have the leisure to hem and haw, or really think twice about anything. I had to do it without flinching, sweat pouring down my brow in the blazing summer heat with the burners on the stove cranked up, washing pans until my hands were literally raw. Perhaps by sheer luck, only two dishes failed – a sixteenth-century chickpea fritter from Bartolomeo Scappi, which fell apart because I tried to cut a corner rather than trust the author's directions. Always a mistake. The other was a medieval beer for which I sprouted the barley, malted it, ground it, fermented it, etc. It was turbid, sour, and flecked with bran. It almost worked, but didn't taste much like beer. It didn't matter in the end, considering everything else worked fine and in many cases was spectacular. I mean so good that the crew and I devoured the food with giddy glee.

Barring the sink or swim approach, there are a few other tactics that will help you face difficult and dangerous tasks. For example – using a sharp knife with confidence. First, you absolutely must respect your tools. Become intimately acquainted with them. Hold them often and practice until the blade feels like an extension of your own hand. I know this is much more difficult than it sounds. Sometimes, for kicks, I will try to chop vegetables with my left hand. It is almost impossible, but with a little practice, it can be done, just like you can learn to write with your non-dominant hand. I have a Japanese pottery wheel that goes counterclockwise, and that's a really interesting exercise. It forces you to rethink exactly what you're doing, and then when you go back to the direction you're used to, you suddenly gain a greater appreciation for what has probably become pure rote muscle memory. It helps refine your technique, adjust to become more efficient, and in the end be more confident. Any intentional challenge you throw in your way will ultimately make you more nimble-footed.

You must never think whether you can do something, just how you will do it. Whether it's stretching filo dough by hand, throwing twenty pounds of clay on a wheel, or chopping through what seems an enormous width of tree trunk – having the 'courage

Figure 5. A mad collection of spoons resting on the windowsill, Stockton, CA, USA, 2024.

of your convictions', as Julia Child said, is essential. I hear you ask, what if you really have no skill whatsoever? I can say that no matter how hard I try I will not be able to solve a problem in calculus. Nor can I speak Japanese, however much I want to. But if the desire is there, small steps most certainly will add up eventually from simple equations or sentences to increasingly more complex ones and ultimately with persistence mastery. There is nothing worse than thinking you can never do a task simply because you haven't tried. All this is merely to say, if you set realistic goals for yourself and put in the time, confidence is the natural and inevitable outcome.

As with anything, getting good at something can only happen with maniacal repetition – only in madness can there be method.

About the Author

Ken Albala is Tully Knoles Endowed Professor of History at the University of the Pacific in Stockton, California. He has published twenty-nine books and won the 2023 Outstanding Faculty Award at the university.

2
How Does Your Justice Grow
Gardens and Gardening in Agatha Christie's Fiction

Tatiana Alekseeva

'But your gardens, you English, I admire. I sit at your feet! The Latin races, they like the formal garden, the gardens of the château, the Château of Versailles in miniature, and also of course they invented the *potager*. Very important, the *potager*. Here in England you have the *potager*, but you got it from France and you do not love your *potager* as much as you love your flowers. Hein? That is so?'

—Agatha Christie, *Third Girl*[1]

The first childhood memory in Agatha Christie's autobiography is a tea party in the garden marking the author's third birthday – images of cakes, sugar icing, and candles intertwine with ruminations on the exceptional role of this particular garden in the universe of Agatha's childhood.[2] Mirroring the beginning, *An Autobiography* ends with Ashfield's garden revisited: there is nothing left of it physically, but it remains untouched in the author's memory, being the very garden she keeps seeing in her dreams.[3] Between these two episodes, many other gardens constituted an intrinsic part of Agatha Christie's personal life.

Not surprisingly, gardens make abundant appearances in her fiction, often marked as something quintessentially English. John Hammett in 'The Augean Stables' (1940) represents 'every quality which [is] dear to Englishmen', such as 'his fondness for gardening'.[4] In *Evil Under the Sun* (1941), Rosamund Darnley claims that she had 'a very English childhood', with an inevitable 'neglected garden'.[5] A living image of such a childhood, with an obvious connection to Christie's personal experience, is Angela Warren in *Five Little Pigs* (1942), whose usual occupation at the age of fifteen is 'wandering about the garden, climbing trees and eating things [. . .]. Plums, sour apples, hard pears, etc.'[6] There is also Christie's famous detective, Hercule Poirot, who regularly draws the reader's attention to the phenomenon of the English garden from a foreigner's point of view (as in the epigraph above).

Generally, a neat 'old-world' garden is an image widely associated with Christie's mysteries – it is a familiar background and a convenient topic for conversation.[7] However, its role is not confined to these functions. This paper aims to examine the subject from three different perspectives: social history, plot development, and general symbolism. It is also worth noting that, while the portrayal of gardens in Christie's fiction is a multifaceted phenomenon, this paper aims to focus on food; consequently, some aspects (such as the language of flowers, garden fashion, etc.) are consciously omitted.

Gardens and Social History

Christie's texts bear witness to the changes in the social role of gardens that occurred in the twentieth century. Whereas three main elements of an average country garden stay present – in Christie's books, there is usually a kitchen garden, a lawn, and an ornamental component (flower beds, herbaceous border, rockery garden, etc.) – their importance for a household changes due to historical events. Apart from this, nostalgia is a considerable factor: the author is not an impartial witness, and the way different aspects of garden culture are highlighted in her books is affected by the mere passing of time.

Garden-Hosted Gatherings

'The passion of the English for sitting out of doors' is a source of great annoyance for Hercule Poirot: one is not safe from 'this whimsy' even by the end of September.[8] The 'sitting out of doors' is part of what Christopher Yiannitsaros calls 'Englishness' – a set of stereotypical images thought to constitute national identity and often associated with Christie's fiction.[9] Indeed, when writing of summer events, Christie usually portrays a garden as a suitable place for various social gatherings, even if there is just 'one fine hot week in the summer' and 'one day that the Sunday papers publish articles on How to Keep Cool, How to Have Cool Suppers and How to Make Cool Drinks'.[10]

Firstly, there are regular family meals, such as afternoon tea. Christie's first detective novel, *The Mysterious Affair at Styles* (1920), opens with a scene of afternoon tea served under the shade of a large sycamore, since it is 'too fine a day to be cooped up in the house'. Characters position themselves in basket chairs or directly on the grass, tea is distributed, and a plate of sandwiches is handed around. It is a usual practice at Styles that repeats later in the book.[11]

In *Five Little Pigs*, one of the characters recalls the events that took place sixteen years earlier (i.e. in the 1920s) and speaks of a similar scene, with a big cedar tree as a natural cover.[12] It echoes Christie's personal practices. In 1934, considering the purchase of her future Wallingford house, she carefully assessed its grounds. Noticing 'a particularly fine cedar tree', she immediately pictured how on hot summer days one might have tea under it.[13]

Another practice suitable for summer days is to serve tea on a terrace overlooking a garden, as shown in *The Lemesurier Inheritance* (1923) and 'The Cretan Bull' (1939).[14] However, many of Christie's mysteries draw an entirely different picture, with tea served inside the house regardless of the time of the year. Nonetheless, this meal is often followed or preceded by a tour of the garden (e.g. *Dumb Witness*, 1937).[15]

Secondly, gardens may accommodate more elaborate parties. In 1908, at the dawn of Christie's era, *The Tatler* magazine claimed that garden parties had been 'a favourite form of hospitality' for almost two centuries.[16] In 1999, outdoor parties still could be referred to as 'one of the most delightful ways of entertaining', although they obviously did not stay the same.[17] Agatha Christie had noticed the transformation.

Gardens, Flowers, and Fruit

In the ending of *Nemesis* (1971), we briefly see Miss Marple through the eyes of a side character as a 'young and pretty girl shaking hands with the vicar at a garden party in the country'.[18] This fleeting vision is tinted with nostalgia, probably emerging from the author herself. The ghost of the young Miss Marple belongs to the past, as do the pre-1914 garden parties of Christie's youth. In her memories, these parties remain unparalleled, with food being a foundation of their glory: 'There were lovely ices – strawberry, vanilla, pistachio, orange-water and raspberry-water was the usual selection – with every kind of cream cake, of sandwich, of eclair, and peaches, muscat grapes, and nectarines.'[19]

The garden party in *Peril at End House* (1932) seems to be the closest to Christie's own juvenile recollections. This novel is set in St. Loo, a fictional resort town allegedly based on Torquay, Christie's hometown.[20] The young Magdala 'Nick' Buckley hosts a party at End House on Regatta week and invites guests to watch fireworks from her garden. Dinner is served indoors at eight o'clock, but most of the guests are interested only in the outdoor activities and arrive later: fireworks begin at nine-thirty.[21] Hastings remarks that Nick 'made everyone welcome in an old-fashioned way', not going into details.[22] Later, Christie describes in her autobiography the joy of being invited to watch the Regatta fireworks from the vantage-point of someone's garden in Torquay: 'It was a nine o'clock party, with lemonade, ices and biscuits handed round.'[23]

Thirdly, there are garden fetes. A fete (or *fête*) is a large-scale garden party organized specifically to raise money for charity. In H.R. Humphries's *Fund Raising for Small Charities and Organisations* (1972), garden fetes are promoted as 'good old favourite[s]', both pleasant and profitable. Humphries gives instructions on how to organize a successful garden fete, interestingly corresponding to Christie's fictional fetes.[24]

The most detailed description of such a fete can be found in *Dead Man's Folly* (1956). The setting, Nasse House, bears a strong resemblance to Greenway, Christie's famous Devon estate with its woodland grounds. According to Poirot, the fete proceeds 'in the normal fashion of fêtes'. There is an affordable admission charge, and a minor film star opens the festivities. There is a coconut shy, a skittle alley, and a hoop-la; then, of course, 'various "stalls" displaying local produce of fruit, vegetables, jams and cakes'.[25] Traditional raffles are offered: of cakes, of baskets of fruit, and even of a pig. A Treasure Hunt is another fundraising activity prescribed by Humphries, but in this case, Ariadne Oliver turns it into a Murder Hunt.[26]

An inevitable point of attraction at any fete is a tea tent erected on the lawn. Everyone is supposed to visit it at some point, contributing to the general commotion and stuffiness inside it, and the fete in *Dead Man's Folly* is no exception to this rule.[27] In *The Mirror Crack'd from Side to Side* (1961), a garden fete for the benefit of the St John Ambulance Association is organized at Gossington Hall, recently bought by a famous actress. The tea tent is very crowded, and Mrs Bantry, the former owner of the property, can only hope that there would be enough buns for everyone.[28] The church fete in *The Pale Horse* (1961) also demonstrates 'pandemonium at tea-time' when every patron wants 'to invade the marquee and partake of it simultaneously'.[29]

Consequences of the Second World War

Christie's mysteries set in the interwar period focus mostly on the decorative and recreational functions of gardens: we see characters admiring flowers or having tea on the lawn. The Second World War shattered the established order and made it impossible to keep up appearances. In *Nemesis*, Anthea explains that one could not get any gardeners during the war.[30] The situation did not improve much in the following years: Mrs Marchmont in *Taken at the Flood* (1948) still complains of 'a terrible shortage' of labour in this field.[31] It is much worse than during the First World War when Styles was 'quite a war household', setting 'an example of economy' by reducing the number of gardeners from five to three.[32] In the interwar period, there seems to be no shortage of workforce whatsoever.[33] In comparison with it, the consequences of the Second World War look truly dramatic.

A neglected and 'terribly overgrown' garden becomes a common image that keeps reappearing in Christie's post-war books.[34] In *They Do It with Mirrors* (1952), Miss Marple's first impressions of Stonygates are the badly kept driveway and neglected grounds. Gina admits that it started with 'no gardeners during the war', but later they simply 'haven't bothered'.[35] In *A Murder Is Announced* (1950), Dayas Hall has 'certainly suffered during the war years', and now 'groundsel, bindweed and other garden pests' are showing 'every sign of vigorous growth'.[36] In *4.50 From Paddington* (1957), the grounds of Rutherford Hall are similarly 'overgrown with weeds'.[37]

When resources are limited, one must prioritize. This is where the practical side of gardening comes into the limelight: the neglected gardens in Christie's post-war mysteries are primarily dedicated to growing edible plants. The examples from the previous paragraph have that in common: at Stonygates, the kitchen garden is the only part of the grounds that is 'prosperous and well stocked' because it has 'a utility value'; at Dayas Hall, part of the kitchen garden bears 'evidence of having been reduced to discipline'; at Rutherford Hall, the kitchen garden is 'sketchily cultivated with a few vegetables', though the hothouses are in ruins.[38]

As Ina Zweiniger-Bargielowska writes, fresh fruit and vegetables became luxuries for many households in the post-war years. These food categories were never rationed, and their scarcity made them hard to obtain: they easily disappeared under the counter. There were 'large-scale black market dealings in fruit and vegetables'.[39] Christie's books demonstrate the value of a vegetable garden in such circumstances. *A Murder Is Announced* also shows how it may be an asset in bargaining with neighbours: one could exchange a vegetable marrow for a pot of honey, or lettuces for skimmed milk.[40]

The Old Against the New

This 'shift in power' from flowers to vegetables corresponds to the tastes of a typical gardener from Christie's books. In *The Mirror Crack'd from Side to Side*, Miss Marple is restrained from active gardening by her doctor and relies on the mercies of old Laycock. This gardener does his best 'according to his lights', which usually includes growing cabbages and sprouts in 'inordinate quantities', accompanied by 'a nice Savoy, or a bit of curly kale'. Laycock considers flowers 'fancy stuff such as ladies liked to go in for, having nothing better to do with their time'.[41] Same is the old gardener in *A Murder Is Announced* whose biggest achievement is to 'put in a few cabbage plants'.[42] Ironically, such gardeners are usually not to be trusted with harvesting what they have grown since 'they hate giving you anything young and tender – they wait for them to be fine specimens'.[43]

What they excel at is drinking tea. Old Foster in *Sleeping Murder* (1976) 'has about five cups of tea a day', and Giles comments that 'talk and tea is his speciality'.[44] Old Edwards in *4.50 From Paddington* is doing 'lots of cups of tea and so much pottering – not any real work'.[45] The aforementioned Laycock is also having 'a great many cups of tea, sweet and strong, as an encouragement to effort'.[46]

Noteworthy, all these characters have 'old' added to their names. By the end of Christie's era, they are relics likely to become extinct. The new type of gardener is someone who is hired and sent to you by a big gardening firm – a practice mentioned in *The Mirror Crack'd from Side to Side* and *Nemesis*.[47]

Another novelty is the idea of buying vegetables instead of growing them. In *Taken at the Flood*, Mrs Marchmont's unrealistic hopes of finding a second gardener are contradicted by Lynn's remark that they could buy all the vegetables they need 'for a good deal less than another three pounds a week'.[48] In *The Mirror Crack'd from Side to Side*, Ella Zielinsky expresses her surprise at the English practice to grow one's own vegetables – she presumes that it would be easier to buy them at the supermarket. Mrs Bantry admits that 'it's probably coming to that', although 'they don't taste the same'.[49]

At the same time, there are those with an entrepreneurial mindset considering growing something for the market, although in Christie's books this idea usually does not succeed. Agatha Christie knew the hardships of market gardening from her own post-war experience at Greenway: she turned towards it as a means of maintaining an expensive estate, and the key, she discovered, was to find 'a first-class gardener' – never an easy task.[50] Not surprisingly, when Major Summerhayers and his wife in *Mrs McGinty's Dead* (1952) make attempts at market gardening, they fail since they have 'not an idea of the commercial life'.[51] In *4.50 From Paddington*, Lucy reflects on the possibility of growing mushrooms for the market in the deserted boiler room, but on her part, it is only an excuse for snooping.[52] In 'Strange Jest' (1941), the whole market-gardening idea is a joke referring to the treasure hunt and respective digging.[53]

Generally, the universe of Christie's mysteries is quite a conservative territory. A feeling of nostalgia haunts many of her late works as she reflects on the passage of time. However, Christie appears to embrace inevitable changes and does not see them as something tragic. The real tragedy is to be trapped in the past, like Anthea in *Nemesis* for whom 'nothing is like it used to be – it's all spoilt'.[54] As Miss Marple says in *At Bertram's Hotel* (1965), 'one can never go back [. . .] one should not ever try to go back [. . .] the essence of life is going forward'.[55]

Gardens and Plot Development

The role of gardens in Christie's books is not confined to reflecting social phenomena and constituting a picturesque backdrop – they also serve as an instrument in creating a strong storyline. This function can be divided into several areas.

Firstly, on the basic level, paying attention to gardening details can help a detective to solve a mystery. In 'Ingots of Gold' (1928), Miss Marple reveals a false gardener because 'gardeners don't work on Whit Monday', and in *Greenshaw's Folly* (1956), she only has to look at the contents of a garden basket to say that a person who did the weeding is not a gardener: they have 'pulled up plants as well as weeds'.[56] Hercule Poirot, by contrast, does not require any specific knowledge to find clues in a garden – he relies on his famous love for order and symmetry. In *How Does Your Garden Grow?* (1935), his revelation is based on the fact that the decoration of a flower bed is incomplete.[57] In *The Mysterious Affair at Styles*, he admires the symmetry of flower beds, meanwhile noticing that gardeners must have entered the house to witness the new will.[58]

Secondly, a garden can be a source of physical danger. This location is supposed to be calm and peaceful, but it is not to be trusted. For instance, weed-killer was invented to maintain order here – but it can be turned into a murderous weapon at any moment. So widespread is this practice that for the pathologist in *A Pocket Full of Rye* (1953), 'taxine is a real treat' – he is tired of the 'inevitable weed-killer'.[59] However, taxine, too, comes from the garden (yew berries from the hedges).

Apart from this, a garden is a place where one can hide something related to a crime – be it a piece of evidence (*Hercule Poirot's Christmas, How Does Your Garden Grow?*) or a body (*Nemesis, Sleeping Murder*).[60] Moreover, a victim can be murdered right here, in this picturesque and calm place, like Amyas Crale in *Five Little Pigs*.[61]

Consequently, and thirdly, a garden is a place of metaphysical ambiguity, being both soothing and dangerous. The recurrent metaphor is the Garden of Eden: even there, 'there had been a shadow – the shadow of the Serpent'.[62] In Christie's garden, evil always lurks somewhere beneath the rosy facade, and rural paradise is easily shattered by a cold-blooded murder. Another image from the Garden of Eden is an apple: 'Nothing could be more agreeable than a juicy English apple – And yet here were apples mixed up with broomsticks, and witches, and old-fashioned folklore, and a murdered child.'[63]

Inspector Craddock in *A Murder Is Announced* enjoys the peace and sunshine in the Vicarage garden but at the same time cannot fight 'a nightmarish feeling at the back of his mind'.[64] Not only is there no guarantee of safety in the calmness of any garden, but this calmness generally does not bode well. It contains something ominous in its very nature. Christie repeatedly calls it 'an atmosphere'. Two gardens in her books have a particularly dense 'atmosphere' – the sunken garden in *Hallowe'en Party* (1969) and the garden of the Bradbury-Scott sisters in *Nemesis*.[65] In all these cases, garden descriptions are purposefully used to enhance suspense.

Fourthly, there is what may be called 'a Miss Marple Paradox'. This old lady is a similarly ambiguous figure whose appearance is deceptive. If a peaceful garden is not to be trusted, the same applies to 'an elderly unmarried woman who knits and gardens': under this mask, she is 'streets ahead of any detective sergeant'.[66] Miss Marple consciously uses her harmless appearance: in *They Do It with Mirrors*, she says that weeding is 'about all an old and useless woman can find to do' which is obviously a lie.[67]

Basically, Miss Marple has two main interests in life – gardening and murders. This is why, wanting her to solve an old crime, Mr Rafiel presents her with the tour of Famous

Houses and Gardens.[68] These two interests may look incompatible, but Agatha Christie uses this sort of contradiction as a literary device. Miss Marple is 'the sort of old pussy who would make homemade liqueurs, cordials and herb teas' (this list includes damson gin, cherry brandy, cowslip wine, and chamomile tea) – but in the eyes of Inspector Neele, it automatically makes her an expert on home-brewed poisons, too.[69]

Garden as a Symbol of Society

The symbolism of gardens in Christie's books develops over time. Comparing the first and the last novels featuring Miss Marple, we can observe a striking difference. In *The Murder at the Vicarage* (1930), her vigorous gardening is just a typical old lady's trait, convenient for spying on neighbours; there is no double meaning in this occupation.[70] In *Sleeping Murder* (1976), the whole narrative spins around bringing order to the old garden, and the bindweed with its roots going 'down underground a long way' is directly associated with old crimes.[71]

Comparing weeds to criminals and gardening to the detective work gradually becomes a recurrent motif in post-war Miss Marple books. This character cannot stand the view of a garden taken over by weeds the same way she cannot rest if a murder goes unpunished. In *They Do It with Mirrors* (1952), she still does her weeding in a fairly unobtrusive way, though regarding storytelling, this occupation is already aligned with her investigation.[72] The metaphor crystallizes in *At Bertram's Hotel* (1965), when Miss Marple openly says about the situation that 'It is like when you get ground elder really badly in a border. There's nothing else you can do about it – except dig the whole thing up'.[73]

In *Nemesis* (1971), Miss Marple can hardly restrain herself from pulling up 'the vagrant bindweed' in someone else's garden, and it is the similar urge for order when it comes to disclosing a murderer. The image of flowering *Polygonum baldschuanicum* is another association with a criminal in this novel: 'One of the quickest flowering shrubs which swallows and kills and dries up and gets rid of everything it grows over'. This plant now grows over the destroyed greenhouse where once there was the grapevine, 'the little, small, early sweet grapes' – a metaphor for the murdered girl.[74]

The last of Miss Marple novels, *Sleeping Murder* (1976), is particularly advanced in terms of gardening symbolism. While there is a whole chapter named after the bindweed, Miss Marple's 'weeding' spreads over most of the book. She is regularly seen doing this activity, and when Gwenda says that Miss Marple is 'awfully kind in helping [. . .] with the garden', it is not just the real garden she means, but rather a metaphorical one. The climax of the novel is also symbolic: having saved Gwenda's life, Miss Marple claims it is fortunate that she 'was just syringing the greenfly' off Gwenda's roses.[75]

Another of Christie's detectives, Hercule Poirot, may appear unsuitable for gardening metaphors, at least at first glance. In many ways, he is the opposite of Miss Marple; even speaking of gardens, his tastes differ: 'a well-cultivated neatly arranged kitchen garden' is more likely to bring 'a murmur of admiration' to Poirot's lips than any wonders of nature, whereas Miss Marple prefers flowers and does not 'care so much for vegetables'.[76]

Nonetheless, Poirot has his share of experience in agriculture that becomes a recurring joke: his attempts at growing vegetable marrows are mentioned in at least six books, sometimes commented on as 'never again'.[77] The experience itself falls on *The Murder of Roger Ackroyd* (1926). Previously, Poirot shared his plans to dedicate his retirement to the cultivation of vegetable marrows with the intention to improve their flavour. When it happens, however, he soon 'enrage[s] [him]self' with the plants that refuse to submit to his ideas of order and uses the occurred murder as an excuse to return to detective practice.[78] Within this practice, he knows how the desired order can be obtained, unlike in gardening.

Later, Poirot states that he has abandoned gardening because 'even when you grow vegetable marrows you cannot get away from murder', but basically, the event simply made him realize that he is more effective as an ultimate 'gardener' for society.[79] Inspector Raglan says to Poirot with respect: 'You don't let the grass grow under your feet', which may be applied to Poirot's principles in general – he eliminates perpetrators as if they were weeds, helped by his famous love for order that eventually is not so dissimilar to Miss Marple's urge to ensure justice.[80]

Conclusion

A garden in Agatha Christie's fiction is more than just a picturesque backdrop – it constitutes an important location reflecting historical reality and bearing deep, often dark, symbolism. Generally, the world of Agatha Christie's mysteries is a territory of fiction that abides by its own rules, yet exists in close connection with reality. To some extent, it absorbed and proceeded everything that happened around it, with the author's literary techniques gradually becoming more complex and refined.

Gardens and their portrayal in Christie's books have at least three layers of meaning. Regarding social history, they represent the author's view on phenomena that may be considered characteristic of twentieth-century England. In terms of storytelling, garden descriptions emerge as one of Christie's signature tools for developing a captivating plot and enhancing suspense. On the symbolic level, Christie's fictional gardens serve as a universal metaphor for society in need of order and justice – a meaning particularly evident in the late Miss Marple novels.

About the Author

Tatiana Alekseeva is a Russian food writer living in Armenia who popularizes British heritage and culinary traditions. Her work focuses primarily on the portrayal of food in fiction.

Notes

1 Agatha Christie, *Third Girl* (Collins, 1966), p. 35.
2 Agatha Christie, *An Autobiography* (Collins, 1977), pp. 21–22.
3 Christie, *An Autobiography*, pp. 530–31.
4 Agatha Christie, 'The Augean Stables', *The Labours of Hercules* (Harper, 2010), pp. 113–35 (p. 115).
5 Agatha Christie, *Evil Under the Sun* (HarperCollins, 2001), p. 30.
6 Agatha Christie, *Five Little Pigs* (HarperCollins, 2007), pp. 85–86.
7 Agatha Christie, *Taken at the Flood* (HarperCollins, 2015), p. 17.
8 Agatha Christie, *The Hollow* (Collins, 1946), p. 69.

9   Christopher Yiannitsaros, '"Tea and Scandal at Four-Thirty": Fantasies of Englishness and Agatha Christie's Fiction of the 1930s and 1940s', *Clues: A Journal of Detection*, 35.2 (2017), pp. 78–88.
10 Agatha Christie, *The Mirror Crack'd from Side to Side* (HarperCollins, 2016), p. 45.
11 Agatha Christie, *The Mysterious Affair at Styles* (HarperCollins, 2007), pp. 15, 20, 205.
12 Christie, *Five Little Pigs*, p. 216.
13 Christie, *An Autobiography*, p. 469.
14 Agatha Christie, *Poirot's Early Cases* (HarperCollins, 2016), p. 123; 'The Cretan Bull', *The Labours of Hercules*, pp. 162–194 (p. 167).
15 Agatha Christie, *Dumb Witness* (HarperCollins, 2002), p. 19.
16 'Garden Parties, Past and Present', *Tatler*, 24 June 1908, p. 328.
17 Carolyn Humphries, *The Summer Weekend Cookbook* (Foulsham, 1999), p. 137.
18 Agatha Christie, *Nemesis* (HarperCollins, 2016), p. 297.
19 Christie, *An Autobiography*, pp. 109–110.
20 Andrew Norman, *Agatha Christie: the Finished Portrait* (Tempus, 2006), p. 54.
21 Agatha Christie, *Peril at End House* (HarperCollins, 2015), pp. 33, 68.
22 Christie, *Peril at End House*, p. 77.
23 Christie, *An Autobiography*, p. 109.
24 H.R. Humphries, *Fund Raising for Small Charities and Organisations* (David & Charles, 1972), pp. 40, 40–46.
25 Agatha Christie, *Dead Man's Folly* (HarperCollins, 2014), p. 68.
26 Humphries, p. 50.
27 Christie, *Dead Man's Folly*, p. 142.
28 Christie, *The Mirror Crack'd from Side to Side*, p. 47.
29 Agatha Christie, *The Pale Horse* (HarperCollins, 2017), p. 60.
30 Christie, *Nemesis*, p. 96.
31 Christie, *Taken at the Flood*, p. 61.
32 Christie, *The Mysterious Affair at Styles*, pp. 21, 77.
33 Agatha Christie, *Poirot Investigates* (HarperCollins, 2016), pp. 35, 172.
34 Agatha Christie, *Three Blind Mice and Other Stories* (Berkley, 1984), p. 2.
35 Agatha Christie, *They Do It with Mirrors* (HarperCollins, 2002), p. 35.
36 Agatha Christie, *A Murder Is Announced* (HarperCollins, 2016), p. 74.
37 Agatha Christie, *4.50 From Paddington* (HarperCollins, 2016), p. 45.
38 Christie, *They Do It with Mirrors*, p. 59.
39 Ina Zweiniger-Bargielowska, *Austerity in Britain: Rationing, Controls, and Consumption, 1939–1955* (Oxford University Press, 2002), pp. 126, 31, 162.
40 Christie, *A Murder Is Announced*, p. 144.
41 Christie, *The Mirror Crack'd from Side to Side*, pp. 1–2, 2.
42 Christie, *A Murder Is Announced*, p. 31.
43 Agatha Christie, *The Thirteen Problems* (Ulverscroft, 2010), p. 244.
44 Agatha Christie, *Sleeping Murder* (HarperCollins, 2002), p. 107.
45 Christie, *4.50 From Paddington*, p. 14.
46 Christie, *The Mirror Crack'd from Side to Side*, p. 2.
47 Christie, *The Mirror Crack'd from Side to Side*, p. 47; *Nemesis*, p. 118.
48 Christie, *Taken at the Flood*, p. 110.
49 Christie, *The Mirror Crack'd from Side to Side*, p. 41.
50 Lucy Worsley, *Agatha Christie* (Hodder & Stoughton, 2022), chapter 31, paragraphs 13–18. Kindle edition.
51 Agatha Christie, *Mrs McGinty's Dead* (HarperCollins, 2001), p. 77.
52 Christie, *4.50 From Paddington*, p. 55.
53 Christie, *Three Blind Mice and Other Stories*, p. 74.
54 Christie, *Nemesis*, p. 103.
55 Agatha Christie, *At Bertram's Hotel* (HarperCollins, 2016), p. 188.

56 Christie, *The Thirteen Problems*, p. 64; Agatha Christie, *Miss Marple's Final Cases* (Ulverscroft, 2010), p. 244.
57 Christie, *Poirot's Early Cases*, p. 329.
58 Christie, *The Mysterious Affair at Styles*, pp. 70, 77, 117–118.
59 Agatha Christie, *A Pocket Full of Rye* (Collins, 1953), p. 12.
60 Agatha Christie, *Hercule Poirot's Christmas* (Collins, 1963), p. 132; *Poirot's Early Cases*, p. 329; *Nemesis*, p. 266; *Sleeping Murder*, p. 257.
61 Christie, *Five Little Pigs*, p. 29.
62 Agatha Christie, *A Caribbean Mystery* (HarperCollins, 2002), p. 59.
63 Agatha Christie, *Hallowe'en Party* (HarperCollins, 1994), p. 42.
64 Christie, *A Murder Is Announced*, p. 131.
65 Christie, *Hallowe'en Party*, p. 93; *Nemesis*, p. 277.
66 Christie, *A Murder Is Announced*, p. 43.
67 Christie, *They Do It with Mirrors*, p. 60.
68 Christie, *Nemesis*, p. 56.
69 Christie, *A Pocket Full of Rye*, p. 88.
70 Agatha Christie, *The Murder at the Vicarage* (HarperCollins, 2016), p. 17.
71 Christie, *Sleeping Murder*, p. 280.
72 Christie, *They Do It with Mirrors*, pp. 59–60.
73 Christie, *At Bertram's Hotel*, p. 243.
74 Christie, *Nemesis*, pp. 101, 280, 117.
75 Christie, *Sleeping Murder*, pp. 201, 279, 290.
76 Christie, *Dead Man's Folly*, p. 5; *Nemesis*, p. 12.
77 Agatha Christie, *The Clocks* (Ulverscroft, 2011), p. 174.
78 Agatha Christie, *The Murder of Roger Ackroyd* (HarperCollins, 2007), p. 33.
79 Christie, *Peril at End House*, p. 174.
80 Christie, *The Murder of Roger Ackroyd*, p. 108.

3
Floating Garden Geopoetics
More-than-Human Collaborations in Xochimilca Foodways

Diego Astorga de Ita

¡Jardines polícromos y flotantes
donde el incendio de la tarde brilla...
[...]
¡y se hacen nuestros sueños más fragantes!
—Francisco Villaespesa, *Tardes de Xochimilco* (1917)[1]

Xochimilco is one of the five lakes that covered the Anahuac Basin, over which emerged the three city-states that became known as the Aztec Empire. Today, Aztec seats of power, lakes and all, have been swallowed by Mexico City, becoming neighbourhoods, boroughs, or part of the metropolitan zone of the megalopolis. Likewise, the old food system, reliant on the lakes from which different foodstuffs were obtained through cultivation, foraging, and hunting – from vegetables and maize to salt, crayfish, or aquatic insects' eggs – has slowly faded, giving way to urban life and to the modern capitalist food system. Most lake-foods once eaten throughout Mexico City's history are faded memories or unthinkable foodstuffs, definitely not something associated with what we'd think of as Mexican food: frogs' legs and duck are more likely found in a French bistro than a *fonda*, and axolotl soup would be a scandalous menu item anywhere. Pork and chicken have taken the place of crustaceans, fish, or waterfowl. Yet, in Xochimilco and a few other places the lake-landscape of old still exists.

Xochimilco's networks of canals are the result of chinampas, small islands constructed on the lake by layering organic matter and mud from the lake's bottom, tethered by bonpland willows' roots (*Salix bonplandia*). Chinampas are a pre-colonial agricultural technology that made the lakes an inhabitable space by providing land and food to the large population of the Basin. Though not unscathed, this foodscape of chinampas has survived centuries of hydric policies that desiccated the rest of the Basin – from Spanish colonists' drainage canals to nineteenth-century Mexican modernists' infrastructure projects to neoliberal politicians' privatizations. Xochimilco's long hydrocolonial history, as well as recent urbanization and land use changes in chinampas and in Mexico City at large, has led to worsening water quality in the lake. The water in which canoes and kayaks travel no longer comes from local springs; instead Xochimilco's fresh water is pumped to wealthy neighbourhoods like La Roma and La Condesa, their sewage goes to water treatment plants, and treated water is pumped back into the canals. This is one of the many complex relations that exist between Xochimilco and Mexico City that speak of the contested nature of Xochimilca water and territory. Adding to the tensions between the lacustrine and the urban, the foodscape is contested: different forms of agriculture come into conflict with each other and with tourist uses of the waterways, giving way to questions of gentrification and commoditization as we see emergent projects from outside actors benefit from the cultural and environmental history of chinampas.

Through long histories of hydric dispossession, this waterscape developed into what it is today: a human-made environment which still supplies Mexico City with substantial amounts of food and water, inhabited by diverse species, including humans, migratory birds, and an endemic axolotl. The uniqueness and beauty of this landscape has led authors – from German geographers to Mexican poets – to wax poetic, naming it a 'floating polychromatic garden', though chinampas are not floating, but anchored in place by the aforementioned *ahuejote* roots.

Here, however, I won't delve into the hydrocolonial history of Xochimilco and the Basin of Mexico, nor will I explore the contested nature of this lacustrine territory.[2] Instead I will consider the politics and poetics of a specific everyday practice that takes place in chinampas that are still cultivated using traditional techniques. In this essay I will dig into *chapín*, the agricultural practice where silt is extracted from the lake bottom to build seedbeds. In *chapín* we will find insights that can guide us in the search for more sustainable foodways in a context of complex environmental relations.

This exploration of *chapín* is based upon two field experiences in two chinampas in Xochimilco. Both outings were part of the work of Cocina Colaboratorio, a project that explores issues related to food and food production in traditional food systems with a focus on co-constructing sustainable, resilient, and just food futures. The project works in three sites in Mexico, Xochimilco being one of them. Cocina Colaboratorio uses a participatory action research approach, bringing together scholars, farmers, artists, designers, cooks, activists and collectives, seeking to build transdisciplinary alternatives that are co-designed and co-constructed by all involved, not just those of us in academia. The two specific vignettes regarding *chapín* I present here (gathering mud and planting seeds) come from *tequios* – collective work sessions traditionally practiced by Mexican campesino communities. These two *tequios* were part of Cocina Colaboratorio's 2022 'Archivo Biocultural Vivo' (Living Biocultural Archive) exhibit in Xochimilco and of Cocina Colaboratorio's 2023 Annual Meeting in Xochimilco.[3]

From these two *tequios*, we will consider the poetics, or rather hydropoetics, of doing *chapín*, as well as the relational aspects of this practice. We will explore the ways in which doing *chapín* establishes a relationship with the landscape at large, with the lake, with the chinampas, and with our neighbours, both human and non-human, who inhabit the landscape. From these considerations will arise political questions related to care, collaboration, food, and the environment, which have implications for how we think of nature's Others, especially in relation to food production. However, before looking at these instances of *chapín*, I will outline the notions of hydropoetics and relational sustainability that underpin this work.

Relational Hydropoetics

In recent decades, a poetic turn has emerged in geography. 'Geopoetics', originally proposed in the late 1980s by Scottish author Kenneth White, has gone from a romantic individualist notion rooted in Euro-Western thought to a concept used by critical geographers. Global south geographers, in particular, have used geopoetics to critique hegemonic modes of being-in-the-world and to make visible other ways of being beyond or on the liminality of modernity.[4]

A similar turn has emerged in the blue humanities in response to concerns over rising sea levels and the threat this poses to oceanic ways of life, particularly among Pasifika peoples.[5] From this turn come notions like hydropoetics, hydrocolonialism, and hydrofeminism, among other hydro-logics used to explore how humans relate to water. While these hydro-logics have mainly focused on the sea, and have been largely used by literary scholars, I propose these hydro-logics – especially the notion of hydropoetics – can be useful to understand life in waterscapes beyond (or before) the sea.[6]

Given that hydropoetics refer to ways we relate to water, and through water, we can understand them as a relational way of being-in-the-world.[7] In *Bodies of Water*, Neimanis argues that water makes things fluid, connecting our bodies with bodies of water, both human and non-human, allowing us to enter into relation with the world.[8] Adding Glissant's thinking in *Poetics of Relation*, hydropoetics would be 'latent, open, multilingual in intention, directly in contact with everything possible'.[9]

Xochimilco's relational hydropoetics are not borne out of the same abyss as Glissant's *Relation*, but they are still the product of colonial histories of oppression, resistance, and dispossession. Here, hydropoetics emerge not from the womb-abyss of the slave ship, nor from 'the depths of the sea [. . .] with their punctuation of scarcely corroded balls and chains'. Instead, Xochimilca hydropoetics carry the pain of desiccation, of water swallowed by the city, of ways of life destroyed by hydrocolonial processes. And of resistance. This postcolonial pain is present here: 'feeling a language vanish, the word of the gods vanish, and the sealed image of even the most everyday object, of even the most familiar animal, vanish. The evanescent taste of what you ate'. Here things vanish not because you have been torn apart from your land but because your waters have been dried up and polluted. And yet, hydropoetics remain; as Glissant insists, 'We cry our cry of poetry. Our boats are open, and we sail them for everyone.'[10]

A relational hydropoetics is a possible answer to calls to diversify epistemologies and ontologies coming from the 'relational turn' of sustainability science. Recent works, like the IPBES's recent values assessment, urge us to look at other ways of understanding and valuing nature, and posit that diverse onto-epistemologies are necessary for sustainability.[11] From this perspective our relationship with nature – or in this case with water – has implications for how we understand and value it, depending on whether we live from nature, in nature, with nature, or as a part of nature.[12] Here, I argue that a hydropoetics of relation can make space for other ways of relating to nature beyond the Modern ontology that sees nature primarily as a resource.

Now we plunge into the mud and the relational hydropoetic implications of working with it, but first let us consider the *tequio* and what it means to work *with*. After this, we will delve into doing *chapín*, and what it tells us about working collaboratively with people, with water, and with our non-human neighbours.

To Work With

As mentioned before, *chapín* is the agricultural technique in which seedbeds are built with mud extracted from the lake. The two instances of doing *chapín* I present took place in two *tequios*, which are a form of labour that has its own political and relational implications.

To Work with Others

Tequios are collective work. *Tequio* is the Nahuatl word for this practice, used mostly in the centre of the country; in other places this might be called *faena*, *guetza*, or *manovuelta*. However named, this is a practice that, at least until recently, was common in campesino and Indigenous communities. In a *tequio* a community comes together to do tasks that are needed by the community or to help a member of the community with difficult or time-consuming labours like clearing a field, planting, or harvesting. This is unpaid labour: rather than a market logic, it follows a logic of communality and reciprocity. It is work that needs to be done for the sake of the community, or that will be repaid later on, when one needs help with work in one's own field. While the green revolution and the advent of Modernity have weakened this practice considerably, there are communities and groups that keep this institution alive and see *tequios* as a form of resistance.

Cocina Colaboratorio is a diverse collective working in different sites important for traditional food systems; however, many, if not most, of the members of this project are not part of these communities. So why do we claim to be doing *tequios*, when the work undertaken in these events isn't done by members of a rural community, and when it is not exactly mutual labour? On one hand, our *tequios* are mutual collective work: even if most of us don't have a field that needs planting or weeding, we give a hand to our collaborators in the field – the actual field where food is grown. In the *tequio*, academics, designers, and artists work pulling weeds, carrying mud, or planting seeds; in exchange, we get help in our field when our campesino colleagues co-write or give feedback to our papers. On the other hand, this *tequio* entails another sort of labour for chinampa farmers – they have to teach us to do the most basic things, from explaining what is and isn't a weed to how to use farming implements and how deep a seed should be planted – which raises the question of how successful our attempts at collaboration actually are and whether it is as useful for our collaborators as it is for us, or if, as the Spanish saying goes, *mucho ayuda el que no estorba* – the one who doesn't get in the way helps the most. Still, participating in a *tequio* means upsetting the asymmetry of power-knowledge by giving the role of expert to the farmers who actually have the expertise when it comes to growing food.

We speak of *tequios* because of the radical collaborative political possibility that the *tequio* entails. Cocina Colaboratorio, as the name suggests, seeks to build collaborative alternatives to the current unsustainable food system, and practices like *el tequio* can help construct such alternatives. Naming these events *tequios* is, in a way, an expression of our desire to build alternatives and of our political commitment to work collaboratively. These sort of *tequios* are a way of transdisciplinary collaboration beyond the gallery, the symposium, or the written page.

The *tequio* is the first instance in which a relational ethos is deployed: through *tequios* relationships are established with other people who become our collaborators, coworkers, and friends, but this collaboration extends beyond our human neighbours as we will see in the first vignette of doing *chapín*.

To Work with the Lake

Doing *chapín* is one of the oldest techniques of chinampa agriculture; it entails dredging mud from the canals and using it to build seedbeds. That, in a nutshell, is doing *chapín*. However, behind this seemingly straightforward activity lie numerous skills, knowledges, and relations.

We embark in the pink-and-aqua canoe and meander our way into one of the lagoons at the heart of the canals. Gaby, the chinampa's owner, directs the canoe while those of us that have been previously instructed on the basics of punting take turns helping her propel the boat in a zig-zagging motion across the canals. Except for her, none of us are experts rowing or doing *chapín*. You can tell. Our oars get tangled with overhead cables, we bump against the banks of the chinampas, and we have much wider zigs than zags, or vice versa. Slowly, we make our way to the place designated by Gaby for us to dredge the mud. The spot was chosen because of its location: mud here is cleaner and easier to extract, which is good, given that we are mostly neophytes. We park the canoe by sticking one of the oars into the mud, the other oar will become the *zoquimaitl* – the 'mud hand'.

To extract mud, one needs a *zoquimaitl*, which is made by tying a circular mesh to the end of the oar. The mesh is stretched loosely over a metal circle made of rebar. The oar goes through a hole in the mesh and is tied to the circle making a sort of long spoon or backscratcher that will be used to extract mud from the bottom of the lake.

Gaby ties the *zoquimaitl* together. She explained how to use it before we embarked (Figure 1), but she explains again now, to make sure we get it: you take the *zoquimaitl* and push it to the bottom of the lake, careful to keep the circle flat; you grab a giant spoonful of mud and using the border of the canoe as a leverage point, slowly bring the mud out. This step has to be done carefully; do it too fast and you risk losing the mud or capsizing the canoe. Once the mud is out, you let some of the excess water drip off, then pour it into a bucket.

We are making a small *chapín*. If we were doing this for a bigger *chapín* in a bigger chinampa, we'd do it without buckets, covering the floor of the canoe with mud instead. One sometimes sees people moving in mud-covered boats through Xochimilco, as if truly floating islands were adrift in the canals. This requires further skills and brings about its own challenges. Knowledge of how much mud a canoe can carry, and of how to move in the mud, is necessary. But our *chapín* is small and our skill is, let's say limited, and so we use buckets. While one person grabs the mud, others clean it, taking out branches and stones, throwing them back into the lake (Figure 2).

Figure 1. Gaby explains doing chapín *before we start work. Photo by the author.*

Figure 2. We extract mud from the lake bottom with the zoquimaitl. *Photo by Rubén Garay, courtesy of Cocina Colaboratorio.*

Once we've filled our buckets and other receptacles with mud, we make our way back to the chinampa. We disembark with our muddy boon. At the chinampa the process of *chapín* continues. We raise a little border out of earth, marking the space where the mud will go, and pour it in. At this stage, the mud is still liquid enough that it can be poured. It's then spread, like cake batter in a tin, and left to dry (Figure 3). When the mud cake is dry, in a couple of days, it will be sliced into small squares (Figure 4); a seed, or two, or three, will go into each slice, depending on what is being planted. The seedlings, when grown enough, will be transplanted into a bed in the chinampa (Figure 5).

We can see in doing *chapín* a form of relational hydropoetics. Relational, not only because of the collective work entailed by the *tequio* but because by doing *chapín* we establish and foster a relationship with the lake and the landscape. *Chapín* is one of the ways in which this landscape is cared for: it adds organic matter and nutrients to the soil, but it also adds soil to the chinampa – although fixed in place and guarded by willows, chinampas are eroded over time, so doing *chapín* is an important part of chinampa maintenance. Furthermore, taking mud from the bottom of lagoons and canals in the lake keeps the waterways navigable. But the relational aspect of doing *chapín* goes beyond a mere janitorial upkeep of chinampas and canals. By doing *chapín* we enter into a relationship with the lake as a body of water, the lake as someone with whom and in whom we exist. Doing *chapín* is a collaboration with the landscape itself. We move across its body,

Figure 3. Gaby spreads the mud. Photo by the author.

Figure 4. A chapín *is cut. Photo by the author.*

Figure 5. A seedling from the chapín. *Photo by Rubén Garay, courtesy of Cocina Colaboratorio.*

winding in its waters, walking in its bottom as we step in the mud-covered canoe or in the chinampa built from its silt. If we think of the lake not just as a natural element from which or in which we live, and think of it in a cosmopolitical way – that is, if we extend the *polis* to include not just human bodies, but bodies of water beyond our own – then doing *chapín* becomes a form of relation and care. There is a mutuality and commensality to it: we scratch its back, almost literally with the *zoquimaitl*, and it scratches ours, by nurturing our crops and giving us water.[13] But the cosmopolitics of this hydric inhabiting are diverse as the second *chapín* vignette will demonstrate.

To Work with Critters

Doing *chapín* is a collaboration with the world beyond the human. With the lake, on the one hand, as we've just seen. But also, on the other, with the seeds and plants we grow, and with the critters that inhabit the landscape.

We make our way to Gaby's second chinampa: this time we have two canoes and two rowers who know what they are doing. We eat breakfast – tamales and atole, both made with maize – as we meander through the canals. The warm tamales, filled with wild greens or chicken in red sauce, have been prepared by Xochimilca cooks, and some of the ingredients have come from chinampas like the one we're going to.

The landscape and its inhabitants nurture us as we float across it. When we get to the chinampa we talk of care: of what labours of care entail, of care for territory, for one another, and for the self. We're asked to work mindfully and are given post-it notes to jot down our thoughts and intentions. We post them on a sapling on the edge of raised beds brimming with vegetables and marigolds. The post-its carry our intentions for today's collective work: 'To make deep connections', 'To be able to pay attention to our care for the non-human', 'To work as a team'. The group divides into teams, each with a task: some will clean raised beds, others make a compost pile, my group will sow the *chapín* using some seeds from the project's 'Living Biocultural Archive' collected through seed exchanges. We move to the spot where the mud has been poured, with a plastic sheet covering it from the sun to avoid over-drying.

As someone unveils the shiny dark sheet of congealed mud, we find something unexpected, someone has left a mark in the *chapín*. A few, small hand-like prints speak of an unseen presence in the chinampa (Figure 6). 'That's maybe from an opossum, or, more likely, a *cacomixtle*,' says Gaby. The *cacomixtle* (ringtail or civet cat) is a tropical cousin of the raccoon. Its presence speaks of the non-humans that inhabit this human-made landscape. We admire its handiwork in the mud for a minute and then get on with our work. We cut the *chapín* into squares carefully, trying to maintain the marks of the *cacomixtle*.

Once we've cut the *chapín* we start sowing seeds into each slice. We follow the *cacomixtle*'s example; with our fingertips we make a small indentation in the centre of each mud parcel. Children help – this is more like play than work. We sow each hole, making sure to note what seed is planted in each fingerprint. When we get to the *cacomixtle*'s handprint, instead of undoing it we leave it be: it's deep enough that we can put the seeds there. It's saved us some work (Figure 7). We cover the seeds with a sprinkling of dirt, hiding all handprints, human or otherwise, and water the seeds we just planted.

Figure 6. Cacomixtle *pawprints in the mud. Photo by the author.*

Figure 7. Sown chapín. *Photo by the author.*

Just like that, the *cacomixtle* has become our collaborator. We're now working not only with each other and with the lake, but also with the *cacomixtle*. Its stroll across the mud has become labour. It toils with us to cultivate this landscape. The *cacomixtle* brings the more-than-human collaboration beyond the theoretic or metaphoric: it is literally a collaboration with our labours of care for the seeds, the chinampa, and the landscape at large. By cultivating the landscape, we build a relationship with our non-human neighbours who partake in the sowing and growing.

Towards a Hydrofeminist Cosmopolitics of Care

Relations established by doing *chapín* in this watery foodscape bring forth a cosmopolitics where lacustrine Others become collaborators in our labours of care. When discussing care after the *tequio*, the conversation goes from the immediate labours of care undertaken for plants and seeds and soil, to caring for this lacustrine territory. Territory is seen by many as an extension of our body: our bodies are part of our territory, and our territory is another body we inhabit. This line of thought makes me think of ecofeminist perspectives, and particularly of Neimanis's notion of hydrofeminism, given that Xochimilca territories are also bodies of water.[14]

Neimanis argues that 'water is a conduit and mode of connection'. This rings true for Xochimilca waterways, not only because Xochimilco is an aquatic landscape where everyday life occurs through rowing and cultivating the lake, but because in Xochimilco we see the multiplicity of water that Neimanis puts forward: 'Water as body; water as communicator between bodies; water as facilitating bodies into being'.[15] Neimanis's work posits water as a means of relation where the boundaries of our bodies are blurred, frontiers made fluid.

We see this hydrofeminist fluidity in the notion of the body-territory, and in the labours of care enacted in it, such as doing *chapín*. Care for the territory, from a hydrofeminist perspective, acquires a cosmic quality, a certain topsy-turvyness resulting from the fluidity of it all. When doing *chapín* we care for the body-territory; for the lake and the landscape and the seeds that we plant. At the same time, through *chapín*, the lake cares for us, giving its body over to be sown. From this hydrofeminist perspective, despite willow roots, chinampas become floating gardens, set adrift by the fluidity of relations that hydropoetic inhabitings entail.

In this watery network of relations, hydropoetic inhabitings make space for cosmopolitical collaborations. But in a context of contested spaces, desiccated lakes, and polluted waters, what does this mean? Why does this matter? The cosmopolitics of care that come from doing *chapín* are important because from this praxis of care emerge diverse values and narratives of nature, beyond those that have made hydric dispossession the norm in the Basin of Mexico. Hydropoetics, like Nassar's geopoetics, allow for an interplay of knowledges and practices that challenge the idea of mastery, though here it is not through language, but through embodied experiences of mud and food.[16]

A relational hydropoetics unsettles the notion of water-as-resource and transforms our relationship to nature into a way of being with nature, where we're peers and not masters; we exist as part of a collective that includes animals, and waters, and seeds. While this relationship takes place by doing *chapín*, less so by writing about doing *chapín* – or reading about doing *chapín* – the presence of this practice in our academic spheres is also important, since it speaks of the existence of other ways of being and opens up avenues to think and implement research (and other forms of work) as a collaboration with nature, a 'working with'. Working with allows for a politics of care to inform the way we work and eat, in order to construct, together, more just and sustainable foodways in Xochimilco and beyond.

About the Author
Diego Astorga de Ita is a doctor in Geography, researching questions of food, cultural ecology, and sustainability. He is currently a postdoctoral researcher at Durham University and has previously worked at Mexico's national University (UNAM) where he was part of Cocina Colaboratorio at Xochimilco.

Acknowledgements
I would like to thank my colleague and friend Gaby Alejandra Morales Valdelamar, for teaching and welcoming us into her chinampas, her *barrio*, and her territory in Xochimilco. I am also thankful to Adriana Cadena Roa, Patricia Balvanera, and the whole Cocina Colaboratorio team, particularly the team at Xochimilco. This text presents my thoughts on our collective work and wouldn't be possible without the collective; any errors are my own. Lastly I thank UNAM's Centre for Scientific Research for the funding that made my research possible.

Notes
1 Francisco Villaespesa, 'Tardes de Xochimilco' (Herrero Hermanos, 1919), p. 37.
2 For an exploration on these topics see Diego Astorga de Ita, Gabriela Morales Valdelamar, Adriana Cadena Roa and Patricia Balvanera, 'Axolotl soup: Hydrocoloniality, contested foodscapes, and plural values of nature', *Environment and Planning D: Society and Space*, in press (2025).
3 The exhibit can be found at <https://archivo.bio> [accessed 13 May 2024]. For more on the project see Cocina Colaboratorio, 'Colaboratory Kitchen', 2024 <https://colaboratorykitchen.com/> [accessed 13 May 2024]; Ana Kooi and Mariana Martínez Balvanera, 'CoLaboratory Kitchen', in *Design for Global Challenges and Goals*, ed. by Emmanuel Tsekleves, Rachel Cooper, and Jak Spencer (Taylor & Francis, 2021), pp. 183–203 DOI:10.4324/9781003099680.
4 Diego Astorga de Ita, 'Grassland Geopoetics: Son Jarocho and the Black Sense of Place of Plantations and Pastures', *Antipode*, 56.3 (2023), 872–95, DOI:10.1111/anti.12999; Antonis Balasopoulos, 'Nesologies: Island Form and Postcolonial Geopoetics', *Postcolonial Studies*, 11.1 (2008), 9–26, DOI:10.1080/13688790801971555; Federico Ferretti, 'From the Drought to the Mud: Rediscovering Geopoetics and Cultural Hybridity from the Global South', *Cultural Geographies*, 27.4 (2020), 597–613, DOI:10.1177/1474474020911181; Angela Last, 'We Are the World? Anthropocene Cultural Production between Geopoetics and Geopolitics', *Theory, Culture & Society Geo-Social Formations*, 34.2–3 (2015), 147–68, DOI:10.1177/0263276415598626; Sarah De Leeuw and Eric Magrane, 'Geopoetics', in *Keywords in Radical Geography: Antipode at 50*, ed. by Antipode Editorial Collective (Hoboken &: Wiley Blackwell, 2019), pp. 146–50, DOI:10.1002/9781119558071; Eric Magrane, 'Situating Geopoetics', *GeoHumanities*, 1.1 (2015), 86–102, DOI:10.1080/2373566x.2015.1071674; Aya Nassar, 'Geopoetics as Disruptive Aesthetics: Vignettes from Cairo', *GeoHumanities*, 7.2 (2021), 455–63, DOI:10.1080/2373566X.2021.1913436; Aya Nassar, 'Geopoetics: Storytelling Against Mastery', *Dialogues in Human Geography*, 11.1 (2021), 27–30, DOI:10.1177/2043820620986397.

5   For example, Alice Te Punga Somerville, 'Where Oceans Come From', *Comparative Literature*, 69.1 (2017), 25–31, DOI:10.1215/00104124-3794579.
6   Diego Astorga de Ita, 'Musical Hydropoetics: Fluvial Inhabitings, Son Jarocho, and Anthroposcenes', *GeoHumanities*, 8.2 (2022), 435–56, DOI:10.1080/2373566X.2022.2045208.
7   Diana Alexandra Bernal and Eduardo Marandola Junior, '*Presença-ausência da água na cidade: narrações hidropoéticas no habitar urbano contemporâneo*', *Ateliê Geográfico*, 12.2 (2018), 98–113, DOI:10.5216/ag.v12i2.45807; Ana Patricia Noguera de Echeverri and Diana Alexandra Bernal Arias, '*La Naranja Azul: El Agua en la Era Planetaria*', *Geograficidade*, 5.3 (2015), 4–17, DOI:10.22409/geograficidade2015.52.a12940.
8   Astrida Neimanis, *Bodies of Water: Posthuman Feminist Phenomenology* (Bloomsbury, 2017).
9   Édouard Glissant, *Poetics of Relation* (University of Michigan, 2010), p. 32.
10 Glissant, pp. 6, 7, 9.
11 IPBES, *Methodological Assessment Report on the Diverse Values and Valuation of Nature of the Intergovernmental Science-Policy Platform on Biodiversity and Ecosystem Services*, ed. by Patricia Balvanera and others (IPBES Secretariat, 2022) DOI:10.5281/zenodo.6522522.
12 Unai Pascual, and others, 'Diverse Values of Nature for Sustainability', *Nature*, 620 (2023), 813–24, DOI:10.1038/s41586-023-06406-9.
13 Bruno Latour, 'Whose Cosmos, Which Politics?', *Common Knowledge*, 10.3 (2004), 450–62, DOI:10.1215/0961754x-10-3-450; Eduardo Viveros de Castro, *Cannibal Metaphysics* (Univocal, 2014).
14 Astrida Neimanis, 'Hydrofeminism: Or, On Becoming a Body of Water', in *Undutiful Daughters: New Directions in Feminist Thought and Practice*, ed. by Henriette Gunkel, Chrysanthi Nigianni, and Fanny Söderbäck (Palgrave Macmillan, 2012), pp. 85–100.
15 Neimanis, pp. 97, 98.
16 Nassar, 'Geopoetics: Storytelling Against Mastery'.

4
The 'Apple Isle'
Tasmanian Apples as a Weapon of Ecological Colonization and Icon of Botanic Imperialism

Carla Baker

Lutruwita Tasmania, also known as the 'Apple Isle', is part of Australia. It is a smaller island of 68,000 km² (about the area of Ireland or South Carolina) lying about 200 km south of the Australian mainland.[1] Apples provided an early and tangible expression of British ecological imperialism. The first apple tree was planted in Tasmania (then called Van Diemen's Land) in 1788, when William Bligh anchored in Adventure Bay on Bruny Island and planted a selection of fruit, including three apple seedlings.[2] These were most likely sourced from South Africa. Jan van Riebeeck, founder of the Dutch East India Company trading post at Cape Town, took apples to South Africa in 1654, making fruit-growing compulsory among settlers.[3] In 1804 at York Town, the first settlement in Northern Tasmania, a garden and fruit trees, including apples, were established almost immediately before the community officially moved to Launceston in 1808.[4] This paper de-centres imperialism by recognizing the idealized cartographic distortion between the 'core' and 'periphery' within the British Empire through the study of traditional infrastructure (transportation, refrigeration) alongside the 'media and social networks that create and transfer cultural meanings about food', what Jeffrey Pilcher terms 'culinary infrastructure'.[5] Tasmanian apples provide a domesticated lens to ecological colonization and are an icon of botanic imperialism, thought to provide a healthy body and mind to sustain a 'white Australia'. It also specifically links the historical iconic status of apples to the recent classification of Launceston, Australia's third oldest city, as a UNESCO City of Gastronomy.[6]

This paper focuses on Tasmanian apples from the nineteenth century to the start of the Second World War. After this period, changes to imperial sentiment and technological advancement modified perceptions of the Tasmanian apple even though it retained its iconic status. Analyzing apples as a botanic imperialist icon opens a multifaceted historical dialogue. Through textual sources, such as advertising, we can see the ways in which apples were used to construct knowledge about the relationship between food and place, food and identity, food and race, and food and purity. This paper builds on the work by historians such as Anne McClintock, Antoinette Burton, and Catherine Hall, and follows the concept that anthropologist Claude Lévi-Strauss famously introduced: food is 'good to eat' because it is 'good to think'. Additionally, Sidney W. Mintz's seminal

Sweetness and Power provides the methodological scaffolding of using iconic comestibles as a lens refracting multiple cultural and historical influences. This paper also draws on the comprehensive report compiled by Anne McConnell and Nathalie Servant in 1999 on the history and heritage of the Tasmanian apple industry. By moving beyond an administrative narrative, this paper highlights the historical importance of gastronomy to Tasmania's trade, economic, and regional identity.

Tasmania, out of all the Australian colonies, was most frequently portrayed as bearing numerous similarities to England. An early colonial settler commented that 'Hobart Town looks like a country village in England [. . . .]he gooseberry, currant, apple and pear trees all give promise of extraordinary abundance of fruit [. . . that] inspire me with delight'.[7] The island's temperate climate, green pastures, and lack of open, sunburnt country provided a stronger ideological link between Tasmania and England than did the mainland of Australia. Following its establishment as an independent colony in 1825, free settlers arrived in search of land grants, following the expropriation of land and the massacre of many Indigenous peoples. Orcharding success was used to justify the theft of Indigenous land.

Domestic apple orchards played a central role in the ecological colonization of Tasmania. From the time of American colonization, apples have symbolized, and were thought to have contributed to, a settled and productive landscape.[8] An icon of British domesticity, the swift acclimatization and flourishing of apple trees were perceived to legitimize British settlement. Alfred W. Crosby's *Ecological Imperialism* details the importance of physically and visually transforming the colonized landscape to be more like 'home'.[9] Stolen land was considered 'claimed' as unfamiliar landscapes were transformed with invested British traditions through the planting of apple orchards.[10] Land grants were offered to those free settlers and civil servants who would farm the land 'productively' and were later sold to 'Anglo-Indian' soldiers as an investment or retirement opportunity, again based on appropriate development to enhance the broader British Empire. Apple orchards presented visible results of the inherent British nature of the land and were utilized to stabilize 'settler colonial power'.[11] Apples were a physical representation across the island that Tasmania's ties with Britain remained strong and, consequently, civilized.

Apples as a botanic symbol of British purity held multifaceted importance for this far-flung colony inhabited with countless convicts, both freed and captive. Civilizing the landscape was inherently important due to Van Diemen's Land perceived convict stain; purity and moral purpose were considered vital. Gardening was considered to bring 'one closer to God and was good for the moral fiber of the individual and the colony or nation'.[12] This sentiment was highly valued due to the penal colony's 'morally deficient' convict population. In 1822, 58% of the white populace of Van Diemen's Land was convicts.[13] Van Diemen's Land received convicts until transportation was ceased in 1853, then was renamed Tasmania in 1856 in an attempt to distance itself from its convict heritage, one which it has become proud of only in recent decades. Cultivating the soil was believed to be a God-given directive, a crucial part of Britain's 'civilizing mission'.

The colonial orchard not only supplied comestibles to stave off hunger, but foods like apples were considered 'safe' to the culinary xenophobic British settlers. By consuming foods from 'home', settlers could maintain an adequate level of civilization and British character, guarding against deterioration by the savage elements on the frontier. Marilyn Lake claims that Tasmanians still tended to think of themselves as transplanted Englishmen rather than Australians, up until at least the 1890s.[14] As Rebecca Earle insists, diet was central to the colonial endeavour to combat the fear that living in an unfamiliar environment and climate might alter the physical bodies of settlers.[15] The incorporation of foods with inherent Britishness within the settler-colonial diet was central to developing colonial identity in Australia. Apples were initially grown for local consumption only in Van Diemen's Land; however with the settlement of colonies in Victoria and the continued growth of New South Wales, seaborne apple trade began.

The technological innovation of steam further enhanced Australia's transport links within the empire. The one hundred miles of Bass Strait, the body of water between Tasmania and Australia's mainland, put Tasmania in a category of its own with shipping facilities more important, and more expensive, than anywhere else in Australia.[16] Sydney and Melbourne developed as important markets in the latter half of the nineteenth century once steamers could provide regular services and much shorter voyage times of down to only a few days.[17] The international shipping of apples predated the development of steam or refrigeration: it was trialled early in the nineteenth century. In 1828, Daniel Stanfield became one of the first Tasmanian colonists to export apples to Britain. Stanfield sent three seedling apples to the Royal Caledonian Horticultural Society in September 1828, which reported 'one of these apples was a foot in circumference, and of beautiful appearance, but the specimens having been plucked on 15th April last [...] were too much decayed to afford a correct estimate of their flavour'.[18] By 1850, Van Diemen's Land had the largest cultivated acreage in Australia, with Van Diemen's Land and South Australia the main exporters of produce to the other colonies.[19] It was not just nostalgia which prompted colonists to agriculturally anglicize the Tasmanian environment: they wanted Van Diemen's Land to be more profitable.[20] Land represented status, wealth, and power. Technological developments throughout the nineteenth century meant it was not long until perishable commodities were able to remain fresh and feed empire's bodies.

Apples became commodified as part of the global circuit of exchange, blurring the lines between the distant 'periphery' and 'home'. In 1887, the first commercial shipment of refrigerated apples left from Tasmania.[21] By 1891, at least 80% of Tasmanian apples sent to Britain were arriving in sound and marketable condition leading to an orcharding boom.[22] William Shoobridge, a successful settler and Tasmanian produce export advocate, presented a paper to the Royal Society of Tasmania in 1892 detailing the successful development and implementation of 'cooling apparatus' specifically for long-haul apple transport.[23] The introduction of steam and the opening of the Suez Canal further enhanced transport links and promoted the connection of Australia to a 'wider, transpacific Anglo world'.[24] The ability to source fruit world-wide, breaking down climate and seasonal barriers, shows how refrigeration technology became 'the empire's instrument'.[25] As Susanne Freidberg notes, refrigeration and the extensive cold chain reach revolutionized the geography of fresh food: 'the idea of the durable perishable no longer seemed as paradoxical as it once had'.[26] By the end of the nineteenth century it was possible to carry frozen meat and chilled butter and fruit cheaply and efficiently from Australia to Britain with minimal wastage.

By the twentieth century, apples were available to urban dwellers within the metropole year-round, providing nutritional benefits popularized by the eugenic push for healthier bodies following the South African War. Although large numbers of men volunteered for service in South Africa, only two of every five volunteers were fit to serve.[27] The working-class volunteers' diet relied heavily on bread and jam; high in calories but hardly nutritious. Fresh fruit and vegetables formed only a small part in the diet of the majority of the working class; the nutritional necessity of fruit and vegetables were not yet understood.[28] 'Home' apple supplies were only available from August until the end of December, but otherwise the market was filled with imported apples. A report by A. Hocking in the 1960s, almost a century after commercial British Empire apple exports began, indicated that the United Kingdom market for apples was still supplied, almost exclusively, by Australia, South Africa, and New Zealand in the months from March to July, with stocks of home-produced apples exhausted by March.[29] This was the unique competitive advantage enjoyed by the Antipodes with the inversion of the seasons. London also bought foreign fruit to supplement its supplies as its distribution of domestic fruit was disorganized and expensive to ship.[30] Apples consumed in the UK from 1934 to 1936 included 53.3% from 'home' production; just under half the apples purchased were imported.[31] Geography was no longer considered a significant barrier to comestible consumption, creating a world-wide British Empire food bowl.

Technological advances and successful export to metropole markets fuelled the Tasmanian apple boom at the *fin de siècle*. Across Tasmania, orchard acreage more than tripled between 1900 and 1914. The *Maitland Daily Mercury*, a regional newspaper in New South Wales, reported that 'the apple trade is certainly one of the wonders of Australia'.[32] Christopher Cowles and David Walker suggest that the foundations for the 'Apple Isle' tag were laid in these years.[33] I argue that it was earlier. In 1895, R.E. McNaughten, writing in the British literary periodical *Pall Mall Magazine* described the Southern Tasmanian Huon region as 'Apple Land'.[34] The connection between Tasmania as 'Apple Land' received further world-wide coverage when huge 'Apple Land' arches were crafted as decorations during a royal visit in 1901. As both mainland Australia and Tasmania grew apples, 'Apple Land' evolved to 'Apple Isle' to differentiate Tasmanian apples both domestically and internationally. British food writer and apple tree nursery owner Edward Bunyard compared the quality of home fruits against those imported varieties in 1929: 'as spring approaches we find the later apples mellowing to ripeness, and many of them can hold their own against the excellent fruit now coming from Australia and Tasmania'.[35] As the apple industry rapidly expanded and matured in the early twentieth century, so did competition between growers within Tasmania, Australia, and the world. Tasmanian apples needed to stand out, turning to advertising for assistance.

At the core of Tasmanian apple advertising was the imagined location of the 'Apple Isle', an idyllic, pure, English farming utopia. Apple crate branding and multifaceted advertising campaigns were employed to support this distortion of Tasmania's geographical and cultural remoteness, seeking to preserve 'a veneer of agrarian fantasy even while the way of life it presented was being irreversibly altered'.[36] During the interwar years, efforts were made to strengthen trade ties between members of the British Commonwealth and to boost markets abroad for British goods by promoting emigration.[37] These imperial networks of trade and consumption benefited from the emotive encouragement of the 'global chain of kith and kin'.[38] One of the most prominent ways this was achieved was through an empire-wide advertising campaign, managed by the Empire Marketing Board (EMB). The EMB, active between 1926 and 1933, was the strongest advocate for empire comestibles, most visible through its visually stunning poster campaign. It displayed around 800 posters throughout its existence with specially designed frames placed at 1800 different sites in 450 British towns.[39] Five posters were used at a time appearing in a narrative sequence, similar to a cartoon strip, with each of the posters telling a part of the story.[40] Around 100 sequences were produced: more than 30 were devoted to the Dominions, white settler colonies such as Canada and Australia, with dependent colonies such as India featured only four times.[41] The traditional apple suppliers of empire are depicted as Australia, New Zealand, and Canada. (In Empire Marketing Board imagery, South Africa is more consistently aligned with citrus fruits.) Tasmanian apples are explicitly localized in Empire Marketing Board posters.[42] This reveals that Tasmanian apples were, at the time, recognized in Britain as a separate marketable commodity available in exportable numbers. Tasmania prided itself on the quality of its produce and capitalized on different harvesting and shipping arrival times into the UK compared to mainland states due to its cooler climate. The Empire Marketing Board inextricably connected specific foods and the ongoing health of the British Empire.

Tasmanian apples provide an example of how moral consumerism was considered central to imperial identity during the interwar period. By providing a geographical distinction between empire primary production and appropriate 'white' commodities, a racialized narrative was built based on purity, expertise, and quality.[43] Dominion identity constructed in EMB advertising had distinct elements; dominion primary production was familiar 'home' produce such as dairy, apples, and wheat. Dominions were also depicted most commonly as farms which 'smoothed the strange edges of diverse dominion landscapes'.[44] Conversely, 'exotic' tropical comestibles were sourced by 'dependent' colonies such as pineapples from British Malaya (now part of Malaysia). The construction of Dominion farmlands as 'British' fields served to balance out the increasing metropolitan urbanization and industrialization at 'home' as well as to reinforce Britain's own imagined identity as rural. Empire foods were essential as these products enabled imperial propaganda to infiltrate the daily routines of the British public.[45] England's garden was considered endless, across the wide landscapes of the dominions and its dependent colonies, connected to the metropole through the red lines of empire shipping.

Tasmanian apples provide a unique angle of how this commodity was idealized and imbued with a strong sense of 'Britishness' due to its imagined pure, utopian location separate from the mainland of Australia and other apple-producing Dominions around the world. In the hands of commodity marketers, being British meant being modern, masculine, rural, and white.[46] Felicity Barnes extends this insight of Britishness within goods in her monograph *Selling Britishness*, coining the term 'commodity Britishness'.

Barnes argues that commodity Britishness was a Dominion invention that legitimized the idea of white settler colonies as their power was both camouflaged and naturalized, 'hidden in plain sight, passing unnoticed in the form of a dominion farmer or metropolitan housewife'.[47] Tasmanian apples held this civilizing power not just by taking over the landscape, but also by providing morally pure work for settlers and a healthy diet for empire families across the world. This concept is based on Anne McClintock's idea of commodity racism.[48] McClintock's seminal work *Imperial Leather* details how commodity racism became distinct from scientific racism; a small, cheap, portable commodity such as soap tangibly showcased the presumed superiority of whiteness through cleanliness. I argue that apples held a particularly potent level of commodity racism, similar to that of soap, due to their shared links to health and purity and the construction of whiteness.

The EMB's romanticization of an idyllic Dominion white settler orchard brought imperial power and scientific racism, such as eugenics, together, and 'then weaponized them via mass marketing'.[49] Promoting stylized imagery of imperial Britishness through comestibles symbolically linked the centre and peripheral British world with the significance and normalization of whiteness.[50] The promotion of neat, thriving orchards further consolidated the imperial position of the successful 'civilizing' mission through Australia's colonization and legitimization of Britain's ownership through a policy of *terra nullius*. *Terra nullius* has two different meanings: it means both a country without a sovereign recognized by European authorities and a territory where nobody owns any land at all.[51] As Henry Reynolds states, the doctrine of *terra nullius* had been based on the assumption that the indigenous peoples were primitive with no traditional system of land ownership at all, and therefore could not claim possession of it.[52] Commodity racism associating cleanliness and purity with whiteness was ultimately part of this attempted process of erasure of First Peoples.[53] This ideology played a significant part in ecological imperialism, expropriating unceded Aboriginal lands through the planting of crops, such as apple orchards.

The embodiment of 'commodity Britishness' within apples rationalized racial discourses that influenced Australian immigration policies world-wide such as the *Immigration Restriction Act 1901*. This legislation, also known as the 'White Australia Policy', restricted migration to Australia from 'non-white' countries and remained in place until 1973. Commodity advertising advocated that a 'white' colony supplied appropriately 'clean' produce, connecting whiteness, cleanliness, and race.[54] Empire fruit grown in the Dominions was implied to have been grown and packed by 'competent and clean' people with their race offering proof of their cleanliness and standards of production.[55] Comestibles grown and produced across Australia were expected to exude this same proof of 'whiteness', a central element of Australian identity. For example, cane sugar producers in Queensland overcame the ideological connections between slave labour and sugar by solely employing European 'white' workers.[56] This motivation for cleanliness was heightened not only by race anxieties but also by the creation of industrial foods in the early twentieth century and the removal of the urban public from the growth and production of their food.

At its peak, much of Tasmania was covered in apple orchards, reaching three million commercial trees by 1950.[57] Following the Second World War, the apple export market began to contract. Attitudes towards imperial preference, a waning relevance of empire, and rising nationalism in Australia diluted the fervour with which the imperial metropole was regarded. Maintaining character through consuming 'commodity Britishness' no longer held the same relevance as it had during the interwar period. Continued development in refrigeration technology also lengthened the time that apples could remain in good condition. When Britain joined the European Common Market and tariff preference ended in the 1970s, apple exports massively declined. Tasmanian apple production was virtually halved as Tasmania lost competitiveness in its traditional markets.[58] As a result the apple industry became increasingly regionalized and contracted to the Huon Valley in Southern Tasmania.

Despite the disappearance of the physical orchards from across the state, the cultural significance of apples did not wane; Tasmania retained its title as the 'Apple Isle'. Tasmania's branding as the 'Apple Isle' has remained persistent in mainland and international imaginings until today, even after an enormous reduction in Tasmania's apple orcharding industry. While there has been a significant reduction in the apple export industry, we have also seen the rise of the Tasmanian apple cider industry. In November 2021, Launceston was designated a UNESCO City of Gastronomy, one of only forty-nine around the world, recognizing it as a major food region of the world. Launceston is also proud to be hosting the Australian Symposium of Gastronomy this July, another event proclaiming the excellence of Tasmania's produce and culinary superiority. Tasmanian apples, introduced in the convict era and then entangled with the eugenics movement through Empire Marketing Board advertisements, remind us of how integral ingestion was to imperialism and as well to Tasmania's gastronomic future.

About the Author

Carla Baker is currently a History PhD candidate at the University of Tasmania after graduating with First Class Honours. Alongside her research, Carla also works in the family-owned mower and chainsaw shop and is mum to two beautiful girls.

Notes

1. William Metcalf, 'Utopian Communal Experiments in Tasmania: A Litany of Failure?', *Communal Societies*, 28.1 (2008), pp. 1–26 (p. 1).
2. Nathalie Servant, 'Apple Industry', *Companion to Australian History* <https://www.utas.edu.au/tasmanian-companion/biogs/E000048b.htm> [accessed 4 October 2021].
3. Erika Janik, *Apple: A Global History*, The Edible Series (Reaktion, 2011), p. 25.
4. Coultman Smith, *Town with a History: Beaconsfield Tasmania* (Beaconsfield Museum Committee, 1978), p. 7.
5. Jeffrey M. Pilcher, 'Culinary Infrastructure: How Facilities and Technologies Create Value and Meaning around Food', *Global Food History*, 2.2 (2016), pp. 105–31 (p. 109).
6. Gastronomy Northern Tasmania, 'Sometimes You Need an Outsider to Tell You Just How Special You Are' <https://www.cityofgastronomy.com.au/how-launceston-won-the-designation> [accessed 14 April 2024].
7. G.T.W.B. Boyes, *The Diaries and Letters of G.T.W.B. Boyes: Volume 1 1820–1832*, ed. by Peter Chapman (Oxford University Press, 1985), p. 267.
8. Michael Pollan, *The Botany of Desire: A Plant's-Eye View of the World* (Random House Publishing Group, 2002), p. 20.
9. Alfred W. Crosby, *Ecological Imperialism: The Biological Expansion of Europe, 900–1900*, 2nd edn (Cambridge University Press, 2004).

10. Katie Holmes, 'Gardens', *Journal of Australian Studies Special Edition: Imaginary Homelands*, 61 (1999), pp. 152–62 (p. 152).
11. Jessica Neath, '*Visions of Nature: How Landscape Photography Shaped Settler Colonialism* AND *Colonization, Wilderness and Spaces Between: Nineteenth Century Landscape Painting in Australia and the United States*', *Australian Historical Studies*, 54.1 (2023), pp. 179–81 (p. 181).
12. Katie Holmes, Susan K. Martin, and Kylie Mirmohamadi, *Reading the Garden: The Settlement of Australia* (Melbourne University Press, 2008), p. 17.
13. Robert Hughes, *The Fatal Shore: A History of the Transportation of Convicts to Australia, 1787–1868* (Vintage Books, 1987), p. 371.
14. Marilyn Lake, *A Divided Society: Tasmania During World War I* (Melbourne University Press, 1975), p. 1.
15. '"If You Eat Their Food . . .": Diets and Bodies in Early Colonial Spanish America', *American Historical Review*, 115.3 (2010), pp. 688–713 (p. 688).
16. John Bach, *A Maritime History of Australia* (Thomas Nelson, 1976), p. 247.
17. Rex Cox, 'The Tasmanian Fruit Trade', *Maritime Times of Tasmania*, 2020, p. 18.
18. [Anon.], news item, *Hobart Town Courier*, 27 June 1829, pp. 2–3.
19. Bach, p. 23.
20. Sharon Morgan, *Land Settlement in Early Tasmania: Creating an Antipodean England* (Cambridge University Press, 1992), p. 111.
21. Christopher Cowles and David Walker, *The Art of Apple Branding: Australian Apple Case Labels and the Industry since 1788* (Apples from Oz, 2005), p. 14.
22. Cowles and Walker, p. 10.
23. William Ebenezer Shoobridge, 'Tasmanian Apples in London', *Papers and Proceedings of the Royal Society of Tasmania*, 1892, p. iv.
24. Frances Steel, 'Re-Routing Empire? Steam-Age Circulations and the Making of an Anglo Pacific, c.1850–90', *Australian Historical Studies*, 46 (2015), pp. 356–73 (p. 373).
25. Nicolo Paolo Ludovice, 'The Ice Plant Cometh: The Insular Cold Storage and Ice Plant, Frozen Meat, and the Imperial Biodeterioration of American Manila, 1900–1935', *Global Food History*, 7.2 (2021), pp. 115–39 (p. 119).
26. Susanne Freidberg, *Fresh: A Perishable History* (Harvard University Press, 2009), p. 47.
27. Rachel Duffett, *The Stomach for Fighting: Food and the Soldiers of the Great War* (Manchester University Press, 2012), p. 36.
28. Duffett, p. 49.
29. A. Hocking, *The United Kingdom Demand for Southern Hemisphere Apples* (University of Tasmania, 1969), p. 1.
30. Angeliki Torode, 'Trends in Fruit Consumption', in *Our Changing Fare: Two Hundred Years of British Food Habits*, ed. by T.C. Barker, J.C. McKenzie, and John Yudkin (MacGibbon and Kee, 1966), p. 121.
31. Charles Smith, *Britain's Food Supplies in Peace and War* (Routledge, 1940), p. 156.
32. 'Tasmanian Apple Trade', *Maitland Daily Mercury*, 9 April 1912, p. 2.
33. Cowles and Walker, p. 11
34. R. E Macnaghten, 'Through Apple Land', *Pall Mall Magazine*, February 1895, p. 194.
35. Edward A. Bunyard, *The Anatomy of Dessert* (Dulau, 1929), p. 15.
36. Jessica Martell, *Farm to Form: Modernist Literature and Ecologies of Food in the British Empire* (University of Nevada Press, 2020), p. 7.
37. Christine Boyanoski, 'Selective Memory: The British Empire Exhibition and National Histories of Art', in *Rethinking Settler Colonialism: History and Memory in Australia, Canada, Aotearoa New Zealand and South Africa*, ed. by Annie E. Coombes (Manchester University Press, 2006), p. 157.
38. Felicity Barnes, *Selling Britishness: Commodity Culture, the Dominions, and Empire* (McGill-Queen's University Press, 2022), p. 10.
39. Uma Kothari, 'Trade, Consumption and Development Alliances: The Historical Legacy of the Empire Marketing Board Poster Campaign', *Third World Quarterly*, 35.1 (2014), pp. 43–64 (p. 45).

40 Felicity Barnes, 'Bringing Another Empire Alive? The Empire Marketing Board and the Construction of Dominion Identity, 1926–33', *The Journal of Imperial and Commonwealth History*, 42.1 (2014), pp. 61–85 (p. 65).
41 Barnes, 'Bringing Another Empire Alive?', p. 65.
42 Cowles and Walker, p. 31.
43 Barnes, 'Bringing Another Empire Alive?', p. 74.
44 Barnes, 'Bringing Another Empire Alive?', p. 68.
45 Troy Bickham, 'Eating the Empire: Intersections of Food, Cookery and Imperialism in Eighteenth-Century Britain', *Past & Present*, 198 (2008), pp. 70–109 (p. 74).
46 Barnes, 'Bringing Another Empire Alive?', p. 71.
47 Barnes, *Selling Britishness*, p. 164.
48 Anne McClintock, *Imperial Leather: Race, Gender, and Sexuality in the Colonial Contest* (Routledge, 1995), p. 209.
49 Barnes, *Selling Britishness*, p. 68.
50 Kate Darian-Smith, Patricia Grimshaw, and Stuart Macintyre, 'Introduction: Britishness Abroad', in *Britishness Abroad: Transnational Movements and Imperial Cultures*, ed. by Kate Darian-Smith, Patricia Grimshaw, and Stuart Macintyre (Melbourne University Publishing, 2007), pp. 1–16 (p. 11).
51 Henry Reynolds, *The Law of the Land* (Penguin, 2003), p. 15
52 Reynolds, p. 2.
53 Max Haiven, *Palm Oil: The Grease of Empire* (Pluto Press, 2022), p. 41
54 Barnes, *Selling Britishness*, p. 76.
55 Frank Trentmann, 'Before Fair Trade – Empire, Free Trade and the Moral Economies of Food in the Modern World', in *Food and Globalization: Consumption, Markets and Politics in the Modern World.*, ed. by Alexander Nützenadel and Frank Trentmann (Berg, 2008), p. 259.
56 Stefanie Affeldt, 'The Burden of "White" Sugar: Producing and Consuming Whiteness in Australia', *Studia Anglica Posnaniensia*, 52.4 (2017), 439–66, (pp. 445–46).
57 'Too Many Old Orchards for Good of Industry', *Mercury* (Hobart), 12 July 1950, p. 13.
58 S. Grosvenor and L.J. Wood, '"Bye, Europe. Buy Asia": Reorientation of Tasmanian Apple Production', *Geography*, 81.2 (1996), pp. 170–73 (p. 170).

5
More Than Wine
Vineyards and Grapes in Medieval and Early Modern Dutch Gardens

Mariëlla Beukers

As soon as the word '*wijngaard*', which in medieval or early modern Dutch can mean either vineyard or grape vine, is used in period sources, the association is with wine and wine production. Countless researchers, finding mentions of vineyards in Dutch medieval and early modern sources, have assumed a long-lost production of wine, sometimes even bringing into the discussion the possibilities of warmer temperatures in medieval times.[1] But indications of serious wine production in the Netherlands, the northern part of the Low Countries – outside of the province of South Limburg – have not been found so far, and conditions for making wine were not very favourable. The Netherlands are positioned above the fiftieth parallel, which runs roughly through Louvain (in Belgium) and Maastricht (the extreme south of the Netherlands). There, from the end of the fourteenth century until the late twentieth century, average temperatures simply did not reach above the ten degrees Celsius needed for viable wine production. It is therefore in agreement with this modern knowledge that only in those two areas, around Louvain in Belgium and around Maastricht in South Limburg, is actual proof of viniculture available.[2] Still, many medieval and early modern sources mention vineyards in the part of the country far above the Louvain-Maastricht line. They range from garden manuals to poetry, from estate maps to account books. From these sources it is clear: vineyards were an integral part of many gardens in the northern part of the Low Countries for centuries, in the vicinity of medieval castles and other noble residences as well as on the seventeenth-century country estates of rich merchants and burghers, and even in towns. And this in times when the Little Ice Age extended its grip on the Netherlands! If wine production was not likely, and we do not have further evidence of it, we are invited to consider other uses of the vines to explain those vineyards.

In this paper, I will present two examples that demonstrate that vineyards were an important part of gardens in the Netherlands that were never large agricultural plots of land intended for commercial grape production or viniculture – at least not until the late twentieth century. In laying out these examples, it will become clear that grapes in the northern part of the Low Countries, the present-day country called The Netherlands, were planted for other economic reasons than wine. And there were also social reasons to plant vines, motivated by demands of society or new ways to spend time, whether religiously inspired or not. First, we will delve into the account books of the counts of

Blois, lords of the small town of Schoonhoven at the end of the fourteenth century. Second, the garden manuals of the seventeenth century will illustrate the role vineyards played in the estate gardens of the wealthy, rich burghers and nobility alike. From these two examples we will draw our conclusions.[3]

A Medieval Castle Garden

John of Chatillon, in Dutch history often referred to as Jan van Blois (*c.* 1342–1381), was second cousin to Albrecht of Bavaria (1336–1404), count of Holland. John grew up in France, a son of the count of Blois. From his maternal grandfather, he inherited several lordships in the Low Countries, among them the town of Schoonhoven, in the present-day province of South-Holland. His overlord for Schoonhoven was the count of Holland. The town lies in the middle of vast, flat polders, and the countryside does not resemble any vineyard landscape we are used to: flat, green pastures, lots of water, willow groves. In Schoonhoven Jan restored and expanded the castle his grandfather had already used, which would become John's favourite residence. Around 1365, John's gardener started laying out a garden. The garden was positioned in close proximity to the castle, although determining its exact position has proved impossible. From the entries in the account books, it is clear that this was a noble garden like elsewhere in Europe: part pleasure garden, part source of fruit for the kitchen and the table. A vegetable garden was also present. The garden had low fences, a gate with a lock, and grass benches; roses and daisies adorned the plot, along with fruit trees, red currant bushes, and vines. For the maintenance of those vines, vast amounts of wooden beams, in all kinds of forms, were bought, as were slats, willow wickers, and iron nails to hold the beams together. Sometimes the slats were bent to form arches. The willow wickers were used to tie the slats together, and to bind the shoots of the vines to the slats.

Building a picture based on this wooden material, it is very likely that the vines were not laid out in neat rows and trained along stakes or poles, but instead they were guided over strong wooden structures resembling, for example, the present-day pergolas in Alto Adige used for training the vines, or the *treilles* of the French medieval gardens: covered alleys with arches, to sit and walk under, out of the sun. This image of vines over large wooden structures is also sketched for the English royal gardens in the fourteenth century.[4] This image for Schoonhoven is confirmed by many other anecdotal references in the accounts: at one point trestles were ordered for Hanssen, the gardener, to stand on, 'while he could not reach the vines from the ground'. Or we read that the vineyards had fallen and needed to be erected again, whereupon more wooden beams were bought.

Having these kinds of vineyards over wooden structures also means that another product of the vines comes into view: verjuice, the sour juice of unripe grapes. In the gardens of the French king in fourteenth-century Paris, the *treilles*, arched wooden structures to train vines over, were used to grow grapes for verjuice; actual vineyards for wine were in other parts of the grounds, outside of the gardens. The same evidence is available for the gardens in Germolles, a property of the dukes of Burgundy.[5] An image in a fifteenth-century copy of the *Tacuinum Sanitatis* confirms the use of large wooden structures for the growing of grapes for verjuice (Figure 1). The caption mentions '*agrestum*' (verjuice), and two men are seen crushing the grapes in a mortar and making verjuice below a large wooden structure, built with horizontal beams and vertical slats.

Of course, pergolas and other wooden support systems were (and are) used in viniculture since Roman times, but in fact verjuice is exactly what is mentioned in the accounts of John of Chatillon, and later also of those of his brother Guy, who succeeded John in 1381. Until 1377, verjuice enters the accounts as a commodity bought in Antwerp or Dendermonde, in the south of the Low Countries. From 1377, however, verjuice is homegrown, produced in the Schoonhoven castle gardens, and there are no more entries of verjuice bought elsewhere. From 1377 on, Hanssen the gardener yearly

Figure 1. Making verjuice below a wooden structure covered with vines. From Tacuinum Sanitatis, *BnFr Latin ms 9333, Bibliothèque Nationale de France, Paris.*

ordered several empty 128-litre barrels for his verjuice. The amount produced differs from year to year but adds up to 1040 litres in 1394, the last year we have evidence of verjuice production. After the death of Guy, the castle was occasionally used by the counts of Holland until it burned down in 1518.

The verjuice produced in Schoonhoven was used in the kitchens, for which we have charming evidence. In the accounts of 1365–1366, when the vines were first planted near the Schoonhoven castle but of course did not yet produce grapes, the cook, called Cattenborn, paid the local baker, Hein Witten, for the grapes that grow in front of his house, 'to daily make fresh verjuice'. And in the last years of the lordship of Guy, when Hanssen the gardener was dead, it was a cook who took care of the vines.

Status at the Table

John and Guy were important nobles in their time, John even ruling the county of Holland for a few years in absence of his uncle Albrecht of Bavaria. They had connections to all the relevant noble courts in the late fourteenth century. Guy was a hostage at the

English court in his youth and later served the French king; John joined several of his peers in a crusade to Prussia, winning his knighthood on the occasion. Others have concluded that the court John held at Schoonhoven was lavish, a place where literature and music were promoted, tournaments were held, and hunting parties were organized.[6] Dining and banquets were part of this culture and augmented John's status. Guy did not reside often in Schoonhoven, but all the same expanded verjuice production from 1382 onwards. It is likely the verjuice was transported to wherever Guy was residing (mostly in Hainault), together with other local foods, such as huge sturgeons, already cooked and prepared by the cooks. The accounts mention several trips of cooks laden with such foodstuffs to Hainault.

To serve the right dishes at banquets, verjuice was a necessary condiment to make the appropriate sweet-and-sour sauces. The French food historian Jean Flandrin has estimated that, in the fourteenth century, sixty to seventy percent of the recipes from cookbooks from northern France made use of sour ingredients like wine, vinegar, or verjuice.[7] Since the courts of John of Blois and his uncle Albrecht of Bavaria were closely linked to northern France, we can safely assume that the same applied to the dishes served in the courts of Holland. For a three-day dinner held in honour of the betrothal of Albrecht's daughter Catherine to Eduard of Gelre in January 1369, over 900 litres of verjuice were ordered, to be used in dishes like green *brouet*.[8]

For nobles in regions further south, having vines surrounding their castle was an expression of status, as Roger Dion argued in 1952.[9] Vines along your castle walls or vines in your gardens expressed hospitality, the idea that at this place you would be received with good wine. For the northern part of the Low Countries, making wine from vines around 1370 was not an option; certainly, in some years you could try to make wine, but the quality was never as good as imported wines. In fact, the accounts of John and Guy of Blois testify to the import of large amounts of Malvasia, Greek wines, and wines from the Rhine region. But still, having your own vines was important, maybe even to express that same image of hospitality. The product, however, was not wine, but verjuice, which was at that moment just as important for the table as wine, since it was needed to serve the right dishes at table. Verjuice was so important that John and Guy ordered the production of it in the garden of their Dutch possession, Schoonhoven. Their vineyards served more than one purpose: as producers of verjuice for the table, but also as embellishment and status-enhancing greenery near their castle.

Vineyards for Pleasure

All through the sixteenth century, vineyards keep appearing in Dutch source material, mostly in the gardens of castles and country estates. Also, within city walls vines grew regularly. In 1567, Ludovico Guicciardini, an Italian merchant living in Antwerp, published a famous work describing both the southern and northern part of the Low Countries. On vineyards he notes: 'Concerning vineyards, those you will find enough, of different kinds, in cities and villages but very little in the field; because it seems time and season are not suited to it.'[10] As in medieval times, the vines were more often trained over wooden structures (Figure 2).

Gardens, Flowers, and Fruit

Figure 2. Wooden archways supporting the vines (visible on the right) were part of sixteenth century gardens. Engraving by Pieter van der Heyden, 1570. Rijksmuseum Amsterdam.

In the seventeenth century, most vineyards were found in the gardens of the newly created country estates, the summer homes of the rich merchants of Amsterdam and other cities. They feature for example in a special kind of poetry, the so-called estate poetry, inspired by works from Antiquity such as Virgil's *Georgics* and Horace's *Beatus Ille*. The estate poems extolled the virtues of living in the countryside and speak of inviting your friends to your gardens to dine there on the vegetables, fruit, and other foodstuffs the estate produced (the so-called *dapes inemptae*, unbought gifts). For this same clientele, a collection of garden manuals and household books appeared: how to lay out your estate, how to plant and maintain your garden, and in one instance even how to prepare food from the estate gardens. The oldest cookbook originating in the northern part of the Low Countries, *The Sensible Cook* from 1667, is part of a multipart manual for estate owners.

In most of those manuals, how to prune and maintain vines is extensively explained. Never do the authors elaborate on wine making or techniques to produce the best quality of wine grapes. Only manuals that are translations of French or German works mention wine making. One such work was written by Charles Estienne or, in Dutch, Karel Stevens (1504–1564), a French doctor of medicine. In 1554 he published *Praedium rusticum*, which deals with agriculture and horticulture. Many translations of this work appeared, including in Dutch in 1588. The second, extended, Dutch edition was published in 1594, and more editions followed. Several chapters in this manual deal with vineyards, and even grape varieties are mentioned. The content of original Dutch works is different, however. Lutheran minister Pieter van Aenghelen, the author of the first edition of *De verstandige hovenier* (*The Sensible Gardener*), from 1662, wrote that the vine needs special attention, but he deems it unnecessary to discuss things like pressing, cleaning barrels, filtering, and so on, 'because in the Netherlands vineyards are only planted for pleasure'.[11] And in 1700, another author, Hendrik van Oosten, states: 'The vineyard in these cold lands is never or seldom planted to make wine, but mostly to eat [the fresh grapes] or to make verjuice. The hotter the original country the vines come from, the lesser it is likely they become ripe'.[12] Another work by Pieter van Aenghelen, published in 1663 and titled *Herbarius, kruyt en bloem-hof* (*Herbarium, Herb- and Flower Garden*), was intended for 'gardeners, for the sick, for curious enthusiasts to herbarize and for people with small purses who cannot buy large herbaria.' The minister gives a definition under the lemma '*Wyngaert*': 'That is a noble crop, with long tendrils and broad leaves, [it] is well known everywhere'. What follows is a description of the place the vine should be grown and the time of year the fruit ripens: 'She asks for a warm place, well placed in the sun. The wine grapes become ripe in the autumn months, when it is a warm summer here in the Netherlands.'[13] Then comes a long list of uses for the grapes, the leaves, and the young shoots. The tendrils or leaves are cooling and astringent. The crooked shoots of the vineyard bruised, and the juice ingested stills the bellyache. The water that drips from the tendrils is good for kidney stones. It also cures scabies, spots, and leprosy, but one must rub the spot with saltpetre first. Vine ashes mixed with vinegar cures haemorrhoids. Verjuice made from unripe grapes cools and strengthens the stomach. Drunk warm, it is good for shortness of breath. Among all fruits picked in autumn, wine grapes are the best; they also nourish more than others, except for figs. Grapes that have been cut off and hung long are good for the stomach. It is striking that wine is not mentioned by Van Aenghelen: fruit, leaves and shoots all had very practical uses other than wine!

The presence of vineyards in the gardens of Dutch estates in the seventeenth century is confirmed by other sources. For example, on 2 March 1680, Margaretha Turnor, wife of Godard Adriaan van Reede, lord of Amerongen, wrote her husband that she was busy restoring the gardens near their castle in Amerongen. In the years before, French troops had burnt the castle and laid the grounds to waste. Now, the castle was rebuilt, and it was time to fill the gardens again. She wrote that the gardener was preparing the soil for the vineyards, but that she was still trying to find 'good sorts of grapes'.[14] On a 1744 map documenting the fruit trees in those gardens, the 281 fruit trees include 16 '*wijngaards*'.[15] They are positioned separately, one by one, along the walls of the garden, intermixed with the apple, pear, and plum trees. And when estates are offered for sale in seventeenth- and eighteenth-century newspapers, the vineyards are often mentioned as extra incentive to buy the available house, gardens, and grounds.

Fascination and Study

In the fourteenth century, the grape vines in Schoonhoven delivered verjuice but also status. Practical reasons went hand in hand with social aspects. In the seventeenth century, something similar occurred: not only did the vineyards have practical uses, but they were also planted for social and maybe even religious reasons as well given the intense botanical curiosity of the time. Studying nature, in all its fascinating aspects, started in the sixteenth century with famous botanists like Clusius, but spread rapidly to rich burghers and nobles. A source of obtaining vines in the Netherlands was surely the network of interested collectors of exotic plants. Rich burghers had contacts all over the world to acquire new plant material. Contact with wine growing countries was extensive; the import and promulgation of vines must have been present but remains understudied at the moment. A first look at the many advertisements from arborists tells us that arborists sold vines.[16] In 1639, arborist Willem van der Stoop from Utrecht delivered not only elms but also pear and plum trees and vines for the castle in Buren, owned by stadholder Frederik Hendrik, son of William of Orange.[17] The description on the 1680 map of the estate Randenbroek near Amersfoort mentioned a 'vineyard hill provided with more than 40 types of the best grapes'.[18] The presence of many different varieties in one garden gives a clear indication of the interest of the estate owner in the diversity of vines and grapes. Modern architecture historian Erik de Jong has pointed to the religious inspiration of garden art in Protestant Holland.[19] In the gardens, nature and art worked together to create the most perfect result possible: a recreation of Paradise. God's greatness was expressed in nature as it was shaped and moulded in the gardens. Thereby, the garden was an expression of God's Second Book, the Book of Nature, and needed to be studied. Collecting plants helped in the study of that Second Book.

Being able to maintain your garden, to prune the vines, also added to your status as estate owner. By pruning, you were able to create order in chaos, to keep your gardens in order. A vineyard left untended becomes tangled and unproductive, and pruning is highly necessary. Of course, the parallels with keeping your house and business in order are clear. It is with this in mind that the garden manuals focused on explaining how to prune. It is also relevant that most of the work in the gardens was done by the gardener, but pruning the vines was considered an activity worthy of the owner himself.[20] In a letter from March 1654 to a fellow poet, the famous Dutch author Jacob Cats writes, 'While I plant trees and prune my vines, I think of the house consisting of seven planks.' He is meditating on his death.[21]

Conclusion

From the fourteenth century onwards, vineyards did not produce wine in the northern part of the Low Countries, but nevertheless were planted extensively. The grapes were used for verjuice, or were eaten fresh, and all other parts of the plant had a purpose as well, illustrating a root-to-shoot approach to foodstuffs. Other reasons for planting and maintaining vines were present: the verjuice production in the Middle Ages made sure the right dishes could be produced, thereby adding to the lustre and status of the court. In the seventeenth century, studying nature or the Book of Nature were stimuli to plant different kinds of grape varieties. Pruning the vines kept the gardens in order and was deemed a fitting activity for a gentleman. The order it created in the gardens reflected on the man and his household. In the seventeenth century, the fruits of the vine were eaten, verjuice was still used in the kitchen (although not much longer), and leaves and shoots all had other applications.

The immediate association of vineyards with wine by researchers who come across vineyards in the sources is understandable, but unfortunately has obscured our view of other uses of the vine, especially in countries where the climate was far from favourable for producing good wine. Vines in the northern Low Countries grew in gardens and along walls, not in the open fields. They were part of castle and estate gardens, adding another source of fruit to the table, in non-fermented form, and giving pleasure and meaning to working or residing in those gardens. They were also part of city gardens, for similar reasons. Viticulture was present in the medieval and early modern Netherlands, in several forms and places, but viniculture had to wait to the late twentieth century, when climate conditions and new grape varieties made wine production a possibility at last.

About the Author

Mariëlla Beukers is a Dutch historian and wine specialist. She is working on her PhD at the University of Utrecht, focusing on the history of vineyards in the Netherlands between 1300 and 1800.

Notes

1. This attitude is not only encountered for the Netherlands, but also for countries like Norway and Scotland. Mentions of grapes, vines, or vineyards in the Middle Ages have been interpreted as 'proof' of viniculture. See for example T. Unwin, *Wine and the Vine: an Historical Geography of Viticulture and the Wine Trade* (Routledge, 1996), p. 150. It would be interesting and worthwhile to test those mentions against what I argue in this paper.
2. The south of the present-day Netherlands, the province of South Limburg, is the only region where indications of medieval and early modern viniculture are present. Due to the extremely complicated political history of the region, no study has as yet been made of the nature and character of this viticulture. This paper focuses on the regions north of South Limburg.
3. This paper is part of my ongoing PhD research into the nature and position of medieval and early modern vineyards in the northern part of the Low Countries. Part of this research has been published: Mariëlla Beukers, '*Boven de 50ste breedtegraad. De middeleeuwse wijngaarden van Schoonhoven en Gouda*', *TSEG* 20.2 (2023), pp. 5–32, DOI:10.52024/tseg.11481.
4. Howard M. Colvin, 'Royal Gardens in Medieval England', in *Medieval Gardens*, Dumbarton Oaks Colloquium on the History of Landscape Architecture (Dumbarton Oaks, 1986), IX, pp. 7–22.
5. Gerdi B. Krebber, '*Koninklijke Tuinen in Parijs*', in *Tuinen in de Middeleeuwen* (Verloren, 1992), pp. 203–25; Corinne Beck, Patrice Beck, and François Duceppe-Lamarre, '*Les parcs et jardins des ducs de Bourgogne au XIVe siècle. Réalités et représentations*' ('*Aux Marches du Palais*'. *Qu'est-Ce qu'un palais médiéval? Données Historiques et Archéologiques*), Actes du VIIe Congrès International d'Archéologie Médiévale (Société d'Archéologie Médiévale, 2001), pp. 97–111.
6. F.P. van Oostrom, *Het woord van eer. Literatuur aan het Hollandse hof omstreeks 1400*, 5th edn (Ooievaar, 1996).
7. Cited in Perrine Mane, '*Raisin, vin, vinaigre, verjus dans les traités culinaires... ou "Dans la vigne tout est bon"*', *L'Atelier du Centre de recherches historiques. Revue électronique du CRH*, 12 (2014), DOI:10.4000/acrh.6000.

8 J.M. van Winter, 'A Wedding Party at the Court of Holland in 1369', in *Spices and Comfits: Collected Papers on Medieval Food* (Prospect Books, 2007), pp. 303–18.
9 Roger Dion, *Histoire de la vigne et du vin en France: des origines au XIXe siècle*, Repr. (CRNS Editions, 2011).
10 *Descrittione di tutti i Paesi Bassi, altrimenti detti Germania Inferiore*, in 1612 translated into Dutch as *Beschryvinghe van alle de Nederlanden* (Willem Jansz., 1612), p. 7, translation by the author.
11 Pieter van Aenghelen, *De verstandige hovenier, over de twaelf maenden van 't jaer* (Marcus Willemsz. Doornick, 1662), p. 82.
12 Hendrik van Oosten, *De nieuwe Nederlandse bloemhof, ofte de sorgvuldige hovenier* (Joh. du Vivie en Is. Severinus, 1700), p. 35.
13 Pieter van Aenghelen, *Herbarius, kruyt en bloem-hof* (Ian Ioosten Appelaer, 1663), pp. 356–57.
14 Letter of 2 March 1680 in Het Utrechts Archief, Archief Huis Amerongen, inv. no. 2725.
15 Het Utrechts Archief, Archief Huis Amerongen, inv. no. 1572.
16 Lenneke Berkhout, *Hoveniers van Oranje: functie, werk en positie: 1621–1732* (Verloren, 2020).
17 Lenneke Berkhout, 'De Utrechtse boomkwekersfamilie Van der Stoop', *Tijdschrift Oud-Utrecht* 4 (2023), pp. 4–7.
18 W.J. van Hoorn, *Een hofstede genaamd Randenbroek: van leengoed tot stadtspark* (Bekking, 1991), p. 59.
19 E. de Jong, *Natuur en kunst: Nederlandse tuin- en landschapsarchitectuur, 1650–1740* (Thoth, 1993).
20 Willemien B. de Vries, *Wandeling en verhandeling: De ontwikkeling van het Nederlandse hofdicht in de zeventiende eeuw (1613–1710)* (Verloren, 1998), p. 49.
21 Letter from Jacob Cats to Jacob Westerbaen, 16 March 1654, printed in Jacob Westerbaen, *Ockenburgh* (Anthony Tongerloo, 1654).

6
Some Mediterranean Culinary Isoglosses
Perry's 'Olive Line' in the Broader Context
of the Umami-Enhancer Isogloss Bundle

Anthony F. Buccini

For Charles Perry

Perry's 'Olive Line'
The olive as a fruit is peculiar, though not unique, in a number of respects: it is naturally unpalatably bitter but to many delicious when cured, and its flesh has a relatively high content of particularly salubrious oil which is flavourful and well suited to consumption both raw and in cooking. In addition, its environmental requisites are such that the olive tree thrives best in precisely those conditions which obtain around the Mediterranean Sea – and not elsewhere in that part of the globe – where summers are long, dry, and fairly hot and winters are wet and cool but largely frost free. These requisites render higher elevations unsuitable to olive trees, and consequently the olive can be cultivated for the most part only in areas fairly near the sea itself, with but limited exceptions. Given the biological exigencies of the olive tree, the characteristics of the olive fruit, and the millennia-long interplay of these natural conditions with the development of related cultural behaviours, it is not surprising that, across the entire Mediterranean region, there is a considerable degree of unity not only in the basic methods of cultivating the olive and exploiting its fruit but also a measure of commonalities in pairings with other foods and seasonings: Both in the curing and consumption of olives themselves and in the extraction and culinary uses of olive oil, we find *grosso modo* very similar traditions from east to west and north to south. This general unity is unsurprising given the nature of this food and, moreover, the known diffusion of the culture of olive cultivation and exploitation from essentially one small zone at the eastern end of the Middle Sea.

There is, however, one noteworthy – though generally unnoticed – divide in the use of the cured olive fruit across the Mediterranean area. One of very few food historians who has noticed this divide is Charles Perry, who refers to it as the 'Olive Line', and it was, indeed, he who first called my attention to this feature of culinary geography in a conversation at the Oxford Food Symposium some years ago. In a very brief discussion appearing in the *LA Times*, Perry (1997) suggests that this Olive Line 'runs slantwise through the olive-raising world from the Balkans to North Africa'. To the east of this line lies an eastern Mediterranean zone, in which 'olives are eaten raw' and otherwise are used as a garnish in salads and 'occasionally even cooked dishes but rarely, if ever, are they cooked'. To the west of the line, there is an area comprised of Italy, southern France, Iberia, and the Maghreb, where olives are regularly included as a cooked ingredient in a wide array of prepared dishes. In Perry's view, the Balkans represent 'a mixed zone'.

It is remarkable that, despite the great interest in Mediterranean foodways on the part of both popular and scholarly food writers and their audiences, this culinary divide between eastern and western Mediterranean zones has gone almost completely unnoticed. In my title I refer to this dividing line as an 'isogloss', a term I borrow in this context from the linguistic subfield of dialectology, where it denotes a geographical boundary between two differing realizations of some specific linguistic feature, be it different words referring to the same thing, different meanings of the same word, or contrasting phonetic realizations of a given phoneme; these are all examples of relatively minor, even superficial, dialectal differences, but of course there exist also geographic boundaries that involve differences of features of more fundamental structural import. Where a number of isoglosses run together in close proximity to one another, they can be referred to collectively as an 'isogloss bundle', and, depending on their structural significance, they may together represent a minor or major dialect break or even a division between what we consider to be two ('genetically' related) languages. In a culinary context, an isogloss such as the division between areas where olives are used in cooked dishes or not so used might analogously represent a relatively superficial or a structurally significant contrast in regional foodways.

The Olive Line seems to make no sense in reference only to the culinary properties of the olive itself, nor does it seem to find any ready explanation in terms of broader cultural meanings that the olive might bear. Considered in broader perspective, however, the Olive Line is found to be but one culinary isogloss in the Mediterranean zone involving ingredients that can be added to savoury dishes in order to enhance greatly their level of umami flavour. Though this bundle of umami-enhancer isoglosses roughly divides the Mediterranean zone into western and eastern halves, the division is especially sharp between southern Italy, where not only olives but also fermented anchovies, cured pork products, aged cheeses, and other umami-rich seasonings are used with great frequency in a wide array of traditional dishes, and Greece, where these ingredients are at most seldom used in cooked dishes. That the division is so stark between these two cuisines, which share not only very similar environments and thus very similar sets of foodstuffs but also historically deep and long-standing cultural connexions, is surprising, to say the least. Indeed, this hitherto unnoticed bundle of culinary isoglosses represents a fundamental structural difference in what I call the 'grammars' (Buccini 2016) of southern Italian and Greek cuisines and raises important and difficult questions regarding the history of Mediterranean foodways. This paper is a preliminary report on ongoing research, and definitive answers to the questions raised by the recognition of the umami-enhancer isogloss bundle – insofar as they *can* be found – are yet to be found. Nonetheless, it is clear that some basic assumptions we have long held concerning Mediterranean cuisines are challenged.

The Olive Line Considered Further

Among the peculiarities of the Olive Line is the fact that this isogloss has only to do with the use of the cured olive as an ingredient – and I would say as a flavouring agent – in cooked dishes. In both halves of the Mediterranean zone, olive oil is, of course, a staple in cooking and also used extensively in uncooked form to dress both raw and cooked foods, as well as to help preserve some foods. And likewise, across the entire Mediterranean area, cured olives are consumed regularly as a food unto themselves, most especially as a snack, usually with bread, or as one of a more elaborate array of simple dishes preceding the main elements of a meal, that is, as an *antipasto*, *tapa*, or *meze*. Though there may be some differences in levels of consumption between the various Mediterranean lands, this use of the cured olive is ubiquitous, popular, and undoubtedly very, very old.

As with olive oil, cured olives vary to a degree in their flavour according to the cultivar in question, and there are many cultivars used across the Mediterranean lands with noteworthy variation in taste and texture. Though new cultivars can be bred to emphasize or combine certain qualities over time, there are also a number of short-term decisions which a producer of cured olives can make that affect flavour, appearance, and texture. Most obvious of these is the timing of the harvest – green, semi-ripe, or black/ripe – which impacts all basic qualities, as does the next decision in importance, the method of curing, the overarching purpose of which is to remove the chemical compounds naturally present in the olive fruit responsible for its intense, unpalatable bitterness. Wet-curing typically is done by soaking the raw olives in brine, though alternate methods are also used, such as the less common use of very frequent changes of plain water or the use of a lye bath, favoured in modern industrial processing in some olive-producing regions. The alternative to wet-curing is the method of dry-curing by packing the olives in salt, a process which renders the fruit shrivelled but with very intense flavour profiles. All traditional methods lead to a process of fermentation of the olive which introduces to its flavour profile a fairly strong umami element which stands alongside the usual residual bitter element and whichever secondary flavour 'notes' (fruity, nutty, sweet, etc.) are particular to a given cultivar's fruit.

Further addition of flavour can occur through the means of preservation of the cured olives. The simplest method is to lay them up in brine, which introduces no new taste elements, but in some regions and countries acidic agents are also used, with the addition of vinegar being well-known from such popular varieties as the Greek Kalamata olives and with lemon being a widespread addition, especially in the Maghreb and Middle East. While these acidic additives help with preservation, they obviously also contribute a further pleasant sour note to the flavour profile. In the case of dry-cured olives, placing them in a bath of olive oil is used, which aids preservation and flavour and also helps to restore some plumpness. Finally, cured olives can be further seasoned in myriad ways through the addition of spices and herbs.

Given the great popularity of cured olives as a food on both sides of the Olive Line, it is clear that there is no basic or broad objection to the flavour profile(s) of this fruit anywhere in the Mediterranean world. One therefore naturally wonders whether the general abstinence in the eastern Mediterranean cuisines from the use of the olive in cooking might reflect some higher order culinary principle. Perry (1997) considers one such possibility, observing first that, even if we normally do not think of the olive as a fruit on account of its primary function as a source of oil and further its inherent bitterness, it is from a biological standpoint just that, a fruit, and in the eastern Mediterranean, the omission of the olive in cooked dishes might be seen as following from this status:

> [The olive is] treated as a fruit in a way you might not expect. Turkey and the eastern Arab countries – Egypt, Syria and Lebanon – generally don't go for the idea of cooking meat with fruit. This is really a new thing, because there was no such taboo in the Middle Ages, but it extends to olives.

Especially with regard to the culinary history of the eastern Mediterranean, one should think twice about disagreeing with Perry, but here he seems only to suggest this association as a possible reason for the existence of the Olive Line. Perhaps medieval Arab and Turkish scholars discuss the matter, but my inclination is to think it likely that biological status in matters culinary is at best of secondary import to taste and function, a matter often discussed in connection with the status of the tomato in modern times, though we might note that in earlier times, when many common fruit varieties were far less sweet than they are now, the current strong association of fruit with sweetness that we have was less strong or even absent.

Another approach to explaining the Olive Line is to consider whether there is some non- or extra-culinary cultural association that the olive bears which lies behind the eastern Mediterranean treatment. Such a cultural association is in my view most definitely connected to a major difference between the treatment of olive oil in the rules for fasting in the Roman Catholic tradition on the one hand and the Orthodox tradition on the other (Buccini 2012: 71ff.). Briefly, in the Orthodox Church olive oil is treated as a sacred substance much as wine is (and blood as well), and like wine its consumption is forbidden on the most solemn days of fasting/abstinence, whereas in the Roman Catholic Church, abstinence rules revolve in simpler fashion around the opposition of meat and other terrestrial animal products versus aquatic and vegetable products; olive oil, as the product of a plant, is permitted on even the most solemn days and in the western Mediterranean lands was the primary lipid for such occasions when pork fat and butter were obviously to be eschewed. As I discuss in the 2012 paper, however, what is perhaps most striking and culturally significant about the Orthodox treatment of olive oil in the rules of fasting is that it is *only* olive oil that is treated as a sacred substance: the olive itself enjoys no such status and is, indeed, one of the primary foods permitted on those most solemn holy days. In light of this fact, the coincidence of the Olive Line and the division between Roman Catholic and Orthodox spheres which fall together in the Adriatic between Italy and Greece on the European side of the Mediterranean appears to be coincidental and clearly has no explanatory power, considering the fact that on the southern side of the Middle Sea, the Olive Line divides a western group of Muslim lands from an eastern one.

The Umami-Enhancer Isogloss Bundle

On first inspection, the Olive Line represents an extremely interesting but puzzling aspect of Mediterranean culinary geography, given a) the fact that the eastern area generally rejects the olive as an ingredient in cooked dishes while at the same time being an area where this same food is otherwise extremely popular, and b) the fact that the isogloss runs roughly north to south and thus at once divides both European and Muslim lands and so joins together opposing sets of Christian and Muslim culinary cultures.

In linguistics, analogous situations are known, where we find isoglosses running in similar patterns, that is, across borders between unrelated or only distantly related languages with unity regarding the feature in question being between unrelated languages (or dialects of unrelated languages). An example of such a situation is found in the southern Low Countries, where a group of isoglosses runs *across* the Germanic/Romance linguistic boundary, separating on the one side the Dutch dialects of Brabant from the Dutch dialects of Limburg and on the other side of the boundary separating the Walloon dialects of Brabant from the Walloon dialects of Liège; in other words, in certain specific respects the Dutch of Limburg and the Walloon of Liège together are in agreement against the Dutch of Brabant and the Walloon of Brabant. We would naturally expect the speakers of the Dutch Brabantic and Limburgish dialects to form more of a speech community with each other than with their Romance-speaking Walloon neighbours to the south, and, in a general historical sense, they have, but in a certain period of history, political, ecclesiastic, and economic factors drew together speakers from the two sides of the Germanic/Romance boundary into habitual interactions that led to more or less widespread bilingualism (at least among culturally influential circles) and the development of the seemingly counterintuitive isoglosses. In other words, for a time, there was a degree of linguistic convergence across that Germanic/Romance boundary reflecting a period of intense social interaction, a convergence (albeit here on a limited scale) which gives rise to what linguists refer to as 'areal phenomena'.

The situation described just above under point a) is sufficiently peculiar that it seems the geographical situation under point b) is quite unlikely to have arisen through coincidence. And if that inference is correct, it seems most likely that the Olive Line is, as we find described in linguistics, an areal phenomenon, and that, at some time in the past, cultural contacts – and specifically culinary discourse (Buccini 2016) – drew together western Christians and western Muslims or eastern Christians and eastern Muslims and thus brought about the east-west divide.

A question we have yet to address is whether the Olive Line represents a culinary contrast of any significance, and it must be said that on first inspection, though very curious, it seems to involve a minor element, not some deeper or broader 'grammatical' difference. If we are, however, correct that it must have gained its present form through areal convergence in one or the other half (or both halves) of the Mediterranean, the Olive Line should not be an isolated isogloss, and I believe it is not.

To render the argument more concrete, let us consider briefly a recipe which appears in Barron's cookbook *Flavors of Greece* for a dish which is named there in English '*Macarónia* with Olive, Anchovy, and Sweet Tomato Sauce' and in Greek '*Macarónia me*

Eliés ke Sáltsa Domátas' (1991: 314–15). This recipe stands in apparent violation of the Olive Line: a Greek recipe which includes cooked olives and, indeed, within this cookbook, it is the only such recipe other than one for fried eggplant which is, in effect, just garnished at the end with olives (a practice which Perry in his formulation of the Olive Line takes into account) and another for bread with chopped olives incorporated into the dough; all other recipes including olives here are salads. The brief description preceding the list of ingredients and cooking instructions in Barron's *macarónia* recipe is worth quoting (p. 314):

> Strong flavors and bold color contrasts make this dish exceptionally attractive and appetizing. The cooked *macarónia* is coated with an anchovy-garlic paste, and then a sweet tomato sauce is poured over it. Interestingly enough, although basil is seen growing in pots all over Greece, this is one of the few basil-flavored Greek dishes I know.

The comment regarding the great rarity of basil in the cookery of Greece is echoed by many knowledgeable chefs and students of the cuisine and, especially in tandem with the inclusion of cooked olives, one must conclude that there is an Italian background to this preparation, which is indeed suspiciously similar to *pasta alla puttanesca* and its ilk. Yet, Barron's recipe is unquestionably a Greek dish, a Hellenized take on a presumable Italian ancestor, for it differs from *puttanesca* in part in cooking technique and also in the addition of a very Greek ingredient in savoury dishes, namely, honey, which is added in sufficient quantity to render the tomato sauce sweet and quite unlike the related Italian version.

There is, however, one further observation to be made regarding this dish: not only are the basil and cooked olives atypical for Greek recipes but so too is the inclusion of the salted anchovies. Just as is the case with olives, salted anchovies are a beloved and commonly eaten part of the Greek diet, but – also like olives – they appear as snacks, alone as a *meze*, and as an element in salads but at most only rarely as a flavouring agent in a cooked dish. From this fact, it seems clear that the Olive Line might well run together with other isoglosses between Italy and Greece and further that the omission or rejection of olives in cooked dishes has not to do with something specific about the olive *per se* but rather more likely with a common flavour element (beyond their saltiness) that cured olives and salted anchovies share: like many fermented foods, they are both bearers of umami.

To test this hypothesis, we might consider the degree to which cured pork products are used as flavouring agents in Greek cuisine, and, though I have come across some recipes where they are used, they are not especially common across much of present-day Greek territory and the recipes are in many cases of northern Greek origin. Again, Greek charcuterie appears to be most often enjoyed on its own and as a featured food, like olives and salted anchovies.

The contrast between Italian and Greek cuisines in this regard is quite sharp: throughout Italy, umami-enhancing ingredients in cooked dishes are widely and frequently used.

This practice can perhaps be most easily observed in the myriad pasta dishes which feature one or more umami-rich foods. For 'lean' days (Catholic holidays requiring abstinence from meat or all animal products), cured olives and anchovies are particularly favoured but mushrooms, especially umami-bearing porcini, are popular throughout the regional cuisines when in season, and the even more concentrated umami flavour brought to bear by dried mushrooms are used the year round. On ordinary and 'fat' days (when no abstinence rules obtain), these same umami-enhancers are employed but, in addition, many regional cured pork products, high or very high in glutamate content, are used, of which the most notable are *prosciutto*, *guanciale*, *pancetta*, and *lardo*. That the intention in using these ingredients is to permeate the dishes with umami flavour is clear from the fact that they are normally added early on in the dish, to the *battuto* or *soffritto* – i.e. more or less gently fried in olive oil, butter or pork fat together with aromatics (onion, garlic, carrot, etc.) – so that the lipid helps carry the umami flavour in the course of cooking throughout the final dish. This umami-enhancement in the initial stage of preparation is bookended, at least with many non-lean day pastas, by means of a final addition of umami-enhancing grated aged cheese, most commonly *pecorino* or *parmigiano*, one of the strongest umami-bearing foods in the world (Mouritsen and Styrbæk 2014: 148–49).

While this general Italian love of umami-enhancement is easily seen in the vast national and regional repertoires of pasta dishes, it is no less a part of all other areas of savoury cookery – meat, fish, and vegetable preparations also frequently include the same set of ingredients and methods of application just discussed. In this regard, we must further call attention to the Italian use of tomato, a moderate umami-enhancer, and its derivative, tomato-paste, a more concentrated, stronger umami-enhancer. Though foreign views of Italian and especially southern Italian cookery imagine a cuisine in which the tomato is omnipresent, a considerable exaggeration, it is of course true that the tomato is a fundamental ingredient and one which is hardly limited in expression to the lavish use of full-blown tomato sauces. For example, there are many dishes in which only a touch of tomato is used, often in the form of paste and included in the initial *battuto/soffritto* stage where its purpose is clearly in part as an umami-enhancer. Along similar lines, it does well to note that one of the most famous Italian preparations, *bolognese* sauce, is not a tomato sauce but a meat sauce with a bit of tomato which, along with grated parmesan cheese, enhances the umami savouriness of pastas made with it.

Tomato sauce, a cover term for an almost infinite variety of tomato-based preparations in Italian cookery, itself features a measure of umami flavour without enhancers, and I believe part of its great success in southern and central Italian regional cuisines is due to this characteristic: it functions as the poor man's substitute for the meat broth that was so much a part of elite and bourgeois cookery as represented in medieval and early modern Italian cookbooks but was not affordable for frequent use by the poorer classes.

Greek and other eastern Mediterranean cuisines today all make extensive use of tomatoes in various forms, including paste, and tomatoes are no less an indispensable part of Hellenic cookery than Italian. Surely this fruit's natural umami element is part of its popularity in Greece, but one of the virtues of the tomato is the flavour complexity it provides, as well as its textural and colour contributions. In light of the Greek predilection for not using strong umami agents as flavourings in cooked dishes, perhaps the moderate umami contribution is just a (secondary?) part of the overall gustatory value of the tomato for Greeks; its sweetness, its acidity, as well as its colour and texture, have made it a beloved part of the Greek culinary palette. In this regard, however, we might also note that, whereas Italians often augment their tomato sauces with the aforementioned umami-enhancers, the Greeks often brighten their tomato sauces with the addition of honey or 'sweet-spices', such as cinnamon. In Italy, some of the same spices are found in tomato sauces as well as in tomato-less dishes, but their use is relatively marginal in Italian regional cuisines and largely limited to preparations more festive in nature, likely a reflection of the almost exclusively elite use of those spices in earlier times. Sweeteners in Italian savoury dishes are encountered here and there but, again, with far more limited occurrence than what we see in Greek cuisine.

Preliminary Conclusions

In my Oxford paper of 2010 I discuss the centrality of umami flavour in Italian cuisine(s), arguing that a feature of ancient Italian cookery – the extensive use of the quintessentially umami-bearing fermented fish condiments – which has seemed to many food writers (including some Italians) as something alien to modern Italian cookery, is in fact an element of deep culinary-grammatical continuity (Buccini 2011: 71–74.). In this connexion, it must, however, be noted that widespread use of fermented fish condiments was also a feature of Hellenic cookery in classical times and the practice continued on well into the Middle Ages, where at least the elite of Byzantine society continued to use fish sauce extensively, a practice then alien to western Europeans and objectionable to some commentators (Dalby 1996, 2003; Anagnostakis 2013; Dalby and Dalby 2017). I strongly suspect the objection was not to the umami element itself but the other flavour elements of fish sauce, as those westerners were themselves making extensive use of other umami-enhancers, if we can judge from the use of cured pork products in roughly contemporaneous western cookbooks, such as the *Liber de coquina* (Maier 2005).

I maintain that there is at a deep level a good measure of culinary continuity between classical and modern times in Italy. Italians (and other westerners) may have largely given up fish sauces as a habitual umami-booster, but they did not give up the practice of habitually enhancing the umami element in their cookery: anchovies, cured pork, aged cheeses, olives, and more recently tomatoes, are essential seasoning elements in Italian and, to varying degrees, other cuisines of the western Mediterranean. There appears, however, to be a basic discontinuity in the eastern Mediterranean, where the Greeks have left behind along with their fish sauce the more general practice of using umami-enhancers in cooked dishes. Perhaps there is a parallel in the Middle East, where the strongly umami-bearing fermented barley sauce, *murri*, of medieval Arab cuisine – according to Charles Perry, similar to soy sauce in flavour – was also abandoned and with it perhaps the general principle of umami-enhancement of cooked dishes (see Perry 1988).

As a preliminary conclusion, pending completion of continuing research (including on areas left wholly aside here, such as the Maghreb and Iberia), I suggest that over many centuries of cultural contact, the cuisines of the eastern Mediterranean largely gave up the practice of umami-enhancement and in its place developed a strong preference for brighter flavourings, in particular through the addition of sweeteners and spices in savoury dishes. In contrast, in the western Mediterranean and more generally in western Europe, the addition of sweeteners and spices to cooked dishes may at first glance seem to have been widespread already in classical times, judging from the recipes of the *Apicius* collection, and continued on in the cookery attested in the medieval cookbooks, but in reality these practices were largely limited to the foodways of the elite. In the west, the use of sweeteners and spices trickled down the social scale but only to a rather limited degree, appearing primarily in the festive cookery of the non-elite (where their use survives till today to a degree). In the eastern Mediterranean, though, the incorporation of expensive spices and sweeteners came to be nativized in progressively lower socio-economic circles. In Greece, culinary influences from the east began in very remote times, but in light of the Olive Line and the discontinuity of umami-enhancing in the non-replacement of fish sauce, it seems likely that the crucial period in the development of this culinary areal feature of the eastern Mediterranean was that of the Ottoman domination.

About the Author
Anthony F. Buccini (PhD Cornell University) is an historical linguist and dialectologist who formerly taught at the University of Chicago in the departments of Germanic Languages and Literatures and Linguistics. As a food historian, his research focuses on the Mediterranean and Atlantic World. He is a two-time winner of the Sophie Coe Prize in Food History.

References
Anagnostakis, Ilias (ed). 2013. *Flavours and Delights: Tastes and Pleasures of Ancient and Byzantine Cuisine* (Armos).

Buccini, Anthony F. 2011. 'Continuity in Culinary Aesthetics in the Western Mediterranean: Roman *Garum* and *Liquamen* in the Light of Local Survival of Fermented Fish Seasonings in Japan and the Western Mediterranean', in *Cured, Fermented and Smoked Foods: Proceedings of the Oxford Symposium on Food and Cookery 2010*, ed. by Helen Saberi (Prospect Books), pp. 66–75.

———. 2012. '*Chi vuol godere la festa, digiuni la vigilia*: On the Relationship between Fasting and Feasting,' in *Celebration: Proceedings of the Oxford Symposium on Food and Cookery, 2011*, ed. by Mark McWilliams (Prospect Books), pp. 66–75.

———. 2016. 'Defining 'Cuisine': Communication, Culinary Grammar and the Typology of Cuisine', in *Food and Communication: Proceedings of the Oxford Symposium on Food and Cookery 2015*, ed. by Mark McWilliams (Prospect Books), pp. 105–21.

Dalby, Andrew. 1996. *Siren Feasts: A History of Food and Gastronomy in Greece* (Routledge).

———. 2003. *Flavours of Byzantium* (Prospect Books).

Dalby, Andrew, and Rachel Dalby. 2017. *Gifts of the Gods: A History of Food in Greece* (Reaktion).

Maier, Robert (ed and trans). 2005. *Liber de coquina. Das Buch der guten Küche* (F.S. Friedrich).

Mouritsen, Ole G., and Klavs Styrbæk. 2014. *Umami: Unlocking the Secrets of the Fifth Taste*, trans. by Mariela Johansen (Columbia University Press).

Perry, Charles. 1988. 'Medieval Near Eastern Rotted Condiments', in *Taste: Proceedings of the Oxford Symposium on Food and Cookery, 1987*, ed. by Tom Jaine (Prospect Books), pp. 169–77.

———. 1997. 'The Olive Line.' *Los Angeles Times*, 24 September 1997 <https://www.latimes.com/archives/la-xpm-1997-sep-24-fo-35628-story.html> [accessed 15 May 2024].

7
Cultivating Liberation
Gardens as Agents of Connection and Empowerment in Havana, Cuba

Mallory Cerkleski

Cuba is constantly in a state of change. As historian Louis Peréz states, 'Change recurs with such frequency in Cuba that it assumes the appearance of a changeless condition.'[1] Therefore, studying Cuba can feel as though you are trying to grasp onto air; once you feel you understand something, a new experience shifts this perspective, and you can never quite catch the truth. Paying attention to materiality can make this process a bit easier; people's direct interaction with and influence on material realities, such as gardens, can reveal complex problems and solutions at play.

While urban gardening has long been recognized as a transformative practice in Cuba, fostering sustainability, diversity, and educational opportunities,[2] recent global events, such as the COVID-19 pandemic and the Russian-Ukrainian war, have brought about new challenges and opportunities for urban gardens in Havana.

Urban gardening in Cuba is a communal activity beyond merely producing food. The practice and outcomes of urban gardening reflect diverse values and realities, such as the depth of communal knowledge, the relationships between people and the land, the availability of materials, and cultural preferences in design and food choices. Additionally, through the exchange of locally grown produce, urban gardens serve as alternative forms of currency during economic instability.[3] They provide a practical means of trade and solidarity when either people cannot access currency or the official currency loses its value.

Case Studies

Since greenery permeates everywhere in Cuba, a lush, tropical island, defining what constitutes an urban garden can be challenging. While scholars have proposed various precise definitions, I have approached this definition by prioritizing the lived experiences and societal contexts of gardening spaces in Cuba.[4] Rather than relying solely on these precise and formal definitions (sometimes coming from the outside), I have included spaces based on whether the practitioners identified the spaces specifically as tools for resistance, liberation, and community building. These definitions have emerged from interactions with the people themselves, providing a nuanced understanding of the significance of gardening spaces beyond technical categorizations.[5]

To create an order of the urban gardening spaces observed, I divided them into three main types, each emphasizing communal aspects. The first category encompasses private homes where gardening activities facilitate public and community interactions, blurring the lines between private and public spheres. The second category includes private spaces that openly welcome public engagement, often hosting community organizations or events. These spaces foster a sense of openness and accessibility, inviting community participation and collaboration. The final category comprises spaces that lack residential presence and resemble more formal or business-like setups. Despite their non-residential nature, these spaces remain integral to community well-being, embodying a tension between private ventures and more revolutionary ideals.[6]

Private Homes
The first example is from Desirada, who lives in Los Pocitos.[7] I got to know Desirada a couple of weeks prior to our interview at a site visit in her neighbourhood, where she is the co-founder and coordinator of a community project. I remember the first time I met Desirada because she had intense energy and was very dedicated to the community. She explained to me how determined she was to wake up every day to do her work because many were relying on her.

After a large tour of the neighbourhood, Desirada showed us her backyard. She had many different plots and even some pens for pigs. She had different raised beds that used old glass bottles as borders. The typical Cuban practice of turning trash into treasure (known as *resolver*) can always be observed in a garden space. She had strawberries, corn, beans, pineapples, and more, all intertwined and flourishing. Desirada explained that even though this plot was in her personal backyard, her experiences there and her experiences in the community could not be separated. It was an intimate relationship, an ebb and flow between the two, allowing each to exist in harmony. The experience she gained in her garden went directly into helping and maintaining the other fifty-eight gardens in, with, and for the neighbourhood. Nothing could be considered a completely personal endeavour around here.

The second example is from Vinita, also a resident of Los Pocitos. Desirada introduced me to Vinita some weeks after our interview when I was volunteering at the community centre. Vinita had a large and diverse home garden where she raised pigs and bees and an extensive fruit orchard and vegetable garden. We knocked on her door to see if we could walk around, and she brought out fresh *mamey* juice for us. Outside her house, she and Desirada were chatting about getting milk for their kids, which was a typical conversation. I heard many women talking about getting their children milk and how they would figure it out the following week. This interaction, where community members discussed and brainstormed solutions to their problems, happened often around me and was their backbone for survival. After speaking outside her house briefly, I asked if she would mind being interviewed. She quickly agreed and invited me inside. I was pleased to be in her house; she had loving and caring energy. Vinita was fifty-two years old and a *campesina* (farmer); she had her home plot and space in a cooperative. She told me she is originally from a humble peasant family from Santiago de Cuba (in the east), from the Sierra Maestra, but she moved and has lived in Havana since 1998. Vinita was a powerful example of the hope that can be had when one takes their food supply into their own hands. As a home gardener, she felt calm because she had access to an abundance of food grown by her hands; if she had nothing else, she would have food. Through her actions, she wanted to inspire others to gain that autonomy and empower themselves by growing their food. Vinita explains how and why she has done that:

> With the Covid issue [. . .], it forced me to plant a little more. I didn't plant this much [before], it was just one fruit tree for the year. And [since] Covid I said: in any little piece [of land] I put a plant and we have encouraged many women to reflect on the food issue of sowing your little piece, of sharing with your neighbour, of exchanging experiences [. . .]. So Covid has taught many people today to have everything cultivated and to share [. . .]. At least among us [who have] *patios* we do that.[8]

Vinita's reflection contributes significantly to the overarching theme of cultivating and sharing food and experiences, while also highlighting the disparity in food shortages and experiences between urban and rural areas. Vinita's experience epitomizes urban gardening's role in empowering individuals and communities within Cuban society, showcasing self-sufficiency amidst economic challenges. Through cultivating her own food and sharing resources with neighbours, she embodies resilience and solidarity, bridging urban and rural divides. Her narrative underscores the transformative potential of urban gardening in fostering autonomy, community cohesion, and resistance to external pressures.

Private/Public Spaces
The first person I will discuss in this section is Nalda, who lives in a neighbourhood of Havana where many houses are considered 'illegal settlements'. I later discovered people call this *llega y pon* (arrive and put). When I got there, Nalda's husband invited me into her home, where I was very unexpectedly greeted by twelve children aged nine to thirteen from the neighbourhood. They gave me speeches about Cuba and their desire to cure the earth and its people. I soon understood that Nalda, with the help of other community members, ran a project centred around creating food and energy sovereignty in their neighbourhood, which included an after-school programme for youth and teenagers on various topics related to sustainability, agroecological food production, and much more. They started the project in 2010 illegally, and it was given official approval in 2014. Nalda had an aura around her of love that allowed her to have faith in the future. She spoke of the issues and frustrations clearly but did not let these thoughts discourage her from fighting for her community's survival. She was clear that she and her family did not speak about politics. She expressed she was in no place to speak of them; I respected this wish and asked little about socialism or the government. Through their actions of gardening and organizing, though, I observed they were consciously disconnecting: trying to not rely as much on the promises of socialism and instead create for themselves what they needed. Rather than openly criticizing the government, they expressed their dissatisfaction through actions – demonstrating what they were doing and why, speaking volumes through silence.

Overall, Nalda and her community displayed more faith and hope than I had observed in my previous interactions. They were convinced that, despite their lack of access to material wealth, they could achieve self-sufficiency through gardening and mutual support, believing this would be their path to survival. It was an inspiring micro-community solution. Below is a quote from Nalda, who, before the interview, gave me a tour of her *patio*. She had many plants in her yard, all with a purpose and intricately intertwined to create an oasis of care. She stated:

> As you see, this [mango] is also produced in the yards, not in my backyard, but since we are thirty-five families together, this is produced by a friend, others produce other things [. . .] we exchange. 'Ah, I have mango, [here is a] mango, ah, I have bananas. [Maybe] I don't have [anything], [but someone comes and says] look here's a banana,' and so. And that helps us breathe a little bit. Because the food situation is no secret. It's a little bit difficult for us.[9]

This aspect of community support and solidarity is expressed in the words above. Her explanation of how she can get food for the week through her community's support and her community's ability to get access through her is a direct example of the reciprocity economy. Here, urban gardening serves as a concrete action to try to overcome the vulnerability of the crisis. Nalda was very passionate about providing for herself and not relying on the government or the market for her survival. This reflection exemplifies the collective memories of support these thirty-five families share. When other members of the community were asked how they imagined their future, they wrote me two letters, which I translated and transcribed below:

> I hope that it will be a better world, I hope it will be a better world [. . .] if we all work with forces we can achieve armies, without many words. And the shortages that exist will disappear, especially in the capital. We want our community and our family to have clean streets and healthy food. We are sometimes in power outages, but [with] the project we want to make renewable energy [. . .]. Those of us who are in the communities want our children and grandchildren to have a better future, which is possible if we have the desire and organization.[10]

> If we all take advantage of the land, help each other, and cultivate together, we can offer food between families, [and] communities. All the communities have the right to their own development, freedom, equipment, tools, cages, medicine for animals [. . .]. We are not [just] fishermen; we are surrounded by land. We can exchange. Those of us who cook, make food to feed the family and ourselves. That's what we want to see food development in the place where we live. Resources and freedom to produce, that's what I want.[11]

Recently, I texted Nalda to check in on her and ask her how the project is going. Unfortunately, since I have left Cuba the situation there has worsened with rising inflation, increased migration, global increases in gas prices, and the ever-present blockade from the United States. She stated:

> Our orchards and gardens are becoming less and less. We continue with the planting of clean, good and healthy food. The seeds here are not easy to find and the state prices are prohibitive. We have a program called: The Seed; which consists of recovering all the seeds of all the fruit trees. We sowed chard from some seeds that some European students brought with them and we ate them and some of them were left in the field for seed, this is the seed that is in the process of drying to sow again. The celery that is in the orchards and gardens today is already sowing, these will be the seeds that we will use and so on. The fertilizers are natural: excrement from rabbits, goats, etc., as well as compost from all the kitchen waste, whether it is eggshells, food collected from the yard, etc. And to fight pests we have the neem tree and the Cardona and other natural repellents. We take leaves and branches and put them to ferment and with that we make fertilizers. Right now, we have Californian worms and we produce our own humus. We don't have many working tools but the exchange between all of us this month helps us and we lend [to each other]. That solidarity among neighbours helps us a lot.[12]

Nalda's story illustrates the impact of urban gardening on community resilience and self-sufficiency in Cuba. Despite socio-economic challenges, she and her neighbours have created a network of mutual support and sustainability that persists even in the harshest of times. Their gardens not only provide food but also strengthen community bonds.

Lastly, and perhaps most importantly, this example illustrates that to truly understand these spaces, one must engage with the people who cultivate them. The true cultural and relational significance of urban gardening emerges through their stories and the observation of micro-interactions. Nalda's ongoing efforts highlight not only the necessity of community cooperation in overcoming setbacks but also the celebrated ability of Cubans to adapt, persevere, and maintain resilience.[13]

The second example in this section comes from Duaro. Duaro and his wife, Lily, live across the bay in a section of Havana called Guanabacoa, far from the central city. Guanabacoa is one of the oldest municipalities located in the east of Havana. Once I reached their house, Lily offered me coffee; a must when entering a Cuban household (even if it is something they have difficulty getting their hands on these days). I accepted. She sat beside me and joined in discussing their lives and what they do. Her energy was very loving and warm. She showed me a small porch where she kept her bees, and her cat who she picked up and asked me to give it cuddles.

After a little while, they took me out to their project grounds. Together, they run a community project that uses gardening as a tool to support and empower people with mental and physical disabilities. This approach not only helps individuals but also fosters awareness and education about sustainability and disability issues. Their space was beautiful – filled with vegetables, fruits, medicinal plants, worm compost, and their germination greenhouse. They toured me around, showing me the new baby kittens that had just been born. I remember this space as an abundant paradise within the city, and as they spoke to me I felt their passion and love for what they do. Duaro was a strong man, as I felt from when I first met him: he was hard-headed and independent but also very loving. He was a dedicated socialist and loved Cuba deeply. He truly embodied the idea that life was supposed to be lived with and for others. He told me:

> We try to educate the children to take care of nature by sowing a plant and watering a plant so that tomorrow the plant will not be [vulnerable]. There are two different visions of life, the one who thinks of being and the one who thinks of having. The one who thinks about being has the possibility to look for the path to happiness. The one who thinks about having will never be happy. That is the consumerist world that we detest. We think of the conservation of natural goods to give to future generations in which we can breathe and live in harmony.[14]

Their community project offers a compelling example of how urban gardens can function as more than just sources of sustenance; they become avenues for social change. By intertwining education, therapy, and environmental stewardship, their initiative creates a holistic approach to community development. Through their dedication to fostering connections with nature and each other, they not only cultivate vegetables and fruits but also sow the seeds of resilience, self-reliance, and collective well-being. In doing so, they embody a vision of urban gardening as a tool for liberation, connection, and empowerment within Cuban society.

Businesses

The first example in this section comes from Yosary, who lived in the same community as Nalda, yet in a different configuration of space. Yosary had lots of land, and she expressed that the main thing she wanted from the land was to make a profit from gardening to benefit the community. She had a guava orchard, many mango trees, coconuts, and small garden plots everywhere. She expressed wanting to have a fruit stand on the side of the road as well as create an agritourism destination to generate profits for reinvestment in the community project.

The second space, located in Vedado, was a garden run by a friend of our professor. He drove us there and introduced us to the caretaker of the garden, Floriano. This garden was fully dedicated to producing medicinal plants, showing the pivotal role of urban gardens in meeting healthcare needs. Traditional remedies are popular amidst medication shortages, with urban gardens serving as essential resources, especially for Santería practitioners. Moreover, integrating urban agriculture into the public health system reflects efforts to address healthcare disparities and promote alternative medical practices amid economic challenges.[15] Floriano explained to us the different plants and the remedies they were used for. He had a small board showing the offers of the day which listed *caña santa* (lemongrass), *menta* (mint), *guacamaya francesa* (candle bush), *masito manzanilla* (chamomile chewing gum), *hierva de la sangre* (blood grass/weed), *verbena* (lemon balm), *halbaca morada* (holy basil), *sabila* (aloe), *rompe camisa* (lantana), *abre camino* (thoroughwort), *siguaraya*, *pasi flora* (passionfruit flower), moringa, and oregano.[16] Every single item was ten pesos which was about ten cents (USD), highlighting that while this is technically a business, the emphasis is more on support and exchange rather than profit, as the pricing suggests minimal profit motive.[17]

Conclusion

I want to conclude with some reflections. The first is the blurred line between private and public spaces in Cuba. As evident in each section, no space is easily defined as either public or private – regardless of certain laws or perceptions. Part of this is cultural, rooted in the Cuban island culture of hospitality, where doors are left open, and people flow freely in and out. The second reason is necessity: in a system marked by struggle, people increasingly rely on each other, creating an expectation of cooperation. And the third reason? It may very well be the elephant in the room: the Revolution. Yes, there is that – a foundational reason for mass solidarity. Briefly touched upon in the quote from Duaro, the idea of living for others has been ingrained in the Cuban consciousness for over sixty years now. But how has this evolved over time? And how can gardens illuminate this?

Gardeners often navigate complex relationships with government bodies, Cuban NGOs, and community entities to secure and sustain their plots. This process is often precarious, requiring gardeners to continually demonstrate allegiance to revolutionary ideals and communal welfare.[18] Additionally, tensions arise from the differing motivations of older and younger generations. While older gardeners may see their work as a continuation of revolutionary principles, younger participants may be driven more by economic necessity and personal advancement.[19] This intergenerational divide reflects broader shifts in Cuban society, where the legacy of the revolution intersects with modern economic realities. Despite the challenges, Cuba's transition can be seen as a testament to the possibility of moving beyond an exploitative capitalist model toward a more holistic form of well-being.

Every day, as I monitor the exchange and inflation rates on the informal market, I watch the Cuban peso's value fluctuate against the US dollar and Euro, a harsh reminder of the worsening crisis in Cuba. The numbers may signal a decline, but they also reveal the evolving nature of the situation. Since my departure, I've noticed other changes like new cars filling the streets and stores appearing more stocked. These shifts hint at a transformation in the city's aesthetics, suggesting a landscape in flux. To some, my observations seem overly optimistic, but I cannot deny the magical quality of Cuba – forcing you to be filled with hope of an abundant future. Gardens, while a small component of this broader narrative, represent a history of resistance. They are not just green spaces but symbols of an enduring struggle that has shaped and will continue to shape Cuban society – from colonization to independence to the revolution. As I prepare to return to Cuba to further explore and document the evolving role of the food system, gardens being a part of that, I am reminded of a valuable lesson from a mentor: the more you think you understand Cuba, the more confident you can be that your grasp is incomplete.

About the Author
Mallory Cerkleski (she/her) is a doctoral candidate in History at the Scuola Normale Superiore in Pisa, specializing in the comparative history of communist food systems in Cuba and Kerala. Her research focuses on food justice, sovereignty, and cultural heritage through oral histories.

Notes

1. Louis A Pérez, *On Becoming Cuban: Identity, Nationality, and Culture* (University of North Carolina Press, 2008), pp. xi.
2. Carey Clouse, *Farming Cuba: Urban Agriculture from the Ground Up* (Princeton Architectural Press, 2014); Gustav Cederlöf, 'Low-Carbon Food Supply: The Ecological Geography of Cuban Urban Agriculture and Agroecological Theory', *Agriculture and Human Values*, 33.4 (2016), pp. 771–84, DOI:10.1007/s10460-015-9659-y; Mickey Ellinger and Scott Braley, 'Urban Agriculture in Cuba', *Race, Poverty & the Environment*, 17.2 (2010), pp. 14–17; Susan Anne Mansel Fitzgerald, *Havana: Mapping Lived Experiences of Urban Agriculture*, 1st edn (Routledge, 2022), DOI:10.4324/9781003201410; Marina Gold, 'Urban Gardens: Private Property or the Ultimate Socialist Experience?', in *Cuban Intersections of Literary and Urban Spaces*, ed. by Carlos Riobó (State University of New York Press, 2011); Sinan Koont, *Sustainable Urban Agriculture in Cuba* (University Press of Florida, 2017); Sinan Koont, 'Urban Agriculture in Cuba: Of, by, and for the Barrio', *Nature, Society, and Thought*, 20 (2007), pp. 311–26; Sinan Koont, 'A Cuban Success Story: Urban Agriculture', *Review of Radical Political Economics*, 40.3 (2008), pp. 285–91, DOI:10.1177/0486613408320016; Friedrich Leitgeb, 'Increasing Food Sovereignty with Urban Agriculture in Cuba', *Agriculture and Human Values*, 33.2 (2015), pp. 415–26, DOI:10.1007/s10460-015-9616-9; Ola Plonska and Younes Saramifar, *The Urban Gardens of Havana: Seeking Revolutionary Plants in Ideologized Spaces* (Springer, 2019), DOI:10.1007/978-3-030-12657-5; Adriana Premat, 'Small-Scale Urban Agriculture in Havana and the Reproduction of the 'New Man' in Contemporary Cuba', *Revista Europea de Estudios Latinoamericanos y Del Caribe*, 75 (2003), pp. 85–99; Adriana Premat, 'Cuban Counterpoint of the Public and the Private: Reflections on the Making of Urban Agriculture Sites in Havana, Cuba' (York University, 2004); Adriana Premat, *Sowing Change: The Making of Havana's Urban Agriculture* (Vanderbilt University Press, 2012); Marina Gold, 'Peasant, Patriot, Environmentalist: Sustainable Development Discourse in Havana', *Bulletin of Latin American Research*, 33.4 (2014), pp. 405–18, DOI:10.1111/blar.12175.
3. Cuba has faced significant economic challenges since the early 1990s, beginning with the 'Special Period' following the collapse of the Soviet Union, its main ally and economic supporter. This sudden loss plunged Cuba into a severe crisis, marked by acute shortages of food, fuel, and other essentials. The situation has worsened in recent years due to several factors. Long-standing U.S. economic sanctions have tightened, further restricting Cuba's access to international markets and financial resources. Additionally, the COVID-19 pandemic severely impacted tourism, a vital source of revenue, leading to further economic strain. Global events like the Russian-Ukrainian war have disrupted supply chains, increasing the cost and difficulty of importing goods. These compounded challenges have resulted in high inflation, widespread shortages, and increased hardship for the Cuban population.
4. Clouse, pp. 83–134; Koont, *Sustainable Urban Agriculture in Cuba*, pp. 1–29.
5. Fieldwork for this piece was carried out May-August 2022 in Havana. To clarify, my primary objective in Havana was not specifically focusing on the study of urban agriculture. Instead, I was engaged in the collection of oral histories using the subject of food as a tool for relationship building. However, due to this approach I frequently found myself guided into people's backyards where cultivation spaces were commonly encountered. As such, the findings presented in this paper emerged as an incidental outcome of my primary research. It is important to note that this paper is not a formal or exhaustive update on urban gardening or urban agriculture in Havana. Numerous urban planners are actively addressing such things. Rather, it serves as a testament to the first-hand experience of a food researcher who interacted with practitioners in the field, supplemented by observations and interactions while immersing oneself in the urban landscape of Havana. Although the method for my broader research was fairly complex

and theoretical, placing itself in the field of contemporary history, for the sake of this paper I will simply explain the logistical facts. To get data for the findings and analysis, I mainly utilized an oral history methodology based on the field of food history and memory studies; this meant that although the interviews were not directly related to gardening, the aspects related that came out will be used in this paper. Additionally, there are two spaces included in the last section (Yosary and Floriano's garden) where I did not formally collect an oral history, but instead spent time with practitioners in their gardens; for this reason, no quotes will be used. Each interview and visit were done in Spanish and later transcribed and translated by two Cuban classmates.

6   Gold, 'Urban Gardens', p. 42.
7   For the sake of safety, all original names or identifiers have been changed. 'Los Pocitos is a marginalized neighbourhood of Havana that has been especially hard hit by the COVID-19 pandemic. The community is afflicted by an array of social issues that include squalid living conditions, economic insecurity, and high rates of unemployment. Most of its residents work in the informal market and are not college-educated. Economic insecurity has forced many families to work on the streets to earn a living despite early government recommendations to stay home' ('Proyecto Akokán: A Local Initiative Building the Los Pocitos Community', *EEAbroad* <https://www.eeabroad.com/blog/proyecto-akokan-the-los-pocitos-community-project> [accessed 6 September 2024]).
8   Personal interview with Vinita, Havana, Cuba, 22 June 2022. *Patios* are backyard gardens.
9   Personal interview with Nalda, Havana, Cuba, 16 June 2022.
10 Text from drawing created during personal interview with Silviana, Havana, Cuba, 16 June 2022.
11 Text from drawing created during personal interview with Verita, Havana, Cuba, 16 June 2022.
12 Text message exchange with Nalda (Whatsapp), 14 April 2024.
13 This sentiment is not meant to over-romanticize the ability that Cubans have to adapt. In general, the idea of resilience in Cuba is over-done, but it also should not be ignored. For further exploration of this idea, see Mallory Cerkleski, 'Resolver and Rebusque: The State of Cuban Land and Food Sovereignty in 2022', in *Food Sovereignty and Land Grabbing*, ed. by Gabriele Proglio (Cambridge Scholars Publishing, 2023).
14 Cerkleski, Mallory. Personal interview with Duaro, Havana, Cuba, 29 June 2022.
15 Gold, 'Urban Gardens', p. 12.
16 *Francesa* is known for being antifungal; *hierva de la sangre*, as insinuated by the name, is used for blood problems in general, respiratory diseases (cough, bronchitis), skin (erysipelas, syphilis, pimples) and gastric diseases among others; *rompe camisa* is bought by many Cubans for spiritual cleaning purposes, mostly people who practice Santería; *abre camino* is also bought for spiritual purposes (gaining power), but also it can be used for treatment of circulatory problems, colds and fevers, and diabetes; people practicing Santería believe *siguaraya* to be an *Oricha*, a force of nature, that can have powers for healing the body but more so for healing the spirit and soul as well.
17 In Cuba, there is an official exchange rate and an exchange rate on the black market, if one were to exchange pesos on the black market this would've been the exchange rate. Now (as of May 1st 2024) it would be about 2 USD cents.
18 Gold, 'Urban Gardens', p. 43.
19 Gold, 'Urban Gardens', p. 42.

8
The Blight of #Cottagecore
Past and Present S/Place-Making in the Garden and Home

Jolin Chan

In #cottagecore, young women frolic across blossoming fields and phone screens.[1] Some bake homemade bread in their rustic – yet perfectly curated – kitchens, while others bring you into their backyards to pick fresh berries. Some are even in modern kitchens in urban apartments, emanating the cottage lifestyle through flowers and homemade scones.

Cottagecore is a fairy tale wrapped neatly in a bow, served on an antique platter. As an aesthetic, it is founded on slow living, idyllic lifestyles, pastoral clothing, and homemade foods, all made beautiful through an abundance of natural imagery and hues of pink, green, and brown. Thriving plants in vintage pots, ceramic cups of tea, and woven baskets decorate cottagecore spaces. Individuals dress in flowy, floral midi dresses with puffy sleeves while wearing wide-brim straw hats with bows. Homemade baked goods, fresh out of the oven, line rustic tables: scones with jam, crusty baguettes, and a latticed apple pie. This world is simultaneously a paradise, escape, and refuge from life in the twenty-first century, blissfully free of sitting in traffic, pollution, processed foods, and (any apparent) health issues. And it is largely young women on social media who are running towards this paradise – and running it.

The aesthetic, lifestyle, and online space make up more than a frivolous trend. Rather, they reveal legacies of expansion and exclusion. Though cottagecore is often associated with European histories and motifs, I expand its history by situating it within the legacies of the American frontier. Early twentieth-century American gardening and home magazines constructed the garden and home as the new frontiers. Mass media helped to maintain expansionist values of space-making, control over land, and ideals of whiteness, thereby overlooking the past and present involvement of people of colour in gardening and homemaking.

This paper reckons with cottagecore's roots in frontierism and follows them into present-day print media and conversations on social media. By romanticizing a nostalgic yet exclusionary past, #cottagecore as a white space complicates relationships to land, the garden, and the home, particularly for people of colour. However, its growing popularity has transformed #cottagecore into a digital site for place-making. Contrasting space-making with place-making, I define the latter as turning space into place by imbuing it with meanings, values, and a sense of community. Participants of cottagecore are increasingly looking toward a future of inclusivity and directly challenging histories of space-making.

#Cottagecore encapsulates tensions we must acknowledge, critique, and navigate as social media use becomes more intertwined with our lifestyles. It bridges the gap between the natural world and the digital realm; it simultaneously reflects the past, present, and future; and it both excludes participants and fights for inclusivity. Our goal, now, is to stop scrolling and instead critically engage with cottagecore's roots and implications.

Cottagecore's Roots

Cottagecore emerged as a popular subculture in 2017 and comes from a long tradition of aesthetics often popularized on platforms such as TikTok, Instagram, Tumblr, and Pinterest. Recent aesthetics have been the bow-obsessed 'coquette' and the stylish 'coastal granddaughter/grandmother'. The 'cottage' part of 'cottagecore' refers to a shared pastoral ideal, while the '-core' part is a suffix that stems from 1980s 'hard-core punk music' and denotes a genre of content.[2] Other types of '-cores' fill up the 'For You Page' (a personalized page or feed of content determined by an algorithm) of platforms like TikTok and Instagram: grandmacore, angelcore, goblincore, and even corecore. What makes cottagecore unique, however, is not only its longevity but also its transformation from an aesthetic into a lifestyle.

The COVID-19 pandemic allowed cottagecore to thrive due to easy access to the rural landscape, even if digital. Users continue to share their experience with handcrafting art, making baked goods from scratch, and pressing flowers, to name a few examples. As of May 2024, #cottagecore on TikTok contained 1.7 million posts, while Instagram's #cottagecore contained 5.5 million. In this paper, I focus on the cottagecore garden and home, though it is important to note the wide variety of activities that make up the aesthetic and lifestyle. The site of the garden and home are only the starting points for idyllic dreams.

While cottagecore's popularity peaked in the past several years, it reflects recurring desires for a rural escape. The earliest evidence of pastoral desires is found in Arcadia in Ancient Greece, which became a romanticized representation of an idyllic utopia in the Hellenistic period.[3] Pastoral desires continued into the eighteenth century with Marie Antoinette's *Hameau de la Reine* ('the Hamlet'), a village meant to be a return to nature and escape from royal life. It included a farmhouse, a dairy, a barn, a mill, and more to evoke the imagery and style of a rustic village.[4] Other writers and scholars have tied cottagecore to Romanticism, the Arts and Crafts movement, and Aestheticism.[5] Common among these examples is an invitation to escape the present day and a longing for an open, rural landscape.

Space-Making: The American Frontier to the American Garden

While much of cottagecore's historical inspirations has roots in Europe, I expand its histories to draw a connection to the legacies of the American frontier. The American frontier and its expansionist values have shaped the space of the garden and home in the twentieth and twenty-first centuries, making them the new frontier. These frontier values later inform the principles of cottagecore, as well as give insight into representations of the garden and gardener, and home and homemaker. We must ask: who is in the frontiers, the rural landscapes, and the garden, both in the past and in the present? Who is not?

The American frontier was an unknown yet alluring space, and, as Manifest Destiny argued, Americans were destined to conquer it. In his 1893 paper 'The Significance of the Frontier in American History', Frederick Jackson Turner argued that the 'true point of view in the history of this nation [was] not the Atlantic coast, it [was] the Great West'.[6] Reflecting the closing of the frontier, Turner constructed the frontier as a space where savagery met civilization and a space that represented values such as democracy and individualism.

Even with its closing, the frontier was not gone. Rather, its location simply changed. In his acceptance speech for the Democratic National Convention in 1960, then-Senator John F. Kennedy proclaimed:

> Today some would say [. . .] that there is no longer an American frontier. But I trust that no one in this vast assemblage will agree with those sentiments [. . .] We stand today on the edge of a New Frontier – the frontier of the 1960's.[7]

Kennedy continues to describe the New Frontier as 'uncharted', 'unsolved', and 'unconquered', a place where one could breathe 'the fresh air of progress'.[8] Since the closing of the frontier, the United States expanded its borders from the American West to places even further west, such as Southeast Asia, and to places not even on Earth, toward the moon and beyond.

The American frontier, whether in the United States or not, represents untapped resources and power, as well as individualism and ingenuity. Richard Slotkin, however, identifies this as the myth of the frontier, what he terms 'regeneration through violence', which embodied utopian ideals that concealed horrendous acts of violence.[9] Literature on the frontier from the seventeenth to nineteenth centuries formed a unified vision that attempted to justify and reimagine violence through narratives on the frontier's abundance of opportunities.[10] After all, it was a bloody conquest involving the ethnic cleansing and removal of Native Americans from their homelands. Gardens were no exception to space-making and myth-making. Writers portrayed the nature of the New World as a garden for discovery and as a possibility for renewal.[11]

Frontiers, then, are constantly shifting for potential space-making. In the early twentieth century, the American frontier – or at least, its values – shifted into gardens and homes. Though not the expansive American West, these spaces nevertheless represented the same principles of self-reliance and growth, reflecting the persistent desire for space-making.

The start of the twentieth century marked a 'flourishing era of garden writing' for America, reaching millions of households and inspiring amateur gardeners to develop their skills.[12] With the closing of the frontier, magazines such as *Country Life in America*, *House and Garden*, and *Ladies' Home Journal* illustrated that building idyllic gardens could fill in the gap. Their content invited Americans to participate in space-making in the comfort of their home. For instance, Neltje Blanchan in *Country Life in America* wrote about the joy of gardening:

> Your true gardener realizes that he is permitted, invited, nay, entreated, by nature, as no other of her children is, to cooperate with her in multiplying beauty on

> earth [. . .]. Nature's favorite child, her partner, lets his imagination ride with loose rein through the field of garden knowledge [. . .]. He dreams dreams, sees visions, and plans for the coming year months before he can strike a spade into the earth [. . .]. He vows that the plot of ground he controls shall awake to a greater glory than it ever before attained.[13]

Blanchan's description evokes a sense of exploration and destined power as man is 'entreated' to transform the earth. The gardener embarks on a journey of expansion and conquering: 'loose rein', 'strike a spade', 'the plot of ground he controls', and 'greater glory'. Novelist Christopher Grant La Farge in *Country Life* characterized the garden as a wedding of 'conquering a wilderness' and 'a refinement that belongs to leisure and settled conditions'.[14] With a vocabulary that resonated with descriptions of the 'Wild West', magazine writers created a new frontier myth in the garden. This imagery of conquering and control transformed the garden by bringing the wild frontier into one's home, where one could 'tame' it.

As with Slotkin's 'regeneration through violence', gardening and home magazines' version of conquest similarly emphasized rejuvenation, planning, and control to enjoy – quite literally – the fruits of one's labour. *House and Garden* magazine in 1912, for example, taught gardeners how to '[restore] vigor' and 'bring the old orchard back into fertility' through practices such as pruning, spraying, and grafting.[15] Another article on greenhouse fruits discussed the benefits of the 'well-ordered' greenhouse and ensured readers that the gardener had 'absolute control'.[16] Whereas the frontier is often a masculinized and rugged space, gardening also gave women control over their land. In 1907, Frances Duncan in *Ladies' Home Journal* guided the reader through planning a garden. Referring to the female reader as both 'the owner' and 'the gardener', Duncan designated the garden as a women's space – not to confine her but to give her freedom of choice and freedom to conquer.[17] Articles emphasizing projects that transform the land put control, power, and endless possibilities into the hands of the gardener and homemaker.

Though these magazines suggested that anyone could 'conquer' nature, they also wrote for a specific audience: the white middle class. In an essay for *House and Garden*, Emma Paddock Telford wrote that her new home and garden must have 'refined neighbors as homogenous as possible'.[18] Magazines on gardening and homemaking produced content that marked who and who did not belong as garden- and homeowners, reflecting racial (as well as class) boundaries. They thus constructed the gardens as enclosed spaces only accessible to the white well-to-do. Articles, stories, and images featuring gardeners of colour are absent, despite their involvement in gardening. There were, however, publications that did reflect Black gardening practices and knowledge, such as the *Handbook of the Negro Garden Club of Virginia*. The club, founded in 1932, included a chapter on their history and values: beautifying the home and community, improving race relations, and expressing oneself creatively.[19] In addition to Negro Garden Clubs, histories of slavery and sharecropping speak to African Americans' longstanding involvement in caring for land, farms, and gardens. Control over land, then, also came with a desire to control who could enter and who could benefit from the literal and metaphorical fruits of gardening.

Representing American gardens and homes, these magazines demonstrated new ways of space-making and controlling land. As people sought new spaces to conquer, the frontier expanded domestically into people's gardens. They became a more accessible way to continue the frontier legacy, and print media articulated and perpetuated ideals of freedom and control. Space-making also meant enclosing spaces, as gardening was depicted as an activity reserved for the white middle class. While cottagecore takes inspiration from European history, the American garden and home as a new, romanticized, and exclusionary frontier also laid the foundations for the aesthetic. This shifting of the frontier demonstrates the flexibility of where space-making can take place. Cottagecore, then, moves the frontier into the digital realm.

(Digital) Place-Making: New Frontiers, Performing Place, and Futures of Inclusivity

In creating the new frontier, American gardening and home magazines illuminated the role of mass media in maintaining the frontier myth. The centrality of mass media has only intensified with #cottagecore's popularity on social media. Defining itself as a refuge where one has freedom in nature and content creation, cottagecore continues this frontier legacy through its core values. Cottagecore's digital nature, however, makes it an aesthetic of contradictions. Part natural and part digital, it relies on nostalgia and rests on histories of space-making, yet its openness also allows participants to envision a more inclusive future through performance and place-making.

The internet has become an ever-expanding landscape. Its vast knowledge, combined with social media's never-ending 'For You Pages', invites users to explore the boundless digital frontier – hence why we fall into Wikipedia rabbit holes or get stuck 'doomscrolling' before bed. Though part of this larger frontier, cottagecore creates its own frontier through magazines, cookbooks, and social media content. At first glance, cottagecore is a concept in contradiction with itself. In emphasizing connection to nature, it lives online through digital content, from blog posts to Instagram photos. Platforms like TikTok, Instagram, and Pinterest display a stream of cottagecore content: a young woman twirling in a milkmaid dress, a woven basket of blueberries, and a rustic kitchen with a freshly baked cake. Viewers on their phones and computers scroll through idyllic scenes to (digitally) explore the pastoral frontiers.

However, much like the expansion of the physical American frontier relied on the frontier myth, the digital cottagecore lands 'grow' through users romanticizing a frontier-like lifestyle 'through a rose-colored lens'.[20] *Country Living*'s guide to cottagecore describes the style as '[pursuing] your wildest pastoral dreams', using phrases such as 'fairy tale-esque', 'quintessential storybook look', and 'whimsy'.[21] Cottagecore on the internet hosts a powerful combination of dreamy tunes, whimsical gardens, and well-decorated homes to build a bucolic space, even if only online.

Alongside this fairy tale imagery exists continuing frontier values. Self-sufficiency and a sense of control over one's life are fundamental values to the cottagecore lifestyle. Participants engage in skills that laud the hand- and homemade. *The Spruce* writes that the aesthetic encourages self-sufficiency through gardening, foraging, and raising one's own animals.[22] The online blog and magazine *Cottagecore Dream* calls on readers to 'embrace self-sufficiency' and '[regain] a sense of peace and security' through hands-on activities like bread-making and raising chickens and describes the garden as essential for fresh foods, saving money, and feeling food secure.[23] Other participants might also use seeds, roots, and flowers to create their own natural herbal medicine rather than rely on 'overly processed, manufactured, and oftentimes dangerous chemical alternatives'.[24] As with early twentieth-century American gardens, mass media appeals to readers by presenting them with a sense of control over their lives and space.

This present-day romanticization also relies heavily on nostalgia to evoke the perfect cottagecore scene, as the desire for an idyllic space becomes intertwined with the desire for the past. Present-day print and social media disseminate nostalgic messages about idyllic pastoral life untouched by industrialization. *The Cottagecore Baking Book*, for instance, paints pie baking as a way of romanticizing time in the kitchen, with the aroma of freshly baked goods evoking a sense of nostalgic comfort. Meat pies might be 'reminiscent of medieval tavern food', while baking bread from scratch allows readers to 'step back in time'.[25] More recently, social media figures – both implicitly and explicitly – emphasize traditional roles, as seen with burgeoning 'trad wife' content (sometimes seen as emerging from or related to cottagecore).[26] Whether baking in the kitchen or decorating a bedroom, cottagecore looks toward the past for inspiration.

Behind the romantic activities and aesthetic food is a bucolic yearning for a past world seemingly free of present-day issues, from health to politics. People have addressed the dangers of nostalgia and histories of exclusion. *Cottagecore Dream* mentions the historical 'harsher realities' and 'how much worse it was for most in the past' but notes that 'cottagecore casts aside the pessimistic realities and incorporates only the best'.[27] *The Spruce* also notes that the aesthetic '[skews] towards a more patriarchal era'.[28] Critically engaging with cottagecore requires understanding that, as its romantic imagery emanates an innocent utopian life, it is simultaneously rooted in an expansionist history. Photos and videos of white women often dominate #cottagecore, thus prompting conversations and protests about #cottagecore as a white space. For instance, cottagecore's desire for the past has prompted critics to rename the aesthetic 'colonizercore'.[29]

Social media, from Twitter to TikTok, has long been used to create networks of dissent and protest.[30] With social media algorithms determining whose content is promoted, questions of who we see on our screens (and who we do not see) encourages us to move beyond the rose-coloured lens. Cottagecore content creators of colour, for example, might use hashtags like #representation and #cottagecoreblackgirls (alongside the regular #cottagecore) to discuss the lack of representation and bring the conversation to more users. Another example is *Cottagecore Black Folks*, a site and 'thriving home for black lovers of cottagecore, sustainability, community and simplicity'.[31] *Cottagecore Black Folks* presents cottagecore not as a romanticization or rejection of an exclusionary history but rather a reclaiming of it – and in the process creating a community as well.[32] Similar sentiments, not exclusive to cottagecore, have prompted viral trends such as the 'black men frolicking' trend in 2023 on TikTok.[33] As cottagecore grows, so do questions regarding the aesthetic's foundations and representations.

Reckoning with cottagecore's roots also involves more subtle tactics than direct conversations. Challenging #cottagecore as a white space often happens through sharing and watching content. Cottagecore cooking and baking videos, for example, have adopted the popular 'hands-and-pans' video style.[34] These videos show an up-close glimpse into the baking or cooking process, focusing on just the hands and tools used and condensing perhaps multiple hours of work into a video that is less than thirty seconds. However, these types of videos – often showing slender, fair-skinned hands – also reflect the lack of diversity in these trendy food-making videos.[35] Furthermore, with personalized For You Pages, content creators often face algorithms that privilege whiteness and generate 'algorithmic invisibility', in which attempting to win against the algorithm (and thus becoming visible or even going viral) is a game of power.[36] In this algorithm-directed digital frontier, whose hands one sees making bread or frosting a cake has larger reverberations.

#Cottagecore as a digital space on social media makes participating in it a performative act. After all, most individuals do not have easy access to open greens and pastoral land, or even large houses as showcased in magazines. They make do with what they do have: their clothing, their kitchens, their home furnishings and decor, and their time (for hobbies). That is not to say that the lifestyle is fake or performative in the negative sense. Rather, it is an active process in which, as Judith Butler puts it, social reality is constituted 'through language, gesture, and all manner of symbolic social sign'.[37] Kathryn Ryan and Antoaneta Tileva apply Butler's concept of performativity to queer women on TikTok who are involved in cottagecore content creation. In performing cottagecore activities online, queer TikTok users demonstrate 'a constructed performance of rurality' that also reclaims the frontier and makes queer women more visible on social media.[38]

For marginalized groups who have not seen themselves represented in the aesthetic, performing cottagecore protests the normalization of #cottagecore as an exclusively white space. Whether digitally, in urban settings, or through food or fashion, performing the pastoral life allows the community to reimagine and redefine what cottagecore can be and what it should be. By embracing the cottagecore lifestyle, participants transform the digital space into a meaningful place of community and belonging, directly challenging expansionist space-making. #Cottagecore provides a platform within a platform in which marginalized groups can be visible and have a sense of place. Individuals uncover a shared – and no longer exclusive – refuge. While cottagecore is nostalgic for the past, it increasingly looks toward a future of inclusivity.

Despite cottagecore's worldwide reach, it is as if participants all exist in a shared pastoral setting that extends beyond the hashtagged space. The openness of social media thus parallels the openness of the frontier and pastoral lands. Place-making through performance on the internet is what makes the digital kitchen feel real, what brings nature closer, or what makes the open fields feel less isolating. It allows users to not just consume content but also feel empowered to produce and share content. The growing community of cottagecore imbues the digital frontier and online space with new meanings and values that empower a more inclusive 'pastoral' land.

Conclusion

The aesthetic, then, is not a frivolous excuse to dress up and play pretend, and nor is it just a trend. Cottagecore is a useful starting point in looking at past and present representations of the garden and gardener, and the home and homemaker. In extending the history of cottagecore to highlight its ties to the American frontier, I analyze how the frontier shifted into the garden and home through mass media, thus maintaining expansionist ideals of space-making.

Representations of the garden and the home, however, often excluded those who were not part of the white middle class. These frontier legacies, including its romanticization and nostalgia for a pastoral past, inform the core values of cottagecore and continue to portray the garden and home as new – and now digital – frontiers. Yet, cottagecore's growing popularity has also inspired conversations about #cottagecore as an exclusive white space. To challenge this expansionist history, performance and place-making become powerful strategies to envision a more inclusive future.

Frontier, then, takes on a new meaning. This opening of the cottagecore 'frontier' – dismantling the barriers – reveals how gender, race, and class intersect and interact in our ever-changing social media landscape. Though cottagecore is rooted in the past, its community is grappling with this history while cultivating a new digital frontier. As cottagecore straddles and blurs the boundaries of nature and technology, past, present, and future, and space and place, it calls for a more conscientious engagement with social media and continues to shift the frontier to new places for protest and empowerment.

About the Author

Jolin Chan is a senior at Harvard College, pursuing a joint concentration in History & Literature and Social Anthropology. She is interested in food studies, Asian American studies, and archival and ethnographic research.

Notes

1. I use #cottagecore to name the online s/place for content-sharing and discourse (where the hashtag allows for efficient categorization), and I use cottagecore to refer to the aesthetic and lifestyle people can adopt.
2. Isabel Slone, 'Escape into Cottagecore, Calming Ethos for Our Febrile Moment', *The New York Times*, 10 March 2020 <https://www.nytimes.com/2020/03/10/style/cottagecore.html> [accessed 10 May 2024].
3. Angelica Frey, 'Cottagecore Debuted 2,300 Years Ago', *JSTOR Daily*, 2020 <https://daily.jstor.org/cottagecore-debuted-2300-years-ago/> [accessed 10 May 2024].
4. 'The Queen's Hamlet', *Palace of Versailles*, 2018 <https://en.chateauversailles.fr/long-read/queens-hamlet> [accessed 10 May 2024].
5. See Rebecca Jennings, 'Cottagecore, Taylor Swift, and Our Endless Desire to Be Soothed', *Vox*, 2020 <https://www.vox.com/the-goods/2020/8/3/21349640/cottagecore-taylor-swift-folklore-lesbian-clothes-animal-crossing> [accessed 10 May 2024]; see also Edgar James Ælred Jephcote, 'The Significance of Victorian England for the Cottagecore Aesthetic', *English Studies at NBU*, 9.2 (2023), pp. 293–311, DOI:10.33919/esnbu.23.2.8.
6. Frederick J. Turner, 'The Significance of the Frontier in American History', in *Annual Report of the American Historical Association* (American Historical Association, 1893) <https://www.historians.org/resource/the-significance-of-the-frontier-in-american-history/> [accessed 10 May 2024].
7. John F. Kennedy, '"The New Frontier" Acceptance Speech of Senator John F. Kennedy Democratic National Convention, July 15, 1960' (John F. Kennedy Presidential Library and Museum, 1960), pp. 1–8 (p. 6) <https://www.jfklibrary.org/asset-viewer/archives/jfksen-0910-015> [accessed 10 May 2024].

8 Kennedy, pp. 6, 7.
9 Richard Slotkin, *Regeneration through Violence: The Mythology of the American Frontier, 1600–1860* (Wesleyan University Press, 1973).
10 Slotkin, pp. 18–19.
11 Slotkin, p. 30.
12 Virginia Tuttle Clayton, *The Once & Future Gardener: Garden Writing from the Golden Age of Magazines, 1900–1940* (David R. Godine, 2000), p. xi.
13 Neltje Blanchan, 'The Joy of Gardening', *Country Life in America*, March 1910, pp. 541–44 (p. 541).
14 Christopher Grant La Farge, 'Our Country and Our Gardens', *Country Life in America*, May 1920, pp. 42–46 (p. 42).
15 Stephen N. Green, 'Rejuvenating the Old Orchard', *House and Garden*, October 1912, pp. 210–12, 245–48 (p. 210).
16 William C. McCollom, 'Greenhouse Fruits with Outdoor Flavors', *House and Garden*, September 1917, pp. 46–47, 58, 60, 62, 64 (pp. 46–47).
17 Frances Duncan, 'How to Fit a Garden to a House', *Ladies' Home Journal*, February 1907, p. 24.
18 Emma Paddock Telford, 'What We Learned When We Built Our House', *House and Garden*, January 1914, pp. 25–27, 55–58 (p. 25).
19 *Handbook of the Negro Garden Club of Virginia*, ed. by H. Hamilton Williams (Hampton Institute, 1943), p. 83.
20 Kristen Hohenadel, 'What Is Cottagecore? A Simple Guide', *The Spruce*, 16 July 2024 <https://www.thespruce.com/cottagecore-design-style-5095952> [accessed 10 May 2024].
21 Anna Logan, 'Here's Everything You Need to Know About Cottagecore', *Country Living*, 2024 <https://www.countryliving.com/home-design/decorating-ideas/a60604002/cottagecore-aesthetic-vibe-guide/> [accessed 10 May 2024].
22 Hohenadel.
23 Lindsey Knight, 'How to Embrace Self-Sufficiency with Cottagecore', *Cottagecore Dream* <https://cottagecoredream.com/article/how-to-embrace-self-sufficiency-with-cottagecore> [accessed 10 May 2024].
24 Emily Kent, *The Little Book of Cottagecore: Traditional Skills for a Simpler Life* (Adams Media, 2021), p. 81.
25 Kayla Lobermeier, *The Cottagecore Baking Book: 60 Sweet and Savory Bakes for Simple, Cozy Living* (Page Street Publishing, 2024), pp. 42, 70, 145.
26 See Jessica Grose, 'Opinion | "Tradwife" Content Isn't Really for Women. It's for Men Who Want Submissive Wives', *The New York Times*, 15 May 2024 <https://www.nytimes.com/2024/05/15/opinion/tradwife-tiktok.html> [accessed 25 September 2024].
27 Leelee Sedlacek, 'How the Cottagecore Aesthetic Turned into a Lifestyle', *Cottagecore Dream* <https://cottagecoredream.com/article/how-the-cottagecore-aesthetic-turned-into-a-lifestyle> [accessed 10 May 2024].
28 Hohenadel.
29 Sydney Gore, 'Life After Colonizercore', *The Cut*, 2022 <https://www.thecut.com/2022/06/colonizercore-cottagecore.html> [accessed 10 May 2024].
30 See Sarah J. Jackson, Moya Bailey, and Brooke Foucault Welles, *#HashtagActivism: Networks of Race and Gender Justice* (Cambridge, Massachusetts: The MIT Press, 2020).
31 'About Us', *Cottagecore Black Folks* <https://cottagecoreblackfolk.com/pages/about-us> [accessed 10 May 2024].
32 'Cottagecore Isn't Monolithic', *Cottagecore Black Folks*, 2021 <https://cottagecoreblackfolk.com/blogs/cottagecore-isnt-monolithic/cottagecore-isnt-monolithic> [accessed 10 May 2024].
33 Starting around May 2022, the 'Black men frolicking' TikTok trend consisted of Black men posting videos of themselves frolicking in large open fields. It emphasized mental health among Black men, sparked the hashtag #BlackBoyJoy, and challenged stereotypes. See Tomas Kassahun, 'Black Men Are Frolicking and Bringing Some Much-Needed Joy To The Internet', *Blavity News & Entertainment*, 2022 <https://blavity.com/black-men-are-frolicking-and-bringing-some-much-needed-joy-to-the-internet> [accessed 10 May 2024].

34 Alexandra Jones, 'Hands and Pans: The Boiled-down Recipe Videos Cooking up a Storm', *The Guardian*, 11 April 2018 <https://www.theguardian.com/lifeandstyle/2018/apr/11/hands-and-pans-the-boiled-down-recipe-videos-cooking-up-a-storm> [accessed 10 May 2024].

35 Alison Saldanha, *The Subliminal Racism of Instagram Food Videos*, Medill Reports Chicago (Northwestern University, 25 June 2020) <https://news.medill.northwestern.edu/chicago/the-subliminal-racism-of-instagram-food-videos/> [accessed 10 May 2024].

36 Safiya Umoja Noble, *Algorithms of Oppression: How Search Engines Reinforce Racism* (New York University Press, 2018); Taina Bucher, 'Want to Be on the Top? Algorithmic Power and the Threat of Invisibility on Facebook', *New Media & Society*, 14.7 (2012), pp. 1164–80, DOI:10.1177/1461444812440159.

37 Judith Butler, 'Performative Acts and Gender Constitution: An Essay in Phenomenology and Feminist Theory', *Theatre Journal*, 40.4 (1988), pp. 519–31 (p. 519), DOI:10.2307/3207893.

38 Kathryn Ryan and Antoaneta Tileva, 'Taking the Past out of the Pastoral: TikTok's Queer "Cottagecore" Culture and Performative Placemaking', *Queer Studies in Media & Popular Culture (Online)*, 7.3 (2022), pp. 165–76 (p. 171), DOI:10.1386/qsmpc_00077_1.

9
Preserving Fruit at the Nexus of 'Lay' and 'Expert' Knowledge

Danille Elise Christensen

In 1914, American universities and the United States Department of Agriculture (USDA) joined forces to take science to the people via the Cooperative Extension Service (CES). This federally supported system built on agricultural and homemaking associations, including the boys' corn clubs and girls' tomato canning clubs established to combat poverty in the rural South. Armed with new degrees in domestic science, USDA Home Demonstration Agents promoted everything from housefly eradication to vitamin-filled menus in their assigned counties.[1] They also encouraged 'every suburban and country household' and 'provident housekeeper' to preserve foods using 'modern' methods: bottles sterilized and sealed at home would maintain the fresh taste of raw foods and counter the dubious quality of commercially canned goods. Although home agents advised on jam and pickle production, they emphasized preserving plain fruits and garden produce, such as bottled peaches or canned green beans only lightly syruped or salted. In the rhetoric of the day, sugar-saturated preserves and fermented vegetables were nice old-time accompaniments, but not the foundation of efficient self-sufficiency.[2]

Today, these CES specialists are called Family and Consumer Science agents, and their casual conversations about home bottling ('canning') sometimes include horror stories, such as tales of people using a dishwasher's relatively low, quick steam heat to seal jars of repackaged snack sausages without sterilizing them. At the same time, practitioners question institutionally generated knowledges. In 2024, one meme pictured shelves filled with home-bottled goods, declaring 'As for me and my family? Whatever the gov recommends, we'll do the opposite'.[3]

Framing bottling techniques as 'lay' and 'expert' thus pits tradition against science in ways that feed broader polarizations.

Here, I explore several American home bottling practices alongside unfolding and contested knowledges within the US scientific establishment, including the early twentieth-century move to bottle foods beyond fruits and preserves. In historical records household food producers largely present as people working to understand changing official recommendations and reconcile them with their own experience, resources, and aesthetic standards. Today, as reasonable suspicions and unfounded assumptions persist across the board, credentialed experts and 'rebel canners' alike could learn much from the nuanced contexts of their predecessors.

Figure 1. 'Wife of FSA (Farm Security Administration) borrower [right] discussing pressure cooker with Home Supervisor'. Photo by John Vachon, Mille Lacs County, Minnesota, 1941. FSA/OWI Collection, Library of Congress Prints and Photographs Division. https://www.loc.gov/pictures/item/2017812917/

Current Recommendations and Vernacular Practice

In 2024, USDA representatives teach that foods bottled safely at home must be heated in gasket-closed jars at temperatures high and long enough to kill or disable pathogens and create a seal that keeps new microbes out. A major concern is *Clostridium botulinum*, a bacterium whose spores love canning jars: they thrive in room-temperature, moist, low-oxygen, low-acid environments, creating a potent neurotoxin as they grow. Officials promote two heat/seal processes. A boiling waterbath is appropriate for most filled fruit bottles, since fruit's naturally high acidity stymies *C. botulinum*.[4] However, low-acid foods such as meats and vegetables need to be 'pressure canned' in vessels that push temperatures beyond boiling; if hardy *C. botulinum* spores aren't constrained by acid, they need extra-high heat instead. For both methods, specific food acidities, densities, and other factors determine the length of heat processing.

A third method – one used for centuries, and in much of the world – is known as 'open kettle' or 'cooked-in-the-kettle' canning.[5] Boiling-hot sugared fruit preserves are ladled directly into clean jars and capped, in the past using alcohol-soaked paper or paraffin wax, today with a gasketed lid. Jars are then stored without further heat processing.

In the 1920s, US food safety experts warned against open-kettle processing. In the 1980s they disavowed the process even for high-acid, high-sugar foods, noting that moulds can enter jars while they are being filled, or if the weak vacuum produced by the open kettle method breaks. Because moulds can lower acid levels, they open the door to botulism. Thus, for decades American home canning manuals have directed readers to submerge capped preserves in boiling water for ten minutes or more, demanding that home cooks 'Forget Grandma's Open Kettle!'[6]

Nevertheless, a survey conducted at the turn of the millennium found that twenty-one percent of respondents used some open-kettle procedures.[7] Much of this output surely involves fruit, including the apple butter described below. Historically, home bottlers have also treated tomatoes as fruits, and some who've learned canning from experienced elders apply the open kettle process to tomato-based soups and salsas – despite researchers' warnings that tomatoes are less acidic than they used to be and that additional ingredients can raise pH beyond safe levels for waterbath canning.

Another vernacular departure is using a boiling waterbath for foods that the USDA says should be pressure-canned or not canned at all. In June 2024, a Facebook canning group circulated photos of bulk cream cheese rebottled for pantry storage. When someone observed that it was 'a botulism party waiting to happen', a defender asked, 'What [do you think they did] before refrigeration? People have been canning dairy products for centuries.' A later post featured a galvanized washtub boiling over an open fire, filled with dozens of bottled green beans. Scores of group members recalled similar sights from childhood, some recommending outdoor boiling as a cooler, more efficient, less explosion-prone practice than pressure canning. Others simply wanted to can 'the old-fashioned way'. One champion of waterbathed green beans referenced a beloved book series about nineteenth-century homesteading: 'How do you think people [bottled beans] before pressure canners were invented, or before they had power. I don't think Little House on the Prairie was digging out the pressure canner!'[8]

These apologists assume that country people have always bottled everything; in fact, home canning didn't take off until the early twentieth century, alongside USDA encouragement. In *Little House* times, people dried or pickled beans. And cheese itself is the older way to preserve dairy: unsweetened evaporated milk was not canned in the US until the mid-1880s, using high-pressure factory retorts.[9] Then again, academically trained food safety advocates may not know how current 'expert' recommendations relate to rescinded prior guidance. There is need all around for a fuller understanding of precedent.

'Safe and Sound': Changing Guidelines

When officials first promoted bottling food beyond fruit and condiments, many home cooks doubted that the effort was economically or aesthetically sound. Sensible people knew that (starchy low-acid) vegetables were difficult to store in a plain, moist state. Ethel Downen, born *c.* 1885 in Indiana, recalled that around the time of the First World War, 'My mother-in-law came up, and I told her that I'd got to canning peas. "Oh," she said, "they won't keep."'[10] Indeed, one USDA employee later recalled that products tinned and bottled at home in the 1910s were 'discredited' because of high spoilage rates. When she began working with community canning clubs in 1911, she aimed to teach young women and their mothers 'safe and sound canning procedure and give them enough supervision to insure good standards'.[11]

However, what was considered 'safe and sound' changed rapidly during the first decades of the twentieth century. American home canners had to navigate not only diverse experimental vessels, closures, and equipment, but also myriad expert and promotional discourses. Researchers pursued many questions: Was it better to sterilize a batch in a single, long cook ('one-period') or shorter periods on successive days ('fractional' or 'intermittent')? Could standard boiling temperatures eventually kill bacterial spores, or were higher temperatures required? How did factors like food acidity and density, container size, and geographical location affect spore germination and thermal death point? Was canning meats and vegetables at high heat under steam pressure a necessity, or did it simply produce a fresher product? In 1922, botulism research wasn't clear on whether the toxin could develop in opened food, how salt and sugar affected spores, and why lemon juice didn't always prevent toxin development. One report even implied race-based differences in symptoms.[12]

Failure: Social and Material Variables

The spread of this emerging, partial knowledge was fraught. Between 1913 and 1916, USDA's O.H. Benson demonstrated *cold packing*, a confusing shorthand for 'one-period processing in the jar, usually at the temperature of boiling water'. Youth club leaders were taught to pack food 'cold' into jars (some raw fruit and all vegetables were blanched and cold-dipped first), cover it with boiling syrup or a weak brine, then cap and heat (Figure 2).[13] But Benson 'encounter[ed] considerable opposition' from educators; at Purdue University, for instance, home economics staff maintained that the older 'open kettle method [. . .] was the only safe method for home use'.[14]

Their reticence was understandable. As the First World War escalated in Europe, instructions for processing fruit and vegetables 'in the jar' were widely variable. Publications from the mid-teens make little distinction among vessels that boiled, steamed, or cooked under pressure: these included the patented four-chamber Mudge Canner (1886), 'water-seal outfits', and lidded washboilers normally used for clothes. If timetables were keyed to different technologies, they emphasized overall cooking time rather than achievable temperatures.[15]

Further, the USDA was slow to update home canning bulletins. Although microbiologists began investigating food spoilage in the late 1890s and the National Canners Association recruited scholars for a research committee in 1913, the USDA resisted industry-linked findings. In 1917, its representatives denied a Stanford professor's assertion that government cold-packing instructions promoted inadequate boiling times and temperatures for vegetables; instead, the USDA attributed vegetable spoilage to home canners' slack attention.[16]

Laypeople experienced this confusing expert guidance as failure. Loss rates were especially high as war-effort canning ramped up. German sabotage was suspected, but the actual trouble proved to be a confluence of new methods, material shortages, minimal

Figure 2. Mandel Brothers store circular, 1917, linking academic credentials to the 'cold pack' ('cooked-in-the-jar') method. National Archives and Records Administration–Central Plains (Kansas City, MO), RG4: Records of the US Food Administration, 1917–1920, Series 58: Correspondence and Data of the Women's Central Committee on Food Conservation, 1917–1918, Box 223, Folder Corr 1917.

refrigeration, novice enthusiasts, and an impulse to salvage windfalls and rejects. Thinner wartime rubber rings failed when subjected to the hours of boiling recommended for vegetables processed using the new cold pack method.[17] In 1917, when the Women's Central Committee on Food Conservation opened a community cannery in St. Louis, they prided themselves on teaching cold pack methods, including processing some foods at low (5 psi) pressure (108°C [227°F]). In two summer months, volunteers produced 7361 jars; twelve percent spoiled by September, including much bottled corn and most okra, squash, and wax beans. This waste undermined the credibility of the pre-suffrage woman-run organization.[18] USDA Home Demonstration Agent Jessie W. Harris encountered a different problem at Texas A&M University: tin shortages meant her team began using glass jars in 1917, but experience with metal cans and pressure retorts didn't transfer to the new equipment. Quick pressure releases siphoned liquid from jars before they sealed; then agents would reopen and fill the jars with boiling water. Many spoiled.[19]

New methods are often at odds with the means at hand. Material objects and the techniques and logics linked to them persist in use, their life cycles affected by physical durability, consumer behaviour, and cultural expectations.[20] When lid-and-band caps became popular after 1915, their 'made-on' rubber gaskets failed when subjected to long-recommended techniques for tightening rubber rings, the technology they replaced.[21] Indeed, instructions prove difficult to call back. In 1918, home canners were taught to keep bottle closures above the waterline during processing; almost fifteen years later educators were still trying to undo that advice.[22]

Navigating Expert Recommendations: Fruit vs. Veg

During these rapid shifts, home canners could easily slip off the cutting edge. Around 1908, an Indiana girl named Agnes Bell tried canning green beans with her sister while their parents were away; despite '[using] the right directions', every jar developed 'flat sour', a non-toxic but unpleasant accumulation of lactic and acetic acid. 'We thought we were real smart,' she remembered. 'And we were disappointed. We [had been] so proud of ourselves'.[23]

The girls may have used what was variously called a 'staged', 'intermittent period', or 'fractional' sterilization method, which involved heating filled jars for an hour or so in a waterbath on several consecutive days. This strategy applies the logic of eighteenth- and nineteenth-century (high-acid) whole-fruit preserves, heated multiple times to ensure sugar saturation and prevent fermentation. Into the mid-twentieth century, the USDA promoted low-heat staged processing as a way to maintain taste and appearance for fruits (and tomatoes). In 1909, government experts extended multi-period processing to vegetables.[24] Although people died as a result, the method remained an approved option in the USDA's 1921 revamp of canning instructions. What didn't make the cut? Open kettle.[25]

Grandma's Open Kettle?

Open-kettle processing is time- and resource-efficient, and it had been a common preserving strategy among credentialed experts. North Carolina's Jane McKimmon remembered that in 1911, when home agents were asked to teach club girls how to seal food in metal cans, 'not one of us had mastered the art of canning in tin. All of us had canned fruit in glass jars, largely by the open kettle method, which meant pre-cooking, pouring into the jar at boiling temperature, and sealing [i.e. securely covering] immediately' (Figure 3).[26]

As microbiologists learned more about pathogens, manuals promoting open-kettle canning became obsessed with sterilization. Whereas a 1907 bottling handbook assured readers that 'germs do not incubate in a vacuum' and were awakened only by oxygen, by 1915 the precursor to the *Ball Blue Book* confirmed that a food's pH mattered, and that one deadly microbe even thrived without oxygen. 'It is dangerous to let the jar stand open for even a moment,' the bottle company warned, 'as spores are floating about in the air at all times, and even one [. . .] may be enough to spoil a whole jar.' Revised open-kettle instructions said to pack a single scalded jar at a time, overfilling the very hot fruit with boiling syrup, then to immediately cap and invert the jar so as to kill bacteria on the lid.[27] At the same time, early directives produced by commercial interests recognized consumer

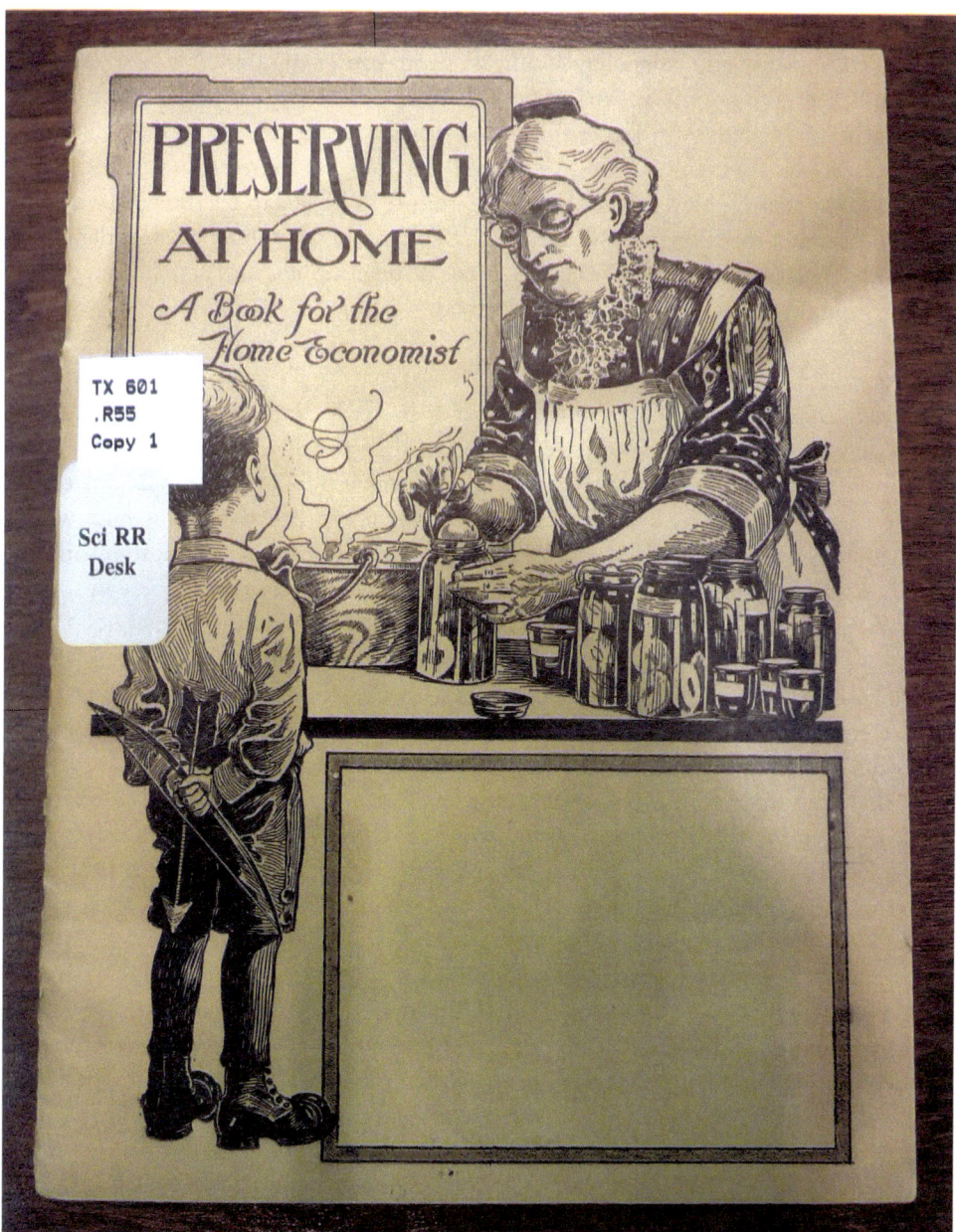

Figure 3. Shortly before USDA introduces the cold pack method, the cover of Emily Riesenberg's Preserving at Home: A Book for the Home Economist *depicts an authoritative elder packing hot fruit alongside jelly jars, no waterbath boiler or pressure canner in sight (Rand McNally, 1916).*

agency. The 1915 Ball manual, for instance, described three open-kettle variations, each keyed to fruit acidity and balancing colour, shape, flavour, and safety.[28]

By 1918, however, experts designated open-kettle methods 'old-fashioned', 'uncertain', and 'largely superseded' by in-jar sterilization. Open-kettle processes were deemed 'troublesome', not only for involving messy scalding syrup, but also for requiring 'considerable intelligence to guard against contamination'.[29] Failure to fill jars to overflowing was 'one of the commonest mistakes in home canning.' Because reusing rubber rings or damaged jars were other common missteps, many open-kettle practitioners may actually have been thrifty to a fault, not careless or poorly informed.[30]

'Yes, but . . .'

Negative twentieth-century characterizations of open-kettle canning align with gendered and ageist stereotypes about 'lay knowledge': food industry insiders often assumed housewives to be distracted or 'irrational and endlessly suggestible', while media portrayed older women as comforting but out of touch, offering potentially dangerous 'guesswork' governed by rote tradition. Home cooks have pushed back, calling attention to shifting official recommendations and demonstrating knowledge of processing times and temperatures, acidity levels, heating strategies, and pathogens. In oral history interviews they often note when apparently outmoded strategies were effective in context, tempering their acknowledgements of procedural dangers with, 'Yes, but . . .'.[31]

One common assertion is that older methods and more accessible technologies were successful with fruit. When Hattie Wells (b. 1919) recalled her mother's food work in Kentucky, she suspected that preserving powders (e.g. salicylic acid) kept waterbathed green beans and sweet corn free of botulism, but worried whether the additives had contributed to family cancers. Fruit preserves, however, successfully skipped heat processing in the jar altogether: 'Of course jams and jellies, she just made that open kettle on the stove. And she made plenty of them.'[32] A Montana couple who married in 1920 never canned vegetables due to 'too much talk along about then, about people getting that poisoning'; they had no pressure canner. But using a boiling waterbath, 'We had a lot of fruit canned all the time. Jellies and jams and syrup'.[33] And Agnes Bell – the girl whose beans had soured in Indiana – affirmed, 'My mother would can tomatoes, peaches, apples, and cherries. Not like a dozen, more like a hundred of each kind'.[34]

'No Additional Process Required': Open Kettle Apple Butter

Bell's mother's cellar likely held apple butter, a highly condensed fruit spread traditionally made outside in giant copper kettles – 'open kettle' in the most obvious sense. Although batches today generally use fresh apples (or applesauce), sugar, and sometimes cider, historically apple butter might also be made from dried apples, or with the leavings from maple syrup or sorghum molasses evaporating pans.[35] Despite its name, the condiment contains no fats, aside from clove or cinnamon oil flavouring.

Batches must be stirred constantly, and the kettle's contents caramelize during all-day events that harness community muscle, tools, and good humour against discomforts like wood smoke and wasps.[36] Few families make apple butter anymore, but the practice has become linked to churches, service clubs, and other civic organizations, and it helps recruit members and raise funds.

At almost any apple butter making today, the finished product is taken hot off the fire and poured directly into warmed glass jars, then capped with gasketed metal lids that vacuum-seal from the product's heat. The USDA Department of Chemistry recommended that method in 1917: 'When the [apple] butter is thick as desired, place in hot containers and seal immediately'. As in commercially produced apple butter, 'No [additional] process is required'.[37]

Open kettle techniques are what permit large-scale apple butter events to continue today, since quantity of product and location of manufacture are not limited by access to boiling waterbaths and the time they require. In Woolwine, Virginia, for instance, apple butter is the star of the thirty-year-old October Festival, which funds volunteer emergency services in this rural mountain community. The finely tuned operation includes century-old, carefully maintained cast-iron apple peelers; six copper kettles on custom-made propane stands; and near 200 bushels of locally grown Golden Delicious apples (pH ~3.6). Volunteers peel, pare, and quarter the fruit and toss it into dozens of clean 32-gallon plastic trash cans.

At 4 am the next day, apple butter cooking commences. Food-grade grinders chop the apple quarters, and volunteers add them to the gas-fueled kettles inside the old fire station's open bays. Each pot cooks for 12 to 14 hours and produces about 35 gallons. Teenagers once joined older men to stir the apple butter through the night, a rite of passage involving coffee and beer, maybe a little moonshine. Population decline prompted organizers to purchase electric stirrers, though volunteers still monitor propane jets and skim seeds that float to the surface (Figure 4). Cornbread and pinto beans feed the rotating workers.

When the batch is nearly done, participants 'pull the heat' and switch to hand-stirring with wooden paddles before ladling the apple butter – 'like liquid magma', as hot as 232°C (450°F) – into quart-size Mason jars. Then volunteers wipe down each jar and its rim, seat the lids, and screw on the bands. Woolwine's product is legendary and trusted: in 2022,

Figure 4. Eddie Harbour (left) and David Goode tend apple butter in Woolwine, Virginia, 6 October 2022. Photo by Laura Saunders, for the Southern Foodways Alliance.

the group presold nearly all of the 1400 quarts it produced, retaining just 36 quarts for festival sales.[38] In the end, it is the apple's particular composition and apple butter's high sugar content that creates specific conditions for asserting community-based expertise in the face of official recommendations.

Humility and History

Food safety educators know contexts matter: thermal death times are affected by altitude; *C. botulinum* exists in more and less virulent strains, distributed unevenly; today's tomatoes and vinegars are less acidic due to consumer preference, and thus poor candidates for open-kettle methods. But it's also important to consider the work processes, technological systems, ethical norms, and knowledge networks in which fruit, microbes, and humans are embedded.[39]

In the oral histories cited here, women narrate how they learned new techniques alongside their mothers and peers in the early twentieth century and reflect on which rapidly shifting recommendations their families chose to adopt. They emphasize how they applied official terms and procedures while claiming their own agency.

Today, lay knowledge of canning's convoluted histories is known only second-hand, if at all, which may explain why contemporary canning talk can be past-facing and contrarian. In many rural places, transfer of gardening and food preservation knowledge has been disrupted by long commutes, homeplace loss or pollution, and outmigration. In these contexts, even turning to outdated canning manuals can feel like reclaiming a lost past.

Anti-elitist and anti-corporate discourses also emerge in canning forums, warning those with institutional power against mistaking culturally important practices for backwoods ignorance.[40] Requiring specialized equipment or forbidding people from reusing spaghetti sauce jars, they say, gives rise to the adage 'hipsters can / poor people can't'. Criticizing people for reportioning bulk minced garlic or nacho cheese bought on sale feels punitive to those doing their best to get by. Others assert that capitalists stand to gain by promoting food safety regulations that discourage thrift and community-based production.

In these physical and rhetorical contexts, food safety educators could build trust by acknowledging that what is now deemed incorrect may once have been officially sanctioned, and that best practices are often best suited to specific constellations of material factors and human intentions. Home canners could draw on the truths of experience while also conceding the dynamism of past practice and the emergence of new variables. In short, a richer understanding of canning's histories can help address information gaps and communicative misfires on both sides of the apparent divide between lay and expert knowledge.

About the Author

Danille Elise Christensen is Associate Professor in the Department of Religion and Culture at Virginia Tech, a research-intensive university in the heart of Appalachia's Blue Ridge Mountains. A folklorist by training, she studies the politics of knowledge production and is completing a monograph entitled 'Freedom from Want: Home Canning in the American Imagination'.

Notes

1. Wayne D. Rasmussen, *Taking the University to the People* (Iowa State University Press, 1989); Franklin M. Reck, *The 4-H Story* (National Committee on Boys and Girls Club Work, 1951); D.B. Williams, *Agricultural Extension* (Cambridge University Press, 1968); Jorge H. Atiles, Caroline E. Crocoll, and Jane Schuchardt, 'Extending Knowledge, Changing Lives: Cooperative Extension Family and Consumer Sciences', in *Remaking Home Economics*, ed. by Sharon Y. Nickols and Gwen Kay (University of Georgia Press, 2015), pp. 36–53; Elizabeth Engelhardt, 'Canning Tomatoes, Growing 'Better and More Perfect Women': The Girls' Tomato Club Movement', *Southern Cultures* 15.4 (2009), pp. 78–92; Carmen Harris, 'Grace under Pressure: The Black Home Extension Service in South Carolina, 1919–1966', in *Rethinking Home Economics*, ed. by Sarah Stage and Virginia B. Vincenti (Cornell University Press, 1997), pp. 203–28.
2. Janet McKenzie Hill, *Canning, Preserving and Jelly Making* (Little, Brown, and Co., 1915), pp. 1, 2. Sugar limits bacterial growth by reducing water activity. On canning as modern, see Danille Elise Christensen, '"The Father[s] of Canning"? Narrating Nicolas Appert/American Industry', *Journal of American Folklore* 136.539 (2023), pp. 16–47.
3. Meme credited to Barefoot Mimosas, a 'crunchy mom' homesteader aligned with anti-vax and prepper politics.
4. The pH of figs, sweet tomatoes, and white-fleshed peaches may exceed 4.6, so they should be acidified with lemon juice, made into jam, or frozen (National Center for Home Food Preservation <https://nchfp.uga.edu/> [accessed 25 January 2025]).
5. *Open kettle* is also historically called *hot pack*, not to be confused with the current method of the same name, in which foods are blanched or simmered before further heat processing in jars.
6. J. Orvin Mundt, 'Effect of Mold Growth on the pH of Tomato Juice', *Journal of Food Protection* 41.4 (April 1978), pp. 267–68; USDA, *Home Canning of Fruits and Vegetables*, Farmers' Bulletin 1211 (US Government Printing Office, 1921); Martha Zepp, 'Avoid Open Kettle Canning: Always Process Canned Goods' (PennState Extension, 30 April 2024) <https://extension.psu.edu/avoid-open-kettle-canning-always-process-canned-goods> [accessed 25 January 2025]; *Kerr Kitchen Cookbook: Home Canning and Freezing Guide* (Kerr Glass Mfg., 1990), p. 9.
7. Zepp.
8. People have indeed survived eating bottled green beans boiled for hours over unsteady heat at standard atmospheric pressure – perhaps because they generally reheat jarred beans before serving, which would neutralize botulism toxin developed during storage. E.g. Jimmie Dean ('Jim') Williamson, interview by Katherine N. Bills (24 May 2012), Berea College Special Collections & Archives, Berea, KY, Appalachian Foodways Oral History Collection (SAA 164), Box 1, Folder 28.
9. Evaporated Milk Association Collection, David M. Rubenstein Rare Book & Manuscript Library, Duke University <https://idn.duke.edu/ark:/87924/m1889x> [accessed 25 January 2025].
10. Ethel Downen, interview by Dee Ann Cabell, Montgomery Co., IN, in *Going to Club: Memories of Hoosier Homemakers*, vol. 5, ed. by Eleanor Arnold (Indiana University Press, 1985), p. 83.
11. Jane Simpson McKimmon, *When We're Green We Grow* (University of North Carolina Press, 1945), p. 3.
12. Jacob Casson Geiger, Karl Friedrich Meyer, and Ernest Charles Dickson, *The Epidemiology of Botulism*, Public Health Bulletin 127 (US Government Printing Office, 1922), p. 50.
13. Katharine Blunt, Florence Powdermaker, and Elizabeth C. Sprague, eds., *Food and the War: A Textbook for College Classes* (Houghton Mifflin, 1918), p. 212. The *cold-pack* waterbath process is called the *boiling water method* today; it was known as *appertization* in the nineteenth-century and *bottling as for gooseberries* in the eighteenth; Danille Elise Christensen, '"Already acquainted with the methods of Mr. Appert": Bottled Gooseberries, Imperial Science, and Vernacular Knowledges in the Early Modern Anglophone Atlantic' (forthcoming, *Global Food History*).
14. Reck, pp. 126, 165.
15. O.H. Benson, 'Home Canning by the One-period Cold-pack Method', USDA Farmers' Bulletin 839 (States Relation Service, 1917).
16. Elizabeth L. Andress and Gerald D. Kuhn, 'Critical Review of Home Preservation Literature and Current Research', (University of Georgia, Cooperative Extension Service, 1988); J.H. Young, 'Botulism and the

Ripe Olive Scare of 1919–1920', *Bulletin of the History of Medicine* 50.3 (1976), pp. 372–91; Anna Zeide, *Canned* (University of California Press, 2018).

17 A.C. Boomslater to Herbert Hoover (23 July 1917), National Archives and Records Administration–Central Plains (Kansas City, MO), RG4: Records of the US Food Administration, 1917–1920 (hereafter RG4), Series General Correspondence, 05/1917–08/1917, Box 19, Folder Rings; Boston Woven Hose & Rubber Co. to Herbert Hoover (17 August 1917), RG4, Series Supplemental Correspondence and other Records, 8/1917–11/1918, Box 785, Folder Glass Jars & Rings.

18 'Spoilage Sheet' (6 September 1917), RG4, Series 58: Correspondence and Data of the Women's Central Committee on Food Conservation, 1917–1918, Box 223, Folder Corr 1917.

19 Jessie W. Harris (19 June 1917), report appended to Alexander Bruce Bielaski's letter to Herbert Hoover, 14 July 1917, RG4, General Correspondence 5/17–8/17, Box 19, Folder Jars.

20 Sarah H. Hill, 'An Examination of Manufacture-Deposition Lag for Glass Bottles from Late Historic Sites', in *Archaeology of Urban America*, ed. by Roy S. Dickens (Academic Press, 1982), pp. 291–327.

21 USDA Bureau of Human Nutrition and Home Economics, Poster set on home canning (1943), Library of Congress, Prints and Photographs Division, PR 06 CN 183 (unprocessed collection).

22 Blunt, Powdermaker, and Sprague, p. 215; 'Lecture Demonstration by Miss Gladys Kimbrough' (Hartford, CT, 2 February 1932), Minnetrista Heritage Center, Muncie, IN (hereafter MHC), Ball Associates Collection (2000.81.1), Folder 29, p. 15.

23 Agnes Bell, interview with Wanetta Edgerly, Hamilton Co., IN, in *Feeding Our Families*, vol. 1, ed. by Eleanor Arnold (Indiana University Press, 1983), p. 35; Gregg Steven Pearson, 'The Democratization of Food: Tin Cans and the Growth of the American Food Processing Industry, 1810–1940' (unpublished doctoral dissertation, Lehigh University, 2016), p. 376.

24 Edith Bradley and May Crooke, *The Book of Fruit Bottling* (J. Lane, 1907), pp. 6–8, 31; A.W. Bitting, 'The Canning of Foods', USDA Bureau of Chemistry Bulletin 151 (US Government Printing Office, 6 June 1912), p. 14; J.F. Breazeale, *Canning Vegetables in the Home*, USDA Farmers' Bulletin 359, Bureau of Chemistry, 1909.

25 Young, 'Botulism'; P.J. Sanders, letter dated Feb 1916, National Agricultural Library Special Collections, Elsie Carper Collection on Extension Service, Home Economics, and 4-H, O.H. Benson's Canning Papers, Box 5, Folder 68; USDA, Farmers' Bulletin 1211.

26 McKimmon, p. 23.

27 *The Ball Canning and Preserving Receipts and Spraying Calendar*, no. 14/Edition F (Ball Bros. Glass Mfg. Company, 1915), pp. 6, 12, MHC, Ball Corporation Collection (98.37), Box 1, Folder 7: Ball Blue Books: Nos. 14, 15.

28 Bradley and Crooke, p. 8; *Ball Canning and Preserving*, edition F, pp. 6, 10–11.

29 Blunt, Powdermaker, and Sprague, pp. 212, 366–67; Mary E. Creswell and Ola Powell, *Home Canning of Fruits and Vegetables*, Farmers' Bulletin 853 (US Government Printing Office, 1917), p. 6.

30 *Ball Canning and Preserving*, edition F, p. 12.

31 Zeide, p. 118; Danille Elise Christensen, '"Still Working": Performing Productivity through Gardening and Home Canning', in *The Expressive Lives of Elders*, ed. by Jon Kay (Indiana University Press, 2018), pp. 106–37.

32 Anne Van Willigen and John Van Willigen, *Food and Everyday Life on Kentucky Family Farms, 1920–1950* (University Press of Kentucky, 2006), p. 213.

33 Rebecca T. Richards and Susan J. Alexander, 'A Social History of Wild Huckleberry Harvesting in the Pacific Northwest', General Technical Report (Pacific Northwest Research Station: USDA Forest Service, February 2006), p. 13.

34 Bell, interview, p. 31.

35 Violet David, interview by Garnet Parsley, Brown Co., IN, and Beatrice Shuel, interview by Ruby Rumble, Gibson Co., IN, in Arnold, vol. 1, pp. 18, 20; Ralph Rinzler and Robert Sayers, *The Meaders Family: North Georgia Potters* (Smithsonian Institution Press, 1981), p. 101.

36 Williamson interview; Rose Moore Ramsay, interview by Chelsea Bicknell (12 July 2012), Appalachian Foodways Oral History Collection (SAA 164), Berea College Special Collections & Archives, Berea, KY, Box 1, Folder 20.

37 Creswell and Powell, *Home Canning*, 29; Arvil Wayne Bitting and Katherine Golden Bitting, *Canning and How to Use Canned Foods* (National Capital Press, 1916), p. 43.
38 Kenneth Belcher and Connie Belcher Goode, interview by Danille Elise Christensen (18 November 2022, Blacksburg, VA), Southern Foodways Alliance Oral History Program, Patrick County Public Events Baking Project.
39 Centers for Disease Control and Prevention, *Botulism in the United States, 1899–1996* (Centers for Disease Control and Prevention, 1998); Ruth Schwartz Cowan, *More Work for Mother* (Basic Books, 1983); Anna Lowenhaupt Tsing, *The Mushroom at the End of the World* (Princeton University Press, 2015).
40 E.g. Fred W. Tanner, 'Home Canning and Public Health', *American Journal of Public Health and the Nation's Health* 25.3 (1935): 301–13.

10
Syl Anagist Was a Garden
Food and Vitalism in *The Broken Earth*

Sara Clugage

Syl Anagist is the massive empire at the root of the problem in N.K. Jemisin's *The Broken Earth* trilogy – its apocalyptic self-destruction has shattered the world, leaving a planet rocked by earthquakes and volcanic eruptions. As characters travel through the last days of the present-day Sanze empire, where nothing can grow on the Earth's surface beneath a sky full of ash, their movements are motivated primarily by the search for nourishment. The novels' central preoccupation with food, rare in science fiction, is enriched by a conception of nourishment that encompasses flows of energy between people, plants, and minerals. *The Broken Earth* envisions a world where climate disasters have made human survival so tentative that the only way to live is through cooperation with life in all its forms, organic or inorganic. Jemisin's novels reckon with the long history of humanism, critiquing ideas about nationhood, scientific racism, and vitalism that supported and encoded the imperial desires of Enlightenment-era Europe – ideas that still lie beneath the world today. Behind *The Broken Earth*'s vision of empire is a rich material history of food production and consumption practices at the dawn of modern empires in the eighteenth century, especially in idealized landscapes and, I argue, tablescapes.

In post-apocalyptic Sanze, the empire that succeeds Syl Anagist, gardens will soon cease to grow. Villages quickly convert any available green space within their walls to food production. The only pleasure gardens are found in Sanze's central city of Yumenes, which includes the Fulcrum. The Fulcrum trains (and also imprisons) 'orogenes', those born with an acute sensory awareness ('sessing') of the Earth's lifeforce and the dangerous potential to channel and control it. In peripheral Sanze, orogenes struggle in a mineral wasteland pockmarked with 'dead civ' elements of fallen and forgotten civilizations. One curious dead civ artifact is the collection of giant crystal obelisks that Syl Anagist created, now floating loose from their sockets and wandering the sky with no discernible purpose. They are the shattered debris of Syl Anagist's own apocalypse, caused by the revolt of orogenes' enslaved ancestors, who were transformed by the cataclysmic energy their revolution released into immortal beings who live in the epistemological gap between rock and flesh – 'stone eaters'.

Like obelisks in gardens and urban plazas all over Europe, the obelisks function as reminders of the conquering empire's reach and power. In *The Broken Earth*, however, they are not dead stone monuments but living conduits for magic. As the story progresses, orogenes learn to 'enter' the obelisks, phasing their lifeforce into crystalline alignment. The obelisks' lattice structures are perfectly patterned, but rigid and static – if the alignment becomes too smooth, the orogene will turn to stone, a phase of matter that Jemisin treats as dead but also newly, strangely, alive.

Jemisin builds a speculative world where minerals, plants, and people are all hungry for life, feeding off each other in a vitalist food chain that turns our ecological pyramid on its head. In a post-humanist twist, it is the magical transfer of energy between incommensurable substances (crystals and plants, earth and flesh) that creates Jemisin's pathway out of both racial classification and the organic/inorganic dichotomy. Power, in *The Broken Earth*, flows through everything – and the flows can be reversed.

At stake in these novels are our ideas about the human, the world that we design and inhabit, and the genealogies of previous systems of knowledge. As one rapacious empire gives way to another, the future organization of life turns on the questions of whether consumption will always demand the suffering of others and what (or who) is left to be consumed. A comparison to the experiences of racialized Others in our own Western empires is easy to discern – Jemisin uses fictional empires to examine how Black people, and Black women in particular, have suffered the real and epistemic violence of being denied humanity.

Sanze

We open the first novel just before the end of the world, caused by someone with very good reason to end it. Alabaster, the world's most powerful orogene, surveys Sanze's capital city of Yumenes. He is standing on a hill outside the city, cynically marvelling at audacious architecture like towers, bridges, and balconies. Most architecture in the Stillness is low, solid stone – the kind of buildings that won't collapse in frequent earthquakes. But Yumenes enslaves orogenes to divert quakes and volcanic eruptions away from the city, allowing it to build ambitiously.

In the middle of Yumenes stands its biggest and bravest structure, a massive obsidian pyramid with a star-shaped base. At the apex of the pyramid balances a golden-hued geodesic sphere: the imperial palace where the staid and fearful emperor lives 'preserved in amber'. A sphere on top of a pyramid, an arrangement that looks precariously balanced but whose exact geometry is actually quite stable, keenly conveys the rigid anxiety of Sanze's power structure. In the shadow of the pyramid is a garden:

> There's a group of young women walking along one of the asphalt paths below; the hill is in a park much beloved by the city's residents. (*Keep green land within the walls*, advises stonelore, but in most communities the land is fallow-planted with legumes and other soil-enriching crops. Only in Yumenes is greenland sculpted into prettiness.)[1]

The spatial order of this opening scene, a monumental stone structure surrounded by a garden, is echoed throughout the novels. The pyramid in Yumenes sets the hierarchy for the landscape, dominating by virtue of its spectacular height and its unusual geometry.

Gardens, Flowers, and Fruit

In the Western imagination, pyramids are exotic and mysterious, the ruins of powerful ancient civilizations. By putting pyramids and obelisks at the centre of her world, and indeed by choosing the word 'obelisk' to describe these massive crystal shards, Jemisin asserts a correlation with the visual strategies of European empires that conquered North African territory and took these structures back home as trophies.

The other signal that this is an imperial landscape is that the garden is designed for pleasure. It is pretty but it does not produce – with the parenthetical advice about fallow-planting, Alabaster draws a comparison to less-wealthy communities that need their land to produce food. But Yumenes takes what it needs from other places and other people. This opening scene establishes the order of the Sanze empire – beautiful, rational, and extractive.

The real-world references for a 'foreign' monument set in a park are found all over Europe. As burgeoning industrial empires, France and Britain were particularly fascinated by the rise and fall of other empires. Eighteenth-century empires resonate with *The Broken Earth* because they mark the turn to economic and political modernity through the transatlantic slave trade. This historical rupture transformed the category of Blackness, ripping Black people from their cultural contexts and redefining them as the negative of whiteness. In the seat of empire, the world-making projects of gardens and dining rooms contested the meanings of these breakages by assembling the fragments in arrangements that favoured the empire's perspective.[2]

Eighteenth-century gardens proliferated with obelisks. Through the previous century, productive farms and kitchen gardens on large estates had been kept proudly close to the manor house, but by the 1760s gardens were imagined not as practical spaces to be worked in but as beautiful spaces to be looked at.[3] A new school of thought, spearheaded by English garden designer William Kent, shifted its priorities from the experience of the person strolling on the ground to the elevated point of view of the manor house, providing sweeping vistas punctuated with obelisks and other fanciful architecture like pagodas and Palladian temples.[4]

These are gardens not for producing food but for consuming it, an order reproduced in miniature on dining tables. At the beginning of the century, the fashionable dining style was *service en pyramide*, in which piles of fruit and pastry were structured to give height and visual impact to a dining table. It was especially used for displays of sweet dishes eaten at the end of a banquet, often set up on a separate table.[5] As the century wore on, the confectioner's *office* was tasked with creating ever-more-elaborate architectural towers. Constructions of gum and sugar paste that copied ancient temples and statues provided focal points in miniature landscapes. Built on raised parterres (a term used in both garden and confectionary design), the tablescapes combined edible and inedible elements to model an ideally ordered version of nature.[6] Much like Yumenes's pleasure garden, these are landscapes for consumption.

In one of the most popular volumes of table designs from the period, *Le Cannameliste Français*, confectioner Joseph Gilliers designs a tablescape with a central fountain made out of spun sugar (Figure 1). The fountain's high central froth, on a square stepped base over the watery surface of a mirrored parterre, recalls the Obelisk Grove at Versailles,

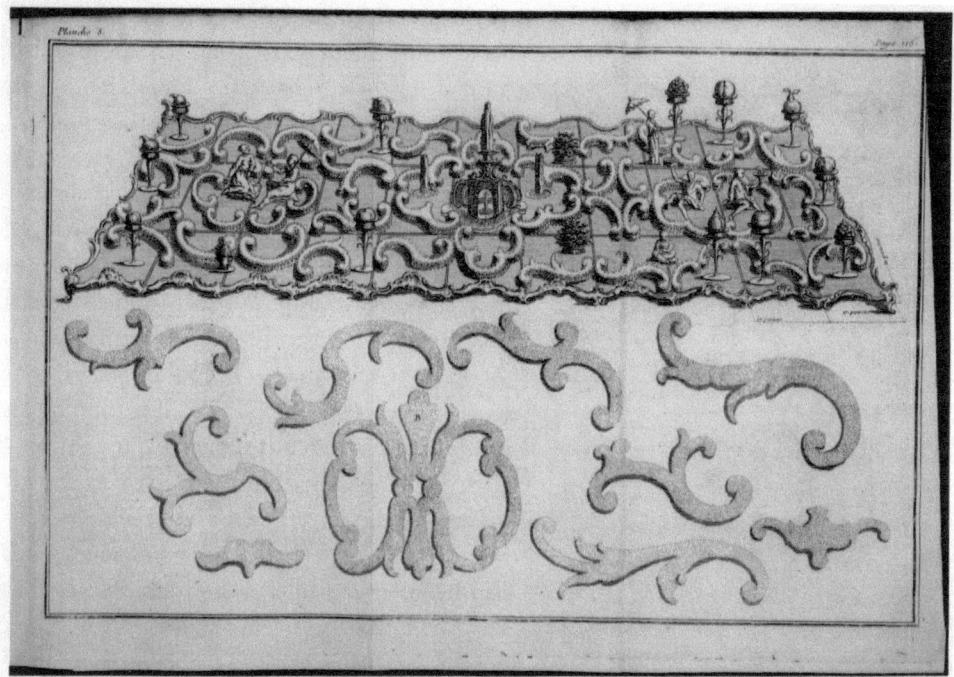
Figure 1. Engraving from Joseph Gilliers, Le Cannameliste Français, *1768. (Public domain.)*

whose powerful water jet sprays a liquid obelisk into the sky. Specimen fruits and berry pyramids on fanciful footed dishes line the edges of the scene. To each side of the central fountain lounge Orientalized figures made out of gum sugar and caramel, pastry parasols held high.[7]

The engraving uses an elevated three-quarter profile that gives us a god's eye view of its elaborate design. It is the perspective of the dinner guest who looks down at the display of miniaturized confections and the reduced-scale perspective of a palace overlooking the grounds. Among its exotic monuments, Gilliers's garden features 'exotic' people.

In the early eighteenth century, driven by the craze for Chinese export porcelain, manufactories like Meissen and Sèvres began offering huge ceramic dinner services that included buildings and people along with the dishes. Other potteries followed suit, and by mid-century, the figurines in a tablescape were more likely to be ceramic than sugar-work.[8] These included ceramic monuments that mimicked the full-size models. Wedgwood, for instance, made 'pyramid moulds' comprised of an inner core around which clear jelly could be set and the outer mould to contain it: 'The cores were usually tall obelisks, either round or square in section, or flat-sided rectangular wedges tapering up to a narrow top strip, similar in shape to the Egyptian pylon.'[9] Once unmoulded and set on a table, diners could see the interior painted design on the cores through the layer of clear jelly, which caught the candlelight in glowing visual effect (Figure 2).

Moulded ceramics also replaced the temporary gum and sugar-paste human figures. A Derby pottery shape-book from 1756 includes these figures representing the 'four quarters

Gardens, Flowers, and Fruit

Figure 2. Core mould, Wedgwood, c. 1780. Encased in lemon jelly and photographed by the author.

of the globe' (Figure 3). While there had long been ways of representing the visual variety of humanity, these tablescape designs and their figural elements encode the classification systems of eighteenth-century scientific racism. As Europe's colonial and imperial projects intensified across Africa, Asia, and the Americas, natural scientists and philosophers generated theories that supported European belief in their natural superiority and sorted the rest of humanity into subgroups (most notably in Carl Linnaeus's 1735 text *Systema Naturae*). Early anthropologists like Johann Friedrich Blumenbach investigated traits like skin colour and skull shape, and by the end of the century they had built the scaffolding for intellectual theories of race. (This pernicious episteme is also evident in *The Broken Earth*, where orogenes supposedly get their powers from abnormally enlarged organs in the

Figure 3. Figures, The Four Quarters of the Globe, *made by Derby Porcelain Works. Derby, Derbyshire, England, 1760–1770. Porcelain (soft-paste); Lead glaze. 1958.2262.001–.004 Bequest of Henry Francis du Pont. (Courtesy Winterthur Museum, Garden & Library).*

brain called 'sessapinae'). What is striking about the 'four corners of the globe' figures is their very ubiquity. The types show up everywhere from tables to gardens to fountains, often holding up the celestial sphere between them: the whole world, in order.

The totality of this visual order is mapped onto physical reality through these miniature gardens. Each tablescape is a description of their rightful place in the world that the designers make of themselves, to themselves. To more precisely define that place, they include for comparison the things that they are not: Chinese men at leisure, ancient Egyptians. These tablescapes matter because each time this order is inscribed, it becomes more naturalized, until it seems objective, until it seems to reflect the world 'out there'. The sedimentary layers of meaning build up until the ground seems unshakable.

Perched on a hill, Alabaster sees Yumenes's self-representation from his position above and outside the city. He is about to end the world because he knows it is impossible to fully live on its terms of order. Alabaster's outside perspective is doubly inscribed – he is the first character in *The Broken Earth* to develop what W.E.B. Du Bois called 'double-consciousness': as an orogene in a world of stills, like a Black person in a world defined by whiteness, he lives with the tension of understanding himself as a human being while everything around him tells him that he is not.[10] As Sylvia Wynter tells us, such a world cannot be merely 'unsettled' (hearing that word with seismic resonance) by rearranging its terms.[11] It can only be ruptured.

So Alabaster sends his orogenic ability down into the bedrock below the city, and up into sky to the obelisks that hover above it. He grasps the world and he breaks it – a deep gash in the Earth that neatly bisects the continent ripples out from Yumenes. It will spew magma and ash into the atmosphere for millennia to come.

Syl Anagist

In ancient Syl Anagist, every city was a garden. Buildings were grown from cellulose, their patterns hidden by layers of leaves and clusters of fruit. Vehicles scuttled through the streets on arthropod legs. And at the centre of each city large enough to bear the name, a monumental obelisk sat firmly socketed in the Earth.

Descriptions of Syl Anagist at first seem like a green and growing paradise, a welcome relief from the rigid structure of Sanze's imperial park or the desolate landscape of the Rifting. With its buildings grown from plants and its bioengineered vehicles (vehimals, they're called), Syl Anagist achieves a state of complex living harmony that is so sophisticatedly engineered as to seem wild. It sprawls in organic patterns that cover the world, 'much in the manner of mold forking and stretching along the rich veins of a growth medium'.[12] Its understanding of nature is characterized by intuition and precision. Unlike Sanze's orogeny, which mechanically forces the heat and pressure of the earth, Syl Anagist's magic is a cultivated energy network that covers the whole earth: 'Webs of life, if you want to think of them that way. Life, you see, is sacred in Syl Anagist.'[13]

We learn about Syl Anagist from Hoa, a stone eater who will befriend Essun, *The Broken Earth*'s protagonist. As Hoa tells his story to Essun in his own narrated chapters, he was a Tuner. Millennia ago, the empire of Syl Anagist bioengineered the Tuners to 'tune' the Plutonic Engine, a worldwide network of crystal obelisks that would allow them to draw energy from the Earth. But Hoa discovers the empire's shocking secret: its gardens are fed by the lifeforce of Tuners – their bodies kept alive but incapacitated, their energy siphoned through implanted vines and stored in the obelisk at each city's centre. The garden that Syl Anagist's subjects enjoy is only made possible by the violent enslavement of another race – not a garden of paradise at all, but a plantation. The empire values life not spiritually but economically: 'Life is sacred in Syl Anagist – sacred, and lucrative, and useful.'[14]

This realization brings Hoa and his fellow Tuners to political consciousness; they take control of the obelisks that have been draining their life force and use them to power a slave revolt. The Tuner Rebellion channels so much energy that the Engine catastrophically fails (as Jess A. Goldberg writes, it is impossible to hear Tuner Rebellion without thinking of the most successful slave revolt in US history, 1831's Nat Turner Rebellion).[15] The power the Tuners channel also crystallizes them, described as a re-alignment of their basic particles into a lattice that can seamlessly integrate with the magical network of energy in all living things, including the Earth itself. This is life, although not a kind of life we previously recognized as such.

Tirimo

In present-day Sanze, Alabaster's own world-destroying slave revolt duplicates and reasserts Hoa's rebellion. His 'Rifting' of the continent destroys Yumenes quickly but is a slower-moving cataclysm for its periphery. People across the Stillness spend their lives preparing food for an ecologically disastrous Fifth Season, never knowing what kind of disaster it might be. Each comm fills its 'cache' with supplies in a nutritional ratio of protein and carbohydrates prescribed by law. When Seasonal Law is declared, the comm closes its gates and tries to wait the disaster out. This Season, however, might never end.

Essun lives in a comm called Tirimo in the middle of the continent. Essun is an orogene 'passing' as a still, working as a creche teacher and raising two young children with her husband Jija. On the day Alabaster breaks the world, her own personal world is shattered when Jija sees their toddler Uche use orogeny to quell the shake and kills him in a fit of rage. Essun knows she has to leave – she been found out, along with her children, as an orogene, and she needs to find her daughter Nassun. As she departs, she sees a group of people building a paddock in the corner of the village green to prepare to bring livestock within the walls. Right now the green is an unplanted grassy field, but today everyone has started to look at it as *'cropland to be'*.[16]

Essun's epic journey in search of her daughter is fuelled by food she can scrounge up or trade for: 'cachebread' (a sparsely described dry loaf made for long-term storage), dried meat, and what fresh fruit and vegetables are able to grow under a rapidly darkening sky. The food is simple, portable, and often unpalatable: 'Food is that which nourishes, the lorists say.'[17] Stores are often hard-won – Essun braves abandoned tradeposts and hostile comms in search of supplies.

It is unusual in science fiction to describe a hero's quest as a search for food. But in this world, no one can survive alone, and besides, Essun is a mother (another rare thing in science fiction).[18] She has been shaped by the care she gives others and the care she has not been allowed to give.

The Fifth Season tells parallel stories of Essun at two periods in her life when she went by different names, the child Damaya and the young woman Syenite (the reader might not catch on that they are the same person until near the end of the book). Damaya is born to non-orogene parents who send her away in fear and disgust. She is fetched to the Fulcrum by Schaffa, a Guardian (a violently paternal figure who reflects slavecatchers and overseers). There she takes Syenite as her namesake (all orogenes go by the names of minerals). After learning from Alabaster about the Fulcrum's crimes, she escapes with him. They have a child, Corundum, during a brief period of safety in a maroon colony. But Schaffa finds them. Unable to bear watching Corundum be captured and enslaved, in a gut-wrenching scene, Syenite suffocates him. The tragedy echoes the famous tale of Margaret Garner, an enslaved woman in the US who self-emancipated with her children in 1856; when the Garners were tracked down by slave catchers, Margaret killed her daughter rather than see her returned to slavery. Corundum's death destroys Syenite and prompts her to flee – she hides in Tirimo as Essun for the next ten years.

Essun experiences successive frustrated attempts at maternity. As Syenite, she biologically bears a child but does not have maternal rights. As Essun, she tragically doubles the intimate pattern of violence with Nassun that she had learned at Schaffa's hands. Joy James aptly coins the phrase 'captive maternal' to describe the Black female protagonist in science fiction who overcomes the intimate violence of a world that despises her – Essun strives for her own freedom and the flourishing of those she loves within a matrix that enslaves her, but as she resists that matrix she 'reproduces resistance detrimental to the conditions from which [she] emerged'.[19] Essun resists through mothering. And so she sets out from Tirimo to find her daughter.

Obelisk

On the way, however, she finds someone else to mother. Hoa, the stone eater, takes the form of a human boy (albeit a boy with stark white skin, hair, and eyes) to approach Essun. Her instinct is to feed him as she would any other child, but he always refuses food, clutching a bundle of rocks that are the only supplies he carries. Toward the end of the first novel, Essun learns why.

> There's a crunching sound, surprisingly loud in the silence of the room. Several more crunches, not as loud, but leaving no doubt that what he's chewing on is by no means food. And then he swallows, and licks his lips.
> It's the first time you've ever seen him eat.
> 'Food,' you say.
> 'Me.' He extends a hand and lays it over the pile of rocks with curious delicacy.[20]

It is through the stone eaters that *The Broken Earth* conceives of food beyond our conception of it, as nourishment writ large. By asserting that rocks are alive, *The Broken Earth* mounts a challenge to the ordering hierarchy of mineral, vegetable, animal. This classification system is built on a scale (a food pyramid) from things are the least like humans and those that are most like humans. Stone eaters throw the whole scale into chaos – they are minerals who eat people. As Alabaster and Essun strive to harness the obelisks for one more world-changing event, each attempt to channel the obelisks' power gradually turns their bodies to stone. This flesh-turned-rock, it turns out, is the particular kind of stone that nourishes stone eaters. In scenes as tenderly sensual as they are cannibalistic, Essun offers her crystallized flesh to Hoa's diamond-sharp teeth – an arm, then her left breast, which she recalls was Nassun's preferred food source too.[21]

Essun rations her orogenic activity so that she will survive long enough to learn how to manipulate magic and reach Nassun. Meanwhile Nassun, under the tutelage of Essun's old Guardian Schaffa, has grown into a powerful orogene herself. The two of them both want to use the obelisks' power for drastic change: Essun wants to restore the Earth, and Nassun wants to end the world. They both learn to 'enter' the obelisks, phasing themselves into alignment with their lifeforce. Essun visualizes this as falling up: 'she's falling, up, and this somehow makes sense. All around her, the place she's falling through, is color and faceted flickering, like water'. This phasing of matter is not quite physical but not quite metaphysical. Essun's experience of the obelisks is a mix of sensation that exceeds rational thought:

> It's beyond her, too complex to perceive in full. Something pours through somewhere, warms with friction. Someplace inside her smooths out, intensifies. Burns. And then she is elsewhere, floating amid immense gelid things [. . .].[22]

As Essun falls up, the viscosity of the obelisk admits her, but not fully – she retains her sense of self and does not dissolve into it. This obelisk is perhaps textured more like a Wedgwood jelly mould than a granite monument. The evocation of a gelatinous interior is particularly evocative because its movement brings the shape to life. Jelly coheres into masses that hold their boundaries – in the loosest sense of the term, they are bodies. On the table, they also wobble and wiggle like bodies do, and this can give us the delightful or even horrifying idea that they might be just a little bit alive. Perhaps we can think of this animated gelatinous imaginary as akin to Jemisin's generative misdirection, away from the rigid hierarchies of inanimate/animate and animal/human that have defined our conceptualizations of 'the human'.

For Essun, the sensation of living inside the obelisk is dizzying but freeing. Zakiyyah Iman Jackson, writing about Nalo Hopkinson's novel *Brown Girl in the Ring*, describes Hopkinson's Black female characters' own magical 'second sight' as a kind of vertigo: 'vertigo – that sense of unhinged reality, a communion with death and that realm which exceeds life – [that seems] to threaten a total loss of self as incommensurable metaphysical frameworks and sensory maps meet'. Jackson's conception of vertigo acknowledges the 'unsettling' sensation of becoming something dangerously new, outside the territory and even the cosmology that map the character's world. But there is no freedom to be had within the boundaries. Hopkinson's 'second sight' and Jemisin's 'sessing' of obelisks are both versions of Du Bois's 'double-consciousness,' sense-abilities that come from being attuned to a world order that will not fully admit you.[23]

Becoming stone allows Essun to broaden her awareness of the things from which she can take energy. This radically large conception of nourishment, encompassing the freely flowing magic between all things, is the path N.K. Jemisin threads out of the world that cages her characters. Essun breaks free of the world-ordering hierarchies that have defined her as non-human by discovering a kind of life beyond the human. By the time that Essun finally meets Nassun again, she has been fundamentally altered.

In the final showdown, Essun and Nassun wrestle for control of the Obelisk Gate. Even while fighting, Essun tries to protect Nassun from the power of it, which threatens to turn her to stone: 'And then it happens. You cry out as you feel a change, a snapping-into-line: Nassun. The magics of her substance are fully aligned; her crystallisation has begun.'[24] Essun cannot kill another one of her children, and so she makes the most earth-shaking choice she can: she lets go. The excess power of the obelisks shunts into Essun, and she turns to stone.

Her transformation is a kind of death. The scene refracts Essun's previous choice to kill her son Corundum rather than submit him to slavery, and Margaret Garner's choice to kill her child for the same reasons. If Essun died, it would be a true reversal of that choice.[25] But the story cannot be confined to the rules of our reality – Essun does not die, she 'aligns'. This is not true death but a hopeful stasis, as Essun awaits her rebirth as a stone-eater in geologic time. In time, she will eat people too. The transformation keeps the long drumbeat of the novels – every time the world ends a new one begins.

Jemisin narrates a post-apocalyptic conception of gestation, mothering, and nourishment so distant from us that we do not yet have the words to describe it. The energy flows in *The Broken Earth*'s abstracted lifeforces map a world beyond ecological pyramids or taxonomies of the human and less-than-human. They dream of a kind of food that does not require the domination and suffering of others. These abstractions are purely speculative, but they are the feats of imagination necessary to express the feeling of being on the verge of a new world. It is a post-humanist move, redolent of other post- moves (post-colonial, post-revolutionary). We do not know what the new world will feel like, but we can sense something out there. Will it be any better? It has to be.

About the Author
Sara Clugage is a New York-based artist who focuses on political issues in craft and food. She is Editor-in-Chief of *Dilettante Army*, an online magazine for visual culture and critical theory, and is currently at work on a book project about Jell-O, animacy, and abstraction.

Notes
1. N.K. Jemisin, *The Fifth Season* (Orbit, 2015), p. 9.
2. My gratitude to Rebecca Ariel Porte for this idea and for her insightful comments on this essay.
3. Tom Williamson, 'Production, Power, and the "Natural": Differences between English and American Gardens in the Eighteenth Century', *Huntington Library Quarterly* 84.3 (September 2021), pp. 467–90, DOI:10.1353/hlq.2021.0032.
4. 'Kent, William,' Grove Art Online <https://doi.org/10.1093/benz/9780199773787.article.B00098060> [accessed 10 May 2024].

5 Roy Strong, *Feast: A History of Grand Eating* (Houghton Mifflin Harcourt, 2003), p. 240.
6 Barbara Ketcham Wheaton, *Savoring the Past: The French Kitchen and Table from 1300 to 1789* (Touchstone, 1983), p. 186.
7 Louise Conway Belden, *The Festive Tradition: Table Decoration and Desserts in America, 1650–1900* (Norton, 1983), p. 73.
8 Strong, p. 239.
9 Peter Brears, *Jellies and Their Moulds* (Prospect Books, 2010), p. 99.
10 W.E.B. Du Bois, *The Souls of Black Folk* (A.C. McClurg, 1903), pp. 3–5.
11 Sylvia Wynter, 'Unsettling the Coloniality of Being/Power/Truth/Freedom: Towards the Human, After Man, Its Overrepresentation – An Argument', *CR: The New Centennial Review* 3.3 (September 2003), pp. 257–337, DOI:10.1353/ncr.2004.0015.
12 N.K. Jemisin, *The Stone Sky* (Orbit, 2017), p. 12.
13 Jemisin, *The Stone Sky*, p. 13.
14 Jemisin, *The Stone Sky*, p. 294.
15 Jess A. Goldberg, 'Living After, and Before, the End of the World: Toni Morrison's *Beloved* and N.K. Jemisin's *Broken Earth*', *Women's Studies* 52.2 (17 February 2023), pp. 173–91 (p. 179), DOI:10.1080/00497878.2022.2156507.
16 Jemisin, *The Fifth Season*, p. 40. Emphasis in original.
17 N.K. Jemisin, *The Obelisk Gate* (Orbit, 2016), p. 711.
18 For more on gender, see Joanna Russ, *The Image of Women in Science Fiction* (Warner Modular Publications, 1973).
19 Joy James, 'Captive Maternal Love: Octavia Butler and Sci-Fi Family Values', in *Literature and the Development of Feminist Theory*, ed. by Robin Truth Goodman (Cambridge University Press, 2015), pp. 185–99, DOI:10.1017/CBO9781316422007.014.
20 Jemisin, *The Fifth Season*, p. 294.
21 Jemisin, *The Stone Sky*, p. 117.
22 Jemisin, *The Fifth Season*, p. 125.
23 Zakiyyah Iman Jackson, *Becoming Human: Matter and Meaning in an Antiblack World* (New York University Press, 2020), pp. 111, 105.
24 Jemisin, *The Stone Sky*, p. 359.
25 For an analysis of Essun's crystallization as a true death, see Goldberg.

11
Towards a Gastronomic Taxonomy for Fruits and Flowers
Beyond Linnaeus's Garden

Len Fisher and Anders Sandberg

Introduction: Linnaeus and his Garden

Mankind has been attempting to classify plants and their uses since ancient times. The earliest extant text, compiled during the Chinese Eastern Han Dynasty (25–220 AD), was *The Divine Husbandman's Classic of Materia Medica*, which classified 365 herbs with medicinal properties into three categories – high-grade, medium-grade and low-grade – according to their medicinal effects and toxicity (Zhao and others 2018).

The modern classification of plants may be said to have begun with Carl Linnaeus and his famous botanical garden in Uppsala, Sweden (Stearn 1976). Freshly laid out by him in 1745 after a disastrous fire destroyed an earlier garden on the site, it was populated by thousands of flowering plants sent by collectors from all over the world. It was one of the first to be arranged with an orderly categorization of plants – in fact, several orderly categorizations.

In some beds, plants were divided according to whether they flowered annually or perennially. Other beds contained plants selected for their usefulness in medicine. In a series of three greenhouses, plants were divided according to the temperatures of their native habitats.

Overall, the plants were classified using an anthropomorphic system that had deliberate parallels with human sexuality (Müller-Wille 2007). Flowers were 'beds'. The male stamens and female pistils were referred to as 'husbands' and 'wives'. Flowers were classified into families according to the numbers of stamens and pistils, producing a veritable catalogue of eighteenth-century sexual relationships, from monogamy and polygyny to homosexuality, miscegenation, and incest.

Linnaeus's sexual classification system was artificial, and did not necessarily reflect relationships between different plants – a fact that he recognized himself. It was designed as a mnemonic recognition system for his collectors, and it remained the dominant system until well into the nineteenth century.

Within the families thus characterized, the flowering plants were further subdivided using Linnaeus's now-famous system of *binomial nomenclature*, which has survived the test of time in principle. It subsequently became based on more realistic assessments of the relationships between different plants, either through morphology or through developmental patterns, and more recently through DNA relationships (Paterlini 2007).

Linnaeus was handicapped in his use of the binomial system by a belief that plants which had been bred and interbred by humans were artificial, rather than naturally God-given, and should not be counted as separate species. Edible plantains and cooking bananas, for example, were grouped together as a single species named *Musa paradisiaca*, *Musa* being a Latinization of the Arabic name for the fruit and *paradisiaca* because Linnaeus believed that this, and not the apple, was the fruit that Eve had offered to Adam.

Linnaeus was the first person to successfully grow fruiting banana plants in Europe. He did so in the greenhouse of his friend, the banker George Clifford, by reproducing the banana's tropical growing conditions, which meant placing the plant in rich, fertile soil and keeping it dry and cool for a few weeks, then increasing the temperature and giving the plant plenty of water. A London diarist, encountering this novel fruit for the first time, recorded it as 'a kind of vegetable sausage tasting like marrow flavoured with pineapple' (Stearn 1976: 26).

What Is Classification?

The description of a banana by the anonymous diarist was an early attempt to classify a plant in terms of its culinary properties. Here we discuss how a broader culinary classification might be attempted. In order to do so, however, we must be careful to understand just what we mean by classification, and what we are attempting to achieve by our efforts.

At the most fundamental level, our goal is to get some kind of useful order into our knowledge (Glushko 2016) – a goal that has been a major obsession with philosophers since the time of Aristotle, who wanted to define a complete list of the highest kinds or genera. *Everything* was to be included, and the contents of these categories would be the totality of what exists.

But how was such a list to be ordered? Linnaeus's classification of living things (and even non-living things, such as minerals) and its subsequent development provides an example of one sort of ordering – a nested or tree structure, with ever broader categories encompassing all of those below.

A disadvantage of this approach is its Procrustean nature – everything must be made to fit. An alternative, and less rigid, approach is to use tagging. Instead of finding a perfectly consistent structure, one is content to describe various properties of each item, and using these tags to organize them. This approach has the advantage that it can be done informally by large groups of people without careful instruction, and the tags processed statistically. Such 'folksonomies' can be powerful because they are easy to maintain and extend, while principled classifications often run into trouble with intermediate cases, disagreements, and the changing complexity of the world.

Culinary Classification

What approach might be best suited to culinary classification? None of the classical 'nested' approaches (genetic, morphological, or DNA-based) would seem to be of much value. The sweetness of fructose, or the tartness of tartaric acid, for example, can both be important in cooking, but knowledge of a plant's genetic or morphological classification tells us little about how much of these important compounds might be present, since their concentrations vary widely, not only across fruit families, but also within individual families (Muir and others 2007).

It would seem, then, that some form of tagging might offer better opportunities for conceptually separating plants into culinarily relevant groups. Even this simple distinction, however, is not always clear-cut. Tomatoes are botanically a fruit, for example, but have been legally classified as vegetables (US Supreme Court Tariff Act 1883). The botanical classification is also misleading in culinary terms. Tomatoes may be a fruit, but one does not put them into a fruit salad.

A more fine-grained division for edible plants is to categorize them according to plant type and/or part. Major categories along these lines are grain crops, pulses, roots and tubers, oilseeds, tree fruits, nuts and seeds, leafy vegetables, and spices/condiments (Fuller 2006). Simple inspection of this list shows that, as with genetic and morphological classification schemes, it is not of much use in defining culinary properties, such as the contribution of the plant as an ingredient to the flavour of a dish.

An approach with more culinary relevance is to categorize edible fruits according to the sorts of nutrients that they provide. Pennington and Fisher (2009), for example, find that groupings in terms of colour and part of the plant (dark green leafy vegetables, deep orange/yellow fruits, roots and tubers, etc.) are good predictors of particular sets of nutrient values.

Cultural values can also provide powerful epistemic criteria relevant to the gastronomic contribution of a plant material to a dish. These values range, from simple binary classifications (such as sweet/savoury) to more complex intertwined criteria (Leschziner 2006), including parallels and comparisons with what the diner has been used to (Coe 1994).

Tagging according to plant type, nutrient content, or cultural epistemic culinary criteria, although derived from culinarily relevant criteria, still does not offer much help, however, in answering the principal question that concerns the working chef: what is the functional contribution of the plant material as food ingredients to the qualities of a dish? If tagging is to be useful for chefs, it must address this question.

Functional Tags and Gastronomic Taxonomy

The term 'gastronomic taxonomy' was conceived in 2017 by a group associated with Ferran Adria and his innovative El Bulli restaurant. Their goal was to provide a framework for the 'categorization of unelaborated food products that follows currently accepted culinary criteria yet avoids contradiction with scientific knowledge' (D'Ambrosio and others 2017: 525). The classification scheme that they devised was based on the division of culinary ingredients into five 'worlds' (plants and fungi, animals, microorganisms, waters, and minerals), with various subcategories.

Unfortunately, this attempt at gastronomic taxonomy did not include sensory experience and desirability. An effort to tag foods in such gastronomic terms was the concept of 'flavour pairing' – the idea that food ingredients which share more aroma components would be more compatible. The hypothesis provided an early guideline for culinary innovation by Heston Blumenthal (2008). It was supported by some research into Western foods, but, puzzlingly, it fell down when applied to Asian cuisines, where quite the opposite effect was found, possibly because other food qualities than flavour profile took precedence (Ahn and others 2011).

Blumenthal himself was later sceptical about the idea, and stated in a *Times* newspaper article that 'Looking back at my younger self I'm almost embarrassed at my bumptious enthusiasm, not least because I now know that a molecule database is neither a shortcut to successful flavor combining nor a failsafe way of doing it. Any foodstuff is made up of thousands of different molecules, that two ingredients have a compound in common is a slender justification for compatibility. If I'd known then what I know now, I would probably never have tried this method of flavor pairing: there are simply too many reasons for it not to work. As it was, in my naivety I just got stuck in' (qtd. Spence 2022). As the chef put it in *The Big Fat Duck Cookbook*: 'I soon realised that the molecular profile of a single ingredient is so complex that even if it has several compounds in common with another, there are still as many reasons why they won't work together as reasons why they will' (2008: p. 171).

According to Spence (2022, lines 192–3), 'experimental studies do not tend to support the food pairing hypothesis'. Spence quotes Kort and his collaborators (2010), who put the food pairing theory to the test in an experiment with untrained participants, and found that: 'food pairings with more aroma overlap did not taste better than food pairings with less overlap. For example, chocolate and tomato (43% overlap) did not taste better than cauliflower and pear (no overlap) [. . . . F]ood pairing based on aromatic overlap is not a guaranteed recipe for success' (qtd. in Spence 2022: lines 194–99).

In any case, the idea of 'flavour pairing' only covers one aspect of the full gastronomic sensory experience. According to Barrett and others (2010), the major factors for culinary acceptance of a fruit or vegetable are not only flavour, but also colour, texture, and nutritional quality. Here we propose to amend this list by including flowers as an ingredient, and by:

- dividing flavour into taste (on the tongue) and aroma
- subsuming 'colour' as a subcategory of 'appearance'
- adding the other effects of the plant material as a food ingredient (e.g. the astringency of some materials, the tongue tingling of hot chillis, the conveyance of a feeling of satiation by some foods, and the psychoactivity from the mild (coffee) to the extreme (magic mushrooms)).

We also propose classifying ingredients that have been treated in different ways (e.g. boiling, roasting, fermenting, or fresh) as essentially different ingredients.

Here we consider how all of these qualities may be brought together in an overall gastronomic classification scheme, confining our discussion in the first instance to fruits and flowers, while noting that the arguments presented here may readily be extended to include vegetables and other food ingredients.

A New Gastronomic Taxonomy
Potential Classification Criteria and their Problems

Our set of gastronomic tags thus consists of the contribution of the fruit or flower to:

- taste
- aroma
- appearance
- texture
- other effects

Tagging food ingredients by these criteria could be useful to the working chef in a number of ways, including:

- choice of recipes containing or based on a particular ingredient or combination of ingredients
- achievement of a particular effect
- compatibility with other ingredients within or between any of the criteria categories
- choice of the closest replacement ingredient if a particular ingredient is not available, or otherwise not advisable (as with dietary sensitivities or allergies)

We note, however, that there is a particular problem with developing such a gastronomic taxonomy. That problem concerns the different dimensionalities of the criteria.

Some of the criteria listed under 'other effects' do indeed offer the possibility of a single, one-dimensional measure. The perceived 'heat' of a chilli-containing dish, for example, is related to the concentration of capsaicinoids, expressed in Scoville units (the number of dilutions needed for the heat to become indetectable). Psychoactivity may similarly be related to the concentration of the psychoactive component, and even astringency may be assessed subjectively along a unidimensional scale.

The basic specification of *taste* requires five measures: sweetness, sourness, saltiness, bitterness, and umami (Kramer 1959), each of which is detected by a different set of receptors on the tongue, and each of which may also be assessed subjectively along a unidimensional scale.

Aromas, however, present a bigger challenge. The human nose has around 400 receptors for individual aromas, and the perception of an overall aroma is governed by a balance between the outputs of these receptors to the brain. It has been suggested that the human nose may consequently be able to distinguish up to *one trillion* individual perceived aromas, each made up of a different set of components in different proportions. Perception also varies between individuals; no two noses are the same and almost everybody has a number of anosmias for different aroma components.

Categorising *appearance* offers even greater problems, since there is an infinity of colours, let alone all of the other factors that contribute to the visual appeal of an object.

Texture also offers an infinity of possibilities. Scientifically, it may reasonably be assessed by the viscoelastic properties of a material (i.e. its flow properties and its stiffness or elasticity), but many ingredients are heterogeneous in this respect, and it is also well-known that viscoelasticity properties alone are insufficient to fully categorize the perceived texture of an ingredient.

Finally, there are many crossover effects between the different criteria. In a famous experiment, Paul Breslin and his colleagues at the Monell Chemical Senses Center in Philadelphia had subjects chew mint-flavoured gum until all of the mint flavour had disappeared. The subjects were then asked to sip some slightly sweetened water and then resume chewing, whereupon the mint flavour reappeared at full strength! The concentration of minty aroma in the nose had not changed, but the brain had simply become bored, only to be re-awoken by the taste of sugar.

A Fresh Approach

How might the problem of an essentially infinite number of possible descriptors across the tagging criteria be overcome?

One possibility is to simplify the criteria so as to make them manageable. Texture, for example, could be classified according to the descriptors most commonly found in recipe books (creamy, gelatinous, crunchy, etc.), with each descriptor given a value in the range 1–5 (say). Appearance could similarly be rated as subjective visual appeal of dishes containing the ingredient, again on an arbitrary scale of 1–5. Aromas, however, present a bigger problem, and the classic wine critic's approach ('has notes of . . .') would not appear to be of much value.

An alternative possibility, only available in the last few years, is to enlist the aid of artificial intelligence (AI), which is already finding widespread application in the food industry (Thapa and others 2023), including recipe development (Uçuk and others 2023). Current examples mainly use large recipe databases (Marin and others 2021), where the properties are left implicit in how human cooks combine ingredients (this has been used to propose exchange ingredients in dishes (Yamanishi and others 2015; Pellegrini and others 2021), even weighting by novelty (Liu and others 2018)).

Here we propose to improve on these largely empirical approaches by using the classification criteria that we have listed above (appearance, smell, etc.) as a basis for a machine-learning algorithm (or even just a statistical algorithm) which takes this list of criteria for each ingredient and learns to recognize and quantify these. The system would then be instructed to develop an inherent model for exploring which ingredients are similar to which and can be used interchangeably in recipes.

This, of course, is only a beginning. As those who have worked with AI know only too well, it is quite capable of producing its own surprises. A fairly simple tweak, for example, could enable an AI programme to deduce and produce its own original recipes, and others producing recipes tailored to suit individual preferences. Watch this space!

About the Authors

Len Fisher is a scientist, author, and broadcaster, whose books range from *How to Dunk a Doughnut* to *Crashes, Crises and Calamities: How We Can Use Science to Read the Early-Warning Signs*. He won a spoof Ig Nobel prize for using physics to work out the best way to dunk a biscuit.

Anders Sandberg is a lapsed computational neuroscientist researching emerging technology, global disasters, and the very long-term future at the Future of Humanity Institute at University of Oxford.

References

Ahn, Y.-Y., and others. 2011. 'Flavour Network and the Principles of Food Pairing', *Scientific Reports*, 1: 196, DOI:10.1038/srep00196.

Barrett, D.M., J.C. Beaulieu, and R. Shewfelt. 2010. 'Color, Flavor, Texture, and Nutritional Quality of Fresh-Cut Fruits and Vegetables: Desirable Levels, Instrumental and Sensory Measurement, and the Effects of Processing', *Critical Reviews in Food Science and Nutrition*, 50, pp. 369–89.

Blumenthal, H. 2008. *The Big Fat Duck Cookbook* (Bloomsbury).

Coe, S. 1994. 'New World Produce', in *America's First Cuisines* (University of Texas Press).

D'Ambrosio, U., and others. 2017. 'Classification of Unelaborated Culinary Products: Scientific and Culinary Approaches Meet Face to Face', *Food, Culture & Society*, 20.3, pp. 525–53.

Fuller, D.Q. 2006. *Plant Systematics and Economic Botany: Background Packet* <http://www.homepages.ucl.ac.uk/~tcrndfu/Abot/Systematics%20base.pdf> [accessed 20 October 2024].

Glushko, R.J. 2016. *The Discipline of Organizing: Professional Edition* (O'Reilly Media).

Kort, M., Nijssen, B., K. van Ingen-Visscher, and J. Donders. 2010. 'Food Pairing from the Perspective of the "Volatile Compounds in Food" Database', in *Expression of Multidisciplinary Flavour Science: Proceedings of the 12th Weurman Symposium, Interlaken, Switzerland* (pp. 589–592). Institut of Chemistry and Biological Chemistry.

Kramer, A. 1959. 'Glossary of Some Terms Used in the Sensory (Panel) Evaluation of Foods and Beverages', *Food Technology*, 13, pp. 733–36.

'Food Database.' 2008. In *Expression of Multidisciplinary Flavour Science: Proceedings of the 12th Weurman Symposium, Interlaken, Switzerland*, ed. by I. Blank, M. Wüst, and C. Yeretzian (Zürcher Hochschule für Angewandte Wissenschaften Institut Für Chemie und Biologischen Chemie).

Leschziner, V. 2006. 'Epistemic Foundations of Cuisine: A Socio-Cognitive Study of the Configuration of Cuisine in Historical Perspective', *Theory and Society*, 35, pp. 421–43.

Liu, K.H., and others. 2018. 'Alternative Ingredient Recommendation: A Co-Occurrence and Ingredient Category Importance-Based Approach', *PACIS Proceedings* <https://core.ac.uk/download/pdf/301376268.pdf> [accessed 20 October 2024].

Marin, J., and others. 2021. 'Recipe1m+: A Dataset for Learning Cross-Modal Embeddings for Cooking Recipes and Food Images', *IEEE Transactions on Pattern Analysis and Machine Intelligence*, 43, pp. 187–203.

Muir, J.G., and others. 2007. 'Fructan and Free Fructose Content of Common Australian Vegetables and Fruit', *J. Agric. Food Chem.*, 55, pp. 6619–27.

Paterlini, M., 2007. 'There Shall Be Order', *EMBO Reports*, 8, pp. 814–16.

Pellegrini, C., and others. 2021. 'Exploiting Food Embeddings for Ingredient Substitution', *HEALTHINF*, 5, pp. 67–77.

Pennington, J., and R. Fisher. 2009. 'Classification of Fruits and Vegetables', *Journal of Food Composition and Analysis*, 225, pp. 523–31 <https://www.sciencedirect.com/science/article/abs/pii/S0889157509000192?via%3Dihub> [accessed 20 October 2024].

Spence, C. 2022. 'Gastrophysics: Getting Creative with Pairing Flavours', *International Journal of Gastronomy and Food Science*, 27 (March), 100433 <https://ora.ox.ac.uk/objects/uuid:903db240-800e-4986

-ab37-cc677086c408/download_file?safe_filename=Spence_2021_Gastrophysics_getting_creative.pdf&file_format=application%2Fpdf&type_of_work=Journal+article> [accessed 20 October 2024].

Stearn, W.T. 1976. 'Carl Linnaeus and the Theory and Practice of Horticulture', *Taxon*, 25.1 (February), pp. 21–31.

Thapa, A., and others. 2023. 'A Comprehensive Review on Artificial Intelligence Assisted Technologies in Food Industry'. *Food Bioscience*, 56, p. 103231.

Uçuk, C., M. Doğdubay, and S.S. Özdemir. 2023. 'The Use of Artificial Intelligence in Recipe Development: How Technology Is Changing the Way We Create and Innovate in the Kitchen', in *2023: Impactful Technologies Transforming the Food Industry*, ed. by Şule Aydın and others (IG Global).

Yamanishi, R., and others. 2015. 'Alternative-Ingredient Recommendation Based on Co-Occurrence Relation on Recipe Database', *Procedia Computer Science*, 60, pp. 986–93.

Zhao, Z., P. Guo, and E. Brand. 2018. 'A Concise Classification of Bencao (*matéria medica*)', *Chinese Medicine*, 13, pp. 1–4.

12
Gardens and Ghosts
Arts of Living in a Contemporary Bulgarian Village

Lindsey Foltz

Introduction
This paper examines weekend home-gardening in a village located in the foothills of the central Balkan Mountain range. Despite expectations to the contrary given the end of state socialism and entry into the European Union, gardens like this one continue to be a common feature of contemporary Bulgarian life.[1] They complicate narratives of emptiness in the context of depopulating post-socialist villages.[2] And though they don't eliminate the need or desire to consume industrially produced, globally sourced foods, these gardens facilitate biocultural conservation and provide nutrition, pleasure, and opportunities to flexibly negotiate the corporate food regime.[3] They are foundational to the everyday practice of quiet food sovereignty and are worth examining as powerful, though often ignored, niches of resilience. This paper is the result of in-person ethnographic research (during the summers of 2018, 2019, and 2021) and remote research including conventional and social media (2020 to the present). The formal research built upon two years of living and working in Gabrovo, Bulgaria as a United States Peace Corps volunteer (2006–2008).

From the Market to the Garden Gate
In late May 2021 Irina, Tihomir, and I began our Saturday morning in their apartment in the centre of Gabrovo, Bulgaria. We had been friends since my days as a Peace Corps volunteer when I worked for a local municipality with Irina. The sun was already warm as we walked from their block apartment to the open-air market a few streets away. Gabrovo is a town of roughly 60,000 people, large enough to have a daily market. It is located on a sloping street that empties out next to one of the main bridges over the Yantra river that runs through the centre of town. On this particular Saturday in late May the market was almost double its typical size, filled to overflowing with seasonal vendors. In addition to the usual commercial produce sellers and retirees with their home-produced herbs, vegetables, eggs, and honey there were many people selling plant starts. They occupied the lower part of the market that was uncovered and didn't have formal stalls.

The traditional agricultural season runs from St. George's Day (*Giorgovden*) on 6 May to St. Demetrius Day (*Dimitrovden*) on 26 October and correlates to the reliably frost-free nights for most of the country. So, it was high time to be putting in gardens. Most of the plant sellers were working-aged men and though they had a wide variety of stock, including ornamental foreign plants, the bulk of their inventory was vegetable starts in small individual plastic pots. While there were many plant vendors, there weren't really that many varieties of each type of plant. For example, in the whole market there were only four types of cucumbers for sale: 'Gergana' by far the most common, followed by 'Kornichon', and after that one called 'Byala' and another 'Tasty Green'. In the tomato department it was similar. Everyone had tomato starts, but they were mostly labelled 'Ideal', 'Magiya', and 'Rozov'. One small vendor had 'Bilvosko Tsertse', 'Ranno', and black and red cherry type tomatoes. These are types readily available in the commercial seed stores throughout the country.

The upper part of the market remained unchanged from when I first arrived in Bulgaria in 2006. It had a large cover with about fifty old metal stands in place below. There were also informal vendors on the side aisles of the market, on the sidewalk, and on some abandoned and uncovered metal stands at the fringes of the market. These were mostly tended by older women with only a few things to sell. Irina headed straight to a woman with a small selection of herbs, jars of honey, and tomatoes on the outskirts of the market. Irina asked her for eggs, but the woman said that she had already sold out. Irina told me that this woman's eggs were the best and she usually bought them by the flat of two dozen each Saturday. Out of luck with her usual supplier, Irina settled for eggs from an older man in the formal part of the market who also sold eggs. She exchanged her empty cardboard flat for a full one. He sold other things too like cucumbers and herbs, but we just bought eggs from him. We continued our leisurely stroll and eventually Tihomir chose some cucumbers and tomatoes for lunch from one vendor, and Irina bought some green onions from someone else. I asked how they chose who they bought from and Irina just shrugged and said, 'I don't know. I just look around and eventually choose' – though I noticed they tended to favour the older vendors who didn't have a wide selection of produce.[4]

Against Irina's wishes Tihomir bought a few more cucumber starts for their garden, our eventual destination on that day. This was their first year solo gardening in the village of Armenite, about twenty minutes away from Gabrovo by car. Tihomir's grandparents, who had both recently passed away, used to live in that village. Now Tihomir was the steward of their home and garden there. Tihomir's family had a deep history in the village. His great grandfather was a mayor and had lived in a grand two-story house built in 1937. It was also left to Tihomir but was crumbling, with a yard knee-high in grass and weeds. Irina commented that they were ashamed of its condition, that a grand old house would be in such disrepair. But between the two of them they had access to several village homes that they would eventually inherit and there was simply no way to restore or maintain them all. They didn't really feel like they could sell them either, so they decayed in place, full of the leftovers from the lives lived in them.

This was not terribly unusual. Bulgaria is the most rapidly depopulating country in the European Union and is simultaneously experiencing rapid urbanization. There are also very high rates of emigration, and very low levels of in-migration. These patterns have only accelerated since Bulgaria joined the European Union (EU) on 1 January 2007, which opened up new opportunities for the free flow of both people and capital across previously closed borders.[5] This population decline, however, is not uniform across the country.[6] The capital Sofia, for example, has experienced consistent population growth. Varna, on the Black Sea coast, has also experienced growth in recent years. However, outside of larger cities, the population decline is pronounced.

These statistics are made visible when traveling through the rural areas of the Gabrovo region just north of the central Balkan Mountain range, as I have done many times over the past fifteen years. Driving along poorly maintained, narrow roads, it is common to come across villages that are mostly deserted, a scattering of houses in various states of disrepair, grass growing high while rooflines sink. Even in villages that still have full-time residents, it is common to find many houses that are abandoned, especially in the winter months. In these emptying villages there are the vestiges of more robust community life like shuttered shops and cultural centres, crumbling schools, and abandoned theatres.

But summer can confound expectations of emptiness, particularly in villages where there are gardens. During the agricultural season, villages experience episodic fullness related to gardening and food preservation practices. On this spring day we ourselves were headed to the village of Armenite to garden. Irina and Tihomir also spent time during the summer months with Irina's parents in the nearby village of Mladen. During the summer, that was her parent's primary residence, until the weather turned cold in the late fall.

With low birth rates, each child would become more likely to inherit several pieces of land or various kinds of real estate. In addition, there were many absentee owners living and working abroad. So, for those Bulgarian young people who stay in Bulgaria, like Irina and Tihomir, they likely had access to land. In their case, they owned their apartment in the city and had access to several village homes as well.

We left Gabrovo in their small hatchback with our market purchases in tow, and after a twenty-minute drive we pulled up to the garden gate in front of Tihomir's grandparents' old house. We unloaded the car and went through the gate into the walled yard. Irina and Tihomir basically didn't use the house at all, except to get water from the faucet inside and store things like towels and a few basic cooking tools. Last year they brought their old stove out to the village and put it on the patio so that they could cook outdoors when they come. They also bought a new metal porch swing with a shade cover for sitting on and relaxing.

Ghosts of the Past

The house looked like someone just walked away one day, which is basically what happened. Tihomir's mother took care of her father-in-law and mother-in-law here during their last years. She also helped them tend their garden during that time. They both passed away within the past two years. After that, late last summer, Tihomir's parents moved away from the region to a house with a small courtyard garden. However, they had already planted a huge garden in Armenite which was then left to Tihomir and Irina to manage. They typically commuted to work together during the week between their home in Gabrovo and a nearby industrial town. Armenite is roughly on their way home from work, so they established a new pattern of stopping in several times per week to water and harvest. Last year they harvested and preserved hundreds of pounds of tomatoes from the plants that were left behind in the garden – which required a huge amount of weekend time at the end of the summer. Irina said that she had never done this task on her own before but learned quickly aided by memories of her mother and grandmother putting up tomatoes.

As Irina put it there was just so much 'stuff' left over from the generations that had previously lived in the house. She wasn't exaggerating, there were books and random bits of things scattered everywhere: old empty plastic water bottles, pieces of wood and metal, old machinery for threshing wheat, and a very small portable mill. On the front steps there were bits of pipe and wire, an old undershirt cut into pieces and used as a rag, small dishes with cigarette butts from Tihomir's father. 'Most of it is just garbage,' Irina said, but she was overwhelmed with the thought of throwing anything out or sorting through it. So instead, she moved around and through the debris, using the useful things and ignoring the rest. In the basement, all the contents were left over from Tihomir's parents and grandparents. There were a few dusty jars of preserved food on the shelves. In every dark corner there were potentially useful things like old buckets, water bottles, beer bottles covered in dust and cobwebs, large plastic barrels for fermenting cabbage, a huge wooden barrel, plastic bags, and all sorts of other odds and ends. It felt like a museum rather than a living, used space. Really only the outside garden and makeshift outdoor kitchen was still used. Everything else about this house was just something that they worked around: a place for visiting, but not for dwelling.

Gardening and Inheritances

The garden, however, was very much alive: a combination of perennial plants and trees put in by previous generations, annuals that Tihomir and Irina had recently planted, volunteers plucky enough to emerge through self-seeding, and semi-cultivated plants that were allowed space within the garden wall. As we walked through the garden yard at the start of the season, I counted a dozen tomato plants, cucumbers, zucchini, pumpkin, potatoes (both planted and volunteer), raspberry vines, grape vines, strawberries, spearmint, savory, nettle, dill, celeriac, lovage, and flat leaf parsley. There was also a giant walnut tree in the yard, and fruit trees like cherry, plum, and quince. The fuzzy quince were just starting to swell, their blossoms falling to the ground.

We approached our work at a leisurely pace. Irina and I did a little weeding and Tihomir chose a spot for his two new cucumber starts that he bought at the market. The earth was friable and easy to work in since there had been a soaking rain the night before. There were lots of bugs, but I didn't encounter any worms as I pulled the small and easy-to-remove weeds with my bare hands. I noticed that there were a bunch of volunteer tomatoes coming up, and when I showed Tihomir he was very excited. He had wanted to buy more tomato starts at the market, but Irina had stopped him. 'What are we going to do with all those tomatoes?' she asked him, clearly remembering their labours canning last year. He shrugged and didn't put up a fight, but I could tell he was excited about the bonus tomatoes. I transplanted them next to some poles he put in the ground to stake them up once they grew bigger.

Our backs were warmed by the sun, even though it was still early in the day. Irina started to become worried that the gardening was becoming too much like 'work', and so she insisted that we stop weeding and instead begin our leisurely lunch preparations.

Cooking Across the Binaries

The featured dish for the day was *Kasha sus Kopriva*, a savoury porridge featuring nettles and other fresh garden herbs. *Kopriva* (nettles) come up in the early spring. They are a welcome fresh green after the increasingly beige food landscape of winter. They emerge at roughly the same time as people begin to venture from their cosy homes: when there is still a crisp edge in the wind, but there is a promise of increasingly warm days ahead. They grow wild and abundant in many places throughout Bulgaria along paths and roads, in meadows and along riverbanks. They are also kept as a semi-cultivated plant in gardens and provide early season fresh greens while the garden tomato and cucumber plants are still in their earliest stages of growth, months away from producing anything edible.

Inside the walled garden Irina taught me how to harvest nettles and cook them together with other early spring ingredients into the kasha. She started by getting a small white enamel bowl and a white piece of cloth. She didn't have work gloves, so Irina had to carefully use the cloth to pluck the nettles from the stalk without getting stung by the fine hair-like spines. In her usual graceful way, she slowly negotiated these stingy plants and filled the small bowl with tender, young leaves. These nettles were growing inside the inherited garden. Nobody planted them, but they came up in patches every year and a certain amount were allowed space to grow.

We also walked through the garden and gathered some dill and flat-leaf parsley and chopped it finely. Again, these were not things that Irina or Tihomir had planted. They were the progeny of previously planted herbs put down by previous generations. They re-seeded themselves and were encouraged to grow by watering and removing competing plants around them.

We took the nettle leaves over to the table near the stove in the make-shift outdoor kitchen, and then Irina transferred them into a large enamel pot full of water. She then put the pot on the stove and brought the water and leaves to a boil very briefly. Once the leaves were exposed to the heat, they would no longer sting.

While the nettles were boiling Irina washed and chopped a large bunch of green onions, the white and green part all the way to the top. She saved the top-most green parts for our salad and kept the rest to make the kasha. Once the nettles were softened but not mushy Irina carefully strained them off the top of the water with a spoon and chopped them finely on the wooden cutting board. She saved the bright green broth leftover from cooking the nettles to make the kasha and then told me she would water the plants with whatever was leftover. 'It's very good for the plants' she said.

Irina then took a large piece of unsalted butter and melted it in the pan. Then she added the chopped green onions and herbs and stirred them infrequently with a wooden spoon until they were golden brown. Next, she added the flour: 'The most ordinary kind. Three spoons-full because there are three of us', she said as she measured from the bag. Once the flour was browned a bit, she added the chopped, cooked nettles and several ladles full of the broth that the nettles cooked in. She adjusted the consistency and stirred carefully to make a smooth and creamy mixture. Finally, she added small pieces of *sirene* cheese (similar to feta), that she had mashed with a fork. She thought it was still a little too thick, so she added more broth. The end result was thicker than a cream soup but not quite as thick as mashed potatoes. Irina told me that she didn't typically add any salt to this recipe because the cheese was plenty salty to season the dish.

She dished up three bowls with the kasha which we let cool a bit while Tihomir finished chopping the tomato and cucumber salad. Tihomir added salt to the tomatoes so that they bled a little and then mixed in the cucumbers, a drizzle of sunflower oil and the green onion tops. We finally sat down to the feast, that included the kasha, some sausages that Tihomir had grilled, pork skewers, salad, boxed juice from the store, Coca Cola, beer, and homemade plum *rakia* from Irina's father.[7]

Village Time

After lunch we slowly cleaned everything up, watered the plants, and went for a stroll through the village. It was very green with trees and tall grasses on the sides of the small roads. We didn't encounter any cars driving that day. From most of the village you could see the 'winding wall' which was a rocky cliff-face to the north and the snow-capped *Stara Planina* (Old Mountains or Balkan Mountains) to the south. The sun was hot by this point, and we slowly ambled along the lanes as they narrated the sights of the village. For example, there was an abandoned cooperative distillery that had been overtaken by bramble. On the road back down toward the village square we passed a couple of brand-new houses with manicured lawns and metal gates with automated sensors for cars to drive in and out. The people who built these places were new to the village and Tihomir didn't know them. This type of village home, not associated with a productive garden or yard, is commonly referred to as a villa. They are usually used as weekend and vacation homes for urban dwellers or ex-pats. We passed several other houses where people appeared to live full-time. Some of their gardens were already extremely far along, with tomatoes waist high and huge leafy cucumber plants. Tihomir was very impressed. Irina and Tihomir said 'good day' to people in these yards as we passed.

In many ways Armenite was emblematic of other villages that I have visited over the years. There were a few year-round residents, retired people who usually gardened extensively and often kept small livestock like chickens or goats. There were also younger retirees who spent a lot of time in the summers in the village, also gardening but usually not keeping any animals since they require year-round care. Then there were working-age people, like Irina and Tihomir, who visited on the weekends during the summer and the odd evening or holiday. They didn't always keep up the houses for staying in but used the gardens and outdoor kitchens. And finally, there were the new-comers who bought village land to build vacation homes, usually distinct for their ornamental landscaping and new construction.

The garden in Armenite provided a significant amount of produce, herbs, and nuts for Irina and Tihomir's family. They also enjoyed it as a place for peaceful outdoor leisure in between more work-oriented tasks. During the summer season it kept them in supply of fresh vegetables and fruit, and they preserved some of the excess in jars for the winter. However, they obtained most of their food through shops, supermarkets, and informal gifts from family. They considered themselves discriminating food shoppers and relied on their experience and aesthetic senses to choose reliable brands of processed foods like cheese, butter, and yogurt. In terms of meat, Irina would make a special trip out of her way to buy from the large international supermarket chain in town. They also routinely bought boxed juices or soft drinks made by international brands. Their own production and all of these commercial purchases were also heavily supplemented by gifts of jarred foods from Irina's mother, who kept an extensive garden, and wine and *rakiya* from Irina's father's vineyard and orchard. They did not characterize these various food production, acquisition, and exchange mechanisms as oppositional. Rather, they composed the relatively common, patch-work nature of everyday foodways in contemporary Bulgaria.

Loss, Change, and Re-creation

Though, as one full-time village resident put it, 'The graveyard is full, but the village is empty', gardens in villages complicate this emptiness narrative. They drive multi-generational, multi-species gatherings and are a rich assemblage of inherited plants, animals, microbes, knowledge, embodied skills, tools, interpersonal relationships, and familial histories. Emptiness, ghosts, and loss co-exist with ongoing traditions, new arrivals, and the re-creation of rural spaces. Through the garden gate it is possible to glimpse the entanglement of formal and informal economies, cultivated and wild foods, local and global influences that provide gardeners and their families with resilience in terms of food security but also the ability to pursue something beyond mere survival. These garden foods are prized as clean and reliable alternatives to industrial food, tastes of home and the village, and essential components in both everyday and ritual life.

Ecofeminists such as Donna Haraway and Anna Tsing propose that the logic of the plantation (the biological application of the Fordist and Taylorist modes of industrial production to agriculture) organizes not only agriculture but also modern economies, environments, and social relationships in profoundly destructive ways.[8] The logic of the plantation, in its biological and cultural application, has led directly to precarity, which is exacerbated by broader environmental and political instability. Large-scale, industrial agriculture in Bulgaria followed this pattern as it emerged in the early years of state socialism. Small plots of land were forcibly consolidated into big tracts and converted into large scale, mechanized monocultures like grain and oil crops or industrial scale vineyards and orchards.

However, home-gardens and other forms of food self-provisioning continued to flourish and provided an alternative source of food, medicine, and alcohol outside of the formalized state-run economy. Though some expected these gardens and informal food preservation practices to diminish significantly during the 'transition' after the end of socialism, they have persisted. This is despite the fact that they didn't fit the progress narrative or ideals of the state socialist regime, the transition to capitalism, or the contemporary context of EU member states. Gardens are opportune for examining 'what has been ignored because it never fit the timeline of progress'.[9] Food self-provisioning through gardening does not fit the timeline of progress for those who operate within a modernist, capitalist conceptualization of development. These practices may also not seem 'progressive' through the lens of food-based social movements or public efforts to preserve traditions since these practices are deeply entangled with industrial, global, corporate materials. What gets lost in between these two conceptualizations is the actually emerging, evanescent present, manifesting in living practices that draw on inherited, experimental, prototypical, and novel materials, competences, and meanings.

Looking around, rather than ahead or behind, ecofeminists like Tsing and Haraway notice that collaboration and mutuality are at play in these ignored spaces and that resilience is rooted in entangled, interdependent relationships which defy binaries commonly undergirding Western thought.[10] Multi-disciplinary research that has emerged from this work points to the vital significance of 'arts of living' and the 'possibilities of life' even amidst the ruins, rather than despairing an inevitably unfolding apocalypse or seeking refuge in the hope of a progressive utopia.

Gardening in Bulgarian home-gardens requires being fully present, in the middle of literal and metaphorical ruins, and making a life. Through gardening, the leftovers of the past are re-contextualized and become indeterminant inheritances from which living social practices can emerge, always contingent, always evolving. I argue that these practices are living manifestations of 'quiet food sovereignty' that grow in the garden, even though they are not explicitly motivated by a social movement and occur in less visible domestic spaces.[11] However, these practices are not ineffective or less important politically for these reasons. Atanasoski and Vora argue that grappling with the messy aftermath of the end of state socialisms offers insights into the practices and politics of social change that is non-oppositional, and that may often be illegible, especially to Western observers.[12] Specifically, they identify a politics that doesn't aim to culminate in, or need to wait for, revolution. This is a helpful contribution to the emerging conceptualization of a multiplicity of food sovereignties in general and quiet food sovereignty specifically.[13] It reinforces Chari and Verdery's call to consider post-socialism, like post-colonialism, as a theoretical frame that is 'non-progressive', troubles linearity, and makes space to reimagine 'a politics worth doing'.[14] I believe all of these inheritances, evident in everyday practices like home-gardening in contemporary Bulgaria, expand the possibilities of seeing through the posts and ghosts to identify, imagine, and expand diverse food sovereignties in the present and future.

About the Author
Lindsey Foltz is a cultural anthropologist who lives, works, and gardens in the southern Willamette Valley in Oregon.

Notes
1 Jens Alber and Ulrich Kohler, 'Informal Food Production in the Enlarged European Union', *Social Indicators Research*, 89 (2008), pp. 113–27; Petr Jehlička, T. Kostelecky, and J. Smith, 'Food Self-Provisioning in Czechia: Beyond Coping Strategy of the Poor: A Response to Alber and Kohler's "Informal Food Production in the Enlarged European Union" (2008)', *Social Indicators Research*, 111.1 (2013), pp. 219–34;

Richard Rose and Yevgeniy Tihomirov, 'Who Grows Food in Russia and Eastern Europe?', *Post-Soviet Geography*, 34.2 (1993), pp. 111–26; Philip Kostov and John Lingard, 'Subsistence Farming in Transitional Economies: Lessons from Bulgaria', *Journal of Rural Studies*, 18.1 (2002), pp. 83–94.

2 Dace Dzenovska, 'Emptiness Capitalism without People in the Latvian Countryside', *American Ethnologist*, 47.1 (2020), pp. 10–26; Dace Dzenovska, Volodymyr Artlukh, and Dominic Martin, 'Between Loss and Opportunity: The Fate of Place after Post-Socialism', *Focaal*, 2023.96 (June 2023), pp. 1–15; Ivan Rajkovic, 'Whose Death, Whose Eco-Revival: Filling in while Emptying Out the Depopulated Balkan Mountains', *Focaal*, 2023.96 (June 2023), pp. 71–81.

3 Eric Holt-Giménez and Annie Shattuck, 'Food Crises, Food Regimes and Food Movements: Rumblings of Reform or Tides of Transformation?' *The Journal of Peasant Studies*, 38 (2011), pp. 109–44; Teodora Ivanova and others, 'Enough to Feed Ourselves! Food Plants in Bulgarian Rural Home Gardens', *Plants*, 10.11 (2021): 2520, DOI:10.3390/plants10112520.

4 See also Yuson Jung, 'Ambivalent Consumers and the Limits of Certification', in *Ethical Eating in the Socialist and Postsocialist World*, ed. by Yuson Jung, J. Klein, and Melissa Caldwell (University of California Press, 2014).

5 Eurostat, 'Population and Population Change Statistics', 6 July 2024 <https://ec.europa.eu/eurostat/statistics-explained/index.php/Population_and_population_change_statistics> [accessed 25 November 2024].

6 Chavdar Mladenov and Margarita Ilieva, 'The Depopulation of the Bulgarian Villages', *Bulletin of Geography, Socio-economic Series*, 17.17 (2012), pp. 99–107.

7 *Rakia* is brandy typically made from fruit, often grape or plum.

8 Gregg Mitman, 'Reflections on the Plantationocene: A Conversation with Donna Haraway and Anna Tsing', EdgeEffects, 18 June 2019 <https://edgeeffects.net/Haraway-tsing-plantationocene> [accessed 2 October 2024].

9 Anna Lowenhaupt Tsing, *The Mushroom at the End of the World: On the Possibility of Life in Capitalist Ruins* (Princeton University Press, 2015), p. 21.

10 Tsing, *The Mushroom at the End of the World*; Anna Lowenhaupt Tsing and others, eds. *Arts of Living on a Damaged Planet: Ghosts and Monsters of the Anthropocene* (University of Minnesota Press, 2017).

11 Oane Visser and others, '"Quiet Food Sovereignty" as Food Sovereignty Without a Movement? Insights from Postsocialist Russia', *Globalizations*, 15.4 (2015), pp. 513–28; Hannah Wittman, Annette Desmarais, and Nettie Wiebe, *Food Sovereignty: Reconnecting Food, Nature and Community* (Fernwood, 2010).

12 Neda Atanasoski and Kalindi Vora, 'Postsocialist Politics and the Ends of Revolution', *Social Identities* 24, no. 2: 139–54. 2018. Pg. 147.

13 Marisa Wilson, *Postcolonialism, Indigeneity and Struggles for Food Sovereignty: Alternative Food Networks in Subaltern Spaces* (Routledge, 2017); Petr Jehlička and others, 'Thinking Food Like an East European: A Critical Reflection on the Framing of Food Systems', *Journal of Rural Studies*, 76 (2020), pp. 286–95; Hilda E. Kurtz, 'Framing Multiple Food Sovereignties: Comparing the Nyeleni Declaration and the Local Food and Self-Governance Ordinance in Maine', in *Food Sovereignty in International Context*, ed. by Amy Trauger (Routledge, 2015).

14 Sharad Chari and Katherine Verdery. 'Thinking Between the Posts: Postcolonialism, Postsocialism and Ethnography after the Cold War', *Comparative Studies in Society and History*. 51(1):6–34. 2009.

13
Roman Edens
Virgil's Tarentum and Pompeiian Garden Painting (Re)Considered

Christopher Grocock

Introduction

This paper examines a remarkable digression in Book 4 of Virgil's didactic poem the *Georgics*, in which the poet recalls an (alleged) experience he had of a garden created out of waste land by an old man of Tarentum, a Corycian by birth (Corycia is located in what is now northern Turkey). Bee-keeping, the main topic of this particular book of the *Georgics*, is tangentially a part of his description; but as Virgil himself says at the end of his digression, 'in truth I must pass over these details, prevented by lack of space, and leave them to be recalled by others after me.' Thomas regards the *praeteritio* a product of Virgil's imagination, but not an impossibility.[1]

The translation which follows is my own, but I have referred to others, notably Lewis, Wilkinson, Fallon, MacKenzie, and Chew.[2] Useful commentaries on *Georgics* are by Mynors and Thomas; while these focus largely on grammar and syntax, they also make useful attempts at identifying the plants which Virgil mentions.[3] The text has been subject to a wide variety of approaches, needless to say.[4]

Translation: *Georgics* iv. 109–148

    Let gardens breathy with saffron flowers lure them in,
    and as guard against thieves and birds, with his willow sickle,   110
    let the protection of Priapus, lord of the Hellespont, watch over it.
    He – and none but he – should bring thyme and wild laurel from the mountain-tops
    and plant them widely around the hives if he has a care for such things;
    It's his own hand he should callous with hard toil, he should himself
    plant fruitful cuttings in the ground and give them a welcome watering-in.   115
    And indeed, if I were not actually drawing in my sails as I come to the end of my labours
    and hurrying to steer my prow to land,
    I might also be singing of the rich gardens which the attentive gardener
    decorates, the roses of twice-blooming Paestum,[5]
    and how it is that endives[6] rejoice in quaffable streams,   120
    and riverbanks are green with wild celery[7] and how the cucumber winds

across the grass and swells its paunch;[8] nor about late-flowering
narcissus would I have kept quiet, or spiralling acanthus with its curling stems,
and ivy[9] shading into pale and myrtles[10] which love the seashore.
For I remember that below the towers of Oebelia's citadel,[11]   125
where the dark[12] Galaesus wets the goldening fields,[13]
I saw an old man from Corycus who had a few acres
of passed-over land,[14] not fertile enough for ploughing with cattle,
not fit for pasturing herds, and no good for growing vines.
But here, clearing space in the brambles for cabbages spaced out with white lilies
round about,[15] and verbena[16] and fragile-seeded poppy,[17]   131
he imagined he equalled the wealth of kings,[18] and returning home late at night[19]
he heaped his table with banquets never bought.[20]
First to pluck a rose in spring or fruits in autumn,[21]
and when sad winter was even now shattering rocks with its cold   135
and checking galloping water with its reins of ice,
he was already cutting back the leaves of soft hyacinth,[22]
ticking off summer for being slow in coming, and the west winds behind schedule.
As a result, that fellow was first to be rich in a plentiful swarm of productive bees,
and to force the foaming honey from pressed combs;   140
his lime-trees and wild laurels were most productive,
and every fruit his fertile trees produced in their early blossoming
held firm and ripened when autumn came.
He also set out mature elms in a row,[23]
and hardwood pears, and blackthorn[24] already bearing sloes,   145
and a plane-tree already serving up shade for drinkers.
But in truth I must pass over these details, prevented by lack of space,
and leave them to be recalled by others after me.[25]

We have here part of a 'descriptive poem', in which the eulogized old gardener's achievements stretch the bounds of possibility in the extension of the seasons. As L.P. Wilkinson commented, 'Seneca [*Ep.* 86.15] said no more than the obvious when he remarked that Virgil was interested in what could be said *decentissime*, not *verissime* ["most fittingly" not "most truthfully"], and that he wrote not to teach farmers, but to delight readers', and that the passage is a good example of *praeteritio* – an opportunity to insert a short passage about gardens, but leaving most to others (and the later writer on agriculture, Columella, took the hint, and composed Book 10 of his treatise, on gardens, in hexameter verse instead of prose).[26]

But to what extent *might* Virgil's description have been penned 'most truthfully'? We might, first of all, ask – who this old man was supposed to be? Some recent scholarship (as much recent scholarship does) prefers to concentrate on the obstacles in identification, as Richard Thomas argues that 'identification has resisted any real consensus':

> the Corycian has been seen as a philosopher, Epicurean or otherwise, a model of simplicity, a paradigm of hard primitivism, a golden-age or an impossibly idealized figure [. . .] an ex-pirate, as Orpheus, or as a conflation of P. Valerius Cato and Virgil's own Meliboeus from the first Eclogue [. . .].[27]

There does seem to be some hard, historical reality which can provide a context for the old man. First, his origin. Corycos is the name of a mountain and city in what is now Anatolia, southern Turkey, and the fourth-century AD commentator Servius explains that the old man was a pirate captured by Pompey the Great, in his clearance of the pirate scourge in the Mediterranean of 67 BC, and then deported to Tarentum in Calabria. As Mynors notes, none of this is provable; but it makes the character and context plausible, at least. Second, Wilkinson points out, 'There were also still independent *coloni*, notably in the wide plains of the Po valley, in the mountains where only small plots might be cultivable, and on the southern coast, where Greek traditions persisted.'[28] Thomas suggests that the episode may reflect 'the successful workings of private man within the terms of the age of *labor*', while noting that classical scholars have preferred to see the details as 'literary justification'.[29]

Third, what is this piece doing in a book devoted to apiculture? Although Virgil's focus appears to be on the planting, one of the *results* of that planting is advantageous for the old man in terms of supply of honey: Virgil reminds us in lines 139–40 that he *was first to be rich in a plentiful swarm of productive bees, /and to force the foaming honey from pressed combs*; honey and grape- or fruit-juices were crucial products in those years before sugar became common. His plot is small, but still useful: Wilkinson comments that 'the Corycian gardener is only incidentally a bee-master (139–141), though he serves to emphasize that Virgil is envisaging throughout a smallholder's sideline, not mass-production', and reminds us that, in Varro, 'we hear of a bee-farm of a single *iugerum* only about thirty miles north of Rome successfully worked by two brothers', and he goes on to suggest that Virgil's old Corycian might have been inspired by the brothers Veianus of Falerii mentioned by Varro there.[30] Columella and Varro both include compendious lists of suitable plants to provide for bees. Roman bee-keeping has been sadly neglected, but an extremely useful piece of work was done by Hannah Bochain.[31]

A major part of the ethic Virgil promotes in the *Georgics* is that of hard work. Wilkinson (whose views are echoed by Mynors and Tarrant) sets this out very clearly:

> The kind of farmer Virgil seems generally to envisage [. . .] still works with his own hands [. . .]. But he must have been aware that conditions over most of Italy were very different. [. . . I]t seems more likely that Virgil's farmer is presented as his own ideal, the old-fashioned yeoman – *vetus colonus* – revived. He must work himself: the whole moral fabric of the poem is based on this.[32]

This runs counter to Roman views about manual labour, at least for a high-status Roman; Wilkinson himself cites Cicero *de Fin.* 1.1.3 (digging and farm-work are *illiberalis labor*, 'work not suited to a free man') and 3.2.4. (*agri cultura, quae abhorret ab omni politiore elegantia*, 'agriculture, which spurns all more refined elegance'). Catullus opposes to the *urbanus* or 'city type' the *caprimulgus aut fossor*, 'goat-milker or ditch-digger' (22.10); we might compare Columella 1 praef. 20, *communis opinio rem rusticam sordidam opus esse*, 'the common opinion is that farming is a mucky business'. Wilkinson notes that 'Horace might do a little manual labour up at his farm, to the surprised amusement of his neighbours':

> *rident vicini glaebas et saxa moventem*
> the neighbours laugh at me moving the turfs and rocks (*Ep.* 1.14. 39)

but, as he comments, 'they laughed presumably not because he was doing it badly or because he was fat, but because he was doing it at all.'[33]

There is also a glaring omission in Virgil's idyllic picture. There is no mention of work done by slaves anywhere. It is part of the fabric of farm life in Hesiod and underpins the works of Cato and Varro. Horace had fourteen slaves working on his Sabine farm (*S.* 2.7.118). The scholarly consensus is that this must be intentional. But it does create an impression of 'effortless ease'.

Despite these omissions, Virgil clearly presents a picture which, while stretching to the limits the possibilities afforded by ancient agriculture, is largely credible. In this it differs markedly from the magical aspects of the gardens of Alcinous described by Homer in the *Odyssey*, where 'fruit never failed upon the trees', and no season at all was without some produce.[34] Virgil's picture is one just beyond the bounds of possibility – but not utterly fanciful. This also applies to the plants which Virgil mentions, in gardens generally (the first part of the text, to line 129) and also in the description of the old man's garden itself. Space precludes any detailed discussion of these, but every plant Virgil includes in his text gets a mention in one or more of the Roman agricultural writers (Tables 1 and 2).

Not only are the plants Virgil mentions attested by other ancient sources, but archaeology from Pompeii corroborates the lists of planting he provides. We may begin with the trees. Ciaraldi lists a number of food plants and trees from the Bay of Naples sites, but only apple, pear, and pine match Virgil's list.[35] There is a more comprehensive list given by Annamaria Ciarallo, and useful information on herbs is found in Jashemski's book on herbs from Pompeii.[36] Taking all these into account, we find that only two plant types in Virgil's text are not found in the archaeology: *crocus* (saffron) and *intubum* (endive). This may simply be due to the vagaries of archaeological survival; saffron, for example, has been cultivated for a long time and appears frequently in the recipes in *Apicius*.[37] Thus there is a recognizable realism to Virgil's account.

Gardens, Flowers, and Fruit

Table 1. Plants mentioned in Virgil's description of the 'old man' of Tarentum's garden

Plants mentioned in the 'introduction' to the digression – lines 109–124			
Line ref. in *G.* 4	Latin name	English equiv.	Other classical references
109	Crocus	saffron	Varro *RR*.1.35.1, Columella 9.44, Pliny *HN.* 31.90.
112	Thymum	thyme	Varro *RR.* 3.16.14, Virgil *E.* 7.37, Pliny *HN.* 13.138.
112, 141	pinus (MSS majority)	pine-tree	Varro *RR.* 1.15.1; Columella 9.5.6; Pliny *HN.* 16.36, 23.62, 37.42.
112, 141	tinus (emendation)	wild laurel, laurestinus	Pliny *HN.* 15.128, 17.60
119, 134	rosa	rose	numerous poetic refs! and see Pliny *HN.* 21.4.
120	intubum/ intubus	endive	Virgil *G.* 1.120, Columella 10.11, Pliny *HN.* 19. 123, 129
121	apium	most probably wild celery; *petroseliunum* is the culinary term for parsley	*Moretum* 88; Horace *C.* 4.11.3; Pliny *HN.* 14.105.
122	cucumis	gourd, cucumber	Varro *RR.* 1.2.25; Columella 11.3.17; Pliny *HN.* 9.3, 20.3, 37.55, 57; 19.118, 181.
123	narcissus	narcissus	lots in poetry! Virgil *G.* 4.160; Pliny *HN.* 14.244, 28.72.
123	acanthus	acanthus	Pliny *HN.* 22.16, *Culex* 398.
124	hedera	Ivy	Virgil *E.* 7.25; Pliny *HN.* 12.74, 16.145.
124	myrtus	myrtle	Cato *A.* 82, 133.2; Pliny *HN.* 15.118, 23.165.

Table 2

The old Corycian's garden adds these:			
130	olus/ holus	greens' – cabbages, pot-herbs	Cato *A.* 156.1; Varro *RR*.1.23.2; *Moretum* 72; numerous varieties in *Apicius* Book 3; Pliny *HN.* 19.79, 26.124
131	lilium	lily	Varro *RR.* 1.35.1; Pliny *HN.* 16.153.
131	verbena	verbena	Pliny *HN.* 8.12, 15.119.
131, 142	papaver	poppy	Pliny *HN.* 19.168, 20.205.
134	pomum	fruit, specifically orchard-fruit	Pliny *HN.* 12.22, 13.30, citing Cato *A.* 15.74
136	hyacinthus	hyacinth	Pliny, *HN.* 21.67, 170.
141	tilia	lime-tree	Virgil, *G.* 4.183; Columella 9.4.3; Pliny *HN.* 11.32, 16.65, 24.50
144	ulmus	elm	Cato *A.* 6.3; Varro *RR.* 1.35.2; Virgil *G.* 1.2; Pliny *HN.* 11.14, 18.266.
145	pirum	Pear	Cato *A.* 7.3; Columella 5.10.8; Pliny *HN.* 23.115
145	spinus	blackthorn	here only and Varro in Nonius Marcellus p.112M (so *OLD*).
145	prunus	Sloe	
146	platanus	plane-tree	Cato *A.* 51; Pliny *HN.* 12.6, 15.29.

Gardens at Pompeii

When we turn to the evidence for gardens from Pompeii, we must immediately acknowledge the debt all owe to the extraordinary work carried out by Wilhemina Jashemski, a lifetime's achievement which is scarcely to equalled.[38] Town houses at Pompeii have a variety of garden spaces, including peristyle gardens and vegetable gardens, usually at the rear. There were two types of house at Pompeii (as at Herculaneum): atrium-peristyle, and those more irregular in plan.[39] But Jashemski should be allowed to speak for herself:

> At Pompeii not only public buildings, but hundreds of houses and gardens, were preserved under the *lapilli*. These gardens are especially valuable because they reveal much about the importance of the garden [. . . as] an integral part of the house and [. . .] a significant factor in the development of domestic architecture which, only at Pompeii, can be traced for a period of over three hundred years [. . . . E]ven the very poor built tiny gardens in their small, crowded homes, and I have found that gardens played a role hitherto unsuspected in restaurants, hotels, schools, and various types of shops [. . .]. And there were large vegetable gardens, orchards, vineyards, and commercial orchards. A study of these gardens, because they relate to almost every aspect of life, contributes to a better understanding of the ancient world and brings to light an area of our cultural heritage about which very little has been written.[40]

Numerous houses at Pompeii are of particular interest for their gardens, but limitations of space will necessitate a study of just a few. It is worth listing the most significant, however: they are worth exploring in online sources. The House of Pansa (VI.vi.1) is particularly important: it has a large atrium, a peristyle garden with a pool, and a *xystus* (a long open portico). The large garden at the rear, 26.5 m × 30.5 m, occupied almost a third of the insula. Jashemski explains, 'When the house was excavated in the early nineteenth century, the planting arrangement was perfectly preserved. Such evidence has rarely been reported but, fortunately, the French scholar Mazois made a plan which shows that the garden was systematically laid out in rectangular plots separated by paths that were also used as irrigation channels.'[41] The layout is precisely that recommended by Pliny the Elder, who gives directions for marking out the plots and bordering them 'with sloping rounded banks, surrounding each plot with a furrowed path to afford access for a man and a channel for irrigation'.[42] These were not merely for decoration: as Jashemski notes, food was raised in rather sizeable plots in the city in AD 79.[43]

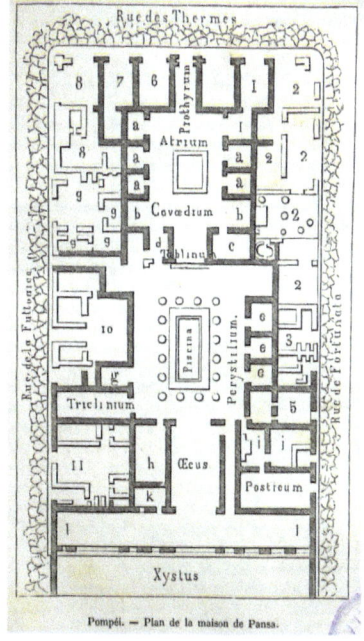

Figure 1. House of Pansa, ground plan. Source: Wikimedia Creative Commons.

Gardens, Flowers, and Fruit

The House of the Little Fountain or House of the Faun (VI.xii.2–5) has two atria and two peristyle gardens – real luxury!

The House of the Vettii (VI.xv.1) is another luxurious example, with two peristyle gardens with statues, ornaments, twelve fountains, frescoes, and marble tables.[44] The House of the Golden Cupids (V.xvi) has a peristyle garden with a large pool (now dry), two fountains, and animal sculptures set amongst plants.[45] Very useful for this study are the House of Julia Felix and the House of Polybius (IX.xiii.1–3): the former incorporates extensive beds to the side and rear (see Figure 4); the latter, excavated from 1965 to 1976,

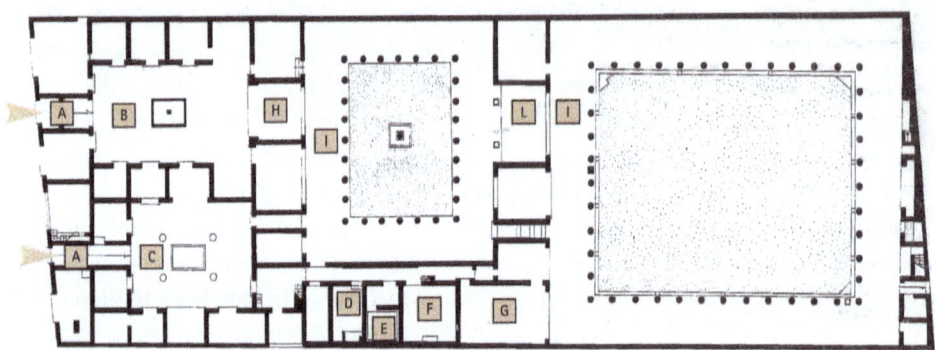

Figure 2. House of the Faun, ground plan. Illustration: C. Grocock.

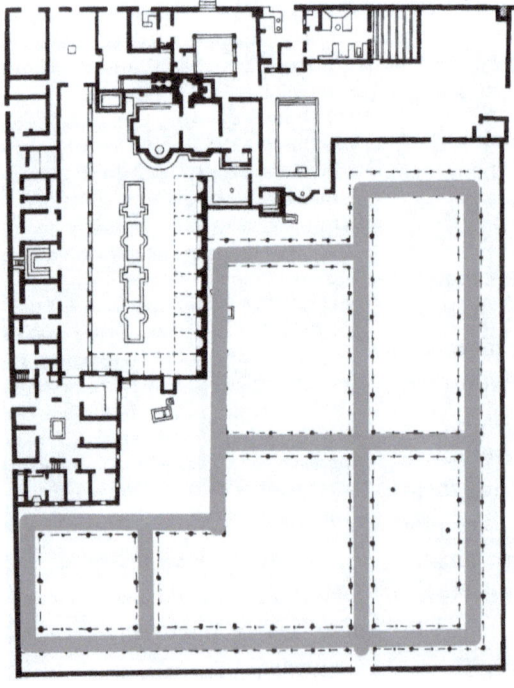

Figure 3. House of Julia Felix, site plan. Illustration: C. Grocock, after Cooley, Pompeii, fig. 11.

136

Figure 4. House of Loreius Tiburtinus, viewed from garden. Photo: C. Grocock.

Figure 5. Garden view, house of Loreius Tiburtinus. Photo: C. Grocock.

Figure 6. Garden view, house of Loreius Tiburtinus. Photo: C. Grocock.

was the first peristyle garden to be discovered in twenty years; garden excavation begun in 1973 benefitted from the use recent developments in archaeological techniques which have found the roots of many trees in this garden, which have been set in concrete.[46]

Of equal importance is the Villa at Oplontis, at Torre Annunziata, on the coast between Pompeii and Herculaneum: here there is a fragment of a once luxurious villa with internal courtyards, surrounded by a garden in which the ancient tree-positions can be seen.[47] More recent advanced scientific archaeological techniques have produced accurate studies of Pompeiian gardens, which have informed our knowledge of what was grown. We have come a long way since Orazio Comes established a catalogue of plants illustrated in frescoes and mosaics back in 1870.[48] The House of Loreius Tiburtinus (also known as the House of Octavius) has a fine, extensive garden-cum-allotment area at its rear. But houses did not have to be large to warrant a garden: even the smallest dwellings had some kind of planting, as a photograph from Jashemski shows.[49] Pompeiians made the most of every space, and mixed ornamental and practical planting.

Garden Paintings

In addition, and more importantly for this study, frescoes were used extensively in houses of all sizes to give the appearance of garden scenes beyond what was actually there – both in the size of what they depict and the plants they show in bloom. This took the form of combining different plants, enlarging their size, and illustrating them in flower when this might occur at different seasons. The aim seems to have been to extend both the size and scope of gardens, and imaginatively to prolong the blooming or fruiting season so that plants were visible at their best even when the season might be at its worst. To appreciate their colour these are better seen on easily-accessible online sites rather than in black-and-white reproductions.[50]

Being open to the elements, these frescoes were and are extremely fragile, as is well expressed by Jashemski's reflections on a specific fresco she wanted to study:

> Because of the transitory nature of these paintings, they date, for the most part, from the last years of the city. Few can be seen by the visitor today, for once excavated and exposed to the weather they rapidly fade and disappear [. . .]. Many years ago, we thought we were finally to see a special garden painting known to us from a drawing made at the time of excavation in 1893. This drawing showed a fruit tree with a nest, in which four little birds eagerly awaited their mother approaching with food. The previous year we had worked our way through the overgrowth of saplings and branches that had choked the garden and had pulled back enough of the vines to make sure that the painting was still there. But when we reached the wall we found that this unique painting had been loosened by the winter rains and was a heap of plaster at the base of a bare wall.[51]

As a result of this fragility, we are dependent on photographic evidence – much in Jashemski's works – for evidence of these frescoes, though some recent restoration work enables us to see how vivid they were.[52]

Of the trees and plants in Virgil's list, pine, rose, cucumber, acanthus, ivy, myrtle, poppy, pear, sloe, and the plane-tree all recur in the frescoes. A fine example is the House of Venus Marina (II.iii.3), a grand house with spectacular wall paintings showing plants (myrtle, rose, ivy, and several others).[53] But like Virgil's depiction, these beautiful images are unpopulated – and the slaves whose efforts without doubt would have made them possible are especially absent. In this sense, they are an unreal ideal, achieved without obvious sweat.

Concluding Remarks

Virgil's text can be interpreted in all manner of literary and political ways, but its contents – unlike its parallel in the gardens of Alcinous in Homer's *Odyssey* – are not at all fanciful, except that they stretch the expectations of what an 'ordinary' gardener might expect (the old man gets the earliest hyacinths, the earliest honey). It shows what might just be achieved. The same is true of the frescoes in Pompeii, to a remarkable extent – even in the smallest Pompeiian houses, there were attempts to extend their gardens – making them larger than they really were through *trompe-l'oeil* effects, or emphasizing lushness and fecundity in combinations of plants which might not flower together, always in bloom, and stretch our expectations of what is achievable. All illustrate the desire – not confined to the Romans – for the good times in our gardens to be extended beyond the normal confines of the seasons, and reveal a delight in the appearance and benefits brought by nature to the human experience.

About the Author

Christopher Grocock has taught Classics for the past thirty years, mainly at Bedales School. He is the author of a number of critical editions and articles in Medieval Latin, most notably texts by the Venerable Bede in *Abbots of Wearmouth and Jarrow* (Oxford Medieval Texts). With his wife Sally Grainger he edited the Roman culinary text *Apicius*.

Notes

1. *Virgil: Georgics vol. 2: Books III–IV*, ed. by R.F. Thomas (Cambridge University Press 1988).
2. *The Georgics of Virgil*, trans. by C. Day Lewis (Oxford University Press, 1947); *Virgil: The Georgics*, trans. by L.P. Wilkinson (Penguin Classics, 1982); *Virgil: Georgics*, trans. by Peter Fallon (Oxford World's Classics, 2009) has Creative poetry more than close translation in places, but conveys many of the nuances of Virgil's poem; *The Georgics by Publius Vergilius Maro*, trans. by K.R. MacKenzie (London, the Folio Society, 1969) renders the Latin quite well, but stilted by modern standards; *Virgil, Georgics*, trans. and comm. by Kristina Chew (Hackett, 2002) is idiosyncratic and interesting, and with some felicitous renderings, and her notes combine scholarly, classical information with agricultural details, jottings from wider reading, and kitchen lore in ways that are frequently a delight to read.
3. *Virgil: Georgics*, ed. by R.A.B. Mynors (Clarendon Press, 1990); *Virgil: Georgics vol. 2: Books III–IV*, ed. by Thomas.
4. For example, a brief search in February 2024 threw up L. Weeda, *Vergil's Political Commentary* (De Gruyter Open, 2015), DOI:10.1515/9783110426427; A.J. Boyle, 'In medio Caesar: Paradox and Politics in Virgil's Georgics,' *Ramus*, 8.1 (1979), pp. 65–86, DOI:10.1017/S0048671X00003994, P. Hardie, 'Political Education in Virgil's Georgics' in *Selected Papers on Ancient Literature and its Reception* (De Gruyter, 2023), pp. 117–41, DOI:10.1515/9783110798852–009. More 'literary' are M.R. Gale, 'Poetry and the Backward Glance in Virgil's "Georgics" and "Aeneid,"' *Transactions of the American Philological Association*, 133.2

(2003), pp. 323–52 <http://www.jstor.org/stable/20054090> [accessed 24 Dec. 2023]; J.S. Clay, 'The Old Man in the Garden: Georgic 4. 116–148', *Arethusa*, 14.1 (1981), pp. 57–65 <http://www.jstor.org/stable/26308077> [accessed 24 Dec. 2023].

5   Thomas sees this as either a marvel, like Alcinous' gardens in *Odyssey* 7. 117–21, or a reference to the autumn-flowering damask rose.
6   'Our curly endive or *frisée*' (Chew).
7   So Thomas – definitely *not* parsley, which I had at first, following Loeb and C. Day Lewis. Mynors points out that celery has been growing in well-watered places since Homer – see *Il*. 2. 776.
8   Mynors has the cucumbers planted round the edge of a plot so they can 'wind at will through the rough herbiage beyond it'.
9   Ivy was associated with Bacchus, and used in the 'poet's garland'; see Virgil *E*. 7. 25.
10 That is, because they are frost-intolerant (so Mynors).
11 Thomas sees parallels with the efforts of Jupiter in book 1, at roughly the same place – here celebrating the success of a private man in the age of *labor*; Mynors p. 244 comments that 'toil, which was a menace (in i. 125–48), has become a liberation'.
12 Mynors: perhaps 'deep and dark' – no pine-woods shade its banks – also recalling Homer's *melanudros* (*Il*. 9.14).
13 Chew provides 'wets the goldening fields'.
14 *Loca relicta* in Roman land-surveying were lands left unallocated outside the boundaries of a *colonia* (Mynors): see B. Campbell, *The Writings of the Roman Land Surveyors*, Journal of Roman Studies Monograph 9 (Society for the Promotion of Roman Studies, 2000), p. 4 line 9, p. 6 line 32.
15 Thomas: not 'the odd vegetable' but 'well-spaced-out'.
16 Plants used in ritual or medicine (Mynors). Celsus 2.33.4 lists eleven shrubs in this category; here, perhaps 'herbs' used for particular purposes; 'any of various aromatic trees or shrubs' used for religious purposes, according to *OLD*.
17 Though the seeds are small, they are edible; 'tiny' might highlight the poor state of the land (so Mynors). They are used too frequently to mention in *Apicius*.
18 Thomas compares Horace's similar view in *S*. 2.6.6.
19 Thomas: like the bees later, lines 185–87.
20 Chew, felicitously again, offers, 'banquets never bought'.
21 His success is within the confines of the real, seasonal world (so Thomas).
22 An odd remark; some kind of early forcing must be intended. Mynors thinks the outer petals are meant by *comam*.
23 The focus now more on making a pleasant place than bee-keeping (Mynors); elms were used for growing vines up – perhaps hinted at here with the *potantibus*. See also the fine mosaic from Saint-Romain-en-Gal <https://commons.wikimedia.org/wiki/File:Saint-Romain-en-Gal.jpg> [accessed 22 August 2024]. There is rhetorical exaggeration, but it need not imply that the trees are fully-grown (i.e. they are not 'whips,' but 3–5 years old).
24 A rare word, according to *OLD* found only here and in a citation of Varro by Nonius Marcellus, p. 112M.
25 At which point Virgil admits he's been digressing and returns to the main theme of 'bees'.
26 L.P. Wilkinson, *The Georgics of Virgil: A Critical Survey* (Cambridge University Press, 1969), pp. 15, 103.
27 Richard F. Thomas, 'The Old Man Revisited: Memory, Reference and Genre in Virg., Georg. 4, 116–48,' *Materiali e discussioni per l'analisi dei testi classici*, 29 (1992), pp. 35–70, DOI:10.2307/40236012.
28 Wilkinson, p. 102. and n. 7.
29 Thomas, pp. 16–70.
30 Wilkinson, pp. 264, 50, 103; Varro, *RR*. 3. 16. 10 11.
31 Hanna L. Bochain, 'The Economy of Beekeeping: Examining an Overlooked Industry of the Ancient World', (unpublished masters thesis, University of Georgia, 2014) <https://getd.libs.uga.edu/pdfs/bochain_hannah_l_201712_ma.pdf> [accessed 21 March 2024].
32 Wilkinson, pp. 53–54.
33 Wilkinson, p. 54.
34 Homer, *Odyssey* 7. 119–141, and n. 11 above.

35 Marina Ciaraldi, *People and Plants in Ancient Pompeii: A New Approach to Urbanism from the Microscope Room*, vol. 1 (Accordia Specialist Studies on Italy Accordia Research Institute, University of London, 2007), p. 39, table 4.
36 Annamaria Ciarallo, *Gardens of Pompeii* ('L'Erma' di Bretscneider, 2000), and *Flora Pompeiana I* ('L'Erma' di Bretschneider, 2004), pp. 67–72; Wilhelmina Feemster Jashemski, *A Pompeiian Herbal: Ancient and Modern Medicinal Plants* (University of Texas Press, 1999).
37 See Wu Mingren, 'Saffron: 'Tracing the Origins of a Treasured Ancient Spice,' *Ancient Origins* <https://www.ancient-origins.net/history/saffron-ancient-spice-0011537> [accessed 27 March 2024]. *Tilia* are attested by Cristiano Vignola and others, 'At the Origins of Pompeii: The Plant Landscape of the Sarno River Foodplain from the First Millennium BC to the AD 79 Eruption', *Vegetation History and Archaeobotany*, 31 (2022), pp. 171–186 <https://link.springer.com/article/10.1007/s00334-021-00847-w> [accessed 27 March 2024].
38 Wilhelmina F. Jashemski, *The Gardens of Pompeii, Herculaneum and the Villas Destroyed by Vesuvius* (Caratzas Brothers, 1979); Elisabeth B. MacDougall and Wilhelmina Feemster Jashemski, eds., *Ancient Roman Gardens*, vol. 7 (Dumbarton Oaks, 1981).
39 Jashemski, *Gardens of Pompeii*, p. 22.
40 Jashemski, *Gardens of Pompeii*, p. vii.
41 Jashemski, *Gardens of Pompeii*, p. 17, citing Mazois, *Les Ruines de Pompei* (Paris, 1812), vol. 2, p. 83 and plate 42.
42 Pliny, *HN*. 19. 60.
43 Jashemski, *Gardens of Pompeii*, p. 24.
44 See 'VI.15.1 Pompeii. House of the Vettii', *Pompeii in Pictures* <https://pompeiiinpictures.com/pompeiiinpictures/R6/6%2015%2001%20plan%202.htm> [accessed 21 August 2024].
45 See 'VI.16.7 Pompeii. Casa degli Amorini Dorati', *Pompeii in Pictures* <https://www.pompeiiinpictures.com/pompeiiinpictures/R6/6%2016%2007%20p5.htm> [accessed 21 August 2024].
46 Jashemski, *Gardens of Pompeii*, pp. 25–26; W. Jashimski and F. Meyer (eds.) *The Natural History of Pompeii* (Cambridge University Press, 2002).
47 See 'Oplontis', *Pompeii* <https://pompeiisites.org/en/oplontis/> [accessed 21 August 2024]; 'Oplontis, Villa di Poppea', *Pompeii in Pictures* <https://pompeiiinpictures.com/pompeiiinpictures/VF/Villa_055%20Oplontis%20Villa%20of%20Poppea%20plan.htm> [accessed 21 August 2024].
48 See for example Ciarallo; M. Mariotti Lippi and Cristina Bellini, 'Unusual Palynological Evidence from Gardens and Crop Fields of Ancient Pompeii (Italy)', in *The Archaeology of Crop Fields and Gardens*, ed. by J.P. Morel, J. Tresserras Juan, J. Carlos Matamala (Edipuglia, 2006), pp. 153–59; Ciaraldi; Clelia Cirillo and others, 'Pompeii-Nature and Architecture', in *Proceedings of the 19th IPSAPA/ISPALEM International Scientific Conference, Naples, Italy*, pp. 719–28. 2015; Daniela Moser and others, 'Archaeobotany at Oplontis: Woody Remains from the Roman Villa of Poppaea (Naples, Italy), *Vegetation History and Archaeobotany*, 22 (2013), pp. 397–408; C. Murphy, G. Thompson, D.Q. Fuller, 'Roman Food Refuse: Urban Archaeobotany in Pompeii, Regio VI, Insula 1,' *Vegetation History and Archaeobotany* 22 (2013), pp. 409–19; Rossella Rinaldi, M. Bandini Mazzanti, and Giovanna Bosi, 'Archaeobotany in Urban Sites: The Case of Mutina,'*Annali di Botanica*, 3 (2013), pp. 217–30.
49 Jashemski, *Gardens of Pompeii*, fig. 152 and see figs. 46–49.
50 There are some very fine photographs on line in Ernesto De Carolis, 'Painted Gardens: Observations on Execution Technique', in *Agrumed: Archaeology and History of Citrus Fruit in the Mediterranean*, dir. by Véronique Zech-Matterne and Girolamo Fiorentino (Centre Jean Bérard, 2017) <https://books.openedition.org/pcjb/2188> [accessed 1 April 2024].
51 Jashemski, *Gardens of Pompeii*, pp. 55–56.
52 Philip Willan, 'Pompeii's Painted Homes Open Frescoes after 40-year Touch-up', *The Times*, 19 February 2020 <https://www.thetimes.co.uk/article/pompeiis-painted-homes-open-frescoes-after-40-year-touch-up-cgrbdohzq> [accessed 1 April 2024].
53 Ciarallo, *Gardens of Pompeii*, pp. 66–72 (plants seen in frescoes in bold).

14
Two Cherry Dishes from Medieval China
A Fruit of Ritual Reciprocity and Sensual Imagination

Zihan Guo

Before European missionaries introduced sweet cherries (*Prunus avium*) into China in the late nineteenth century, Chinese epicures had been relishing their native sour cherries (*P. pseudocerasus*). Different from the Japanese cherry (*P. serrulata*), renowned for its spring blossom, the Chinese cherry is both a flowering tree and a luscious fruit. While the fruit came to be an object of connoisseurship, inspiring artistic presentations and innovative dishes, it was also endowed with ritual, political, and cultural meanings. This essay explores premodern era Chinese cherries (*yingtao*) as represented in archaeological, agricultural, medical, and literary texts. I begin with the Chinese cherry's nomenclature, botanical features, and cultivation practices. The bulk of this essay zooms in on two cherry dishes from medieval China (seventh to thirteenth centuries): cherry dumplings and honeyed cherry, appearing both on elite tables and at daily markets. Their very occurrences reflected the socio-cultural milieux, technological advancements, and linguistic shifts at each historical point. The third section forays into the multivalent symbolisms of Chinese cherries, a fruit of ritual reciprocity and sensual imagination, and concludes with a reflection on the continuities and discontinuities of cherry lore in contemporary China.

Figure 1. Prunus tomentosa. *Photograph by Gidiyorum / Creative Commons / CC BY-SA 2.0 KR.*

'Oriole Peaches'

Nomenclature reveals more than simple designations. In Chinese, there are two terms for cherry: *yingtao* 櫻桃 and *chelizi* 車厘子. The former is a generic name for all cherries, derived from the ancient sense of 'jewel peaches'. The latter is a transliteration of the English word 'cherries', highlighting its imported origin. The foreign-sounding name *chelizi* conjures up an image of big, firm, and crimson cherries that promises sweetness and gustatory satisfaction. Because of their tartness, Chinese cherries almost never appear in markets. In premodern China, however, several kinds of native cherries were recorded, each with their own figurative nicknames: a relatively large and crimson 'Wu cherry' 吳櫻桃 from south-eastern area, a yellow or white 'wax cherry' 蠟櫻, a 'hairy cherry' 毛櫻桃 (*P. tomentosa*, also known as Nanking Cherries, Figure 1), and a dainty bright red 'water cherries' 水櫻桃.[1] Cherries also had two sobriquets associated with birds: *yingtao* 鶯桃 (oriole cherries) and *hantao* 含桃 (sucked cherries). These names relate to the shape of cherries that are so delicate that birds like to hold them in their mouths. The famous Chinese pharmacologist and naturalist, Li Shizhen (1518–1593), kindly advised that one guard the newly ripened cherries, lest they be gobbled by birds![2] For cherry growers today, birds can also be a nuisance as they eat or damage the fruit.[3]

The consumption and cultivation of cherries have a long history in China. In 1973, archaeologists excavated a few dozen seeds in northern China that dated back to *c.* 1300 BC. Among them were several resembling seeds of the aforementioned 'hairy cherries', which might have been used medicinally in antiquity.[4] Descriptions of wild cherry blossoms and grafting techniques have appeared since the sixth century. The earliest extant agricultural treatise, *Qimin yaoshu* (*Essential Techniques for the Commoners*), completed in AD 544, includes brief instructions for grafting cherries: 'In early second month, obtain seeds from mountains. Those from the sunny side should be planted back on sunny land. Those from the shady side should be planted back on shady land'. If the planter does not adhere to this rule, the cherry would either die or not bear fruit.[5] Besides botanical expertise, the terms used to indicate sunny or shady areas, *yang* 陽 and *yin* 陰 respectively, also reflect cosmological beliefs. The principle remains that one should conform to the natural inclination of the fruit, rather than uprooting it entirely from its original environmental conditions. Techniques like this led to the transformation of cherries from a decorative wild plant for visual pleasure to a dish for gustatory enjoyment. By the late seventh century, cherries were highly revered in imperial orchards and, with careful manuring, supplied elite gourmets.

Cherries on the Table of Chinese Epicures

Tang China (618–907) is often depicted as a cosmopolitan era. The newly unified empire prospered with its part-Turkic dynastic family and its multi-ethnic and multi-religious people. The so-called 'Silk Road', a series of transcontinental trade routes, connected China with Central Asia, the Black Sea, and the Mediterranean.[6] New commerce developed with the maritime Southeast Asia as well. Visitors of all sorts – envoys, clerics, and merchants – brought their distinctive foodways to the Tang metropolis Chang'an. Among the imported exotic ingredients, one could find grape wines from the west (native drinks were usually made from fermented rice, millet, or barley), pistachio from Persia, black pepper from India (the native Sichuanese pepper belongs to the family *Rutaceae*), nutmeg (known as 'fleshy cardamom' in Chinese) from Indonesia, bezoar (known as 'ox yellow' in Chinese), human hair from Manchuria and Korea (possibly for magico-medical purposes), and so forth.[7]

Ways of consuming cherries reflected an eclectic taste as well. Eating cherry on its own was already an aesthetic experience that engaged the sensorium for upper echelon consumers. To foreground their bright colour and glowing tint, cherries were typically presented on silver plates and baskets. The famous poet Bai Juyi (772–846) recounted how the fruit was soft and tender enough to be easily pitted and how its succulent juice filled his serving spoon. Apart from artful presentations, people also devised various pairings. Under the influence of Central Asian nomads, dairy products became an important part of Tang diet. Milk was consumed in various forms: cream, yoghurt, kumis, cheese, curds, and butter. Some topped cherry with *lao* 酪, yoghurt or butter.[8] Others, more dietetically minded, paired cherries with sugarcane juice. Medical literature recorded that cherries are 'hot' by nature. To avoid incurring internal heat, the Provision Office at the imperial court prepared sugarcane juice that has a cooling effect.[9] This was a luxury, as the tropical plant from Southeast Asia remained costly in the north through the eighth century.

A most popular yet puzzling dish in the Tang capital was known as 'cherry *biluo*' 櫻桃 饆饠, first recorded in a ninth-century miscellany. What exactly *biluo* 饆饠 was remains a mystery. Some versions of it seem to have included garlic, and ghosts did not like the smell.[10] It must have been a popular street food among Tang urbanites, since there were specialized shops around the Eastern Market in the capital. A certain Han Yue (d. 835), an eminent cook, was known for making cherry *biluo* 'whose colour does not change'.[11] Scholars have speculated that *biluo* could have referred to either a rice or a wheat dish. Based on the phonetic proximity, the Chinese historian Xiang Da opined that it was *pilaf* (also *polo, pilau, plov*, etc.), a rice dish cooked in stock with spice, meat, and other ingredients, common in Central and West Asia and many other regions.[12] *Albaloo polo* (آبالو پلو), a sour cherry rice dish usually served with chicken, is a traditional dish in Persian cookery still favoured among the Persianate societies today.[13] If one considers the Middle Sinitic reconstruction of the written form for *biluo*, *pjit la*, the phonetic resemblance is even more uncanny.[14]

This raises questions of when and where the designation 'pilaf' became widely accepted. The American food historian Charles Perry has noted that techniques for making pilaf first appeared in thirteenth-century Arabic cookbooks and presumably spread in the Near East after the tenth century, possibly even later. The agricultural ecologist Gary Paul Nabhan, however, has pointed out that it was the prominent Persian philosopher and physician, Ibn-Sīnā (c. 970–1037, also known as Avicenna), who left the first record of pilaf preparation during the tenth century.[15] The fact that these textual accounts of pilaf appear later than the Chinese entry does not exclude the possibility that *biluo* could have been pilaf, as culinary practices often predate their formal documentation. Further investigation needs to take into consideration the role of different grains in Tang dietary patterns as well as intertextual references of *biluo*.

Wheat, a Near Eastern cereal first introduced to China near the end of the Neolithic period, had become increasingly popular during the Tang and eventually superseded more traditional staple crops like millet.[16] Grinding wheat flour was facilitated by the development of rotary stone mills during the Warring States period (c. 500–300 BC) and its proliferation in the subsequent Han dynasties (206 BC–AD 220). This technology continued into the Tang and eventually divided China into two dietary zones: granule food in the south and flour food in the north.[17] Since the third century, a type of food called *bing* 餅 became a major dish, referring to wheat dough cooked in different ways (boiled, roasted, baked, fried, or steamed). It is often compared to pasta but can also include bread and flatbread.[18] It had gained such popularity that a third-century poet wrote about rules of seasonal consumption of specific types of *bing* to enhance health.[19] One notable kind often praised by Tang poets was *hubing* 胡餅 (foreign or sesame *bing*), a grilled round flatbread imported from West Asia, likely Persia.[20] Flatbread vendors were common on the streets of the Tang capital. A kind of 'cherry hammer' was recorded, possibly referring to wheat dough kneaded into the shape of a hammer, seasoned with cherries. Because of the prominence of wheat products, some Chinese scholars have proffered that *biluo* most likely referred to a wheat-wrapped food.[21] In fact, there have been archaeological excavations of plum blossom shaped desserts from medieval Turfan in Central Asia. The snack would have been made from flour dough, pressed into the shape of a six-petal plum blossom, and baked, with fruits in its concave centre.[22]

Although cherry *biluo* only make terse appearance in textual sources, references to other kinds of *biluo* are abundant and could corroborate this second view. In the southern frontier (modern Guangdong and Guangxi Provinces), a kind of crab *biluo* was known as a local specialty. A Tang official appointed there recorded the way to make it:

> Inside the shell of the red crab, the yellow and red butter is like the yellow yolk of chicken and duck eggs. Take the white crabmeat and mix with crab butter, fill them inside the shell, sprinkle with seasonings, wrap it with refined flour, and make a crab *biluo*. It is delicious and to be esteemed.[23]

Though this recipe does not indicate the precise cooking method, it does specify the usage of wheat dough as a wrapper. The tenth-century medical treatise, *Taiping shenghui fang* (*Formularies of the Sage Benevolence of the Taiping-Era*), records several *biluo* recipes with fillings of organ meat such as pig liver and goat kidney, which treat problems with stomach and spleen. The physician-compiler instructs to first stir-fry the organs, mix in medicinal spices, use flour to make *biluo*, and bake to finish.[24] Based on these references, one can imagine that cherry *biluo* might have been some kind of cherry dumplings, recalling the sour cherry pierogi and vareniki common in Central and Eastern Europe today.

In the subsequent Song dynasty (960–1279), the cultural milieu changed drastically. The dynastic focus gradually shifted toward the southeast, where aquatic foodways were deeply favoured and dairy products less appealing. This was a time of agricultural, economic, and culinary changes. Tea, rice, and sugar were representative of new dietary patterns, as the thirteenth-century writer Wu Zimu remarked: 'The things that people cannot do without every day are firewood, rice, oil, salt, soybean sauce, vinegar, and tea.'[25] The concomitant commercialization of agriculture led to increases in both production and variety of foodstuffs. Tea came to be consumed as a common beverage at all levels of society (previously it was used mostly for religious and medicinal purposes). New varieties of rice (most notably the Champa rice from what is present-day Vietnam) were introduced. Along with sugarcane, they were produced as important cash crops.

Regional cuisines developed, as reflected in specialized restaurants in the capital cities: northern, southern, Sichuanese, and later Cantonese, forming a simplified overview of the modern pattern.

Whereas the Tang was remarkable for its military prowess and eclectic ambiance, the Song distinguished itself through aesthetic refinement and urbane taste. Culinary abundance was reflected in Song literary sensibilities, bifurcating into paradoxical aesthetics between conspicuous extravagance and principled frugality. Connoisseurship became prominent among writers, some of whom devoted whole monographs to ingredients like mushrooms, crabs, and lychees, cataloguing their varieties and cultivation.[26] While exploring new topics in writing, they also displayed discriminating and experiential knowledge of food. Su Shi (1037–1101), a much-acclaimed Chinese poet, left a long poem, 'Song of An Old Gourmand', describing luxury dining:

> Tasting a piece of fat from pork neck,
> chewing the crab claws before the end of Autumn.
> Boiling pearl-cherries with honey,
> steaming lamb with almond paste . . .[27]

One might be surprised to find the juxtaposition of exquisite snacks like sugared cherries and hearty dishes like steamed lamb, yet such is the contrasting allure of the Song gastronomic world, where every taste could find its own place.

As praised in Su Shi's poem, honeyed cherry became a renowned snack among urban gastronomes. Before the technique of refining sugarcane juice came to China from India in the seventh century, the traditional sweeteners were mainly malt sugar and honey.[28] Honey was used extensively to preserve food, to make mead, and to flavour flatbread during medieval times.[29] Prior to the tenth century, the method of heating honey in a pan was recorded primarily in medical treatises as a way to prepare medicinal substances like ginger, used to treat ailments with digestion. Since the eleventh century, honeyed fruit as a culinary delight came into vogue. A special 'Office for Honeyed Fruits' was established at the imperial court, ensuring sufficient supply for the privileged. Beyond the sweet-toothed elites, honeyed fruits were also street foods widely available in the capital markets.

By the thirteenth century, culinary uses of honey in flavouring fruits had become more sophisticated. One otherwise obscure writer, Chen Yuanjing, active around the mid-thirteenth century, detailed the procedure for preserving fruits with honey. He instructed the gist of this skill: 'Whether it is sour, bitter, pungent, or stiff, one needs to make the preserve in accordance with the nature of the fruits'. He also noted different grades of honey by seasons: honey harvested in the spring is a mixed kind with opaque colour, honey harvested in the winter appears milky white and yet is tart, only the summer honey is distinguished for its pure texture. His recipe for making honeyed cherries reads:

> Cherries. It does not matter exactly how many. Get rid of the pit and place inside
> a stoneware container. First add half *jin* honey [roughly 1.4 pounds] and simmer

> on a slow fire till the juice comes out. Drain them in a bamboo basket and let them dry. Add two *jin* honey again and simmer on a slow fire till they turn into amber colour. Let them cool down. It is best to store them in ceramic containers.³⁰

In general, Chen emphasizes two points: adding honey twice to get rid of the sour juice and storing the product in air-tight containers to avoid insects. This same method can be replicated to preserve other ingredients like lychees, Chinese olives, and bamboo shoots.

A contemporary and equally elusive thirteenth-century writer, Lin Hong, left a slightly different recipe. It is recorded in his cookbook, *Shanjia qinggong* (*Pure Offerings from a Mountain Home*), and reflects an attentiveness to the natural environment and a poetic awareness of taste:

> After rain, worms grow inside cherries, but people cannot see them. Soak cherries in a bowl of water for a while, then the worms will all crawl out. Only then can the cherries be eaten. Yang Chengzhai's [1127–1206] poem says: 'Who has dexterous hands, | smashing ten thousand fragile cherries? | Impress them onto a thin layer of flower shape, | dye with them the purple ice drink. | It is not that northern fruits are scare, | but only this flavour is tempting'. To summarize the gist of making it: simply boil cherries with plum water, pit them, smash them into a paste, and add honey, that is it.³¹

This recipe is quite peculiar. Compared to the previous detailed account, the actual instruction for making the dish in this recipe is quite concise, even nonchalant. Lin's casual tone suggests that his interest lies elsewhere. The first third of the entry provides guidance on properly preparing cherries for cooking, echoing the ideal of 'pure' in the title of his cookbook. Indeed, of the one hundred and four recipes in this work, most are vegetarian, calling for ordinary ingredients like flowers and vegetables and using simple cooking methods. A few even blur the boundary between edible and non-edible, such as boiling pebbles and a string-instrument performance. Another feature of this recipe is the literary allusion to a well-known earlier poet. Lin often references poetry for various reasons: to name a precedent for the dish, to further explain the culinary and medicinal merits, and sometimes even to fault the poetic representation of a dish compared with his own cooking experience. In this way recluses like Lin were able to stay engaged in the literati circle while residing far away from the cultural centre.

The referenced poem illuminates another trait of honeyed fruits – they could be moulded into floral shapes for visual pleasure. Traditionally, cooks would carve on fruits with broad and solid surfaces, such as papaya, and serve them with honey. Besides flowers, they could carve patterns of fish and birds that looked alive. The aforementioned special office had purportedly prepared a papaya carved with a narrative scene from a Chinese legend and presented it to the imperial house. For ordinary consumers, carved fruits were an enjoyment to both the eye and the palate. For elites, they were furthermore an indication of wealth and status to impress their guests. Some of the carved fruits served a purely decorative function and were not to be eaten. The human labour and potential waste behind this practice bespeaks the sumptuousness associated with food beyond its nourishing function.

Gustatory Metonymy and Gendered Fantasies

Because of its dainty shape, red sheen, and portable quality, cherry is commonly featured in artistic and literary creations and speaks to different cultural values. Its symbolism ranges from purity, fertility, and holy blood to eroticism and sexuality.[32] In premodern China, cherry carried complex meanings as well. Compared to other common fruits, cherry matured early in the year, which had earned it the name 'the first fruit of early spring' as well as ritual significance. In mid-summer, the ruler, known as the Son of Heaven, would offer cherries along with young hens and millet to the ancestral shrine, before he himself partook of the new harvest.[33] During the Tang dynasty, the Civil Service Examination flourished as an imperial system whereby the state bureaucracy selected its functionaries. Successful candidates celebrated with Cherry Banquets.[34] Rulers gifted seasonal cherries to ministers as a token of imperial favour and official prospect. Ministers returned the favour through composing poetry, where they elevated cherries to the most wonderful fruit, expressed their gratitude, and displayed their loyalty to the sovereign. For instance, the poet Bai Juyi wrote a long poem about enjoying imperially granted cherries with two of his colleagues in the capital city:

> Like pearls yet unperforated,
> Like flame yet not scorching.
> Apricots are too common to compete,
> Peaches too stiff to match.
> Sneering at loquats' heavy pulp,
> Mocking lychees' wrinkled skin.
> The jewelled ambrosia is filled with tart and sweet flavours.
> Each golden ball is well-proportioned.[35]

The red lustre, sweet aroma, light pulp, smooth texture, mixed flavour of tart and sweet, and exquisite shape together constitute a multi-sensorial tasting experience. Nonetheless, Bai's motivation was not merely composing a panegyric but also in displaying of his loyalty and self-awareness as an imperial subject. He ends the poem in excessive humility and self-deprecation: 'Ashamed most of having not repaid the favour, | this untalented person is fed to satisfaction'. Many contemporaneous poets shared this same repertoire and rhetoric. Writing about cherries came to be a highly symbolic and self-conscious act: it established the poet as a humble and conscientious minister at court and reinserted those exiled outside the capital into the imperial vision.

As Bai's poem stresses, the dainty and scarlet physicality of cherries captured the fancy of Chinese writers, who presented a seemingly paradoxical view between male camaraderie and feminine sensuality. Because of their size, cherries were often associated with mouths, as one might recall Li Shizhen's admonition about birds coveting and sucking cherries. *Jiu Tang shu* (*Old History of Tang*) records that emperor Zhongzong of

Tang (656–710) once took a stroll in the imperial orchard. He invited senior officials and academicians to taste cherries together and asked them to 'mouth-pick' the fresh ones while riding horses.[36] This brief account portrays an ideal monarch-minister relationship and conviviality, embodied in cherries, although the emperor eventually met a tragic death amid political strife.

Where cherries were associated with female beauty, it was often erotic, if not contentious. In the fourth century, the emperor of the state of Zhao was attracted to a singing girl whose name was Zheng Yingtao (Cherry Zheng) and made her his empress. Anecdotes have it that Cherry Zheng twice seduced the emperor to kill two other girls who would be his concubines.[37] Apart from stereotypes of the femme fatale, cherries were also featured in love affairs. The poet Bai Juyi described one of his beloved entertainers, Fan Su, as having a cherry-like mouth.[38] Fan Su excelled at singing 'Tune of Willow Branches'. Bai once composed a eulogy of her performance:

Her mouth moves, the cherry breaks.
Her hair bun hangs, the jadeite ornament dangles.
Like supple twigs is her lissome waist,
Like tender sprouts are her vibrant arms.[39]

Twig-waisted and sprout-armed are classical epithets for female beauty and yet comparing a woman's mouth to a cherry and picturing a dynamic scene were an innovation that continued to hold currency. One recalls how Shakespeare (c. 1564–1616) had Demetrius confess to Helena's 'kissing cherries' and how Édouard Manet (1832–1883) painted his street singer holding cherries toward her mouth, seemingly replacing it.[40] Whereas male elites would mouth-pick cherries, female performers were imagined as having cherry-like mouths. The eroticized dynamic therein awaits further explication.

In contemporary China, people relish the imported sweet cherries in supermarkets, marvelling at their crimson and robust appearance. The native sour cherries have retreated to suburban areas, hidden from general consumers' sight. Their tartness and fragility render them not particularly marketable. One would need to personally pick them in mountainous areas. As the Civil Service Examination was abolished in the early twentieth century, the ritual significance of cherries also gradually waned, but the poetic epithet of cherry-mouth lives on, as it does in other cultures. Rather than an ingredient that seasons different dishes, cherry has become a leisurely fruit to be snacked on its own (Figure 2), satisfying epicures' appetites with juicy sensations and sweet flavour. Just as food is never purely for sustenance but also brimmed with cultural implications, writing about food is never simply hedonist but bears the personal imprint of the writer. The mystery of cherry *biluo*, yet to be fully unravelled, speaks to the cultural exchanges between the East and the West in medieval times. Honeyed cherries, in turn, reflected Chinese connoisseurs' sensibilities to fruit and poetry. The story of Chinese cherries is one of a hidden past, with its paradoxical allure of ritual gravity and sensuous fantasy.

*Figure 2. Chilean cherries (*Prunus avium*). Author's own photograph.*

About the Author
Zihan Guo is a PhD student at Princeton University in the United States, studying pre-modern Chinese literature, food, and medicine.

Notes
1. For most common cherry species found worldwide, such as sweet cherry and sour cherry, see Constance L. Kirker and Mary Newman, *Cherry* (Reaktion Books, 2021).
2. Xu Guangqi, *Nongzheng quanshu jiaozhu* (Zhonghua shuju, 2020), xxx, pp. 1036.
3. Kirker and Newman, pp. 39–40.
4. Geng Jianting and Liu Liang, '*Gaocheng Shangdai yizhi zhong chutu de taoren he yuliren*', *Cultural Relics*, 8 (1974), pp. 54–55.
5. Jia Sixie, *Qimin yaoshu jinshi* (Zhonghua shuju, 2009), iv, pp. 346.
6. The term 'Silk Road' was first coined by a German geographer, Ferdinand von Richthofen, *Seidenstrasse*, in 1877. However, there was not one continuous road but a network of markets, nor was silk the most important traded good. See Valerie Hansen, *The Silk Road: A New History* (Oxford University Press, 2012). For a recent new interpretation of the Silk Road as a diplomatic route, see Xin Wen, *The King's Road: Diplomacy and the Remaking of the Silk Road* (Princeton University Press, 2023).
7. Edward H. Schafer, *The Golden Peaches of Samarkand: A Study of T'ang Exotics* (University of California Press, 1963), pp. 139–54, 176–94.
8. It could also refer to cream. See Shinoda Osamu, *Chūgoku shokumotsu shi* (Shibata Shoten, 1974), pp. 215.
9. Wang Wei, '*Chici baiguan yingtao*', in *Wang Wei ji jiaozhu* (Zhonghua shuju, 1997), iv, pp. 303.
10. Duan Chengshi, *Youyang zazu jiaojian* (Zhonghua shuju, 2015), vii, pp. 607.

11   This is a literal translation of the original terse account. It is unclear as to which substance's colour does not change (the cherry or the *biluo*). Following the interpretation of *biluo* as wheat-wrapped food, I suspect that it is the wrap that does not get tinted by the cherry filling. See Duan, vii, pp. 607.
12   Xiang Da, *Tangdai Chang'an yu xiyu wenming* (Sanlian shudian, 1957), pp. 49–50.
13   Thanks to Nader Mehravari for kindly informing me of this practice.
14   For a philological enquiry into *biluo* and its possible relations with pierogi and pilaf, see Victor H. Mair, 'A Medieval Chinese Cousin of Eastern European Cherry Pierogi?' *Language Log*, 14 August 2024 <https://languagelog.ldc.upenn.edu/nll/?p=65400> [accessed 14 August 2024].
15   Despite general scholarly acknowledgement, these are both paraphrases without actual quotations from the historical records: *The Oxford Companion to Food*, ed. by Alan Davidson (Oxford University Press, 1999), pp. 624–25; Nabhan, *Cumin, Camels, and Caravans: A Spice Odyssey* (University of California Press, 2014), pp. 135.
16   Francesca Bray, *Agriculture*, in *Science and Civilization in China*, ed. by Joseph Needham, 7 vols (Cambridge University Press, 1984), vi, 2, pp. 459–65.
17   Hsing-Tsung Huang and Joseph Needham, *Fermentations and Food Science*, in *Science and Civilization in China*, ed. by Needham, vi, 5, pp. 462–65.
18   Silvano Serventi, Françoise Sabban, and Antony Shugaar, *Pasta: The Story of a Universal Food* (Columbia University Press, 2002), pp. 271–344.
19   Serventi, Sabban, and Shugaar, pp. 292–93.
20   Persian bread (*nan*) reached China probably in the eighth century. Chinese records indicate that Persian refugees from the Muslim conquest were selling *nan* in Tang capital. See Paul D. Buell and others, *Crossroads of Cuisine: The Eurasian Heartland, the Silk Roads and Food* (Brill, 2020), pp. 182–84.
21   Lu Rui, 'Zhuafan haishi bobo: biluo kao', *Xinjiang daxue xuebao*, 43.3 (2015), pp. 140–44.
22   *Secrets of the Silk Road: An Exhibition of Discoveries from the Xinjiang Uyghur Autonomous Region, China*, ed. by Victor H. Mair (Bowers Museum, 2010), pp. 123.
23   Liu Xun, p. 6. For a modern reconstruction of this recipe, see Shiyi boshi Dr Shiyi, '*Tangdai guizu hao zhekou?*' *YouTube*, 7 November 2022 <https://www.youtube.com/watch?v=p5d4wsxiARQ> [accessed 29 July 2024].
24   Wang Huaiyin, *Taiping shenghui fang* (Guojia tushuguan, 1980), xcvii, p. 9772.
25   Michael Freeman, 'Sung', in *Food in Chinese Culture: Anthropological and Historical Perspectives*, ed. by Kwang-chih Chang (Yale University Press, 1977), pp. 141–76 (pp. 146–51).
26   Eugene N. Anderson, *The Food of China* (Yale University Press, 1988), pp. 83–85.
27   Su, '*Lao tao fu*', in *Su Shi wenji* (Zhonghua shuju, 1986), i, pp. 16–17.
28   For an account of how Indian Buddhists helped transmit the technology of extracting sugar from sugarcane juice, see John Kieschnick, *The Impact of Buddhism on Chinese Material Culture* (Princeton University Press, 2002), pp. 249–62.
29   Huang and Needham, p. 246.
30   Chen, *Shilin guangji* (Zhonghua shuju, 1999), vii, pp. 112–14. For a modern reconstruction of honeyed cherry, see Li Xu, Yasheng Zheng and Ran Lu, *Delicacies of the Song Dynasty* (CITIC Press Group, 2024), pp. 94–97.
31   Lin, *Shanjia qinggong*, *Shuofu* (Shangwu yinshuguan, 1927), pp. 18–19. In a later sixteenth-century edition, honey is substituted by white sugar: see Lin, *Shanjia qinggong*, *Yimen guangdu* (Yiwen yinshuguan, 1966), pp. 30–31.
32   Kirker and Newman, pp. 129–67.
33   Sun Xidan, *Liji jijie* (Zhonghua shuju, 1989), xvi, pp. 451–52.
34   Fu Xuancong, *Tangdai keju yu wenxue*, (repr. Zhonghua shuju, 2020), pp. 296–312.
35   Bai, '*Yu Shen Yang er sheren gelao tongshi chici yingtao wanwu ganen yin cheng shisiyun*', in *Bai Juyi shiji jiaozhu* (Zhonghua shuju, 2006), xix, pp. 1558–60.
36   Liu Xun, *Jiu Tang shu* (Zhonghua shuju, 1975), vii, p. 149.
37   The same historical source presents Cherry Zheng in two different identities: a servant boy and a singing girl: see Cui Hong, *Shiliuguo Chunqiu jibu* (Zhonghua shuju, 2020), xvi, pp. 195, 200.
38   Meng Qi, *Benshi shi* (Zhonghua shuju, 2014), p. 88.

39 Bai, *Bai Juyi shiji jiaozhu*, xxxii, p. 2453.
40 William Shakespeare, *A Midsummer Night's Dream*, ed. by Barbara Mowat and others (Folger Shakespeare Library, n.d.), III.2.143 <https://www.folger.edu/explore/shakespeares-works/a-midsummer-nights-dream/read/3/2/> [accessed 15 August 2024]; Janet Beizer, 'Urban Nomads and Vagabond Food: Eating Away from the Table in Nineteenth-Century France', in *Portable Food: Proceedings of the Oxford Symposium on Food and Cookery 2022*, ed. by Mark McWilliams (Prospect Books, 2023), pp. 29–42 (pp. 30–35).

15
Growing Rice, Growing Taste

Chu Hao Pei

Introduction

飯醉集團, also known as *Secret Rice Society*, is an experimental guerrilla rice garden which sources heirloom rice seeds from an informal rice seed exchange network in Southeast Asia. *Secret Rice Society* was started at the height of COVID-19 pandemic lockdown in Singapore, when the trend of home gardening picked up as people were confined to their homes. Most residents grew ornamental plants or herbs to improve their mental health or to supplement their daily food consumption. In contrast to these common plant growers, I had a different purpose in mind – preserving heirloom rice varieties. Hence, I began searching for a space to grow my heirloom rice seeds. Due to the lack of utilitarian spaces and the prevalence of untapped spaces on the rooftops of multistorey carparks in Singapore's public housing estates, rooftop gardens could be built to provide residents a recreational space, while community gardens could introduce a social, communal space to build cohesion among residents.[1] However, the strict policing of planting in public green spaces by the authorities restricted the amount of public gardening spaces available: only designated spaces are allowed for public gardening.[2] The socially engineered design of public gardening spaces combined with the strict landscape control of the state led to the rooftop garden of a multistorey car park in Toa Payoh, an old public housing estate in Singapore where the *Secret Rice Society* took root (Figure 1).[3] I spontaneously started growing the heirloom rice seeds obtained from my informal rice seed network built from my personal connections.

This informal transregional exchange network is centred around the distribution of heirloom rice seeds, which tracks the peer-to-peer exchange and distribution processes using mail. *Secret Rice Society* is the focal point and the overseas custodian of heirloom rice seeds in this informal transregional network, and thus serves as a mode of heirloom rice seed preservation. Commons-based seed practices and initiatives are omnipresent, 'spanning from traditional seed systems (such as seed exchange networks or community seed banks) to recent anti-enclosure movements (such as open-source seeds and organic breeding initiatives) that resist intellectual property rights on varieties'.[4] Sharing similar values, *Secret Rice Society*'s informal networks of heirloom rice seed contributors grew over time with people in the region. My interaction with the contributors embodies the mutual trust system which *Secret Rice Society* relies on, such that we will not share the seeds to anyone without affirming their intentions. *Secret Rice Society*'s rice seeds collection includes heirloom rice seeds from Indonesia, Myanmar, Thailand, and South Korea.

Figure 1. 飯醉集團 Secret Rice Society *before and after. Image courtesy of Chu Hao Pei.*

The name 飯醉集團 (*Secret Rice Society*) shares the same pronunciation (*fàn zuì jí tuán*) as crime syndicate 犯罪集團 in Mandarin Chinese – a wordplay that simultaneously demonstrates the 'illegal' guerrilla activities of the informal rice seeds exchange network and the garden space it occupies. Through the guerrilla act of heirloom rice seeds mail exchange, *Secret Rice Society* circumvents the illegality of undeclared seed movement within state-controlled systems. Ironically, these same systems approved the planting of hybrid or genetically modified seeds on lands that have since lost many heirloom rice varieties. *Secret Rice Society* is an artist movement that challenges and questions the dichotomy of legal and illegal introduction of heirloom seeds in the wider discussions of rice seed sovereignty. In this paper, I will explain the beginnings of the *Secret Rice Society*, the history of seed sovereignty in Asia, and why there is a need to look into heirloom rice varieties. I will then show how the processes of *Secret Rice Society* influenced my artistic practice and incubated subsequent artworks. Finally, I argue that participatory artworks bring to light the issue of seed sovereignty and address the loss of heirloom rice varieties, as well as the loss of agrarian culture and knowledge attached to heirloom rice.

Seeding the *Secret Rice Society*

The seed of *Secret Rice Society* was first planted when I was invited to the project 'Against the Dragon Light' by South Korean curators Eugene Park and Moonseok Yi in 2021. The project evolved into a mail art project as a result of the COVID-19 pandemic, and I was paired with Rice Brewing Sisters Club (RBSC), a South Korean artist collective, for the mail art exchange. We agreed on rice seeds for the mail art exchange as they are most aligned with our respective artistic practices and ongoing interests. They are also small enough to be delivered through mail packages. To circumvent the absence of heirloom rice varieties in Singapore, I sourced patented hybrid rice seeds, SAT-15H, from Singapore's Golden Sunland.[5] RBSC, in return, sent three heirloom rice varieties – 각시나 *Gakshina* (early maturing glutinous rice), 대관도 *Daegwando* (late maturing glutinous rice) and 가투리찰 *Ggaturichal* (late maturing glutinous rice) – with brief historical and geographical information about each variety. The additional information was important because it gave the heirloom rice seeds an identity, and sharing interesting stories behind each variety grew my interest in heirloom rice varieties.

According to Vandana Shiva, seed sovereignty:

> includes the farmer's rights to save, breed and exchange seeds, to have access to diverse open source seeds which can be saved – and which are not patented, genetically modified, owned or controlled by emerging seed giants. It is based on reclaiming seeds and biodiversity as commons and public good.[6]

The concept of seed sovereignty is not new and has gathered momentum over the last few decades in a resistance against the growing corporate dominance in seeds control as championed by environmental activists like Shiva, who has spent more than four decades on the seed sovereignty movement.[7]

Drawing on Shiva's inspiration and my growing interests in seed sovereignty, I decided to start a guerrilla rice garden as a living rice seed library, growing only heirloom rice varieties starting with the heirloom rice seeds from RBSC. The act of growing expresses a desire to further the continued heritage of the endangered heirloom rice. During the process, I picked up invaluable vernacular rice growing knowledge – labour practices, weather observation, moisture level, soil condition, pest control, and the right time to harvest. I adopted a chemical-free method to grow the rice using only food scraps and manure as organic fertilizers. Without any prior knowledge in growing rice, my first attempt was surprisingly successful despite coming with its set of challenges. Pests like sparrows fed on the rice grains while aphids and ants infested the rice plants and roots. *Secret Rice Society* also faced eviction in the process, but was quickly resolved after consultations with the relevant agencies (Figure 2). The rice growing process and the development of the informal heirloom rice seeds exchange network of *Secret Rice Society* is a vital cog in the manifestations of subsequent projects.

The need to reclaim heirloom rice varieties is urgent. Many heirloom rice varieties have been lost since the introduction of high-yielding varieties (HYVs) during the Green Revolution. In India alone, more than one hundred thousand heirloom rice varieties

UNAUTHORISED PLANTING AT PLANTER BOXES

NOTICE 07.03.2023

Bishan-Toa Payoh Town Council

BTPTC has sighted unauthorised planting of plants at the planter boxes by residents. This is an infringement to the Town Council by-laws 12(1) on common property and open spaces.

We seek your co-operation to remove all the plants and items from the planter boxes by **21st April 2023**. <u>Any remaining plants and items will be removed by BTPTC after the stipulated deadline.</u>

For further clarifications, please call BTPTC office at Tel: 62596700. Thank you for the co-operation.

Our Estate • Our Home
Let's work towards a Clean & Green zone!

Figure 2. Warning sign to evict Secret Rice Society. *Image courtesy of Chu Hao Pei.*

have been lost.⁸ The Green Revolution, as the name suggests, 'revolutionized' agriculture because it radically changed agricultural yields and the way farmers have been cultivating for centuries, but it came with its own set of problems with many unprecedented adverse effects that surfaced decades later.⁹ Focusing only on the issue of rice seeds, hybridized and genetically modified HYVs were aggressively marketed and introduced to farmers for higher yields. The seeds are genetically sterilized, making it impossible to grow again after harvest, which puts the farmers at the mercy of rice seed companies.¹⁰ As a result, farmers become permanently reliant on seed companies.¹¹

Rice is also suffering from extreme weather patterns, pest attacks, and polluting farmlands that are all worsened by the current global climate crisis, triggered by centuries of industrialization and unsustainable monoculture agriculture. While we are scrambling to find solutions to rectify these issues using science and technology, an alternative movement – the Indigenous Regenerative Agriculture Movement – searches for more adaptive and less destructive solutions for growing food.¹² Relearning our ancestral sustainable practices and knowledge in rice growing is ever more important as we turn to age-old wisdom to address our food security concerns. We need to identify, source, and start cultivating our heirloom varieties to unlock this knowledge.

Heirloom rice varieties not only provide sustenance to communities, but they also play important roles in spiritual rituals and social currency.¹³ Cultures and rituals developed in rice fields by farmers alongside the heirloom rice varieties in respective lands for centuries; hence, the replacement of heirloom rice varieties not only disrupted the farmers' farming practices but also altered their cultural practices. As a result, the preservation of heirloom rice varieties is not just a food security affair but also extends to cultural preservation.

Incubation: 'Do Seeds Control Us or We Control Seeds?' Workshop

I was invited to be part of the pilot artist residency programme under the Community and Research Residency by Singapore Art Museum in 2021.¹⁴ During the artist residency period, I conducted a workshop with David Chen, co-founder of The Little Rice Company, to explore Singapore's relation with rice and seed sovereignty. The workshop was framed in an interactive setting with a brief discussion of the introduction of HYVs rice during the Green Revolution and an overview of the seed sovereignty movement worldwide. It was followed by a sensorial exploration on the heirloom rice seeds grown in *Secret Rice Society* and rice packaging found in Singapore (Figure 3a–3d). I wrote a reflective piece, '*Air Tajin* (Rice Water): A Speculative Disappearing Recipe', lamenting the disconnection of the participants from the issue of seed sovereignty and rice farming in general due to the lack of agriculture in Singapore.¹⁵ The workshop forced me to rethink the urgency of seed sovereignty and importance of heirloom rice varieties in the context of the urban community. The interesting takeaway from the workshop was that the participants could only relate to rice through rice packets which probably explains the disengagement during the workshop.

Figure 3a. Do Seeds Control Us or We Control Seeds? *workshop at City Sprouts on Farm Day Out. Image courtesy of Singapore Art Museum.*

Figure 3b. A palette of heirloom rice seeds and heirloom rice in Do Seeds Control Us or We Control Seeds? *workshop at City Sprouts on Farm Day Out. Image courtesy of Singapore Art Museum.*

Gardens, Flowers, and Fruit

Figure 3c. Examining profile of rice in Do Seeds Control Us or We Control Seeds? *workshop at City Sprouts on Farm Day Out. Image courtesy of Singapore Art Museum.*

Figure 3d. Details of Do Seeds Control Us or We Control Seeds? *workshop at City Sprouts on Farm Day Out. Image courtesy of Singapore Art Museum.*

From Guerrilla Rice Garden to Art Engagements
Grind, Boil, Smell, and Taste

The workshop 'Do Seeds Control Us or We Control Seeds?' led me to conceptualize the socially-engaged artwork *Grind, Boil, Smell, and Taste*. Participants were asked to bring a handful of rice from their homes which I used to activate the session through a series of gestures in preparing rice tea. They are first invited to observe the profiles of the rice they brought and differentiate it from other types of rice. Under my instructions, each of them grinds their rice in the mortar and pestle and pours it into the clay pot where I will cook it. Building a connection around the rice tea, I shared that I was fed this rice tea as an infant to supplement my diet and that this is an old practice in my culture. I also asked the participants if they had the similar drink as an infant, in an attempt to bridge the participants closer to the rice. The responses were mixed. As the rice cooks, the fragrance of the rice overtakes the air while I scoop the rice tea for the participants. Before drinking, I emphasized first that they should smell the fragrance of the rice tea.

Each of these sensorial elements – touch, smell, and taste – contributes to the relational exchange that the participants have with the rice, together with my storytelling and prompts. The process of preparing the rice tea draws inspiration from the Japanese tea ceremony and is a considered decision to let the participants appreciate the rice more while also tapping into the sensorial elements to enhance the experience (Figure 4a).

The rice packets collage in the installation acts as an intermediary for me to invite the participant to identify the rice they brought (Figure 4b). While we appreciate the rice tea, a mapping exercise is done together to identify the sources based on the rice packet information, which serves as an interactive pedagogical trigger for me to share about

Figure 4a. Preparation of rice tea in Grind, Boil, Smell, and Taste *art installation. Image courtesy of Chu Hao Pei.*

the rice trade, the dominance of HYVs, and the lack of heirloom rice in the Singapore rice market under the context of food security. It becomes an entry point to introduce the topic of seed sovereignty to the agriculture-deprived participants of Singapore. This methodology allows me to weave my long-term research on rice into more personal and customized interactions with participants, thereby unpacking the multiple layers in Singapore's rice sector and complex issue of seed sovereignty.

The failure to build that connection between the participants and rice in the earlier workshop prompted me to improvise my methods to make it more relational. Several prompts were introduced, such as the invitation to bring rice from home, the labour of grinding rice before cooking, and the rice packet mapping exercise. These methods draw the participants' curiosity and also create their relational connection with the common grain. The context is also set in the rice packet mapping exercise where we learn about the types and sources of rice available in Singapore through an interactive

Figure 4b. Grind, Boil, Smell, and Taste *art installation. Image courtesy of Chu Hao Pei.*

setting. Establishing that context and relational connection opens up a space where I could share and unpack complex issues such as seed sovereignty and heirloom rice varieties. The session builds up over several considered gestures and dialogues as it transitions from one phase to another as I seed a certain set of knowledge to the unaware participants in the interactive session.

Seeding Sovereignty

Seeding Sovereignty is an interactive art installation that emerged from the processes of *Secret Rice Society* and especially the nature of the garden as a living seed library. Commissioned for the exhibition *Lonely Vectors* by the Singapore Art Museum, this art installation was exhibited in different public libraries in Singapore. The art installation includes three components – an archival video on the unveiling of 'IR8 Miracle Rice' by former US President Lyndon B. Johnson at the International Rice Research Institute (IRRI) (Figure 5a), scarecrows with placards inspired by different farmers' protests around the world (Figure 5b), and a rice seed library (Figure 5c).

Figure 5a. Display of archival video 'US President Lyndon Johnson visits IRRI, 26 October 1966' in Seeding Sovereignty *art installation at* Lonely Vectors *exhibition. Image courtesy of Singapore Art Museum.*

Figure 5b. Scarecrows with protest placards in Seeding Sovereignty *art installation at* Lonely Vectors *exhibition. Image courtesy of Singapore Art Museum.*

Figure 5c. Details of visitors interacting with rice seed library in Seeding Sovereignty *art installation at* Lonely Vectors *exhibition. Image courtesy of Singapore Art Museum.*

The archival video provided the historical context of the transformation of rice agriculture since the Cold War and the charged history of the introduction of 'Miracle Rice' at the dawn of the Green Revolution. Juxtaposing the archival video with the scarecrow protests placards inspired from modern day protests brought to light the polarized, decades-long debate on the effects of the Green Revolution on food security through the lens of 'Miracle Rice'.

The rice seed library grew out of recognizing the library as a knowledge sharing and learning space that could expand the impact of the living rice seed library in *Secret Rice Society*. The function is to distribute rice seeds and share lesser-known knowledge about rice in the fields of history, culture, politics, and local rituals cited from a diverse range of literary sources from the library. These sources span regional poems, childhood song lyrics, political slogans, cultural rituals, and archival records. The rice seed library takes the form of wooden cabinets where each compartment includes selected and edited write-ups with illustrations and rice seed packets. Visitors are encouraged to open the drawers to learn about the uncommon knowledge of rice and then to take the rice seed packets home to grow them. The content in the drawers is intentionally designed for different age groups where the lower drawers, more likely accessed by kids, contain less text and have more illustrations while the upper drawers are more text-heavy (Figure 5d).

As demonstrated in my rice growing processes in *Secret Rice Society*, knowledge exists in multiple forms and can be disseminated in unconventional ways. Each seed

Figure 5d. Content and rice seed packets inside rice seed library in Seeding Sovereignty *art installation at* Lonely Vectors *exhibition. Image courtesy of Singapore Art Museum.*

encapsulates knowledge, but a knowledge that can only be harnessed by the act of growing. The visitors are encouraged to acquire that knowledge by growing rice following the instructions inside the rice seed packets. A pact with the contributors of the informal heirloom rice seed exchange network, especially given the sensitivity of heirloom rice seeds, led to a considered decision to pack non-heirloom rice seeds in the rice seed packets for distribution.

Distributing these rice seed packets garnered an enthusiastic response from the public. A gallery sitter shared that kids who interacted with *Seeding Sovereignty* in earlier encounters returned to inform them with excitement that they successfully grew rice from the rice seed packets. Some returned to take more rice seed packets to grow again. The rice seeds packets distribution is a carefully planned gesture to motivate the act of growing which has had lasting impacts on the visitors. First, the rice seeds in the packet have broken the material barrier between the visitors and rice. For many visitors, it was likely their first time to touch and feel rice in its unhusked form as rice is only experienced for consumption in Singapore due to the absence of rice fields. Second, the planting of rice seeds, following the planting instructions in the rice seed packets, is an ode to the rice farmers and an invitation to the urban dwellers to venture into unfamiliar grounds to grow their own rice. Therein lies the power of art in convincing visitors to grow rice on such a small scale that bears insignificant or almost no returns for consumption. Finally, the performativity of labour and care on the rice plants through the rice growing process imparts invaluable knowledge to the growers. Regardless of the success rate, the act of growing has not only planted the seeds in the soil but also figuratively in the hearts of the growers. The latter is what *Seeding Sovereignty* aims to achieve – sharing the knowledge and, inevitably, the struggles of rice farmers. In other words, the rice seeds, like books in the library, are also knowledge carriers that require different ways to access that knowledge.

Seeding Sovereignty engages library visitors in multifaceted ways to access the knowledge of rice. The growing of the rice seeds is an important first step in developing a personal connection between visitors and rice, while the other components of *Seeding Sovereignty* bring to light the larger and more complex issue of seed sovereignty. The act of growing should not be underestimated as it creates a more holistic approach in understanding and learning beyond the exhibition premise, similar to my experience in growing heirloom rice in *Secret Rice Society*. While it takes more time and concerted effort to fully apprehend the complex issue of seed sovereignty, distributing rice seeds through *Seeding Sovereignty* aims to shed light on an otherwise heavy topic to a less aware public.

Growing Rice, Growing Taste

Growing Rice, Growing Taste was an event and exhibition in collaboration with Daniyah Az Zahra and Sukabumi Farm. *Growing Rice, Growing Taste* presented a new series of artworks in the medium of rice packaging designs inspired by the historical moments of political rice campaigns which introduced HYVs and displaced many heirloom rice varieties in Southeast Asia during the Cold War. A rice packet design workshop and heirloom rice tasting session (Figure 6) used the recollection of lost taste in heirloom rice varieties to reposition heirloom rice varieties in a hybrid and genetically modified rice dominated market. *Growing Rice, Growing Taste* presented an encounter for participants to discuss the issue of seed sovereignty through the lost taste of heirloom varieties, namely West Java heirloom rice varieties – Cere, Hea, and CN.

Against the backdrop of the rice packaging inspired by rice's visual politics in Southeast Asia, I conducted a rice packet design workshop for local heirloom rice varieties. After sharing the inspiration behind the rice packet artworks, participants were invited to design rice packets for their local heirloom rice varieties. Inspirations were drawn from local cultural emblems, rice architecture, and other rice packet designs as the participants sought ways to represent 'localness'. The rice packet design workshop engaged the relatively young participants in a visual way before the heirloom rice tasting session. Through the rice packaging design workshop, I attempted to raise heirloom rice's profile and to evoke participants' sense of attachment to the forgotten rice.

The tasting session engaged the audience more deeply as their sense of taste was put to the test. For many, it was their first time to taste local heirloom rice varieties. When opinions of the rice taste were solicited, a strange feeling engulfed the room, because the taste of the local heirloom rice varieties felt foreign to the participants. This awkward confusion was felt as most participants found it hard to relate to this forgotten taste. There were, though, some attempts to share their thoughts on heirloom rice varieties based on the taste profiles. One participant shared that the heirloom rice tasted sweeter and had a chewier taste profile than the regular rice found in the market. It

Figure 6. Rice tasting session at Growing Rice, Growing Taste *art installation. Image courtesy of Chu Hao Pei.*

was difficult to get more feedback from participants on the newly introduced heirloom rice varieties.

    I have to consider several factors for the lack of responses in the participants. The rice tasting session involved just plain rice despite the availability of other foods to complement the rice. The spectrum of taste in rice is limited as it does not involve the flavours that our tongue is used to – sweet, salty, sour, bitter, and umami. Hence, one's tastebud has to be very sensitive to distinguish any subtle taste difference in heirloom rice. The other factor to consider is also how the participants process the new taste of the rice. This may take time: participants may find it difficult to offer any thoughts or feedback during the session. There is also a lack of rice knowledge, even though most rice eaters consume rice almost every day. Apart from rice farmers, middlemen, cooks, and rice researchers, most consumers may not be able to identify more than a handful of rice varieties despite the plethora of rice varieties, especially heirloom rice, available. This shortcoming does not only apply to rice but also to other common food products, such as potatoes, and is largely the result of the Green Revolution identifying and hybridizing new uniform varieties that grow fastest with highest yields. This paradigm shift has significantly reduced our food diversity, and, in this case, rice, contributed to the lack of knowledge about rice in the public.

Despite the limitations of *Growing Rice, Growing Taste*, the combination of the rice packaging design workshop and the heirloom rice tasting session has opened new avenues to introduce 'forgotten' local heirloom varieties to a group of local participants. There is more work to be done to spread these heirloom rice varieties, but *Growing Rice, Growing Taste* has seeded some possibilities in engaging heirloom rice to a wider public.

Conclusion

Secret Rice Society, as a guerrilla rice garden growing heirloom rice, has incubated several art projects which attempt to address the loss of heirloom rice varieties and the knowledge encapsulated within a seed. The rice seeds are central themes in the three case studies, which explore different sensory experiences and methods with different groups of participants to share multi-layered knowledge embedded through shared actions. Seed sovereignty remains a large issue that needs a wider collective action to promote, and the art projects incubated by *Secret Rice Society* provide avenues for creative expressions to contribute to this larger movement in the revival of heirloom rice.

About the Author

Chu Hao Pei is a visual artist whose works are primarily influenced by his long-standing interest in the interactions between culture and the environment. Hao Pei's practice explores the shifting physical, sociological, and emotional connections with our natural and urban landscapes.

Notes

1. Koon Hean Cheong, 'Creating Liveable Density Through a Synthesis of Planning, Design and Greenery', in *Dense and Green Building Typologies: Research, Policy and Practice Perspectives*, ed. by Thomas Schröpfer and Sacha Menz (Springer, 2019), pp. 7–12 (p. 10), DOI:10.1007/978-981-13-0713-3_3; Leon H.H. Tan and Harvey Neo, '"Community in Bloom": Local Participation of Community Gardens in Urban Singapore', *Local Environment*, 14.6 (2009), pp. 529–39 (p. 530), DOI:10.1080/13549830902904060.
2. Marvin Joseph F. Montefrio, Xin Run Lee, and Elwin Lim, 'Aesthetic Politics and Community Gardens in Singapore', *Urban Geography*, 42.10 (2021), pp. 1459–79 (p. 1470), DOI:10.1080/02723638.2020.1788304.
3. 'Secret Rice Society 飯醉集團 · 130 Lor 1 Toa Payoh, Floor 8, Singapore 311128', *Secret Rice Society* 飯醉集團 · *130 Lor 1 Toa Payoh, Floor 8, Singapore 311128* <https://maps.app.goo.gl/Ce7gMhs6x4EK3Sn26> [accessed 14 May 2024].
4. Stefanie Sievers-Glotzbach and others, 'Diverse Seeds – Shared Practices: Conceptualizing Seed Commons', *International Journal of the Commons*, 14.1 (2020), pp. 418–38 (p 418).
5. GOLDEN SUNLAND SINGAPORE PRIVATE LIMITED, 'PVP Journal No. 04/2021', 2021 <https://isomer-user-content.by.gov.sg/61/ac6fe424-8ca0-47f8-af38-a442c6653d79/plant-varieties-protection-journal-04-2021.pdf> [accessed 14 May 2024].
6. Vandana Shiva, 'Opinion: The Seed Emergency', *Al Jazeera*, 6 February 2012 <https://www.aljazeera.com/opinions/2012/2/6/the-seed-emergency-the-threat-to-food-and-democracy> [accessed 14 May 2024].
7. 'The Seed Sovereignty Movement: Reclaiming the Seed', 2022 <https://www.chelseagreen.com/2022/reclaiming-the-seed-history-of-the-seed-sovereignty-movement/> [accessed 14 May 2024].
8. Ann Raeboline Lincy and others, 'The Impact of the Green Revolution on Indigenous Crops of India', *Journal of Ethnic Foods*, 6.8 (2019), p. 1, DOI:10.1186/s42779-019-0011-9.

9 Daisy A. John and Giridhara R. Babu, 'Lessons from the Aftermaths of Green Revolution on Food System and Health', *Frontiers in Sustainable Food Systems*, 5 (2021), DOI:10.3389/fsufs.2021.644559; Lincy and others 2019.
10 A.S. Bawa and K.R. Anilakumar, 'Genetically Modified Foods: Safety, Risks and Public Concerns – a Review', *Journal of Food Science and Technology*, 50.6 (2013), pp. 1035–46, DOI:10.1007/s13197-012-0899-1.
11 M Behrokh Mohajer Maghari and Ali M. Ardekani, 'Genetically Modified Foods and Social Concerns', *Avicenna Journal of Medical Biotechnology*, 3.3 (2011), pp. 109–17 (p. 114).
12 Marta Fiolhais, 'The Indigenous Roots of Regenerative Agriculture', *Rainforest Alliance*, 2023 <https://www.rainforest-alliance.org/insights/the-indigenous-roots-of-regenerative-agriculture/> [accessed 14 May 2024].
13 J. Peter Brosius, 'Significance and Social Being in Ifugao Agricultural Production', *Ethnology*, 27.1 (1988), pp. 97–110 (p. 101), DOI:10.2307/3773563; Subir Bairagi and others, 'Preserving Cultural Heritage through the Valorization of Cordillera Heirloom Rice in the Philippines', *Agriculture and Human Values*, 38.1 (2021), pp. 257–70 (p. 258), DOI:10.1007/s10460-020-10159-w.
14 'Pilot for SAM Residencies', *Singapore Art Museum* <https://www.singaporeartmuseum.sg/en/residencies/announcements/pilot-for-sam-residencies> [accessed 14 May 2024].
15 Chu Hao Pei, 'Air Tajin (Rice Water): A Speculative Disappearing Recipe by Chu Hao Pei', *Farm Day Out Singapore Art Museum* <https://samplings.sg/all-posts/air-tajin-rice-water-disappearing-recipe-by-chu-hao-pei> [accessed 14 May 2024].

16
On Gardens, Flowers, and (Possibly) Fruit in Cheese

Ursula Heinzelmann

When the Oxford Food Symposium decided on the 2024 theme of Gardens, Flowers, and Fruit, my cheese-self immediately started thinking about a paper on vegetal rennet: a historic and contemporary overview of the use of plants instead of the enzyme from suckling ruminants' fourth stomachs to coagulate and guide milk into cheese. Then I saw how much research has been done rather recently on vegetal rennet and got interested in why. In short, I fell into the 'vegetarians save the world' rabbit hole – and emerged with some thoughts much larger than thistles and lady's bedstraw (two of the vegetal enzyme alternatives).

The importance of how we humans shape the relationship with our ruminant fellow animals can't be overstated. An erstwhile fundamentally symbiotic connection has veered into a human-dominated, large-scale operation. Refusing these animals their innate way of life, grazing, we are degrading cows, for instance, to 'climate killers'. The contemporary trend towards so-called 'vegetarian' cheeses using vegetal extracts such as thistles, fig, or lady's bedstraw instead of animal rennet tries to block out life's inherent connection with death but does not change anything about the animals' situation.

Ruminants, a group of roughly two hundred species including cattle, sheep, goats, buffalo, and others such as camels, are able to survive and thrive in and from pastures, the humblest, and the largest, of gardens – something we humans are incapable of. An intricate digestive system comprising four stomachs that in turn are the homestead of an enormous microbiome enables them to break down complex carbohydrates. Herding, that is domesticating these animals, allowed our neolithic ancestors to survive in otherwise unwelcoming areas, which were unsuited for food crops, but could be used as pastures (think of wet marshes or mountains).[1] Humans provided food and shelter; in return animals allowed them to partake of their milk. This symbiotic relationship has shaped landscapes and has created some of the most biodiverse environments, environments very good at storing greenhouse gases.[2]

The interaction between ruminants and humans and the resulting milk and cheese can happen at all kinds of levels. At one extreme, without human intervention, milk serves its only natural purpose, nourishing newborns, and there is no cheese. At the other extreme, with humans over-dominating, we are faced with monotony and boredom at best (think of processed cheese) or extinction at worst (think of threatened breeds and the recent 'camembert alert'[3]). Balance is required, which means different things in different situations. Let's look at three cheeses to better understand what is at stake and to hopefully inspire you to think wider and deeper and harder the next time you are standing at the cheese counter, remembering that ruminants' existential need is to be in the garden.

We start in eastern Anatolia, 1000 kilometres east of Istanbul, with *Erzincan Tulum Peynir*, a sheep's cheese. *Erzincan Tulum Peynir* – that is the landscape of the majestic mountain plateaus high above the valley of the Karasu, the black river that further south is called the Euphrates. But it is also the families who come up from villages in the summer with sheep and tents, to translate that landscape into cheese. Sheep and people live in a symbiosis without which, until very recently, it would have been impossible to exist here. It is hard work under tough conditions, but also what the head of one of those families calls 'our factory without smoke'.[4]

In May, the families with their flocks (typically between 200 and 300 animals) move up to the plateaus at over 2500 metres, where the pastures are green and lush, and the people live in tents until September. The men milk the sheep by hand in the morning and evening; the women then immediately slurry the milk with rennet they prepared themselves from their own lambs and scoop the curd into cloth bags to drain. Early next morning, these bags are taken to a central tent that acts as the first dairy. There, the men fill large bags with the fresh curds, sew them up and stack them. Whey continues to ooze out for a week while lactic acid bacteria break down the lactose. In a second step, the now firm, sour curd is emptied into a large tub, broken up with the feet and mixed with local salt. Then it is filled again into large bags to be collected by the cheese merchants driving up from the town.

In some streets in Erzincan, there is one cheese shop next to another, and they all have their contacts to certain plateaus and families. Cheese, here, is a whole culture of life, not only a commodity. In the small urban manufactories, the matured, salted curd is broken up once again and pressed either into large plastic cans (these are the *bidon tulum peynir*) or, traditionally, into cleaned animal skins sewn into sacks, sometimes from sheep, but preferably from goats, whose skin is more robust. The larger these *deri* (literally leather, skin) *tulum peynir* are, the more rustic they appear, due to the unshorn exterior. The interesting thing is that after the four-month maturing period, at just above zero degrees imitating the winter mountain air they would have once experienced, the *deri* version is significantly more complex in flavour than the one from the *bidon* (which itself can be fantastic!) – but the very special quality of the milk is always apparent: extremely concentrated and with a deep umami flavour that in wine we would describe as mineral.

From Anatolia we go to the very south of Germany, and we find a completely different story. *Hofgemeinschaft Heggelbach*, on the western end of Lake Constance, is a farming community of six families, founded in 1987. Back then, it was a run-down estate whose new owners soon realized that selling their biodynamic-certified milk to conventional dairies was a financial fiasco and so decided to set up their own cheese dairy. Soon the first *Heggelbacher Alpkäse* was produced, and the cheese dairy developed into the most profitable part of the farm.

Since 2016 Stephan Ryffel has been responsible for Heggelbach's cheesemaking, and it's indicative of the farm's success that in 2020 he moved – vats, cave, and all – into a new, generously-sized building. Stephan has a degree in agriculture, spent eight summers working in the Alps, wrote his thesis on semihard cheese production, and worked for some time at the Swiss agricultural research centre Agroscope. He combines a great deal of practical experience with a clear vision.

And part of that vision includes the fifty Heggelbach cows, Braunvieh or Swiss brown with proud curved horns. For more than fifteen years they have been living on grass, clover, and hay – exclusively, with no silage, no concentrated feed, no cereals of any kind. 'Ruminants,' Stephan says, 'are specialists in the digestion of grass. This is what we want to promote and not the digestion of grain, which humans can eat directly.' This attitude is very rare in commercial dairy farming today, an attitude generally thought of as economically unviable. But 5800 litres of (excellent!) milk per animal year (or rather lactation) is still a good amount and viable if transformed into good cheese.

The most original of all Heggelbach cheese is *Felsbrocken*, rock. Made from raw milk, it is compact and extra hard like Parmigiano Reggiano, without the washed rind typical of alpine cheeses. Immediately after being brined the young wheels spend ten days at 25°C, to 'fat-sweat'. They mature very slowly; *Felsbrocken* is sold with fifteen months at the earliest, and Stephan doesn't cut the wheels, but rather breaks them open, horizontally, and in fact, just like Parmigiano Reggiano, due to its compact and yet crumbly, loose texture, the rock tastes best when broken.

So, a rock, yes, but the opposite of hard on the palate! It smells of melting butter with a hint of caramel, the milk's lushness making way for a very fine, elegant acidity and (again!) minerality, reminding me of herbs and baked celeriac. At once sweet and savoury, its ageing potential is far from exhausted at fifteen months – those cows and their pastures have much more potential.

So far we have seen a cheese that just clings to its garden, and one that has been created anew to do justice to it, and thus make it economically viable. To finish, I'd like to introduce you to a cheese whose maker had to reclaim – almost reinvent – the garden: let's go to the Netherlands.

The van de Voort family's farm near Lunteren, fifty kilometres east of Utrecht, is situated in an idyllic location, surrounded by heath, pines, birches, and oaks, yet, typical of this densely populated country dedicated to efficiency, close to the motorway. Ninety Jersey cows live here – which is unusual in itself: traditionally, Holland's main milk producers are black and white Friesians, bred in the USA to become the super-efficient Holstein-Friesians, and re-imported as such. Just as unusual, visits to the farm always begin with Jan Dirk van de Voort, in his mid-sixties, calm, friendly, almost a little mischievous, reaching for a spade and leading you to the pasture behind the red-brick buildings. There, he digs out a small square of soil, explaining the many different worms and how important it is to increase the humus content with straw manure and to stop mowing and ploughing the pastures.

Jan Dirk and his family have achieved something extraordinary: in the middle of the Gouda world, they have developed a completely new cheese, Remeker, and thus not only managed to break free from conventional trade and price structures, but to also find their way back to consider animals as partners. To understand what this means, it is important to consider the Dutch cheese psyche and history as such.

Cheese, trade, and building polders are Dutch specialties – almost a quarter of the country's current surface area lies below sea level. Back in the seventeenth century, it was hoped that the reclamation would lead to wheat fields – only to realize that all it was suitable for was pastures. With the industrial revolution the dairy industry began to expand rapidly. Within fifty years from the first milk factories in the 1870s, the milk production doubled. By 2000, it had tripled. Today, there are virtually no more traditional mixed farms in Holland, and dairy farms are becoming ever larger and more efficient. They use over half of the small country's agricultural land and with the help of 1.5 million cows produce over eleven billion kilograms of milk a year. Over a quarter of this is exported in the form of cheese.

The aim with that cheese has always been to extend milk's shelf life and reduce its volume in order to increase its trade potential, and from a technological point of view, the mild, inexpensive supermarket Gouda is a masterpiece of the Dutch food industry. The round wheels are supposed to mature 'dry' and clean. In pre-coating times, this meant a lot of work to keep the wheels free of mould in the humid maritime air, something that is now considered virtually impossible. Which is why coating them with polymers, invented after the Second World War, today is seen as a traditional characteristic of Dutch Gouda.

It requires a great deal of inner independence to think outside of this context and at the same time to live within it. Jan Dirk told me:

> When I took over the farm in 1981, I worked completely conventionally, I was always the first to apply artificial fertilizer in the spring. We switched to organic in 1993, but it wasn't until 2004 that we stopped dehorning and using antibiotics. That was very difficult at first, until I realized that the big balance was missing. We, the cows and the theory, it all has to fit together.

With a large gesture of both hands, he indicated two different levels like an unbalanced scale:

> I had to come down and let the cows up, now we are on the same level, and I understand much better what they offer me. The animals needed more space, so we built a new barn. We also thought about what kind of feed a cow is really built for, grass and hay, not silage or concentrates. And then: everyone always wants more cows, but nobody wants to smell cow dung – how is that supposed to work? Can you be a farmer like that? The cow manure must have the consistency of a fine ointment, then everything is just right!

This is also a fundamental part of a visit in Lunteren: Jan Dirk reaching into fresh cow dung, rubbing the brown mass between his fingers, followed by a gesture towards the visitor, almost an invitation to taste. Colleagues and students now regularly come to Lunteren and want to learn from them, but, he said, such knowledge cannot simply be transferred, everyone has to develop a feeling for their own situation, their own soil, their own manure, their own feeding: 'I always tell everyone: this is what I think, not: this is how you have to do it.'

His father had already switched the herd from the usual black-and-white cows to the beautiful Jerseys who produce less, but much richer milk. Jan Dirk's grandmother still made cheese, but then the milk was collected by the dairy. When Jan Dirk started making cheese again in the mid-1980s, he had to start from scratch: 'Everyone said that you couldn't make good cheese from Jersey milk, that it contained far too much fat. However, they forget that it also contains more protein, which is ultimately crucial when making cheese!'

There were many small steps. Raw milk was a matter of course, but then came the switch to a different salt, and gradually to a new shape, away from the rounded gouda moulds and towards cheeses pressed in hoops in the alpine style, with the Remeker logo embossed on them. And then, the question of coating: 'I always wanted to get rid of the coating, and my wife had been living in India, so we came up with the idea of rubbing the wheels with ghee instead of coating them.' Ghee is dehydrated butterfat, an easy undertaking with Jersey milk. He explained, 'When we started with ghee, we were told again that it wouldn't work. The butterfat would go rancid too quickly. But since our cows have horns again, it's not a problem at all.'

The van de Voorts live biodynamic agriculture like few others. However, Jan Dirk only read Rudolf Steiner's writings a few years ago, he told me to my surprise:

> We do things in a certain way because we observe and go our own way. Everything is connected to everything else, and the cows show you the way. It's all so complex, and the deeper you delve, the more possibilities open up!

On my first visit in 2010, the ghee-rubbed wheels were still the exception; seven years later, the colourful coating survived only as a picture on the kitchen wall. The flavour development has been just as fundamental and almost unbelievable. I had liked Remeker from the start, but it now represents a different world. There are four levels of maturity, culminating in the eighteen-month-old *Pracht* splendour, truly splendid with crystals yet crumbly like the finest shortbread, with a great umami spiciness, so that the sweetness that is often so intrusive with fast-matured gouda is completely integrated.

There we are: three cheeses, three versions of tending the large gardens that carefully managed pastures really are. For saving the one huge garden that is our world – which more honestly should read, ourselves. And no, I don't have a perfect plan for that. All I do know is that if you love cheese, you should definitely ask and know about how the animals lived and what they ate, to try and understand the world and shape it, at least a tiny, crumbly, savoury bit.

About the Author
Ursula Heinzelmann is the author of *Beyond Bratwurst: A History of Food in Germany*. Her work has also appeared in publications such as *Slow Food, Frankfurter Allgemeine*

Sonntagszeitung, *Gastronomica*, *Effilee*, and *Saveur*. She also acts as curator of the annual *Cheese Berlin* festival.

Notes

1. M. Salque and others, 'Earliest Evidence for Cheese Making in the Sixth Millennium BC in Northern Europe', *Nature*, 493 (2013), pp. 522–25, DOI:10.1038/nature11698.
2. My (fleeting) evaluation of the environmental/climate situation is based on an Oxford University study from 2017 and the subsequent reply by the Sustainable Food Trust (T. Garnett and others, *Grazed and Confused? Ruminating on Cattle, Grazing Systems, Methane, Nitrous Oxide, the Soil Carbon Sequestration Question – and What It All Means for Greenhouse Gas Emissions* (FCRN, Oxford University, 2017); 'Grazed and Confused – An Initial Response from the Sustainable Food Trust', *Sustainable Food Trust*, 10 March 2017 <https://sustainablefoodtrust.org/news-views/grazed-and-confused/> [accessed 1 March 2024]), as well as Anita Idel's *Die Kuh ist kein Klimakiller* (Metropolis-Verlag, 2011).
3. Mehdi Harmi, 'French Cheese Under Threat', *CNRS News*, 16 January 2024 <https://news.cnrs.fr/articles/french-cheese-under-threat> [accessed 4 March 2024].
4. I'd like to thank Gamze Ineceli who organized the field trip to Erzincan (and many more) and acted as contact and translator, as well as photographer Hüseyin Tuncer who (together with Arda Ipek) caught it all in pictures and did the driving.

17
Postnatural Apple
Re-Defining Fruit and Re-Thinking Food Systems through Design

Leonie Hochstrasser and Katharina Mludek

The Apple You've Been Dreaming Of. Meet your new favourite apple. The large, juicy and red apple has a perfectly balanced flavour and firm texture, making it ideal for snacking, cooking, baking, and entertaining. The striking colour and shape of the Cosmic Crisp® apple shines in fresh decor like wreaths, floral arrangements and tablescapes.[1]

The apple – symbol of love, fertility, knowledge, wealth; archetype of fruit – has been modified by humans for thousands of years.[2] Nowadays, apple production is an industrialized process where the fruit is treated as a product or even a brand, as the Cosmic Crisp® advertising slogan highlights. When selecting cultivated varieties, the focus lies more on shelf life and profitability than on variety and taste. A complex network of demand and requirements from a wide range of actors spans the entire production process. Not only the apple itself is designed, but also decisions are made in the design of warehouse buildings, tools, logistics processes, packaging, and advertising.

This paper defines the apple as postnatural – which the Center for PostNatural History defines as 'any living thing that has been intentionally altered by people through domestication, selective breeding, or genetic engineering' – with and through design and further introduces experimental design research, on food experience and exhibition design, in order to mediate discourses on the apple and its production.[3] The two authors, product designer Leonie Hochstrasser and design researcher Katharina Mludek, share a passion for object biographies and looking at food from a design perspective. The aim is to inform and stimulate a discourse about complex food production systems, with the help of familiar everyday foods such as the apple. The authors hypothesize that design research, and particularly an experimental approach, can enrich other disciplines around food with hands-on methodologies. They argue that the emerging food design scene and its actors and methods should be taken into account when thinking about the transformation of the food system.

The aim of the first section is to situate the case study and thus the designer's exploration around the apple in the wider context of food design. The project can be positioned as a good example of a new movement in food design that deals with questions of ecology and transformation, and the agricultural system in general. After this contextualization, the exhibition work *PostNatural Apple*, the final outcome of the project, will be introduced by guiding the reader through an exhibition tour. After a brief introduction to the concept of postnaturalness, research methods and results are explained through the exhibited elements. All elements together strengthen the argument of the apple as postnatural and show the various levels of human manipulation of the organism. Finally, the paper aims to serve as a basis for a discussion on how designers can contribute to a possible ecological change in the food system through their approach.

The Future of Design within the Food System – or the Future of the Food System with Design

The young sub-discipline of design is often mistaken for food styling or photography, or packaging design.[4] Design researcher Francesca Zampollo makes clear that food design not only works on topics such as tableware or the next meat alternative, but looks at all aspects of food and food culture: 'Food design is the conscious and deliberate creative process that brings innovation to living beings and the planet on anything related to food and the act of eating: from production, procurement, preservation, transportation, preparation, presentation, consumption, and disposal.'[5]

Food design has been an emergent field for twenty-five years. The initial phase was characterized by a few individual designers, such as the Dutch designer Marije Vogelzang, who introduced the concept of Eating Design. The Austrian studio *honey & bunny* – the duo Sonja Stummerer and Martin Hablesreiter – questions the food culture of today with a more performative, artistic approach. More importantly for the formation of theory, the two of them wrote the initial texts on the concept of food design. The Catalan designer Martí Guixé was also a pioneer in pointing out that food can be used as a material and designed in the same way as traditional product design: he wanted to be the first to 3D-print tomatoes.

From these beginnings, master's programmes, journals, conferences, and a new field of study have developed.[6] Food design has diversified with a new generation of young designers who understand the field much more holistically, as Zampollo's definition underlines. There may be various explanations for this shift: the post-Covid era may be one reason, as communal eating was not possible for many years. Furthermore, a general lifestyle trend towards food awareness can be observed. More importantly, however, is the focus on sustainability in design and thus the attention to organic materials such as fungi, algae, etc. Food as a material is therefore increasingly gaining attention. Moreover, the idea is growing that design, with its sometimes unconventional methods, can contribute to change in the food system or to challenge individuals' everyday eating routines.

Regardless of the still limited collaboration with the industry, many new approaches and projects are emerging.[7] A broadened definition of design and approaches, such as critical or speculative design, challenges existing systems and present possible future scenarios. The examples are diverse: Italian designer Eleonora Ortolani developed vanilla ice cream from plastic by means of enzymatic dissolution; the London-based studio *Cooking Sections* looks critically at the animal-based food production of salmon, oysters, or buffalo; Dutch designer Caroline Niebling reinterprets traditional techniques of preparing sausages in a new plant-based way. Such projects attempt to criticize existing systems or to illustrate new, and often extreme, proposals for future forms and ways of eating and making food production tangible.

How to Make a Postnatural Apple

Having described the dynamic food design scene, Swiss designer Leonie Hochstrasser's case study offers one example of a new approach to food design.

In the course of the initial research on the apple, the idea arose to present the apple as postnatural and thus take a different perspective on the fruit and its production. Recognizing the postnatural creates a de-romanticized and realistic image of the apple and its production, consciously emancipated from the notion of pristine nature. Initiating the discourse of postnaturalness is not about celebrating the concept but, rather, recognizing this phenomenon. It encourages a discussion about our current attitude as humans towards nature.

To reveal the postnatural nature of the apple, the next paragraphs will guide through the exhibition and food experience held at CIVIC, Basel. The project's final outcome was a tangible mediation of the content with all the senses.

The exhibition can be understood as an assemblage of design research findings: information, objects, materials, collages, photos, recipes, and so on. Each exhibit contributed to defining the apple as postnatural and highlighting different arguments of the discourse. The diversity of the exhibition pieces also reflects the different research approaches that were used in dealing with the topic: such as design ethnographic methods, field studies, interviews, and visual documentation, as well as experience-based learning and culinary experiments.

Figure 1–3. Photographs taken in the fields of apple producers in Switzerland. © Leonie Hochstrasser, 2023.

Manipulation by Definition

The exhibition design deliberately played with stereotypes associated with nature and culture, and challenged the image of the seen, untouched, natural, and romanticized apple. To capture visitors' minds, it was important to introduce them to the concept of postnaturalness. The introduction poster can be seen in the photograph of the exhibition (Figure 4).

A short introduction and reflection on postnaturalism and naturalness here is helpful for further understanding of the project. The Center for PostNatural History traces the 'influence of human culture on evolution'.[8] The definition of postnaturalism therefore implies an interpretation of nature or naturalism: in this case, all living beings that have not been intentionally altered by humans.

In times of debates about the age of the Anthropocene and the associated understanding of humans as the greatest influencing factor on the Earth system, the question must be raised whether anything can be considered free from human influence or manipulation. Designer and media culture analyst Katharina Stahlhofen delves into this topic in her essay exploring the concept of the Anthropocene: 'The history of humanity is also a history of its impact on the environment.'[9] As many of the positions mentioned in this article agree, we cannot (and should no longer) clearly separate what is 'natural' from what is human-made.

Despite this understanding, Western society still sees *nature* and *society* as two separate areas. These boundaries are arbitrary, even fictitious, and therefore 'not an indisputable fact, but rather an idea that has evolved over time'.[10] Ailton Krenak, the indigenous

Figure 4. Photograph of the exhibition entrance at CIVIC. © Chiara Marinai, 2023.

activist from Brazil, argues that, at a certain point in history, the 'civilised place of humans' gave a name to something that did not yet have a name: the idea of nature – an invention of culture – came into being. 'Firstly, people create nature and separate themselves from it; then, they idealize it.'[11]

However, the relationship between humans and nature has always been ambivalent: 'Nature is both enemy and friend, murderer and nourisher,' writes Thomas Schramme, a German philosopher whose research interests include bioethics.[12] People have had to and still have to manipulate or even fight the environment in order to meet their needs, such as protection from the cold or animals. Schramme questions 'why the pristine, untouched [nature] should in any way be more valuable than that which has been manipulated by humans'. Nevertheless, it is precisely the image of untouched nature that is still romanticized and associated with an illusory longing. The valuation of naturalness, in this case the positive one, is therefore always influenced by our relationship to it and linked to a specific idea – the 'humanly beneficial'.[13]

The apple industry – or more specifically the advertising and portrayal of apples – also continues to conjure up an image of naturalness and plays precisely on a Western notion of nativeness and authenticity. Over the last 200 years, however, we have seen an increasing standardization and ubiquitous presence of monocultures in today's production landscapes. Apple production in particular can be understood as a planned and organized process and the result – the apples – therefore correspond quite precisely to what is defined as postnatural. And as a glance at history shows, this is nothing particularly new, as apples were domesticated 10,000 years ago and have since been modified, refined, and cultivated.

Manipulation during History
The exhibition poster gave an overview about the (design) history of the apple and the evolution of human manipulation (Figure 5). By emphasizing the silhouettes, the poster deliberately highlights the evolution of the shape and size of the apples. The result corresponds to the human ideal and needs: large, lots of flesh, symmetrical. A brief overview of the apple's history provides some context.

The roots of each apple cultivar lie in the vast Tian Shan Mountains, which cover areas from Kazakhstan through China to Kyrgyzstan, Uzbekistan, and Tajikistan. In fact, entire forests of apple trees still grow there today, with trees reaching up to thirty metres in height. The variety of different apples, in a wide range of colours, sizes, and tastes, is enormous. Over time and with the interest of the bears and wild horses, the small berry-like apples turned into larger fruits. The first domestication by humans – nomads – happened 10,000 years ago. Subsequently, traders transported the apple to Europe through the Silk Road. Around 3800 years ago, the grafting of fruit trees was optimized in Babylon and adopted first in Persia and then in Greece, with wars and conquests driving its spread.[14]

That spread was accelerated by colonialism. 'Although it took around 6000 years for the apple to be brought from the Tian Shan to Western Europe, it reached all other temperate regions of the world in just 300 years' – introduced by European colonial powers, write Juniper and Mabberley in their comprehensive book on the apple.[15] In the course of the

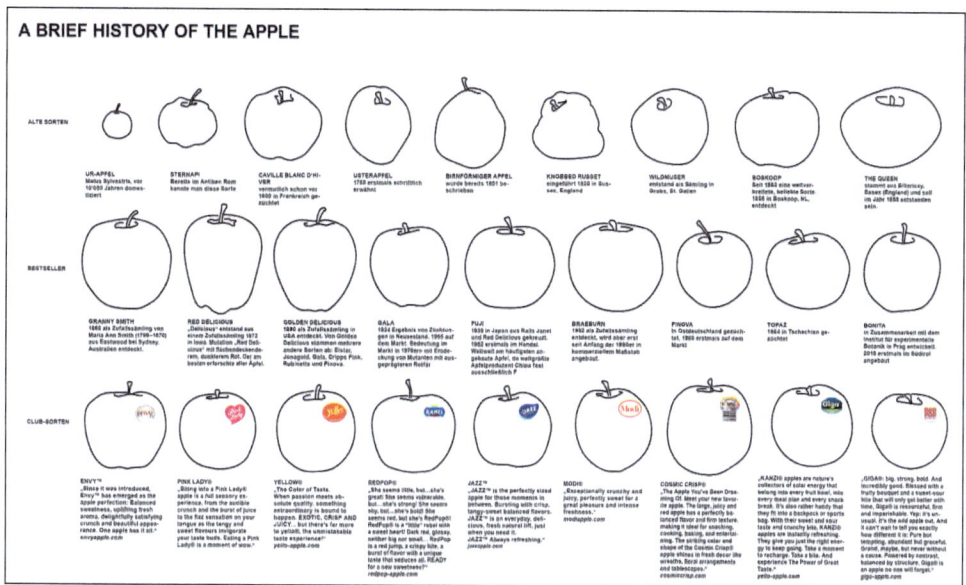

Figure 5. Illustration: 'A Brief History of the Apple'. © Leonie Hochstrasser, 2023.

history of the apple, the proportion of deliberately cultivated fruit plants grew in contrast to wild-growing fruit plants: 'This resulted in cultivated forms that differed more and more clearly from the wild forms and thus increasingly met the wishes of the fruit growers.'[16]

Today, apples are grown all over the world in tropical, subtropical, and temperate regions and are the most widely cultivated fruit in the world after bananas and watermelons. In 2021, China produced the largest quantity with almost 46 million tons, followed by Turkey, the USA, and Poland with up to 4.5 million tons each. In Switzerland, 176,574 tons were produced in the same year, making apples the leader in Swiss fruit production ahead of grapes and pears.[17]

The poster's focus on the developed outline – the shape of the apple – illustrates to visitors of the exhibition with just one glance that today's supermarket apples are the result of a process of design and manipulation over thousands of years.

Manipulation during Cultivation

When visitors look at the exhibition in its entirety (Figures 6–7), the bright colours of red, green, and yellow are particularly eye-catching. The exhibition architecture was designed using collected materials and objects relating to apple production. Standardized green fruit crates served as the basis, used as spatial elements and exhibition platforms for other objects. Red was the basic colour throughout the entire design process, referring to the Swiss flag and the red of the apples. Furthermore a ten-metre-long hail net was stretched across the exhibition space as a frame. The net and other collected objects highlight various practices of manipulation and protection in apple production, some of which have been taken to extremes. Here a few practices found within the research process are presented as examples.

Figures 6–7. Photographs of exhibition and food experience. © Leonie Hochstrasser, 2023.

Today's cultivation in monocultures favours the transmission of diseases such as apple scab to healthy trees. Producers rely on preventative treatments, as otherwise they risk losing up to 80% of their yield. This involves drawing up a spraying plan based on the weather forecast and spraying with chemical-synthetic agents, biological clay preparations, or modern plant strengthening agents with algae.[18]

During winter and early spring, the surrounding grass is mulched to expose the soil, allowing it to warm up more efficiently during the day. This is particularly important in the red bud stage. If the temperature falls below 4°C at this time, fruit growers can lose the entire harvest. They resort to various methods to protect their trees and buds: one farmer, who was visited as part of the research work, operates a heater attached to his tractor, passing through the rows of trees from 1:00 to 9:00 in the morning on frosty nights to protect his harvest. Alternatives are kerosene candles or frost irrigation. Here, water is constantly sprayed onto the flowers at sub-zero temperatures. The resulting layer of ice releases heat into the blossom, protecting it. Not all growers can use this effective method, as it is costly, energy-intensive, and requires sufficient groundwater.[19]

Design decisions can be observed in the arrangement and height of the apple trees in the field, which are adapted to the harvesting processes. Low-stem trees are the standard in current fruit production, as their low height means that maintenance care can be mechanized and carried out from the ground. Further advantages are an early and high yield of marketable fruit, as well as easier installation of weather protection.[20]

During the growing process, the plants are protected from external influences and pests – from nature itself, so to speak. The fragile system can tip over at any time and must be constantly monitored and influenced accordingly by means of fertilizer, heat supply, or hormones. To protect the fragile monocultures, farmers have to go to and beyond the limits of what is tolerable. To show how many machines and devices are necessary for a small producer to achieve apple production to the desired standards, a collage was created (Figure 8).

Manipulation by Standardization

During the research for *PostNatural Apple*, an extensive collection of documents and regulations concerning apple production was uncovered, collected, and analyzed. To illustrate how much bureaucracy, forms, plans, and regulations are involved in apple production, an entire wall was plastered with printed documents (Figure 9).

Regional regulations on apples and specifications from brand licensors, such as 'Pink Lady', serve as the initial and final parameters for the targeted manipulation of apples. They set the standard, the orientation, and the value of the fruit – the more perfect, the higher the profit.

Apple production in Switzerland is carried out in accordance with the Standards and Regulations for Dessert Apples to ensure that the products meet the highest quality standards. This document describes in great detail how an apple should look and perform. If an apple does not meet these points, it is downgraded and can only be sold for less than half the price.[21] The minimum characteristics of all

Figure 8. Collage of the machinery of a small apple producer. © Leonie Hochstrasser, 2023.

Figure 9. Photograph of the wall plastered with printed documents. © Leonie Hochstrasser, 2023.

classes (class Extra, class I, class II) require that the apples are whole and healthy, practically free from pests and damage, abnormal moisture, foreign smell, and taste. Class II includes apples with defects in shape, development, colour, or skin – for example, bruises. The various classes are categorized by diameter in 5 mm increments. They are also divided into colour groups (Figure 10).[22]

The transformation of apples into commercial products becomes especially evident when observing the recurring apple brands featured in various collages throughout the exhibition (Figure 11).

The so-called 'club varieties' are marketed professionally, with their own websites and sometimes even Instagram profiles. In contrast to non-protected apple varieties, which can be grown at your own discretion, club varieties are often backed by privately-financed breeding programmes and work within contractual structures. Their aim is to breed interesting apple varieties for producers, marketers, and consumers. It can take up to fifteen years before a new variety with market potential is developed. If the producer organizations are convinced, a licence agreement is concluded for a fee, whereby they are also obligated, for example, to a prescribed cultivation quantity. Criteria for production, quality, packaging, and marketing are also regulated. Licensees receive a share of the sales proceeds, which they use to finance their research and provide standardized marketing and packaging.[23]

'Pink Lady' is an example of a club brand that advertises with the slogan 'More than just an Apple'. This is made very apparent on the website: in addition to background

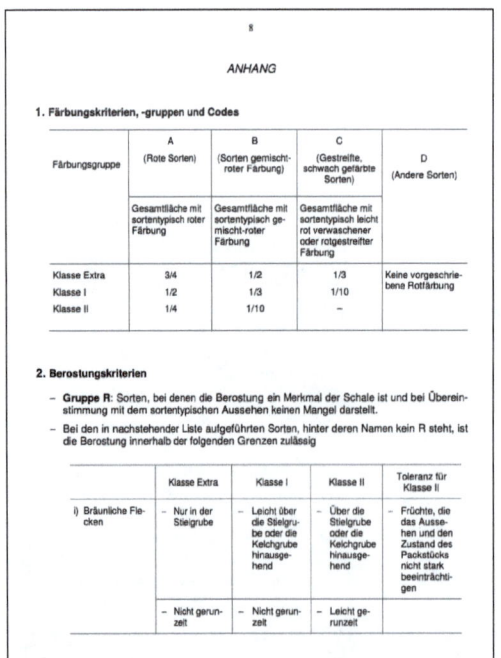

Figure 10. Extract from standards for Swiss dessert apples: Colouring and russeting criteria for Swiss dessert apples.

Figure 11. Collage of licensed apple brands. © Leonie Hochstrasser, 2023.

information on the 'success story', raffles, and recipe suggestions, you can also go directly to the webshop, where branded products such as t-shirts, bags, and kitchen products are advertised. Although the naturalness of the 'Pink Lady' apple product is emphasized, a certain alienation of this ideal image is noticeable.

As a result of these regulations and the licensors, producers are under high competitive pressure to produce apples that look almost perfect and therefore work with extremely regulated production rhythms and storage, as well as the sometimes extensive use of pesticides. If you go to the supermarket today, you will find neatly sorted, identical apples – the same size, the same colour, the same shape – and above all free from blemishes. Certification and licensing leads to supermarket apples being seen as normal and natural by customers, as they are no longer confronted with blemishes and flaws. Customers also develop a relationship with individual apple brands and can expect the same flavour everywhere they shop – even across national borders.

Manipulation from the Fields to the Table
For the designer Hochstrasser, the apple also became a material for experimentation (Figures 12–18). Different apple varieties were compared raw, cooked, dried, and pickled, and a range of flavours were explored through food pairing experiments. This was followed by experiments such as the production of apple leather or the extraction of apple yeast

Figures 12–18. Photographs of food and material experiments (Figures 12 & 13); Apple leather (Figure 14); Apple yeast bread production (Figures 15, 16 & 17); Apple wood (Figure 18). © Leonie Hochstrasser, 2023.

for baking bread. Through these experiments, a menu was conceptualized, reflecting the innovative ideas and insights gained during the process.

Outro

Consumer goods, such as apples, usually do not reveal where they come from and what their origins are. They 'appear as if by magic on the shelves of our shops', and society seems not only to have forgotten their material and practical origins, but also to have 'elevated them to little gods'.[24] Even the apple has been, and continues to be, romanticized and praised as a natural product. The industry thus profitably clings to the marketable naturalness of the fruit. However, as this work shows, the links surrounding the creation and production of the apple are far more complex. Based on the history of domestication and the analysis of current production systems, the postnaturalness of the apple could be proven. The apple has become a product and has adapted to the production and marketing mechanisms of non-edible products, such as an iPhone or a sofa. For consumers, the supermarket apple has become the norm and is no longer scrutinized, as it is considered a regional fruit. Agreeing with the criticism of the desire for naturalness, the German design researcher Johanna Kleinert clarifies how this is opposed to a transformation of the food system: 'The desire for naturalness has so far not resulted in fruit and vegetable products being less designed, but above all in the design interventions being more misjudged and thus less talked about.'[25] She argues that it is not about manipulating plants and food 'ruthlessly and radically' through design, but about recognizing the status quo: fruit are industrial consumer goods shaped by humans and human desires. Kleinert, following design theorist Annette Geiger, describes the potential scope for designers in the food sector: 'To describe fruit and vegetable products as designed therefore primarily implies the idea that they can be designed – and thus also that they can be designed differently.'[26]

Figure 19–21. Photographs of food in the exhibition. © Leonie Hochstrasser, Chiara Marinai, 2023.

Every year, around 50,000 new food products are brought onto the market, and, at the same time, 50% of the world's habitable land and 70% of its fresh water is used for food production.[27] The pressure is increasing to bring innovation and change to the food system. Sonia Massari sees a particular strength in design, which is open to different disciplines, methods, and contexts: 'Design, when applied to the agri-food sector, is a powerful research methodology because it enhances critical transdisciplinarity between people with diverse backgrounds and competences, and helps to develop individual and collective creatives to foster more sustainable innovation processes.'[28]

The work *PostNatural Apple* is a contemporary example of a new tendency in design to focus on food, thereby also broadening the concept of design and revealing new potentials. The project presented in the paper and the exhibition aims to inform about the postnaturalness of the apple and its various manipulations by humans. In doing so, the work does not attempt to moralize, but rather to take visitors on a journey through the research and work process and to arouse their interest through a variety of exhibits that can be experienced with all the senses. The goal was not about developing or presenting an alternative system for apple production or speculating about possible futures. However, the work can be a starting point for such a next step.

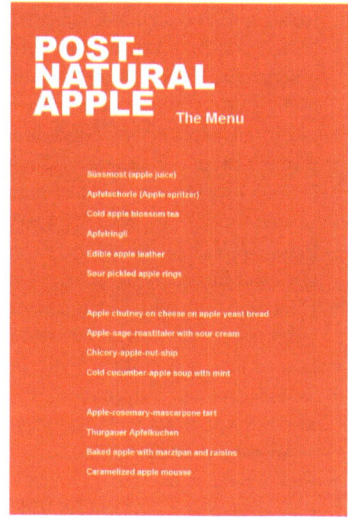

Figure 22. 'PostNatural Apple' menu. © Leonie Hochstrasser, 2023.

Design can utilize unconventional approaches and interventions, such as food experience design, to illustrate and communicate the complex structural challenges in the food system of our time. With this first step, making things transparent, the next step may follow: the actions of individuals to make sustainable decisions in the supermarket or, thinking bigger, a transformation of the food system.

About the Authors

Leonie Hochstrasser is a designer based in Bern who works with an experimental approach to materials, everyday objects and food culture in order to raise questions about society.

Katharina Mludek is a design and cultural researcher specializing in sustainable product design at the Kunsthochschule Kassel and works at the intersection of design, food, and transformation.

Notes

1. Cosmic Crisp® <https://cosmiccrisp.com/> [accessed 20 December 2023].
2. Bernd Brunner, *Taming Fruit: How Orchards Have Transformed the Land, Offered Sanctuary, and Inspired Creativity* (Greystone Books, 2021).
3. 'About', *Center for Postnatural History* <https://www.postnatural.org/About> [accessed 20 December 2023]. By 'with design', the authors mean looking at apple production through the lens of design and

working with design methods within the practical work. Christopher Frayling discusses the categories (1) research into, (2) through, and (3) for design. The first category deals with the analysis of design objects, that is with the thematization of existing artefacts ('Research in Art and Design', in *Royal College of Art Research Papers*, ed. by Christopher Frayling (Royal College of Art, 1994), pp. 1–5). The second category is explained by Nigel Cross's approach of designerly ways of knowing: knowledge is generated through the act of designing (*Designerly Ways of Knowing* (Birkhäuser Verlag, 2007)). The last category is based on the research that designers do to work practically on this basis.

4   LinYee Yuan, 'From Fake Meat to Cheese Architecture, Designing the Future of Food with Katja Gruijters', *MOLD Designing the Future of Food*, 2017 <https://thisismold.com/process/studio-visit/katja-gruijters-food-design> [accessed 26 September 2024].
5   Francesca Zampollo, 'What Is Food Design?', *ResearchGate*, August 2023 <https://www.researchgate.net/publication/377951005> [accessed 15 May 2024].
6   Silvana Juri, Sonia Massari and Pedro Reissing, 'Food+Design – Transformations via Transversal and Transdisciplinary Approaches', in DRS2022: Bilbao, ed. by D. Lockton and others, 2022 <https://dl.designresearchsociety.org/cgi/viewcontent.cgi?article=2774&context=drs-conference-papers> [accessed 26 September 2024].
7   Rick Schifferstein, 'What Design Can Bring to the Food Industry', *International Journal of Food Design*, 1.2 (2016), p. 103, DOI:10.1386/ijfd.1.2.103_1.
8   'About', *Center for Postnatural History*.
9   Katharina Stahlhofen, 'Das Andere', *form – Magazin für Haltung und Design*, 294 (December 2021), p. 69.
10 Stahlhofen, p. 69.
11 Ailton Krenak and Maurício Meirelles, 'Our Worlds Are at War', *e-flux journal*, 110 (2020) <https://www.e-flux.com/journal/110/335038/our-worlds-are-at-war/> [accessed 1 May 2023].
12 Thomas Schramme, 'Natürlichkeit als Wert', *Analyse & Kritik*, 24 (2002), pp. 255–56.
13 Schramme, pp. 250–61.
14 William Mullan, *Odd Apples* (Hatje Cantz Verlag 2021).
15 Barrie E. Juniper and David J. Mabberley, *Die Geschichte des Apfels. Von der Wildfrucht zum Kulturgut* (Haupt Verlag 2022), pp. 191–93.
16 Bernd Brunner, *Von der Kunst, die Früchte zu zähmen. Eine Kulturgeschichte des Obstgartens* (Knesebeck GmbH & Co. Verlag KG, 2022), p. 67.
17 'Crops and Livestock Products', *Food and Agriculture Organization of the United Nations*, 2021 <https://www.fao.org/faostat/en/#data/QCL> [accessed 24 April 2023].
18 Valérie Sauter, 'Pflanzenkrankheiten', *News vom hof Farmticker*, 2020 <https://www.juckerfarm.ch/farmticker/hintergruende/pflanzenkrankheiten-1-schorf/> [accessed 13 May 2024].
19 Personal interview with farmers, Messerli, Schweizer, and Suter (2023).
20 'Obst- und Beerensorten- Inventarisierung Schweiz, Schlussbericht', *FRUCTUS – die Vereinigung zur Förderung alter Obstsorten*, 2005 <https://www.fructus.ch/wp-content/uploads/schlussbericht_nap02-23_inventarisierungobstbeeren.pdf> [accessed 13 May 2024].
21 Schweizer Obstverband, 'Jahresbericht 2021', *Schweizer Früchte* <https://www.swissfruit.ch/wp-content/uploads/2022/04/sov_ueberuns_kommunikation_jahresbericht_2021_d.pdf> [accessed 26 May 2023].
22 'Normen und Vorschriften für Früchte, Ausgabe 2008', *Swisscofel* <https://www.qualiservice.ch/uploads/pdf/Tafel_Aepfel_d.pdf> [accessed 13 May 2024].
23 Beitrag Teilen, 'Club-Äpfel', *LebensmittelPraxis*, 2022 <https://lebensmittelpraxis.de/warenkunden/33158-warenverkaufskun-%20de-club-aepfel.html> [accessed 13 May 2024].
24 David Crowley, 'Foreword', in *The Toaster Project*, ed. by Thomas Thwaites (Princeton Architectural Press, 2011), p. 10.
25 Johanna Kleinert, *Lebendige Produkte* (Verlag, 2020), p. 35.
26 Kleinert, p. 14.
27 Sonja Stummerer and Martin Hablesreiter, *Food Design Small: Reflections on Food, Design and Language* (De Gruyter, 2020), p. 13.
28 Sonia Massari, *Transdisciplinary Case Studies on Design for Food and Sustainability* (Woodhead Publishing, 2021), p. 8.

18
Śiva's Flora

Soham Kacker and Deepa S. Reddy
with illustrations by Chippy Diac Vivekanandah

Kuśasthalī, or the Forest in Time

Nature as botanical material, constituted of plants, flowers, leaves, trees, whole gardens, or entire forests, plays an abundant role in Hindu narrative and worship traditions – but is rarely the focus of specific scholarly attention. Attending to that gap, this essay seeks Śiva's garden or, more accurately, a botanical landscape in which to place Śiva, because forests, plants, and flowers can be called upon to present particular views of Śaiva or Śiva-centric cosmology. Śiva takes many colourful forms in Hindu stories: the wild, fierce Rudra who glows in fire-reds; the great mountain-dwelling ascetic, white and pure; the blue-throated swallower of poisons; the compassionate, beautiful God bedecked with garlands of golden yellow *karṇikā*. What if botanical stories and insights from Śaiva texts could be gathered – what would gardens grown of those plants look like? And what might these tell of Śaiva ways of knowing and being?

No sooner had we posed the question than we found ourselves in a forest. Methodologically, this was simply the immense, dense thicket of Śaiva texts through which we now needed to chart a course. Śaiva literature is comprised of a vast number of works of varying length and complexity, accrued over centuries from different parts of the country, encompassing several lay and initiate traditions, and including devotional expressions as much as doctrinal and practical exegeses. We have tended to rely less on texts expounding Śaiva philosophies and more on literary,

*Figure 1. A single Kaḍamba (*Neolamarckia cadamba*) represents a forest: Koodal and Kaḍamba-vana-kshetra were once names for Madurai where the famous Meenakshi Sundareśwarar temple is now a major pilgrimage center. Also visible are a part of a banyan tree and wild jasmine creepers.*

devotional, and practice-oriented sources, as it is largely there that engagements with the plant world gain importance. We draw mostly from Purāṇic sources, rich in story and mythology, and on Mantramārga ['the path of mantras'] sources within initiate traditions, particularly Śaiva Siddhānta ['the conclusive truth about Śiva'] and (generally South Indian) devotionalism, as each has shaped what we could call 'everyday Śaivism' and each contains, as we read it, a garden. We also reference Vedic and other ascetic traditions where pertinent.

The real forests, Śiva's forests, appear within these texts and story traditions sometimes as a backdrop for a divine pageant, sometimes as though part of the cast therein, sometimes as both. In that last category is Kuśasthalī, an exceptionally heavenly forest, rich in variety and bloom, where trees and creepers are entwined in lovers' embraces or bent over by the weight of flowers and fruits as though in gracious offering.[1] Into this sylvan grove enters Śiva with his *kapāla*: the ash-smeared ascetic-mendicant with a skull as his eternal begging bowl. He enjoys the trees who bow to him, offer obeisances, and ask Him to be present in this forest for all time. Śiva replies that he will indeed be present but the trees may go anywhere, assume any form, and always yield fruit. So saying, Śiva stands in Kuśasthalī for a thousand years, and the forest becomes the Mahākālavana: the great forest beyond time and death.

But then he places the *kapāla* down, and the three worlds shake. Trembling devas and asuras approach Brahma, who instructs them to find refuge in Śiva in the Mahākālavana. When they approach this splendid forest, however, certain of Śiva's presence there – He is nowhere to be seen. Śiva 'cannot be seen by one who is not properly initiated,' Brahma reminds them; once they are in constant communion with him, they will 'know the proper time'.[2]

There are other, prior extolments of Śiva as 'Lord of forests', for instance in the Śatarudriya and the Śri Rudram-Nāmakam of the Ṛg and Yajur Vedas respectively.[3] The Mahākālavana episode stands out for its detailed, lyrical descriptions of how this botanical association came to be and for its placement of Śiva's forest as beyond time. Time in Śaiva texts is itself neither linear nor historicist, and the mingling and melding of ideas and practices across different schools equally have effects of collapsing centuries or transcending them. We therefore regard the 'forest in time' as constituted by a series of moments, the 'when' of which are best known only to Śiva himself.

The Kannada poet of the Vīraśaiva tradition Allama Prabhu offers a riddle of sorts, illuminating this thought:

> Where was the mango tree,
> where the koilbird?
> when were they kin?
> Mountain gooseberry
> and sea salt:
>   when
> were they kin?
> And when was I

kin to the Lord
of Caves?⁴

Spring is when the koel and mango are kin; the bird feasts so eagerly on tender mango blossoms that the poet Kālidāsa imagines that its voice must become clear only after it has done so.⁵ Gooseberries and sea salt become kin in summer pickling jars – and Śiva alone knows the appropriate moment when he should appear, always in the forest for the forest is where Śiva is, to make kin of those who seek him.

This essay finds five such landscapes and five such moments of 'kinship', waiting, yearning, searching, propitiating, speaking of, and speaking to Śiva. Flowers, leaves, plants, and trees become metaphors, media, and means to invoke and discern his presence or determine the import of his emergence.

Dārukāvana, or the Forest of Transformations

Manifesting eternally in Kuśasthalī, Śiva only ever visits Dārukāvana or the forest of pines. The Himalayan *devadāru* (*Cedrus*

*Figure 2. A banyan representing the forests of Thiruvalangādu: the temple and its location take the name of the banyan forests. Malabar glory lilies (*Gloriosa superba*) are at its base and the Himalayan Devadāru (*Cedrus deodara*) forests in the far background as the originary site of Śiva's cosmic dance.*

deodara) forest is a 'place of passage', perhaps also of sport.⁶ But of what kind?

Śiva comes to Dāruvana as Bhikṣāṭana the mendicant, a beautiful unclothed form of the terrifying Bhairava who has decapitated Brahma's fifth head and been fated to expiatory wanderings for committing *Brahmahatya* or Brahminicide – but he bears no sign of Bhairava save the *kapāla*, the skull as begging bowl. So, the sages performing their penances in the pine forest do not recognize him in this form beyond the usual attributes. They are transfixed anyway by Viṣṇu in the equally alluring form of Mohini, who accompanies Śiva, and Mohini destroys their penances. The rishis' wives, for their part, lose themselves seeing Śiva's beauty: their clothes slip from their bodies, and they follow him with such heated desire that the soft rice with which they fill his begging bowl burns to ash. 'There is a very deep ocean that is called "wanting god," and they were drowning in it,' says Kachiyappa Śivachariyar in the *Kanda Purāṇam*, a Tamil telling of the *Skandapurāṇa*.⁷

In the northern *Purāṇic* tellings, the doubly outraged sages curse Śiva to become a eunuch, and the sign that is Śiva's *liṅga* falls down but towers such that neither Brahma nor Viṣṇu can find its beginning or ending.⁸ Here marks the start of *liṅga* worship in temples, and the establishment of four fiercely austere *āśramas* or orders in worshipping the deity: Śaiva, Pāśupata, Kālāmukha, and Kāpālikā.[9] Śiva remains here *niṣkala*: beyond form, beyond measure, beyond time.

In the south, Śiva dances. The outraged rishis cast spells: they send a tiger (which becomes his garment), a snake (a bracelet), the powerful sound of their mantras (his drum or his anklets), and the strange demon Mūyalakaṉ [Apasmāra], the epileptic force of seizure and stupor, on which Śiva simply places his foot and dances. Thus does Śiva emerge to adulation as Naṭarāja, a *sakalarūpa* [with all parts] form that is accessible, comprehensible, suitable for contemplation.[10]

The pine forest is on one level significant simply because it bears witness to all the events of these various tellings – to the transformations of Śiva from Bhairava to Bhikṣāṭana to the formless linga to Nataraja and in one instance also to Kankāla mūrti, the clothed mendicant who bears a *kaṇkāladaṇḍa* [skeletal staff]. The *devadāru* forest is the setting in which all these signs of Śiva are unmade and remade, in which Śiva appears in a form nobody can recognize only to acquire marks that become eternally recognizable: the liṅga, the austere Pāśupata *bhasma* [ash] and *nirmālyam* or the withered taken-down lotus garlands not worth possessing, then the tiger, hatchet, doe, snake, ghouls, a skull, and 'the entire iconographic array' proper to Nataraja.[11] All these fragments of Śiva, all these signs of the Truth that is Śiva, pass through the pine forest, no matter the variations in how the story is told.[12] In each iteration, Śiva returns; '[i]n fact there was never a first time'.[13] His return portents re-emergence.

Yet it would be incorrect to view the pine forest as simply botanical backdrop for what happens in it, or to treat it as a passive but all-seeing witness. 'You are the forest / you are all the great trees / in the forest,' the naked, convention-flaunting Vīraśaiva saint Mahādeviakka later declares, insisting on the singularity of the 'world as god'.[14] The Śaiva Siddhāntins, however, never fully deny the phenomenal existence of the world alongside Śiva's all-pervading presence in it: 'not two and not one', *oṉrirandum il*, in the words of Chidambaram's Umāpati Civācāriyār, for 'if he were two, the world could be utterly distinct from him [. . .] and it is not. If he were one, the autonomy of phenomenal reality would be completely lost – and the Saiva Siddhāntins, in general, refuse to take this step'.[15] Dārukāvaṇa, we might say, then, represents both realities.

The landscape of Śaiva temples in Tamil Nadu most vividly elaborates this logic, far more than other regions. Trees, fruits, forests are botanically and contextually specific instances of phenomenal existence which nonetheless contain Śiva as universal fact. So do we find a proliferation of botanical Śivas, each associated with a local forest or trees growing therein: Jambukīśwarā, Śiva of the Jamun trees; Tirunellikkā Nellivaṉanāthar, Lord of the Sacred Nellikāya groves; Venuvaṇanāthar, Lord of the Bamboo forest; Kurumpala-veesar, Lord of the *kurum-pala* or short jackfruit grove; and so on. Each one discreet and physically distinct, the forests are nonetheless a continuum of sorts, where the devout search and sometimes find, where Śiva visits and sometimes stays.

It is logical, then, that the *Chidambaram Mahatmyam* positions the story of Śiva's appearance in the *thillai* mangrove forest as sequel to the Dārukāvaṇa episode. Ādiseṣa, the thousand-headed serpent on whose body Viṣṇu reclines, hears of Śiva's dance in the

Dārukāvana and yearns to see it. Śiva tells him that to witness it, he must go to a place whose specific coordinates converge on Chidambaram. Ādiseṣa takes the human form of Patañjali and in the *thillai* forest encounters Vyāghrapāda, the saint granted tiger feet and eyes in his palms so that he could climb trees in the early morning dark to collect flowers for prayer unsullied even by bees and insects. On the first full moon day after the winter solstice (*Thaipuṣam*), Śiva reveals the *ānandatāṇḍava*, his blissful dancing form: 'tawny matted locks undulating / as they stretched to the horizon on either side,' 'eyes dancing with compassion / purifying the forest with the lustre of his gentle smile', a *koṉṟai* chaplet of spreading fragrance on He who is the dancer in fire.[16]

Dārukāvana and the Thillaivaṉam, a Himalayan forest and a Tamil mangrove, both dense, dark, somewhat forbidding – these disparate landscapes are conjoined in the story of Chidambaram. Likewise, the Veḷāḷa poet Nellaiappa Pillai's 1829 *Tirunelveli thala purāṇam* or 'story of place' draws the local and equally dense *veṇuvaṉa* of the Nellaiappar temple, the bamboo forest which once named the place before paddy ['*nel*'] did, into the pine forest ambit by equating the two: the bamboo forest *is* where Śiva visits as Bhikṣāṭana. Pillai also alludes poetically to Śiva dancing with Kali in yet another 'forest of pines' which has no obvious referent unless we think of Thiruvālaṅkāṭu, the banyan forest that is another famous setting for Śiva's *tāṇḍava* and, indeed, a performative competition with Goddess Kali. In his imaginary, pine, bamboo, banyan, and perhaps not just these forests but *all* forests become ones through which Bhikṣāṭana may yet pass.

If Śiva is the formless and endless *stambha* [column], the pine forest is where the 'root liṅga' falls and emerges first. If Śiva is the dancing, gathering of vectors, the pine forest and all its corollaries are the axes of movement, the places where strewn fragments are summoned, caught, swirled, and turned inwards.

*Figure 3. The dense thillai mangroves (*Excoecaria agallocha *L.) of Chidambaram are placed in full bloom at the centre as the most significant centre of Śaiva cosmology, one of the most important sites of Śiva's Ananda tandava or cosmic dance. Śiva wears chaplets of Koṉṟai (*Cassia fistula*), a common adornment in Tamil poetic traditions.*

Puśparaṇya, or the Forest of Prayers

Śiva in the forest is constantly transforming, leaving the seeker with an unpredictable deity who may assume the terrible form of Bhairavā, the compassionate Naṭarājā, the unmarked Bhikṣāṭana or the formless *mūla liṅga*. Which Śiva will seekers find? Alongside such wild forest uncertainties simultaneously grew a *Purāṇic* and *Āgamic* garden, in which flowers, fruits, and select botanical offerings become media by which the devoted may commune with Śiva, and invoke his particular forms with greater certainty. From the sylvan witness and symbol of constant transformation, plants as offerings become active agents of that transformation.[17]

The role of flowers as interlocutors between devotee and God is lengthily elaborated in the *Kāmikāgama*, a primarily ritualist text which details the transformation which an offering must undergo in order to 'make it suitable for intimate contact with Śiva'.[18] Only flowers of utmost purity may be offered: those which are unblemished; fully bloomed, but not yet wilted; those with excellent fragrance, but which have not yet been smelled; untouched even by bees and insects such as those the tiger-footed devotee Vyāghrapāda legendarily sought.[19] Such acts of profound devotion suggest that even the blossoming flower's perfection is ephemeral, and the search for the untouched bloom is a process of acquiring purity of faith. Even the choicest blossoms must therefore undergo a further process of ritual purification to imbue them with a 'Śiva-ness'. For 'Śiva does not blend with substance, as water does not blend with a lotus leaf'.[20] Hence flowers, well-chosen and purified, become that substance which facilitates re-integration with Śiva, serving as go-betweens until such time as the devotee can do the same.

Flowers thus make Śiva accessible, and prescriptions of plants to be used in worship become a coded language in which to speak to his particular aspects and to importune specific blessings or outcomes. What emerges from this dialogue is a profusion of botanical offerings – and Śaiva texts go into meticulous detail to provide the seeker with the perfect recipe for attainment.

The *Purāṇic* garden is a methodical arrangement with specific guidelines for time, place, name, form, and number. There is ritualistic reciprocity: you reap as you offer. *Agastya* flowers earn great fame, *datura* blossoms beget sons, and *mallikā* flowers are offered to seek a good woman. Devotees desiring beautiful garments, precious ornaments, and good vehicles should offer worship with *karnikara*, *bandhūka*, and *jātī* flowers respectively. Certain offerings are more esoteric: offerings of *jāpa* flowers bring about the death of enemies, *nirgundī* flowers bring the blessing of a clear mind, and *tulasi* leaves bring both worldly auspiciousness as well as salvation from the world.[21] There are specific flowers for months of the year and for days of the week: *dhaturā* flowers are offered during *pauṣa*, *bilva* leaves during *māgha*, *karavīra* flowers are offered on Wednesday, white lotuses on Thursday, and so on.[22] There are even specifications for flowers offered at different times of day. While *mandāra*, *punnāga*, *arkapuṣpa*, *nāgakesara*, and others may be offered at all times, *kutaja*, *karnikāra*, *kausumbha*, and *mallikā* are best offered at midday, while *mogaraka* flowers are reserved for the night.[23] The Purāṇas are qualitative as well as quantitative in their injunctions: the *Śivapurāṇa* offers a system of measuring flowers by weight, wherein two thousand *bilva* leaves, or twenty full *kamalas*, comprise one unit, and the devotee is urged to use this calculation to measure his offerings to gain the desired outcome.²⁴

Set in this ritualized flower garden Śiva appears much more a domesticated deity than the wild Rudra of the Vedas, one who may be appealed to by means of this floral vocabulary, but only if one knows the correct grammar and the proper etiquette. Amidst the many intricate prescriptions of the *Purāṇas*, however, hide subtle clues of another aspect of Śiva: the Mahākālīśwarā who exists beyond place, time, rules, and conventions. The texts provide the rules – and the rules of circumvention. So, after specifying exact quantities of flowers to offer, the text goes even further to accept a single *arka* flower as the equivalent of ten *karavīra* flowers, but asserts that a single *baka* blossom is dearer to Śiva than even a thousand *arka* flowers.²⁵ Multitudes equal one, and one contains multitudes. The *Āgamas*, after elaborately listing the strict criteria for choosing the appropriate flower, decree that should the devotee fail to find such a specimen, devotion alone is enough to win the heart of the Lord.²⁶ The *Śivapurāṇa*, before highlighting the benefits of offering the choicest celestial bouquet, plainly states that 'If one worships with no specific desires, he will become Śiva himself'. There is a distinct calculus of equivalences that animates Śiva's garden, such that single flowers tend towards infinities much as devotees may tend towards God.

The *Āgamas* place flowers on the fine line between substance and Śiva-ness, dissolving the distinction through flowers. When the devas fail to find Śiva in the forest at Kuśasthali, the forest is revealed to be the physical manifestation of Śiva: 'Formerly, at the outset the entire universe was full of trees. [. . .] All of them originated from the parts of the Lord.'²⁷ Thus, the forest is *aṅgaja*, 'born of the body', literally an embodiment of the Divine. The solution provided to the *devas* is to worship Śiva in the form of the trees – not just as a deity in the forest, but as the deity who is the forest. Hence, in the glorious flame-flowered *palāśa*, 'Hara is always present at its root. The Trident-bearing One himself is present in its stems. Lord Rudra is in the branches, and the destroyer of the Triple-cities is present in the flowers. [. . .] Iśvarā himself is present in the twigs.'²⁸ Śiva is invoked as all-pervasive, yet his different aspects are handled with a botanical and anatomical specificity: he is simultaneously One and many, as the dualists hold. Yet this embodiment further blurs the lines between that which is impure, mortal, and earthly, and that which is pure, eternal, and divine. Śiva is seen in both, and this idea grows and flourishes in later devotional Śaivism.

Figure 4. The bamboo forest of the Nellaiappar temple in southern Tamil Nadu where Śiva is known as Venu-vana-nathar, lord of the bamboo grove, accompanied by Erukkam (Calotropis gigantea) and Dhatūrā (Datura stramonium).

Bilvākṣaraṇya, or the Forest of Devotion

Various everyday Śaiva traditions of worship have adapted botanical myths and practices from the *Purāṇas* and *Āgamas* and turned them into defining devotional practices, premised on the notion that certain plants have arisen from Śiva himself. When Mahādeviakka declares to Śiva, 'You are the forest,' her words pithily summarize Dakṣa in the *Vāyupurāṇa* as he elegizes Śiva in his botanical element: 'You are the creeper, the winding plants, the grass, the medicinal herbs; you are the animals, beasts and birds; you are the beginning of substance, activity and attributes; you are the bestower of flowers and fruits.'[29] In such articulations, plants offer a language of devotionalism beyond ritual forms. Two plants speak best to the variable character of this adoration: *rudrākṣa* and *bilva*.

In the *Skandapurāṇa*, Pārvatī asks Śiva why she only has these dull *rudrākṣa* seeds (pyrenes) with which to adorn herself, and Śiva in response tells the story of their spontaneous creation: 'O Maheśani, once I had been performing penance for thousands of divine years. Although I had controlled it rigorously, my mind was in flutter. . . .' Śiva subsequently opens his eyes to the beauty of the world around him, tears fall from his half-closed eyes, and '[f]rom those tear-drops there arose the Rudraksa plants'.[30] In the often-repeated Tripurasamhāra episode told in the *Mahābhārata*, *asuras* who are Siva-devotees build cities on earth, sky, and air, and win a boon from Brahmā that these can never be destroyed but by a single arrow from Siva, and in their pride and delusion began to wreak havoc. To conjure the beautiful, terrible weapon *aghora*, Siva remains with eyes open for a thousand years – and from a momentary flutter in his great thought and great concentration came three droplets from his three eyes which become the *rudrākṣa*.[31]

The *rudrākṣa* thus becomes emblematic of Śiva's formidable penances. The Purāṇic texts list the effects and extol the (equally transactional) virtues of wearing *rudrākṣa* beads.[32] They mark those with matted hair and *rudrākṣa* ornaments as undoubtedly 'Rudras in human form'.[33] Drawing on this and the Śaiva Āgamas, Nigamajñāna in the sixteenth century sees the seeds as one of five external marks of an *ācārya* or Śaiva preceptor.[34] For lay devotees as for initiates, the *rudrākṣa* is both an identification with Śiva's asceticism and a devotion to its practice.

The *Skandapurāṇa* also tells of the creation of the *bilva* tree from the sweat of Pārvati as she sports in the forest.[35] Born of Pārvati, *bilva* leaves become most dear to Śiva, the foremost of offerings, a prerequisite for worship – a *liṅgam* must be sanctified and cooled with water offered from a bilva leaf before it may be worshipped – and a representation of Śiva himself.

The *Skandapurāṇa* goes on, however, to describe how every holy site in the world is present at the root of the *bilva* tree, deeming the tree itself worthy of worship: 'He who places a row of lighted lamps at the root of the *bilva* tree with reverence becomes endowed with the knowledge of truth and merges into Siva.'[36] The *bilva* tree here is neither offering nor medium, but rather the object of worship itself to be propitiated with lamps, flowers, and scents. The *Vidyeśvarasamhita* of the *Śivapurāṇa* in describing the virtues of the *bilva* says, 'If a slayer of a brahmin circumambulates the trees of Vata and Bilva reciting the verses from Rudrasamhita he will become purified.'[37] The *bilva* is elevated to the level of expiating the sin for which Śiva himself must repent by roaming the earth for a thousand years as an ascetic. The *Bilvāṣṭakam*, an eight-stanza hymn attributed to Śaṅkarāchārya and in praise of the *bilva* extols the numerous merits of offering single *bilva* leaves, concluding each stanza with the refrain '*ekabilvam śivārpaṇam*'.[38] In the balance of singles and multitudes, the exponent which makes the equivalence work is devotion.

Yet all of these legends of the *bilva* tell of a more direct approach to Śiva, one which is unfettered by the specifications of ritual or the injunctions of scripture. Adaptations of botanical images and stories from ritual and mythological texts become the ultimate bases for a personal and embodied relationship with God. If the *rudrākṣa* recalls the fearsome austerities of the ascetic path, *bilva* reminds that any path can be made easier through devotion. In this imagined 'bilvākṣara' forest, Śiva becomes amicable, relatable, accepting of all things offered with or without scriptural directive – the great *yogī* who may at moments flutter, the great God who can be won by the smallest offerings.

Figure 5. The kurumpala or short jackfruit tree represents the forests near the Kuttrālam water falls, which name the presiding deity Kurumpala-veesar or Kuttralanāthar (Lord of Kuttrālam). Alongside are bunches of Ko<u>n</u>rai (Cassia fistula), Erukkam (Calotropis gigantea) shown in larger-than-life form, and Thumbai (Leucas aspera).

Vistarava<u>n</u>a, or the Spreading Forest

Yakśas are the fierce beings who dwell in the roots of trees, tutelary gods of the natural world – but the *yakśaswarūpa* is Śiva's alone, as though to remind that aspects of Rudra endure in the body of a deity now made accessible by devotion. Just as Śiva flouts convention, finding not defilement but the greatest purity in burning grounds, turning ash into the water in which Pāśupata ascetics must bathe in Atimārga traditions – so also does the 'wild and incorrigible [devotee], unquenchable in his yearning' challenge the advantages of wealth and position to turn his body into a temple and offer single leaves and favourite flowers without ritual or ceremony in the *Mantramārga*.[39] Yet, the procedural conventions put in place by the *Āgamas* and the *dharmaśāstra* ideas elaborated by texts of the Śivadharma corpus work in tandem with *Purāṇic*, folk, and bhakti traditions to pull ideas into devotional contexts, creating wide interpretive latitude.⁴⁰ If the *Uttarottarasaṃvāda* [borrowing from the *Umāmaheśvarasaṃvāda*, 'The dialogue between Umā and Maheśvara'] mentions offering *droṇapuṣpī* (*Leucas aspera*) in the month of *Phālguna* as granting a seat beside Indra – that textual referent is accompanied now by the charming story of this devoted little weed's fluster on seeing Śiva, such that it asked for its feet to be placed on his head, when really it meant the reverse.⁴¹ The flower is now among the few that merits placement on Śiva's head in popular southern worship traditions.

Other plants and practices sidle up: the yellow *karavīra* introduced from Central America is a natural addition precisely because of its relatedness to the already-canonical *karavīra* (*Nerium oleander*); the *nagalingapushpam*'s structure, with fertile stamens forming in a ring around the style, and a 'hood' of staminodes with antherodes easily resembles a lingam with a protective snake-like hood hovering above – though the plant was introduced as late as the 1800s. *Bael* sharbat vendors assume logical positions in the vicinity of northern Śiva temples in the hot summer months, even though it is typically the leaves offered and not usually the fruit. It starts to become evident that most of Śiva's flora are inedible, an array of potently medicinal toxicities which must be administered to humans with caution, which are nonetheless the sorts of substances the blue-throated One could swallow so that the *amṛta* [nectar] emerges. Nothing otherwise in Śaiva narrative traditions appears to separate wild from cultivated species; the marking of species for culinary or medicinal consumption arrives largely from Āyurveda.

However classified, the place of botany is paramount and at times heterodox. The opening chapters of the *Śivapurāṇa* declare that 'the merit derived by a person who reads [this text] in the grove of Bilva trees in a temple of Śiva is beyond description in words'.⁴² Leave the temple and read amidst the trees instead, it says, for among the trees and flowers is where merit accrues to those who seek Śiva.

Figure 6. This panorama is a creative interpretation and illustration of the idea of the 'panchasabhai' or the five stages of Śiva as the cosmic dancer, each one associated with a forest in which five major temple shrines once grew in Tamil Nadu, with the Himalayan Devadāru (Cedrus deodara) forests in the far background as the originary site of Śiva's cosmic dance. Other flora associated with Śiva worship and Saiva traditions accompany the trees: Thumbai (Leucas aspera), Malabar glory lilies (Gloriosa superba), Koṉṟai (Cassia fistula), Erukkam (Calotropis gigantea) and Dhatūrā (Datura stramonium).

Glossary of Plant Names

	Sanskrit Name	Botanical Name		Sanskrit Name	Botanical Name
1	Agastya	Sesbania grandiflora	20	Kurumpala	Artocarpus heterophyllus
2	Arjuna	Terminalia arjuna	21	Kutaja	Wrightia antidysenterica
3	Arka/Arkapuśpa	Calotropis gigantea	22	Mallikā	Jasminum sambac
4	Aśoka	Saraca asoca	23	Mandāra	Erythrina variegata
5	Baka	Barleria cristata	24	Mogaraka	Jasminum sambac (var.)
6	Bakula	Mimusops elengi	25	Nāga	Calophyllum inophyllum
7	Bandhūka	Pentapetes phoenicea	26	Nāgakesara	Messua ferrea
8	Bilva	Aegle marmelos	27	Nāgalingapuśpam	Couroupita guianensis
9	Devadāru	Cedrus deodara	28	Nellikaya	Phyllanthus emblica
10	Dhatūrā	Datura stramonium	29	Nirgundī	Vitex negundo
11	Droṇapuṣpī	Leucas aspera	30	Palāśa	Butea monosperma
12	Jāmun	Syzygium cumini	31	Punnāga	Calophyllum inophyllum
13	Jāpa	Hibiscus rosa-chinensis	32	Rudrākṣa	Elaeaocarpus angustifolius
14	Jatī	Jasminum grandiflorum	33	Thillai	Excoecaria agallocha
15	Karavīra	Nerium oleander	34	Tilaka	Wendlandia heynei
16	Karnikā/Karnikāra	Cassia fistula	35	Tulasī	Ocimum tenuiflorum
17	Kausumbha	Carthamus tinctorius	36	Vata	Ficus benghelensis
18	Ketakī	Pandanus odoratissimus	37	Yūthika	Jasminum auriculatum
19	Koṉrai	Cassia fistula			

About the Authors

Soham Kacker is an ecologist and horticulturist based in New Delhi, India. His work looks at plant conservation and ethnobotanical landscapes in the Himalayas and beyond.

Deepa S. Reddy is a cultural anthropologist with the University of Houston-Clear Lake. Her current research is on heritage rice revival, food systems and native forms of ethnobotanical knowledge.

Chippy Diac Vivekanandah is a product designer and an artist who explores biophilic illustrations, visual journaling and captures the world around her as an urban sketcher. Her work is primarily in designing, building, and nurturing creative environments.

Notes

1 *Skandapurāṇa [SkP]*, vol 12, 5.5–36, trans. by J.L. Shastri, G.P. Bhatt, and N.A. Deshpande (Motilal Banarsidass, 2012). Subsequent citations abbreviated as SkP.
2 SkP vol 12, 5.49–72, 6.1–14.

3   Swami Krishnananda, 'The Satarudriya', *Swami Krishnananda: The Divine Life Society* <https://www.swami-krishnananda.org/invoc/in_sata.html> [accessed 21 February 2025]; 'English translation of Sri Rudram – Namakam and Chamakam', Glory of Hinduism, 6 March 2011 <https://gloryofhinduism.blogspot.com/2011/03/english-translation-of-sri-rudram.html> [accessed 21 February 2025].
4   *Speaking Of Śiva*, trans. by A.K. Ramanujan (Penguin, 1973), p. 149; see also Sajjan Shiva, 'Vachana 80: Ettana Maamara', *Vachana-a-Week*, 9 March 2012 <https://vachanaaweek.blogspot.com/2012/03/vachana-80-ettana-maamara-how-are-we.html> [accessed 21 February 2025].
5   *Kumārasambhava of Kālidāsa*, verse 1.4, trans. by M.R. Kale (Standard Publishing, 1917).
6   Stella Kramrisch, *The Presence of Śiva* (Princeton University Press, 1981), p. 292.
7   *Tirunelveli thala Purāṇam of Nellaiappa Pillai/Nellaiappa Kavirayar*, trans. by Don Handelman and David Shulman, in Handelman and Shulman, *Śiva in the Forest of Pines: An Essay on Sorcery and Self-Knowledge* (Oxford University Press, 2004), p. 6. The idea is echoed in Nellaiyappa Kavirayar's 1829 Tamil kavya, the *Tirunelvelittalapurāṇam*: 'There is an ocean called "god's beauty," and they were drowning in it' (trans. by Handelman and Shulman, p. 124)
8   SkP 1.1.6.24, trans. by Shastri, Bhatt, and Deshpande.
9   See Thomas E. Donaldson, 'Bhikṣāṭanamūrti Images from Orissa', *Artibus Asiae*, 47.1 (1986), p. 53, DOI:10.2307/3249979.
10 See *Chidambara Mahatmyam*, verse 14, trans. by E.A. Sivaraman (Bharatiya Vidya Bhavan, 1993).
11 *Pāśupata Sūtra with Panchartha-Bhyasa of Kauṇḍinya*, 1.2–6, trans. by Haripada Chakraborty (Academic Publishers, 1970); Handelman and Shulman, p. 22.
12 Retellings of the pine forest story are a distinct corpus. The tale appears in Kurma, Varāha, Śiva, Skanda, and Vāmana Purāṇas, as well as in several south Indian texts including the Chidambara Mahātmya and the Chidambaram Koil Purāṇam of Umāpati Civācāriyār. For discussions of the various retellings, see Donaldson; Handelman and Shulman; Stella Kramrisch, *Manifestations of Shiva* (Philadelphia Museum of Art, 1981), pp. 153–58, 287–300.
13 Handelman and Shulman, p. 159.
14 Poem 122, trans. by Ramanujan.
15 *Tiruvarutpayan* 8.5; Handelman and Shulman, p. 25.
16 Trans. by David Smith, in *The Dance of Siva: Religion, Art and Poetry in South India*, Cambridge Studies in Religious Traditions (Cambridge University Press, 1996), pp. 182–82; Sambandar, *Digital Tēvāram*, verse 1.45.3, trans. by V.M. Subramanya Ayyar, vol. 13, p. 127 <https://www.ifpindia.org/digitaldb/site/digital_tevaram/U_TEV/VMS1_045.HTM> [accessed 21 February 2025].
17 SkP, vol 1, 5.89.
18 Richard H. Davis, *Worshipping Śiva in Medieval India: Ritual in an Oscillating Universe* (Motilal Banarsidass, 2000), p. 146.
19 *Kāmikāgama*, verse 5.60–61, trans. by S.P. Sabharathnam Sivacharyar (Himalayan Academy, 2020). Subsequent citations abbreviated as KĀ.
20 Nirmalamaṇi, quoted in Davis, p. 144
21 *Śivapurāṇa*, vol 1, 14.26–36, trans. by J.L. Shastri (Motilal Banarsidass, 2014). Subsequent citations abbreviated as SP.
22 *Uttarottarasaṃvāda* 3.12–24, in Nirajan Kafle, 'The Umāmaheśvarasaṃvāda of the Śivadharma and Its Network', in *Śivadharmāmṛta: Essays on the Śivadharma and Its Network* (Unior Press, 2021), pp. 233–54; SkP, vol 3, 7.1–8.
23 SkP, vol 1a, 17.251–254.
24 SP, vol 1, 14.6–8.
25 KĀ, 5.51–54.
26 KĀ, 5.61–63, as translated in Davis.
27 SkP, vol 18, 247.21–22.
28 SkP, vol 18, 249.9–14.
29 *Vāyupurāṇa*, Vol 1, 30.238, trans. by G.V. Tagare (Motilal Banarsidass, 1987).
30 SkP, vol 1, 24.5–7.

31 *The Devi Bhagavata Purāṇa [DBP]*, 11.4.1–11, trans. by Swami Vijñanananda (Allahabad Panini Office, 1922).
32 SP, vol 1, 23–25; SkP, vol 1, 13.50–61; SkP, vol 3, 3.20.1–14.
33 SkP, vol 1, 5.97; Kramisch notes the Tripurasamhara story "assigns to its own setting Rudra's becoming Pasupati," the one who alone can untie "the snares (pāśa) of each paśu" or human animal (pp. 405, 421).
34 T. Ganesan, *Two Śaiva Teachers of the Sixteenth Century: Nigamajñāna I and His Disciple Nigamajñāna II* (Institut Français de Pondichéry, 2009), pp. 3, 15–18.
35 SkP, vol 1, 22.22.
36 SkP, vol 1, 22.23.
37 SP, 2.36.
38 Ādi Śankarāchārya, *Bilvāśtakam*, verse 7 <https://shlokam.org/bilvashtakam/> [accessed 21 February 2025].
39 *Eating God: A Book of Bhakti Poetry*, ed. by Arundhathi Subramaniam (Penguin, 2014), p. 9.
40 See Kafle, p. 240.
41 Kafle, p. 253.
42 SP, vol 1, 2.44.

19
Ancestral Fruits
Strawberries, 'Strawberries' and Acorns in Roman Gardens and Texts

Erzsébet Kovács

Nevana Stajcic writes:

> With food, we are not just buying or consuming a product but a whole system or chain of meanings. An apple is not just the red sweet object that you ingest for nutrition; it is the whole system that contributed to growing the apple: the sun, water, animals, human farmers. Also potentially in the apple is pesticide, transportation issues, Snow White, Macintosh and much more. You are not eating an apple, you are experiencing a system or grammar of food.[1]

In this essay, I would like to give a brief but hopefully all-around overview of what might have gone on in ancient Romans' minds whenever they were having the aesthetic experience of reading pastoral poetry, looking at the garden scenes and still lifes of fresh fruits and frugal meals that decorated the walls of Roman rooms, enjoying dishes served at a banquet, or even foraging for fruit in the Italian countryside. To borrow the analogy of Roland Barthes, like Stajcic did: what was the meaning of certain 'words' in the Roman 'grammar' of food?

I am referring to the fruits of the arbutus tree (*Arbutus unedo L.* and related Old World species such as *Arbutus andrachne L.*) as well as woodland strawberries (*Fragaria vesca L.*) and the nuts of many Old World oak species that bear edible acorns such as the Valonia oak (*Quercus ithaburensis ssp. macrolepis*). It must be noted that for the ancient speakers, genetically unrelated species or parts of plants could have the same name when both looked similar enough or had a similar function. Pliny the Elder discusses both the arbutus tree and woodland strawberries on the same page, while also saying that '[t]he flesh of the ground-strawberry is very different to that of the arbute-tree, which is of a kindred kind: indeed, this is the only instance in which we find a similar fruit growing upon a tree and on the ground'.[2] As for the different oak species known to the ancients, Dioscorides also included sweet chestnuts (*Castanea sativa Mill.*, called here 'Sardian acorns' and 'Zeus' acorns') in his summary of the medical uses of oak, and Pliny the Elder as well as Galen took care to mention that fully ripe dates (*Phoenix spp.*) used to be called 'acorns [of date palms]' in Ancient Greek in general, and in the Roman provinces of Phoenicia and Cilicia in particular.[3]

And so, when we read about these fruits and nuts in different genres of Ancient Greek and Latin texts, both technical and literary ones, we are apparently faced with the traces of a multitude of past realities and ideas, all of which can be reasonably supposed to have belonged together and potentially able to influence each other in the ancient speakers' mental lexicons. In trying to sort out the threads of this tissue of meanings, I would like to build upon the work of Emily Gowers and Jessica Romney, who investigated the literary use of food by certain Roman literary genres such as Roman comedy and satire, and Herodotus, respectively, and found that different foods could represent marked Otherness in a Classical system of dichotomies such as civilized vs. uncivilized, domestic vs. foreign, simple vs. luxurious, and pure vs. mixed or corrupt.[4]

As already discussed by Andrea Maraschi and Corey Straub, acorns of the oak species were held by Ancient Greek and Roman authors to have been the stereotypical food of the people who lived in the mythical Golden Age as well as those Greeks who were living in present Arcadia.[5] Not only does this part of the myth seem to correspond to the lived realities of past human nutrition, it also bears a heavy ideological burden.[6] The Classical myth of the Golden Age was told by and for Ancient Greek and Latin-speaking people in order to make sense of how they themselves and the world in which they were living in came into being. Etymologically, the Greek word *physis* and the Latin word *natura* (from which derive such Latin loans in English as 'natural' or 'physical') originally meant the way, process, and the sum of characteristics of someone or something being born or becoming.[7]

And so, when Greek authors keep recommending to their audience that they should live according to 'nature' or their own specific 'nature' in order to stay healthy, that essentially meant living in the same way, practicing the same habits, and even having the same diet as their ancestors, who were thought to have lived in the Golden Age, before the invention of agriculture, shipping, and other human professions as well as fighting wars. This is what the Classical dichotomy of nature vs. culture or custom/law (in Greek: *physis* vs. *thesis*) means, and it is part of the Classical discourse on food and health. It was also supported by their anthropomorphic understanding of human history, starting with the Golden Age seen as the youth of the world, going on to manhood, and, according to the Classical understanding of the heroic and historic past, finishing with the degeneration of old age, that is, their present age. Additionally, the Stoic philosophers believed in the cyclical nature of world history, meaning that after the end of the world (a universal conflagration, as imagined by them) the same process would start again, and the same Golden Age would return. But a survey and comparison of literary and technical texts will show just how contradictory and controversial recommending practices from before the invention of agriculture to humans who live in the age of agriculture could be.

The first Ancient Greek epic poet to describe the Golden Age was probably Hesiod, who might have been a contemporary of Homer. In his didactic poem that lists peasants' seasonal toils and tasks, *Works and Days* (ll. 109–120), he claims not only that 'a golden race of mortal men' 'lived like gods without sorrow of heart', but also that 'the fruitful earth unforced bare them fruit abundantly'.[8] Since the adjective 'unforced' in the original literally means that the earth produced food for humans automatically (that is, without any human action), it is no wonder that the two Hesiodic criteria for the Golden Age seem to have merged and, in later Greek texts such as Athenian Old Comedy, turned into the Ancient Greek forebear of the medieval European fantasy of the imaginary land of Cockayne. However, the expression translated here as 'fruitful earth' is common with Homer's *Iliad* and *Odyssey*, where the customary epithet of the earth is usually translated as 'life-giving earth', but it literally means 'grain-giving' or 'grain-bearing', and that grain is the Greek name for einkorn wheat (*Triticum monococcum L.*). Even though the *Odyssey* features this archaic wheat with a husk that is difficult to separate from the seed only as fodder fed to horses, we can see that, according to Hesiod, people were eating wheat even before the invention of agriculture. We may also remember that one of the Homeric epithets for Greeks literally means 'those who eat cereals'. It would seem that for the (Athenian) Greeks, even the mythical ancestors of Greeks had to have lived on the typical Greek diet, and the alternative, a potentially Barbarian-looking diet, was yet unimaginable.

On the contrary, in his didactic poem, titled *De Rerum Natura*, 'On the Nature of Things', the Roman poet Lucretius, who lived towards the end of the Roman Republic, not only summarized what the Epicurean philosophic school taught on the human body and soul as well as the world itself, but also wrote that the hardy Golden Age people were not 'to be harmed easily by heat or cold, by unaccustomed foods, or any bodily illness' and that by living their life 'after the roving fashion of the beasts', they 'fed their bodies from the acorn-bearing oaks for the most part, and those arbute berries, which you now see ripe in their scarlet in the winter, then were produced by the earth in greater quantity and larger'.[9] It is almost as if Lucretius could not decide whether to praise or pity the first humans, even when, some twenty lines later, he adds that in those times sex could be bought of a woman 'at a price in acorns, arbute berries or choice pears'. It goes to show the conflicting variety of Greek and Roman opinions on how history began in the course of the five or six hundred years that stand between Hesiod and Lucretius.

In the meantime, a curious new genre also appeared out of, and became distinct from, the scholarly, almost postmodern and intertextual Hellenistic poetry produced by well-read Greek scholars who lived in the newly established Greek royal city of Alexandria. One of them, Theocritus, composed a collection of thirty or so short poems, titled *Idylls*, which, judging by the Greek word's meaning, we can interpret as 'little pictures' or 'small scenes'. Although some of these idylls obviously tried to imitate earlier Greek epic poetry and comedy in a bite-sized form, the bucolic idylls proved more influential and stand at the origin of this new genre which would expand through Antiquity and the Middle Ages to the Early Modern era and even modernity.

Βούκολος (boo-kaw-laws) is Greek for shepherd, and our English adjective, 'idyllic', comes from the term given by Theocritus to his poems, as they describe how shepherds enjoy their way of living in the forest: tending to their animals, but mostly writing poetry in the course of singing contests and courting shepherdesses. In other words, shepherds or, rather, strictly literary shepherds seem to have been understood as being the last remnant of the lost Golden Age. And indeed, Theocritus's shepherds mention the arbutus tree quite often, either the fruit as a delicacy to be chewed on and a thing of pleasure to boast of in a singing contest, or the branches of the tree that goats feed on. When the Roman poet Virgil later imitated the Greek genre and its singing contest between shepherds, and adapted it to Latin, he included this detail, too. In Eclogue 3, the shepherd Menalcas praises his love for Amyntas with these words: 'As moisture to the corn, to ewes with young / Lithe willow, as arbute to the yeanling kids, / So sweet Amyntas, and none else, to me.' And as we remember that Pliny the Elder discusses both the woodland strawberries and the fruits of the arbutus tree in the same chapter, we are not surprised that this eclogue mentions those, too: 'You, picking flowers and strawberries that grow so near the ground, fly hence, boys, get you gone!'[10]

However, in his didactic poem on agriculture, the *Georgics*, Virgil also paradoxically stressed both the need for backbreaking human work and the ever-present possibility of failure of crops and famine. 'Soothly on all must toil be spent' if one cultivates and miraculously transforms nature by grafting: 'But the rough arbutus with walnut-fruit / Is grafted; [. . .] And swine crunched acorns 'neath the boughs of elms'.[11]

And this was the contradiction inherent in the advice of living according to 'nature'. If civilized food is what makes civilized people, and Golden Age people were living like animals, then are their foraged fruits and nuts fit for humans or animals only? What would distinguish humans from animals, if not their habits and, especially, their diet?

And indeed, if we also look at technical literature, they would usually recommend not just the branches (as we have seen with Virgil) but also the fruits of the arbutus tree for animal fodder only. Already the Athenian comedy writer Aristophanes had associated them with bird food, but the Roman agricultural writer Columella included them in two lists. Book 7 (Chapter 9) advises the reader that pigs can be fed with 'oaks, cork-trees, beeches, Turkey oaks, holm-oaks, [. . .] wild fruit-trees like the [. . .] strawberry-trees'. Later on, Book 8 (Chapter 10) gives advice to those who wish to set up an aviary for wild birds to be fashionably fattened and sold for a sizeable profit:

> 'Many people think that a variety of food ought to be provided, lest the thrushes take a dislike to a single food. This variety consists in putting before them seeds of myrtle and mastic, also wild olive and ivy berries and likewise the fruit of the strawberry-tree, for these are the things for which this kind of bird generally seeks in the fields, and so they do away with the distaste for food which they feel in their idle captivity in the aviaries and make the bird population there more voracious, which is a great advantage; for the more they eat the quicker they get fat.'

Just previously, Columella had been discussing dried figs and fine flour for bird food, all of which are originally and mainly food for humans, and apparently also unsuitable for birds, on account of the 'distaste'.[12] So now it is the birds, not people whose 'nature' requires them to feed on the fruits of the arbutus tree – at least for the agricultural writer.

But there had already been a controversy between the literary genres and not just these agricultural manuals but also the medical texts that contained medical theory, recipes, and practical advice on human health. The Hippocratic text titled *On Ancient Medicine* taught that the only reason for inventing medicine and cooking (same thing, the Hippocratic author says) could have been that the beastly diet of Golden Age people, consisting of 'raw foods, unmixed and potent', overpowered their bodies, causing diseases and death. And yet the Hippocratic author's only example for the improved diet is cereals: 'So from wheat, after steeping it, winnowing, grinding and sifting, kneading, baking, they produced bread, and from barley they produced cake.'[13]

No wonder that encyclopedic authors such as Pliny the Elder and medical authors such as Galen indeed warn that the fruits of the arbutus tree are not really fit for human consumption. Pliny the Elder claims in Book 23 that the 'arbutus or *unedo* bears a fruit that is difficult of digestion, and injurious to the stomach' and in Book 15 he even uses this notion to promote his popular etymology of the fruit's Latin name: 'This is a fruit held in no esteem, in proof of which it has gained its name of "*unedo*," people being generally content with eating but one.'

Later, Galen devoted a separate chapter in his book *On the Properties of Foodstuffs* to foraged or 'wild' fruits, which he defines as:

> the plants which spring up in the countryside without any agricultural attention; and indeed they also call wild those vines that no vineyard worker takes care of by digging or hoeing around, or by pruning, or by doing anything like that. Among these plants are the Valonian oak, the common oak, the Holm oak, [. . .] the strawberry tree and other trees like these [. . .].

Interestingly, he also apparently makes the mistake of identifying 'the shrub that bears medlar pears' with the strawberry tree, because he adds that '[i]n Italy they call the fruit of this last shrub "*unedo*"; it is bad for the stomach and causes headache, and is quite sour, with some slight sweetness'. Whichever species he is talking about, Galen claims, nonetheless:

> [the] fruit of the strawberry tree is in every respect inferior to the acorn from the common oak, just as the latter is inferior to what are called chestnuts, for these are the finest of the mast, and some call them 'easily peeled'. Alone of the wild fruits, these give noteworthy nutriment to the body. For [. . .] the fruit of the strawberry tree, [. . .] and the fruit of the wild pear, and others like these have little nutritive value, are all unwholesome, and some of them are bad for the stomach and distasteful – being food for pigs; not the domesticated ones so much as those that live in the mountains. The former, at any rate, are very nutritious.[14]

So, according to Galen, are these foraged fruits and nuts for humans or animals? Galen goes on: 'People in the country regularly eat wild pears, blackberries, mast and *mimaikyla* (as the fruit of the strawberry tree is called), but the fruit of the other trees and shrubs is not eaten very much.' If he might have felt at all compelled to confirm the bucolic ideal, the next sentences make it crystal clear that even real shepherds and rural people would only eat them for famine food and then only after extensive culinary treatment. Galen continues:

> However, once when famine took hold of our land and there was an abundance of mast and medlars, the country folk, who had stored them in pits, had them in place of cereals for the whole winter and into early spring. Before that, mast like this was pig food, but on this occasion they gave up keeping the pigs through winter as they had been accustomed to doing previously. At the start of winter they slaughtered the pigs first and ate them; after that they opened the pits and, having suitably prepared the mast in various ways, they ate it. Sometimes, after boiling it in water, they covered it with hot ash and baked it moderately. Again, on occasion they would make a soup from it, after crushing and pounding it smooth, sometimes pouring in honey, or boiling it with milk.

He might indeed have felt compelled, because he adds the following explanation:

> The nutriment from it [acorns] is abundant, like nothing that has been mentioned in this work up to the present in this section of the book. For mast is just as nourishing as many cereal foods. Of old, so they say, men supported life with it alone – the Arcadians for a very long time – although all present-day Greeks use cereals. But the food from it is slow to pass and has a thick juice, from which it follows that it is also difficult to concoct.[15]

Although the culinary techniques listed by Galen will surely remind us of the instructions requested in the recipes of the only remaining Roman cookbook, the *Apicius*, it may be no coincidence at all that there is not a single *Apicius* recipe for the fruits of the arbutus tree and (as noted by Christopher Grocock) only one with acorns. Book 8, titled *Tetrapus*, of the *Apicius* has ten recipes for dressing wild boar, seven recipes for deer, three recipes for wild goat, three recipes for wild sheep, and thirteen recipes for hare, all of which must have been hunted. The rest of the book is for domestic beef or veal, kid, lamb, and pig, plus there is one recipe at the end for dormice, which was, in all probability, fattened on nuts and acorns. The one recipe with acorns (8.8.3) calls for 'whole pine nuts, chopped almonds, hazelnuts or acorns, whole grains of pepper, meat from the hare itself [. . .] bound together with broken eggs' and stuffed into a hare that would be roasted in the oven. Grocock's translation preserves the semantic choices of the original, yet Grocock adds that '[a]corns were a traditional rustic food for the poor, and we might consider [the word *glandes*, "acorns"] out of place here and take the word to be *iuglandes* and therefore walnuts'.[16] So, what was served on Roman tables?

It may be that satirical texts were the closest to the truth, since humour needs to riff off what its audience holds true, otherwise it will not elicit honest laughter. In the same sense, the Roman satire writer, Juvenal was also probably brutally honest when in his sixth satire he wrote about the wife of the Golden Age cave-dweller, 'one whose breasts gave suck to lusty babes, often more unkempt herself than her acorn-belching spouse (*glandem ructante marito*)'.[17] Belching, we may remember, was considered at the time to have been a sign of indigestion and a digestive disorder itself in medical texts.

In contrast, one of the dishes served at the *nouveau riche* Trimalchio's banquet in Petronius' satirical novel may help us to guess the lost context of Apicius's many recipes for game. Even before this dish comes up, the author prepares us for it as one guest tells a story about his acquaintance who 'used to dine like a prince: boars cooked in a cloth'.[18] Shortly, Trimalchio will have arranged for a fashionable hunting scene:

> the servants came and spread over the couches coverlets painted with nets, and men lying in wait with hunting spears, and all the instruments of the chase. We were still wondering where to turn our expectations, when a great shout was raised outside the dining-room, and in came some Spartan hounds too, and began running round the table. A tray was brought in after them with a wild boar of the largest size upon it, wearing a cap of freedom, with two little baskets woven of palm-twigs hanging from his tusks, one full of dry dates and the other of fresh.

Even the meat-carving slave looked like 'a big bearded man with bands wound round his legs, and a spangled hunting-coat of damasked silk, who drew a hunting-knife'. But the imitation of the hunt does not end there:

> Trimalchio ordered everybody to be given his own portion, and added: 'Now you see what fine acorns the woodland boar has been eating.' Then boys came and took the baskets which hung from her jaws and distributed fresh and dry dates to the guests.[19]

Considering the semantic variety of the word 'acorn' mentioned earlier and the contradictory values attached to the nut, we may conclude that, for the purposes of Petronius' satirical novel, lowly acorns just had to be there for the hunting scene, but also that they just had to be upgraded for the next best thing: palm dates.

So far, we have been able to grasp at the cultural paradox that lay at the heart of Ancient Greek and Roman cultures, and also see that paradox made tangible in a piece of fruit or acorn. In a way, literary descriptions of a barely edible or even unhealthy, but also aspirational food item can provide one with a safer (because spiritual) mode of consumption.

About the Author
Erzsébet Kovács is a Hungarian scholar of Classical philology, who is interested in the intersections of literary and material cultures in Classical Antiquity and the Middle Ages. She researches culinary and medical recipes for ancient and medieval food and medical history.

Notes

1. Nevana Stajcic, 'Understanding Culture: Food as a Means of Communication', *Hemispheres*, 28 (2013), pp. 77–87 (p. 77).
2. Pliny the Elder, *The Natural History*, trans. by John Bostock (Taylor and Francis, 1855), Book 15, Chapter 28. Subsequent references are cited parenthetically.
3. Pedanius Dioscorides of Anazarbus, *De materia medica*, trans. by Lily Y. Beck (Olms – Weidmann, 2005), p 77; Pliny, Book 13, Chapter 9; Galen, *On the Properties of Foodstuffs*, trans. by Owen Powell (Cambridge University Press, 2003), p. 91.
4. Emily Gowers, *The Loaded Table: Representations of Food in Roman Literature* (Oxford University Press, 1993); Jessica Romney, 'Bekos: Food and Notions of Civilisation in Ancient Greek Literature', *YouTube*, 10 November 2016 <https://www.youtube.com/watch?v=yUn2zRTOvo8&t=1040s> [accessed 1 October 2024].
5. Andrea Maraschi, 'The Seed of Hope: Acorns from Famine Food to Delicacy in European History', in *Seeds: Proceedings of the Oxford Symposium on Food and Cookery 2018*, ed. by Mark McWilliams (Prospect, 2019), pp. 177–85; Corey Straub, 'Revisiting the Acorn Eaters: The Case of the Arkadians in Greek Antiquity', in *Seeds*, pp. 281–85.
6. Sarah L.R. Mason, 'Acorns in Human Subsistence' (unpublished PhD thesis, University College London, 1992).
7. Roger French, *Ancient Natural History: Histories of Nature*, Sciences of Antiquity (Routledge, 1994).
8. *Hesiod, the Homeric Hymns and Homerica*, trans. by Hugh G. Evelyn-White (Heinemann, 1914).
9. Lucretius, *De Rerum Natura*, Book 5, ll. 925–44.
10. Virgil, *Eclogues*, trans. by J.B. Greenough (Ginn & Co., 1895), Eclogue 3, ll. 82–83.
11. Virgil, *Georgics*, in *Bucolics, Aeneid, and Georgics*, trans. by J.B. Greenough (Ginn & Co., 1900), Book 2, ll. 61–72.
12. Lucius Junius Moderatus Columella, *On Agriculture*, trans. by E.S. Forster and Edward H. Heffner (Heinemann, 1954), Vol. II, pp. 293, 373.
13. Hippocrates, *On Ancient Medicine*, ed., trans., and comm. by Mark Schiefsky (Brill, 2005).
14. Galen, *On the Properties of Foodstuffs*, pp. 97–98.
15. Galen, *On the Properties of Foodstuffs*, pp. 97–98.
16. *Apicius: A Critical Edition with an Introduction and English Translation*, ed., trans., and comm. by Christopher Grocock and Sally Grainger (Prospect, 2006), p. 353.
17. Juvenal, Satire 6, ll. 9–10 = *Juvenal and Persius*, trans. by G.G. Ramsay (Heinemann, 1928), pp. 84–85.
18. Petronius Arbiter, *Petronius*, trans. by Michael Heseltine (Heinemann, 1913), chapter 38.
19. Petronius Arbiter, *Petronius*, chapter 40.

20
The *Petit Fruit* that Could
How the *Camerise* Captivated Québec

Ivy Lerner-Frank

When I returned to live in Québec in 2018, after over twenty-five years outside the province and Canada, one of my biggest moments of surprise was being at the local farmers' market and seeing something I'd never encountered before. It looked like a blueberry, but longer. Matte and tiny, pendant-like in shape, it was indigo-coloured, lip-puckeringly tart, and a bit sweet.

At first I thought it was a new kind of blueberry, longer than the standard size, and bigger than the wild blueberry that parts of the province are famous for. But these five-quart baskets, the subject of a great deal of attention and excitement, weren't full of blueberries. '*Ah, ce sont des camerises!*' my favourite vendor told me: they're *camerises*!

These *petits fruits* with an almost grape-y texture taste like a cross between a blueberry and a raspberry, and they've taken the province by storm. Starting around 2006 across Québec, farmers took notice of the berry's potential to survive winter and thrive in the province's summer fruit landscape. Recipes for the *camerise* abound on Québec's French-language food websites that emphasize Québec's *terroir*. References to it as our own – *notre petit fruit d'ici* – are not uncommon.

Québec has long been a province with a distinct and enthusiastic food culture, a by-product of long winters, a penchant for preserving, common religious traditions, and government entities keen to support agricultural self-sufficiency which encourage buying local. 'In Québec, we're curious, we're gourmand, we're epicurean,' says Geneviève O'Gleman, a Québécoise nutritionist, best-selling author, and Montréal-based food media personality. She adds, 'We have a culture of curiosity and openness, and at the same time we have a responsibility to local production. These local ingredients represent our tenacity, our desire to always produce more locally and to be more self-sufficient in a climate that's not easy.'[1]

The province was ripe for the production and development of another crop, in part to fully utilize the capacity of processing facilities used for blueberries. But the role of a key government ministry, as well as a lively food scene and an active media to amplify new products, should also be credited with the popularity of this *petit fruit*.

For all the local pride and support around the berry, though, the *camerise* – the *lonicera caerulea* – didn't originate in Québec.

Haskap's Canadian history

Lonicera caerulea is a circumpolar species native to boreal forests in Asia, Europe, and North America, mainly found in low-lying wet areas or in high mountains. Siberian haskap seeds were carried by birds to northern Japan (including Hokkaido Island), where the Ainu people called it haskap (or *haskappu*, among other similar sounding names), meaning 'berry of long life and good vision'. Siberian horticulturists became engaged with the plant, which they called blue honeysuckle, a descriptive translation from the Russian, in the 1950s. Cultivated haskaps were first introduced to Canada that same decade in Beaver Lodge, Alberta, but these were reported as unpalatable.[2]

In the 1990s, Dr Maxine Thompson at Oregon State University worked with Japanese varieties, while Oregon nursery owner Jim Gilbert was exploring Russian versions of the berry, calling those varieties Blue Belle and Berry Blue.[3]

Dr Bob Bors, recently retired from the Department of Plant Sciences at the University of Saskatchewan, became intrigued with the haskap varieties created in Oregon based on the Japanese and Russian varieties. Widely acknowledged as the Canadian father of the haskap, his lab is a centre for its hybridization, knowledge, and development, enthusiastically promoting the suitability of the plant in Canada due to its hardiness, ease of growing, disease resistance, and deliciousness.

Bors began collecting haskap germplasm (plants collected for breeding and research) in 1997, and started a breeding programme, blending the tiny wild Canadian haskap, the Japanese haskap, and the Russian cultivar in 2002. The aim was to develop 'good tasting fruit that can be harvested mechanically'.[4]

While tiny, bitter wild haskaps are found throughout Canada, save for British Columbia, it 'was a crop living in swamps no one wanted to visit'. Bors wrote that the 'berry size of wild North American plants are the size of lentils so it wasn't worth the effort to pick them. But in Japan and Russia, some wild haskap had fruit much larger and in certain areas it was worthwhile to gather them'.[5] The author's research was inconclusive regarding indigenous interest in wild haskaps. Indigenous experts, when asked, told the author the berries were 'as tiny as a mouse poop' and 'anyone relying on these would starve'.[6]

Bors sought to breed berries that would thrive in a challenging environment. 'Hardiness is not just the ability to survive extremely cold winters,' Bors explained, citing that his crops have never seen winter damage, even when his research group forgot to cover plants in small pots with woodchips, their usual procedure for winter stock: 'It may [. . .] involve the ability to stay dormant when warm weather occurs in the middle of winter.' The Japanese types in particular are slower to come out of dormancy, and Bors's breeding plan is to 'develop cultivars with a deeper dormancy for warmer areas and a later season crop'.[7]

As a crop, the berry continues to grow across Canada, including Yukon, with a research centre in Prince Edward Island along with the one in Saskatchewan.[8] But it's in Québec, with its fervent sense of provincial culinary pride, that the *lonicera caerulea* has truly taken off and has been taken on as one of its own.

Why Québec? The Birth of the *Camerise*

Québec has a love affair with berries.

The province is the largest producer of cranberries in Canada, with more acreage than any other province.[9] The Saguenay Lac St-Jean region in central Québec is known as the home of the blueberry, both wild and cultivated, and it's not uncommon to affectionately refer to people from the region as *bleuets*, or blueberries. So it's not surprising to imagine that word of a new berry, easy to grow and suitable for Québec's long, cold winters, would be well received.

In 2006, MAPAQ, the Ministère de l'Agriculture, des Pêcheries et de l'Alimentation (the Ministry of Agriculture, Fisheries, and Food) sent a delegation of farmers from the region to meet Bors at the University of Saskatchewan to learn about the haskap cultivar programme. Twenty farmers and MAPAQ officials went for an exploratory visit. In an interview with the author, MAPAQ's Pierre-Olivier Martel explained that the interest of the Québec delegation was in part because of the commercial opportunities presented by this new crop.

Combining the economic opportunity of freezing facilities otherwise unused at the time of year haskaps ripen with the support that MAPAQ has offered since that first delegation to Saskatchewan made for a winning combination. 'We could sort, clean, and freeze the berries, and then export these *petits fruits* all over the world,' Martel explained. 'The use of existing freezing facilities remains a factor in the berry's success, given its fragility as a fresh fruit with a shelf life of about 72 hours.'[10]

The fact that the haskap is one of the first fruits of the season, before raspberries and blueberries, was not lost on the delegation. 'They immediately saw there was enormous potential for the region', Martel recalled. 'In the beginning, those first cultivars were bitter, and the plants were less productive. But when Bob Bors put the new cultivars like the Indigo series on the market, the taste was so much better, and a lot of producers started growing them.'

The delegates, who continued to be trained by the breeders at the University of Saskatchewan, thought it was important to have a name that would describe the fruit better for Québécois French speakers. They came up with *camerisier*, combining the words for 'at ground level' and 'cherry tree'.[11]

Planting began around 2007 in the Lac St-Jean-Saguenay blueberry growing region. Commercial harvests started in 2012, with the first significant crops produced about seven years later.[12] Hardy to minus 47 Celsius and loving the cold, the plant is relatively insect- and disease-resistant, with modest watering requirements.[13] Depending on the variety, the harvest can go on for weeks after ripening in mid-to late June, just in time for the St-Jean Baptiste provincial holiday.

According to Camerise Québec, as of 2017 (the most recent data on their website), there were more than 200 farmers growing *camerise*, only slightly less than half the farmers growing wild blueberries, with an expected production of over 150 thousand kilos of the fruit that year.[14] In a recent interview, Martel was excited about increases in production, citing the government body's statistics (quoted in pounds): from 550,000 lbs in 2021 (~250,000 kg), to 625,000 pounds in 2022 (~283,000 kg) to over 775,000 in 2023 (~350,000 kg). Martel estimated that the yield in 2024 would be over 1 million pounds.

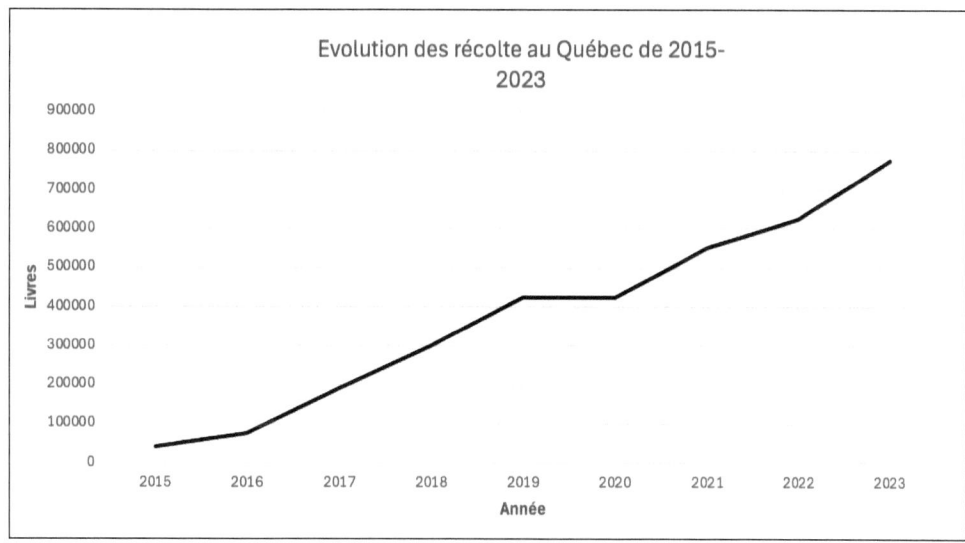

Figure 1. Evolution of the Québec camerise *harvest from 2015 to 2023. Source: MAPAQ/ Pierre-Olivier Martel.*

This yield remains dwarfed by the production of blueberries (approximately 100 million pounds) and cranberries (200 million pounds) in the province. The *camerise* is thus viewed as an emergent fruit, with considerable potential for additional growers to engage with the plant and the development of new markets.

At the time of the most recent Camerise Québec report, about 25% of the *camerise* crop was sold directly to consumers at farm, with an important *auto-cueillette* or you-pick component, the remainder going towards transformation into yogurt, preserves, drinks (including beer, gin, and *hydromel*, or mead), wholesale, industry, and dried or other uses. Berries destined for transformation are generally harvested by machine, as they need not look perfect for the consumer.

About one-third of the *camerises* planted in Québec are in the Lac St-Jean–Saguenay region, estimates Manuel Gosselin, president of Camerise Québec, which represents approximately eighty growers. The reason for this concentration of *camerise* growing in the region is manifold: not only is the region particularly suited to growing the berry, with ample facilities for processing, but MAPAQ's close collaboration with the Lac-St-Jean based Vegetolab, a propagation laboratory, was active in providing training, conferences, and support for *camerise* growers. 'In other words,' said Gosselin, a *camerise* farmer himself, 'there was a market, we were set up, and all we had to was grow it. We were already leaders with blueberries and cranberries; this is how we also became leaders with the *camerise*.'[15]

Access to fresh produce at a reasonable price, especially in season, is a provincial preoccupation. Knowing more about these berries and being able to find them readily – what cider maker Marc-Antoine Arsenault-Chiasson called a 'democratization' of the fruit – is an important consideration.[16] With the ever-increasing number of plants grown

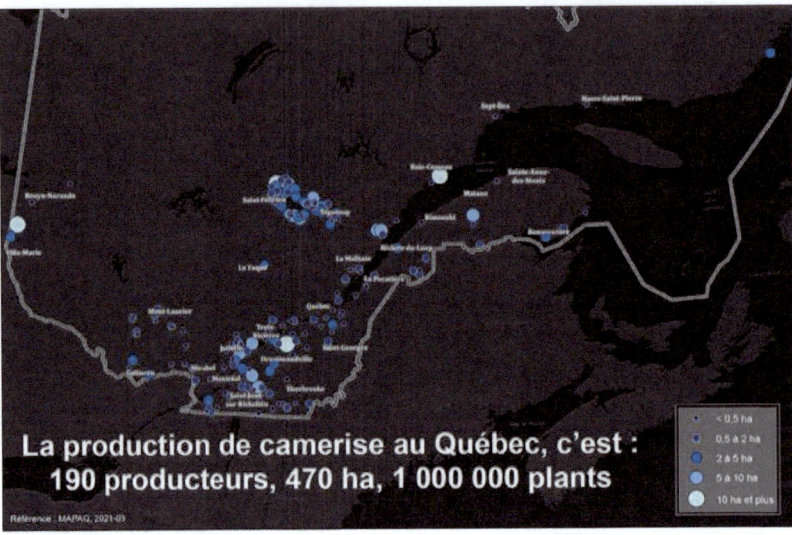

Figure 2. Québec camerise *production: 190 producers, 470 hectares, 1,000,000 plants. Source: MAPAQ.*

across the province, Québécois keen to try this emerging fruit can easily visit an *auto-cueillette* farm, a phenomenon that is more popular here than in any other province.

Institutional Strength and Support

The role of MAPAQ cannot be overstated in Québec. With a mandate 'to promote a quality food supply and support the development of a prosperous and sustainable biofood sector that contributes to the vitality of the regions and the health of the population', it plays a huge role in the province's approach to food systems, and feeds overall into a feeling of provincial pride and identity.[17]

Economic opportunity sustains the vitality of Québec's regions, and with the Ministry's responsibility for everything pertaining to agriculture, food processing, and the food trade, as well as hotels, restaurants, and private and institutional markets, MAPAQ's promotional, training, and support activities can make or break the success of an agricultural endeavour. This is why the Ministry's support for *camerise* growers, through training and their promotion of the crop, has been so key to its status as an emerging Québec food.

In 1996, MAPAQ created Aliments Québec, a consumer-forward NGO with the same mission: promoting Québec-produced ingredients. Membership in Aliments Québec, which includes the *Québec au menu* initiative is open to restaurants and food processors, providing certification for products that contain more than 85% of Québec-sourced ingredients and transformed foods.

Camerise fits right in there.

Farming (and Picking) in Québec

Québec's diverse farming activities make it the leader in numerous commodities, including maple products, dairy cows, pigs, and cranberries.[18] But there are other distinctive aspects to the provincial agricultural landscape.

According to Statistics Canada, Québec has the most farms reporting organic production; in 2021, this represented almost 2500 farms, including those producing maple syrup, and 43.7% of organic production overall in Canada. This represents almost 2.5% more than the next closest province, and a statistic increasing year upon year from the previous census.[19]

While much of the province's organic production is related to maple, the fact that *camerise* lend themselves to being grown organically is not lost on growers like Nathalie Lacroix of Les Petits Fruits St-Louis, whose family farm boasts over 8000 plants in St-Louis-de-Gonzague, an hour from Montréal. 'For any agricultural producer, there's a risk in playing with chemicals,' Lacroix said in interview. 'We're responsible for this land, for the health of our family, and the health of the people who come here to pick. We were close to being organic prior to certification, and we just had to modify a fertilizer. We pushed for that certification, to reassure everyone.'[20]

There's more synergy between maple production and *camerise* production in the province: the phenomenon of going out to the farm to experience it first-hand. In the springtime, when maple syrup sap flows, people flock to sugar shacks to enjoy maple syrup tapping and traditional foods. The same enthusiasm greets the ripening of *camerise* as the first crop of the year. Statistics Canada reports that, in 2020, over 20% of farms in Québec reported direct sales, versus slightly over 13% for the rest of Canada that same year.[21] While this statistic can represent buying produce at a farmer's stand as well as *auto-cueillette*, the interest of Québécois in purchasing directly and connecting with the growers is part of the province's culinary culture. 'It's not just to go get fruit,' Manuel Gosselin from Camerise Québec says. 'It's a societal phenomenon, part of our tradition to pick berries.'

Nutritionist and author O'Gleman echoes Gosselin's comments about the popularity of *auto-cueillette* in the province, especially for *camerise*: 'It's a great activity and symbol of the summer. People say "We're not going to pick strawberries, we're going to pick *camerises*!" That generates excitement, when you help someone discover something new.'[22]

Camerise Québec's Gosselin also cites the ease of picking *camerise* as a factor in its popularity; there is no crouching as with blueberries and strawberries, and there are no thorns as with raspberries or blackberries. The bushes grow to a height easy to pick – approximately 5 feet or 1.5 metres high – so small children can pick at the bottom of the bush and adults from the top. Most importantly, the entire bush generally reaches maturity at once, meaning that there is no rushing up and down rows, trying to find ripe berries on one bush or another.[23]

Before harvesting machines came into vogue, growers used to put hard plastic kiddie pools cut in half underneath the bushes, shake them and harvest from there: it's that easy, Gosselin says. Picking *camerise* by just shaking the bushes is a method still used by Hugo Bourdelais, a specialty grower in the Lanaudière region who sells his perfect-looking fruit to top restaurants.[24]

Promotion of the *camerise* is very much focused on the healthful antioxidant properties of the berry and the fact that it's grown locally by primarily small family farms, providing an alternative to other berries which come to the province from mega-producers in more southern climes. 'And it's blue', says Gosselin, alluding to the bright colour that graces the Québec flag and promotional material for the province.

Enthusiastic Growers of All Sizes

The sweet, tart taste of the *camerise*, ease of growing, antioxidant health benefits, and sheer novelty are enough to keep growers, home gardeners, and hobbyists alike coming back for more.

Camerise Québec divides growers roughly into three groups: less than 2000 plants; between 2000 to 10,000 plants; and more than 10,000 plants, referring to semi-industrial growers. According to Gosselin of Camerise Québec and owner of family *camerise* farm Indigo+, the bulk of Québécois growers are in the middle category. The numbers are hard to pin down, he says, because many growers will have *camerise* as anywhere between their fifth to tenth crop, often as part of a suite of other fruits ranging from strawberries to raspberries and apples. The most recent official statistics from MAPAQ indicate 190 growers over 470 hectares across the province, with approximately eighty of those members of Camerise Québec.

Yield per bush can vary. Saskatchewan's Bors says the oldest plants in his lab produce about seven kilos per bush after about five years. Québec producers interviewed informally indicated yields from one to ten kilos per bush. In his document regarding growing haskap in Canada, Bors raises two points: first, because the crop is produced in the early part of the season, he doubts it would be as productive as cherries or apples; and secondly, that those early crops can be anticipated to be consistent, as the plant has most of the summer to prepare for winter. The longevity of the plant is another plus, says Bors, noting there were 'many productive 30 year old plants at Japanese farms he'd visited, and ornamental blue honeysuckle plants the same age at University of Saskatchewan'.[25]

Nathalie Lacroix and her husband, of Petits Fruit St-Louis, knew that they wanted to have a family farm. When a parcel of land with seventeen arable acres became available, they jumped on it, but weren't sure what to grow. Nathalie read about *camerise* in a catalogue and became obsessed; the couple packed their three young children into their van straightaway, driving five and a half hours to Lac St-Jean to see the plants for themselves.

With training from MAPAQ about growing and pruning, Petits Fruits St-Louis saw annual yields between 5000 to 7000 kilos from their organic bushes nestled at the edge of a forest: less than they'd hoped, but still a comfortable harvest. Lacroix and her husband have divided up their business, with *auto-cueillette* representing a third of their output. During harvest time, they cover the rows in netting to keep the birds out; once they start picking, they rent a refrigerated truck, filling up cases with the fragile berries, and taking laden pallets to a facility where the berries are cleaned, sorted, and frozen. (The same facility handles raspberries and blueberries later in the season; as with raspberries, the *camerise* are cleaned with forced air instead of water.) The remainder of the harvest is transformed into jellies, jams, coulis, and even candles and soaps, sold in boutiques, on-site, and at other markets throughout the year.[26]

Similar to apples, *camerise* need cross-pollination. Successful crops require the help of native insects, bumblebees, and honeybees, so growers (both commercial and small producers) usually plant two or three varieties, ideally in hedgerows to facilitate easy pollination.[27] Given that *camerise* flower in April, well before other berry crops, some growers buy flats of workhorse bumblebees, tolerant of the cold weather that the plants thrive on. (The bumblebees die at the end of the season.)

Growers like Lacroix share the expense of honeybee rental with blueberry growers who need pollination later in the growing season, shipping the hives off once the buzzing has subsided with the *camerise* harvest. That sharing and biodiversity is something that producers appreciate and seek, according to Marc-Antoine Arsenault-Chiasson, owner of Cidricole Équinoxe, an apple cider producer in Farnham, southern Québec, who grows *camerise* alongside apples, sumac, and other berries, diversifying his offerings and promoting the berry at the same time.[28]

Montréal-based home gardener Patricia Gagnon bought her *camerise* bushes as an end-of-year special, starting with three plants. They did so well in her driveway that she bought two more and made an edible hedge out of them. Gagnon, an inveterate experimenter, eats her home-grown *camerises* with overnight oats every day, documenting her urban paradise throughout the year on Instagram. 'They're now about one metre high; they grow surprisingly well despite the fact that I do nothing to them,' she said. Gagnon has blueberries, too, but they don't do as well. 'I always tell people to try to grow things. And with *camerises*, it's so simple.'[29]

Osteopath François Saine divides his time between Montréal and land he inherited in the Mauricie region in central Québec, where he has 150 *camerisiers*. 'I can get a bit obsessive, so when I found plants wholesale at $3 each, I thought "Great! I have a tractor to plant them!"' Like Gagnon, Saine has been delighted with how easy growing *camerise* is and has a particular fondness for the larger fruit of the Aurora variety. He does 'nothing special' to take care of them, even to protect them from birds: 'Right now it's not part of the birds' habit to come find them at my place. But they love the plants I gave my neighbour, so they go over there. They haven't figured out I have them, too.'[30]

Culinary Delights of the *Camerise*

The popularity of *camerise* can be attributed to its taste, with its unexpected tartness a big part of the allure. 'The taste is amazing,' says Marc-Antoine Arsenault-Chiasson of Cidricole Équinoxe, who transforms his *camerise* into hard cider. 'It's like a blueberry in the shape of a banana with a taste of raspberry – but really, the taste can't be described. A *camerise* is like a *camerise*!'[31]

Melanie Aumais, co-owner of Montréal's Fin Soda, makes ready-to-drink non-alcoholic cocktails in tiny cans, using all fresh ingredients. The former bartender wanted to create a new flavour. 'We knew we wanted a berry, and we knew it was sour. Then we went *camerise* picking!' she says, laughing. 'When we found out we could have frozen Québec *camerise* all year long, we mixed it with balsam fir and cedar.'[32] Fin Soda's website describes it as having a 'subtle forest vibe, typical to our terroir – and perfect before, after, or during a ski day'.[33]

Aumais' *camerise* soda is sold across the province, including in high end restaurants like Montréal's renowned Vin Mon Lapin. There, the purple concoction gets mixed with mezcal as a fun pairing. Mon Lapin co-owner and co-chef Jessica Noël cooks with *camerise* during the season, mostly in desserts. 'I love to showcase them without doing too much, like in an olive oil semifreddo,' she says. 'They're much brighter in flavour than blueberries, and when they're simmered with a little sugar, they taste like sour patch kid candies.'[34]

Mon Lapin's *camerises* come from the decades-old Ferme Bourdelais in the Lanaudière region, between Montréal and Québec City. *Camerises* comprise a quarter of the crops on the thirteen-acre family farm known for its high-quality produce. They initially planted the first cultivars, Berry Blue and Indigo Gem, in 2007, and then shifted to the newer, sweeter varieties like Aurora and Boreal Bizarre. Their *auto-cueillette* and restaurant wholesale business make the farm a destination for gourmands in the region, especially a once-a-year *camerise* event at peak harvest time which features a visit to the farm and a gourmet meal in their sugar shack.

Award-winning Montréal-based chef Simon Mathys has made his career focused on local ingredients and takes advantage of the brief *camerise* season. 'The *camerise* has a nice complexity, with good acidity,' he says. 'I find it goes better with savoury than with other sweeter fruits, like strawberries. I might smoke it lightly to dehydrate it a bit, and then put it with something like duck foie gras or beef fat. Or just eating it plain. It's really good.'[35]

With frozen berries available year-round, recipes for *camerises* abound in local media. In mid-winter, the website for national French broadcaster Radio Canada's food vertical, *Mordu* (literally, a play on the word 'bitten'), featured a *Longue Vie* or 'long life' cocktail, named after the Japanese origins of the word 'haskap', an oolong tea mocktail made with a *camerise* syrup. *Mordu* featured almost twenty *camerise* recipes on their site this year so far, with the promise of more coming as the berry season approaches. A sampling includes grilled cod with *camerise* sauce, frozen yogurt with *camerise*, mint, and elderflower, and a real cocktail called *Bisou du Nord* (literally, Kiss from the North) featuring local Québec gin, *camerise*, and sparkling wine, a northern riff on a Kir Royale.

Québec alcohol producers form an important component of wholesale *camerise* buyers, according to Camerise Québec: gins such as a salty *camerise* gin made by the Menaud distillery, along with their *camerise* beer, are but two alcohol-based options, alongside liqueurs like *crème de camerise* and even *camerise* wine.

Conclusion: Saskatchewan-Created Hybrids Become Québec's Pride
The rise of the *camerise* and its adoption by Québec growers and gourmands is a curious yet somewhat inevitable and tasty conclusion to what Bob Bors started in Saskatchewan with Siberian, Japanese, and Canadian haskap. The berry has served as a culinary departure point, embraced wholeheartedly by a province connected to yet unbound by traditions, a forward-looking Ministry, and an adventurous public keen to try new tastes.

Nutritionist and author O'Gleman concurs. 'We're excited when we have one more local product.' she says. 'It's to our credit as a people to be proud of what we produce, and to carve out a place for ourselves in the world's gastronomy.' O'Gleman concludes, 'We're not sitting on our heritage. We're still in pioneering mode, trying to produce better and more, defying the climate to be increasingly self-sufficient and to have more diversity of local products.'

The *camerise* fits that criteria.

About the Author

Ivy Lerner-Frank is a former oratorio singer and Canadian diplomat, now a food and travel writer based in Montréal. She has called Beijing, Bangalore, Delhi, Hong Kong, and Manila home, but loves Montréal – including its farmers' markets – the best.

Notes

1. Genevieve O'Gleman, interview with the author, 25 April 2024.
2. 'Haskaps (Cultural)', University of Saskachewan, 2 May 2004 <https://gardening.usask.ca/gardening-advice/gardenline-nested-pages/food-plant-pages/fruit/haskap.php> [accessed 28 February 2025].
3. Bob Bors, 'Growing Haskap in Canada', USask <https://research-groups.usask.ca/fruit/documents/haskap/growinghaskapinCanada.pdf> [accessed 27 February 2025].
4. 'Haskaps (Cultural)'.
5. Bob Bors, 'Haskap Rumours', University of Saskatchewan <https://research-groups.usask.ca/fruit/Fruit%20crops/haskap-rumors.pdf> [accessed 28 February 2025].
6. Manuel Gosselin, interview with the author, 16 April 2024.
7. Bors, 'Growing Haskap in Canada'.
8. Jennifer Cole, 'How the Haskap Berry Survives Arctic Temperatures', *Modern Farmer*, 19 July 2023 <https://modernfarmer.com/2023/07/how-the-haskap-berry-survives-arctic-temperatures/> [accessed 28 February 2025]; 'Research & Development', Phytocultures <https://phytocultures.com/camerise-haskap-berries/research-development/> [accessed 28 February 2025]; 'Usask Fruit Program', University of Saskatchewan <https://research-groups.usask.ca/fruit/Fruit%20crops/haskap.php> [accessed 28 February 2025].
9. Bernard Brault, 'It's Harvest Season at the Cranberry Capital of Canada', *The Globe and Mail*, 17 October 2021 <https://www.theglobeandmail.com/canada/article-its-harvest-season-at-the-cranberry-capital-of-canada/> [accessed 28 February 2025]; Martin S-Beaulieu, 'Quebec Continues to be the Main Force behind Maple, Blueberries, Cranberries, Dairy Cows and Pigs', Canadian Agriculture at a Glance, Statistics Canada, 3 February 2023 <https://www150.statcan.gc.ca/n1/pub/96-325-x/2021001/article/00005-eng.htm> [accessed 28 February 2025].
10. Pierre Olivier Martel, interview with the author, 19 April 2024.
11. 'Name Origin', Jardin de Camerise <https://camerises.ca/en/name-origin/> [accessed 28 February 2025].
12. '*L'industrie de la camerise*', Camerise Québec <https://camerisequebec.com/a-propos/lindustrie/> [accessed 28 February 2025].
13. Dave Trenholm, 'Why I Love Growing Haskap', *Cold Climate Abundance*, 30 December 2020 <https://www.coldclimateabundance.ca/2020/12/30/why-i-love-growing-haskap/> [accessed 28 February 2025].
14. '*L'industrie de la camerise*'; 'Survey of the Wild Blueberry Industry in Quebec', Le bluet – perle des p'tits fruits <https://perlebleue.ca/images/documents/amenagement/guideanglais/e001.pdf> [accessed 28 February 2025].
15. Manuel Gosselin, interview with the author, 16 April 2024.
16. Marc-Antoine Arsenault-Chiasson, interview with the author, 15 April 2024.
17. '*Mission et mandats du ministère de l'Agriculture, des Pêcheries et de l'Alimenation* (MAPAQ)', Gouvernement du Québec, 5 February 2025 <https://www.quebec.ca/gouvernement/ministere/agriculture-pecheries-alimentation/mission-et-mandats> [accessed 28 February 2025].

18. S-Beaulieu.
19. S-Beaulieu.
20. Nathalie Lacroix, interview with the author, 21 April 2024.
21. S-Beaulieu.
22. Geneviève O'Gleman, interview with the author, 25 April 2024.
23. O'Gleman.
24. Hugo Bourdelais, interview with the author, 6 May 2024.
25. Bors, 'Growing Haskap in Canada'.
26. Lacroix.
27. 'Pollination', Phytocultures <https://phytocultures.com/camerise-haskap-berries/growers-guide/pollination/> [accessed 28 February 2025].
28. Marc-Antoine Arsenault-Chiasson, interview with the author, 15 April 2024.
29. Patricia Gagnon, interview with the author, 23 April 2024.
30. François Saine, interview with the author, 18 April 2024.
31. Marc-Antoine Arsenault-Chiasson, interview with the author, 15 April 2024.
32. Melanie Aumais, interview with the author, 17 April 2024.
33. 'Haskap Berry and Balsam Fir', Fin Soda <https://finsoda.com/products/camerise-sapin-baumier-12-x-250ml> [accessed 18 March 2025].
34. Jessica Noël, interview with the author, 29 February 2024.
35. Simon Mathys, interview with the author, 28 April 2024.

21
#mangowars
Why Are Some Varieties of Fruit More Popular than Others?

Priya Mani

'Alphonso is the most overrated stuff you will ever come across . . . Try Dashahari, Chausa, Safeda and ultimate Langda/Malda.'

—#mangowars

In the early days of April 2021, a seemingly innocuous tweet sparked a fervent discourse on the merits of the Alphonso, the crown jewel of Indian mangoes. What began as a gentle prompt to celebrate regional diversity swiftly evolved into a conflagration of hyper-regionalism, trending as #mangowars, a vivid testament to the dichotomous nature of mango appreciation, transforming into a battleground of condescending clashes. Boundaries, both regional and national, blur and intertwine, defying precise delineation and instead beckoning only to be hinted at.

On the world stage too, mango patriotism is not new. The mango is so ingrained in local memory that people rarely acknowledge that it is the product of early globalization. The humble tropical fruit today is cultivated in no less than eighty-seven countries.[1] Yet, a little-known truth persists: all commercial cultivars trace their lineage to a singular progenitor – the *Mangifera indica*, India's native mangoes meticulously bred for superior texture, taste, and aroma, then propagated across the globe.[2] Indeed, the mango stands as colonialism's greatest, albeit unsung, discovery, its seeds and seedlings freely disseminated and transplanted worldwide in a grand endeavour to commercialize the fruit on a global scale.

The sweet Alphonso mango, once hailed as India's fruit ambassador, has taken an unexpected turn toward bitterness. Cultural conditioning has elevated the Alphonso to an unassailable position in the public palate, casting a shadow over hyper-local varieties, and its consistently high value distorts market dynamics to alarming proportions. Ingrained in the Alphonso mango narrative is a deep-seated classism, perpetuating culinary elitism that caters to a privileged few. In recent years, both scholars and the media have delved into the origins of the Alphonso mango. One prevalent narrative suggests that Jesuit priests introduced the technique of mango grafting to India, in Goa, Portugal's first territorial possession in Asia captured in 1510. They used a mango from Brazil to propagate the extraordinary variety and bestowed upon it the name of their esteemed leader, Afonso de Albuquerque. Could India have developed a reputation for its mango if such knowledge was to be gained only as recently as the sixteenth century? Literature review before this period does not name specific varieties, but mangoes from certain regions had already gained nationwide reputation. To understand this, I delved a little into how the Alphonso came to be, the legacy of its name, and the propagation of mango trees in India.

Whose Alphonso Is It Anyway?

In the annals of botanical exploration, the Portuguese herbalist Garcia de Orta stands as a pioneering figure, chronicling the verdant treasures of early modern India in 1563. Engaging in a fictional dialogue with his esteemed guest Dr Ruano, Orta painted a vivid picture of the local mangoes of his time. 'The mangoes of the kingdom of Guzerat are also very good,' he remarked, 'especially some they call Guzaratas, which are not so very large, but with a splendid scent and taste. The stone is very small.'[3] Yet, curiously, Orta did not mention the Alphonso mango by name, despite his work being composed in Goa.

A century later, Venetian traveller and self-taught physician Niccolao Manucci in his 1698 travelogue of Mughal India offers some idea of how these local mango names came to be:

> I may mention that the best mangoes grow in the island of Goa. They have special names, which are as follows: mangoes of *Niculao Affonso*, *Malaiasses* (? of Malacca), *Carreira branca* (white Carreira), *Carreira vermehla* (red Carreira) of *Conde*, of *Joani*, *Parreira, Babia* [229] (large and round), of *Araup*, of *Porta*, of *Secreta*, of *Mainato*, of *Our Lady of Agua de Lupe*. These are again divided into varieties, with special colour, scent, and flavour. I have eaten many that had the taste of the peaches, plums, pears, and apples of Europe.[4]

Interestingly, Manucci brings up a mango variety named after Niculao Affonso, which shares similarities with today's renowned Alphonso mango. The Alphonso mango, in its journey through language and culture, has been vernacularized in various forms, including alfonso, alphonse, appus, hapus, and more. As the urban legend goes, the Alphonso mango is named after Portuguese statesman and Goa's first Duke Afonso de Albuquerque (1453–1515). However, I have been unable to trace any primary source to establish this fact. More recently, Goan agriculturist Fernando do Rego noted that 'Jesuit priests "baptized" new mango species with the same zeal they renamed converts to Christianity'.[5]

The Alphonso mango, once a coastal variety with a generic name, has metamorphosed into a branded entity. Further, the mango as a fruit exhibits a wide variability in quality, even amongst the clones of the same variety. In 1998, Pandey reported seven clones of 'Alphonso': 'Alphonso Behat' and 'Alphonso Bihar' from Bihar; 'Alphonso Batli', 'Alphonso Black', and 'Alphonso Bombay' from Maharashtra; 'Alphonso Punjab' from Punjab; and 'Alphonso White' or 'Bili Ishada' from the North Canara district of Karnataka.[6] Nearly a century before Kumar, George Woodrow, in his handbook on Indian mangoes, offered a glimpse of the Alphonso diversity present at the turn of the twentieth century. He lists the Afonza of Goa alongside the Alphonse from Kirkee (present

Figure 1. 'Alphonse' in The Mango: Its Culture and Varieties *by G. Marshall Woodrow, 1904. Illustration by R.K. Bhide. The author notes that this is a mango from Kirkee (Pune district, Maharashtra).*

day Pune), Kagdi alphose and Swrawini Alphonse from Bombay and Kala Alphonse (no provenance mentioned). He describes their varying morphological and organoleptic variations, and adds, 'The keeping qualities of this fruit are excellent, and it is generally admitted the best of all Mangoes. The name is applied in the markets to many distinct sorts of greatly varied merit.'[7]

In India, a popular notion persists that mango trees are grown from seeds. Grandparents retell the fables of cherished mango trees in family gardens that began with the serendipitous act of tossing a venerable mango seed into the garden. The 'Mango Seed Trick' is a popular street magic performance. In this mesmerizing display of sleight of hand, the *jadugar* (magician) appears to sow a mango seed and then promptly waters it and covers it with a basket. In the blink of an eye, spectators witness a young plant emerge, adorned with its characteristic rose-tinged leaves.

Figure 2. Performance of the mango-from-seed-trick in colonial India.[8]

But interestingly, mango seeds are either mono-embryonic (single embryo, like *M. indica*) or poly-embryonic (multiple embryos, as in most Southeast Asian mangoes). Only poly-embryonic seeds produce true-to-type (clones) of the parent. Leading experts on mango R.E. Litz and S.K. Mukherjee state that only a few poly-embryonic cultivars occur along the west coast of India; however, they may have been introduced into Goa from Southeast Asia, perhaps by the Portuguese from their colonies of Malacca in the Malay Peninsula or Timor in the Indonesian archipelago, which has later been confirmed by gene analysis. Mango varieties in India are primarily mono-embryonic, so most cultivars do not produce seedlings true-to-type, and vegetative propagation is necessary to develop trees that produce uniform yield, fruit size and quality. So, while the lore of the tree-from-the-seed holds true, the creation of varieties requires deep knowledge of vegetative propagation for consistency of taste, texture, and aroma.

*Figure 3. Magician Cherpulasseri Shamsudheen staging 'Mango Tree Magic Show'.*⁹

In 1920, American plant explorer Wilson Popenoe, who worked for the US Department of Agriculture, noted:

> It has been observed in Florida that mono-embryonic grafted varieties, such as Mulgoba, will, when grown from seed, sometimes revert to polyembryony in the first generation. G.L. Chauveaud has advanced the theory that polyembryony is a more primitive state than mono-embryony, which would seem to be borne out by this observation; for it must be true that the choice mangos of India which have been propagated by grafting for centuries are less primitive in character than the semi-wild seedling races.

Scholarly squabbling is not new either, as Popenoe points out:

> Inarching is an ancient method of vegetative propagation. While several writers have attempted to show that it was not known in India previous to the arrival of Europeans, and that the Jesuits at Goa were the first to apply it to the mango, others have held the belief, based on research in the literature of ancient India, that the Hindus propagated their choice mangos by inarching for centuries before any Europeans visited the country.¹⁰

The legend of the Ekambareshwar Temple in the southern city of Kanchipuram may add to our understanding of early mango grafting in India. Shiva, the presiding deity of this Pallava-era temple is manifested as *eka-amba-eashwar*, or the Lord of the Mango tree. The sacred temple tree or *sthalavruksha*, is a unique mango tree that bears four different varieties of mangoes, in four different parts of the tree, finding mention in various Tamil literary sources as early as 600 CE, including the classical works of the Sangam period.

Orta, too, describes a mango tree in Bombay 'which has two gatherings, one at this season, and another in the end of May. As other fruit may exceed this in scent and taste, so much this exceeds others in coming out of season.'[11]

However, it is possible that the work done by the Portuguese missionaries and the Goan cultivators popularized Goan mango grafts. Nandkumar Kamat and others claim:

> the first reference to the grafting of mango trees is to be found in a 1710 publication by Jesuit priest Francisco de Souza. Father Clemente da Ressureicao in his *Tratado de Agricultura* (1872) describes grafting techniques. Bernardo Francisco da Costa in his manual *Practico do Agricultor Indiano* (1872) writes on mango cultivation methods.[12]

My Mangoes Are Better than Yours: A Case of Reverse Export

Mangifera is a large family of flowering plants primarily found in Southeast Asia and edible fruits are produced by at least twenty-six of its species, in addition to *Mangifera indica*, or the mango as we know it.[13] Recent taxonomic and molecular evidence indicates that the mango likely evolved within a large area of northwestern Myanmar, Bangladesh, and northeastern India.[14] Vivid portrayals of mango cultivation adorn the pages of Indian scriptures and mythology, depicting the lush orchards, sacred groves, and verdant landscapes where these fruits thrive. The Malayan name of *mango mangga* attests its origin outside Malaya, being the same word as the Tamil (south Indian) *mangas*, and produces evidence for introduction of superior races from India into Malaya. The size of the seeds is too great to allow carriage by birds or other animals, but the frequency of its cultivation in Malaya suggests dispersal by human agency. Its introduction into the islands of the Malay Archipelago and other East Asiatic countries was very likely brought about by Indians making voyages to those areas during the Buddhist period (fourth and fifth centuries BC).[15] Maritime trade routes with the Arab world likely facilitated the introduction of mangoes to East Africa around the tenth century AD. By the early sixteenth century, Portuguese explorers expanded its distribution to both East and West Africa, eventually reaching Brazil. The fruit made its debut in Barbados by approximately 1742, and later spread to the Dominican Republic and Jamaica around 1782. Mangoes also reached Mexico through Spanish trade routes from the Philippines and the West Indies.[16]

In 2006, a study led by French scientist Marie-France Duval proposed that mangoes may have reached the French West Indies through two distinct routes: cultivars from Central America (Mexico) and South America (Colombia). These varieties likely originated from Southeast Asia and former French territories in the Indian Ocean.[17] Notably, mangoes thriving in the Caribbean were immortalized in Étienne Denisse's meticulously hand-coloured lithograph masterpiece, *Flore d'Amérique*, alongside numerous native American species, serving as evidence of their integration into new geographical landscapes.

By the dawn of the nineteenth century, the Indian mango had garnered significant favour among colonial tastes, emerging as the focal point of dual commercialization endeavours. One aimed at exporting both fruits and fruit products, while the other focused on exporting the grafted plants themselves for large-scale cultivation elsewhere. In 1889, G. Marshall Woodrow, formerly a Professor of Botany in Pune, orchestrated the shipment of Malgoba plants, initially sourced from Chittor in the Madras Presidency, alongside Alphonso and various other popular varieties to Florida.[18] These saplings were subsequently disseminated among other horticulturists in the region. Although many failed to thrive, Professor Elbridge Gale observed in 1898 that one Alphonso tree was bearing fruit, while two Mulgoba specimens had endured.[19]

In 1910, a new addition emerged in Florida's mango cultivation scene: a seedling of the Mulgoba variety:

> Its fruit had a highly attractive red blush, and appeared to bear more heavily than its parent(s) [. . .]. Based on more recent genetic analysis involving microsatellite markers, it is now estimated that the majority of Florida cultivars are descended from only four mono-embryonic Indian mango cultivar accessions, i.e., 'Mulgoba', 'Sandersha', 'Amini' and 'Bombay', together with the poly-embryonic 'Turpentine' from the West Indies.[20]

In the last century, Florida has become the focal point of modern mango germplasm. The Florida mango cultivars have been found to be highly adaptable to many

Figure 4. 'The Mazagon Mango of Bombay with the Papilio Bolina or Purple-eyed Butterfly', James Forbes, 1813.[21]

Figure 5. Le Mangotier (Plate 61), Flore d'Amérique, Étienne Denisse (1843–1846).[22]

agroecological areas and bear regularly, whereas many of the outstanding Indian cultivars have been unproductive outside their centre of domestication, and are alternate bearing. In the latter half of the twentieth century, plantings of Florida cultivars have been established in many countries and now form the basis of international trade of mangoes.[23]

Mango Identities in Early Modern India

Horticulture provides a revealing perspective on the hierarchical dynamics of land utilization throughout history. During the reign of the Delhi Sultanate and later under the Mughals, the introduction of numerous fruit-bearing trees from regions like Samarkand into India led to the establishment of expansive orchards. Fruits, like sweet confectioneries, portrayed a sense of refined taste and an appreciation of the edible, and perhaps in the hands of nature, even more a sense of awe and blessing. They represented a luxury beyond the basic requirements of grain and salt, a high social currency as much as their economic importance. Orchards thus emerged as lucrative investments for the affluent and the aristocracy, fostering widespread patronage of the art of horticulture. Cultivated fruits, distinct from wild varieties, attained elevated social status, becoming prized commodities for food diplomacy and ceremonial gift-giving.

Abu'l Fazl's history of Mughal emperor Akbar's reign in the *Āʾīn-i Akbarī* describes the culture of eating mangoes but provides no information about how it was propagated.[24] In his memoir, the *Jahāngīrnāma: Tūzuk-i Jahāngīrī*, Mughal emperor Jehangir notes in chronological detail several instances of mangoes being enjoyed in the Mughal courts, after being received as gifts from noblemen.[25]

Beyond their significance as lucrative land use for the royal and their nobility, fruits have been important tokens of gift-giving and diplomacy. The seventeenth-century French traveller François Bernier writes that despite the high cost of these fruits:

> nothing is considered so great a treat; it forms the chief expense of the Omrahs [notables], and I have frequently known my Agah spend twenty crowns on fruit for his breakfast. Ambas or Mangues, are in season during two months in summer, and are plentiful and cheap; but those grown at Delhi are indifferent. The best come from Bengale, Golkonda, and Goa, and these are indeed excellent. I do not know any sweetmeat more agreeable.[26]

Mangoes in the Indian Landscape and Plate

Owing to the antiquity of *Mangifera indica* in India, there is a rich living tradition of mango appreciation. Mango varieties, with their local, often lyrical names, are, for most Indians, the reason to look forward to the hot summer months. Some varieties arrive as early as April, while others thrive in the scorching Indian summer heat. The last few varieties come with the first monsoon showers, marking the progression of summer in the hearts and minds of the people. It is the world's fifth most important fruit, and India still leads its global production. However, in India and in the western perception of Indian food, cultural conditioning has elevated one variety, the Alphonso mango, to an unassailable position in the public palate, overshadowing hyper-local varieties and distorting market dynamics to an alarming proportion.

Although mangoes grow everywhere in India, they are generally divided into Northern and Southern varieties and growing belts. Northern varieties, such as Kesar from Gujarat or Dashehari from Uttar Pradesh, are alternative bearers: if they produced a high yield last year, they will produce a low yield this year. Southern varieties are regular bearers and produce the same volume each year, as is the case for the heavy bearer Totapuri. Alphonso is originally an alternate or shy bearer but behaves more regularly in Southern Indian growing belts, although with lower sugars and pulp quality. Mango tree production depends on flowering, which is often the first indicator of the yield for the coming season. Flowering starts in December–January in the South and lasts till February to the beginning of March in the North and can be distorted by sudden rains or heat waves during this period. Not all trees flower equally in one orchard, especially for varieties like Alphonso. A mango tree can produce from 20 kg up to 200 kgs per harvest season. Mango yield is thus difficult to forecast, as it varies a lot across years.[27]

Figure 6. Mangoes vary in their size, shapes, the feel of their skin, the texture of their pulp, aroma, sweetness and tartness. Here is a selection from June 2023. Priya Mani, A Visual Encyclopaedia of Indian Foods, M/ Mango.

A large proportion of mango trees in India (not orchard cultivated) indeed have stemmed from seeds, and perhaps in their speculative quality and taste landscape lies the proof of Indian culinary ingenuity. More than just a table fruit in India, they hold a revered place in regional cuisines, celebrated at every stage of their existence, from blossom to seed. Mango blossoms are transformed into chutneys, while tender, unripe green fruits are pickled to create *achaars*, India's indigenous oil-preserved pickles. Along the Konkan and in the southern states, the tiny berries that form during fruit setting are also highly prized. Their stones not yet set, these berries are pickled in mustard and brine to produce *maavadu*. As summer arrives, ripe fruits become a staple on every table. Mango seed kernels are sun-dried and are a handy cupboard ingredient used to make sour stews. Mangoes with all the faults that a fruit connoisseur might find – fibrous pulp, sour, astringent, overripe, misshapen and fallen – find purpose in many avatars of regional mango-based curries devoured with great appetite along with rice and rotis.

Mango varieties with loose, juicy pulp are pressed to make *aamras*, a thick mango extract served alongside local breads. Alternatively, the pulp may be spread thin and sun-dried to create *aam papad*, a delightful fruit leather. The seeds of certain varieties are sun-dried to produce *aamchur*, an essential tangy condiment found in Indian street foods. Mangoes find their way into savoury dishes, are preserved in pickles, and contribute to chutneys and relishes that reflect the profound connection between culture and this beloved fruit. As fourteenth-century Sufi poet Amir Khusrau wrote:

> The mango is the pride of the Garden,
> The choicest fruit of Hindustan,
> Other fruits we are content to eat when ripe,
> But the mango is good in all stages of growth.[28]

The Mango Season Is for Mango Wars

The rush for early-arrival Alphonsos has been a long-standing phenomenon, which has propelled their market value to astronomical heights, leaving the average rates of other varieties far behind. Mangoes' cyclical seasonality (they are biennial bearing) has historically fuelled consumer hysteria, incited price wars, and stoked nationalistic tempers. The pursuit to buy the season's first Alphonsoes has transcended into a quest for prestige and privilege, with rarity, high cost, and seasonality becoming the new benchmarks.

Through the last decade, I have encountered Alphonso orchards across South India. In 2023, I travelled from Chennai on the east coast to north Kanara in June, and Alphonso was available for purchase on the most off-grid highways. It sold for a premium compared to other varieties. As an example, Banganapalli sold at INR 190/kilo, Salem gundu at INR 90/kilo, Mallika at INR 75/kilo while the Alphonso sold at INR 220–250 per kilo. In Mumbai I was told the rates started at nearly INR 2000/kilo for early season fruit in April, and the prices simmer to INR 450/kilo in June as the monsoons arrive.

Thus, keen to grow high-value varieties, farmers around India have chosen Alphonso over local varieties. It is indeed a variety that is cultivated in many parts of India. This creates a surplus, flooding Indian markets (and exports) that undermines the exclusivity of GI-tagged Alphonso mangoes from Ratnagiri or Devgad regions of western India. It depresses traditional Alphonso farmers' market values and raises poignant questions about the essence of hyper-regionalism and cultural identity embedded in the nation's favourite fruit.

Mango Wars Cross the Borders: Seasonality All Year Round

Following recent studies of the global mango trade, mangoes from Mexico, Central America, and South America are shipped to the USA at different times of the year, which means that fresh mangoes are available year-round. For example, Mexican mangoes are shipped from late February until September; Peruvian mangoes from mid-November until April; Ecuadorian mangoes from late September until December; and Brazilian mangoes from late September until December. Compared with Mexican, Central American, and other South American mango exporters, Brazil has the costliest ocean freight to the USA, so it concentrates more on the European Union market.[29]

Even in India, where the mango has a deep socio-cultural connection as a summer delicacy, the Alphonso has come to be enjoyed as winter fruit now in an odd way. Grafts of the Geographical Indication (GI) tagged Ratnagiri Alphonsoes were acclimatized in Malawi and reverse exported to India as 'aplhonso' to be sold in winter. Eventually, due to objections from local merchants, regulatory authorities have mandated that they be labelled as Malawi Alphonso mangoes.

Mango Wars Go Local for a GI Tag

It is estimated that, in India, the total number of varieties could be several thousand across the whole country.[30] Only about 25–30 varieties are sold in major cities and about 5–8 varieties are exported, such as Alphonso, Totapuri, Dashehari, Banganapalli, and Kesar.[31] Has mango favouritism led to poor documentation, poorer commercial appeal, and, as a result, allocation of privileges such as the GI?

Today, Alphonso cultivation is spread across Ratnagiri, Sindhudurg, Raigad, and Thane districts in Maharashtra, and Dharwad and Belgaum districts in Karnataka, although the demand from specific pockets in South Konkan region of Maharashtra is high. All these regions, and many others, had queued up to apply for a unique GI tag for their Alphonso. After local squabbles amongst mango cooperatives the GI tag for Alphonsoes was awarded to producers from Devgad and Ratnagiri.

The complex policies of the GI tag programme have exacerbated the decline of diversity in India's mango industry, reinforcing a system of privilege and exclusivity that marginalizes small-scale farmers and regional producers. In the evolution of the Alphonso mango from a coastal variety with a generic name to a branded entity, many regions across southern India have grown Alphonsoes. I am not contesting that all these Alphonsoes are surely different, but this problem has diluted the exclusivity of GI-tagged Alphonso mangoes from the Ratnagiri or Devgarh regions of western India. It has led to an oversupply, flooding Indian markets and exports – erratically, owing to the unpredictable nature of mango yields – depressing traditional Alphonso farmers' market values, and raising poignant questions about hyper-regionalism and cultural identity inherent in the nation's beloved fruit. In the face of climate change and sustainability challenges, it is crucial to advocate for policies that preserve culinary traditions and honour the diverse array of flavours that shape our nation's gastronomic heritage.

This trend has resulted in the disappearance of many unique taste profiles and traditions, replaced by a standardized flavour profile tailored to mass-market preferences and commercial interests. This underscores the importance of addressing the hyper-regionalism and cultural identity associated with the nation's favourite fruit, urging us to confront the underlying realities of our collective culinary consciousness.

The mango industry in India, which accounts for 56% of global production, is a significant economic endeavour. Throughout mango history, novelty, authenticity, and exclusivity have shaped consumer preferences. With the rise of smartphone usage in rural India, farmers now have access to global insights into the fruit market. Interestingly, in response to the widespread availability of regional varieties, farmers are turning their attention to international high-value mango varieties over local favourites. Many Indian farmers have successfully cultivated the highly priced Japanese Miyazaki mango, positioning it as the 'new Alphonso', attracting attention, and fetching substantial values for Indian farmers while becoming sought-after by affluent consumers in India.

The Miyazaki was based on the Irwin mangoes developed in Florida in the 1940s, still popularly sold across America. In the 1980s, a few farmers in Miyazaki (Kyushu prefecture) began farming the Irwin, and soon discovered that the taste of the fruit could vary distinctly depending on how they were nurtured and harvested. To ensure premium fruit quality, farmers selectively remove approximately 80% of the developing fruit, channelling the tree's resources into a chosen few. As the mangoes mature, each fruit is meticulously wrapped with a net attached to an overhead wire suspension system within the greenhouse. This elaborate setup safeguards the individual fruits, allowing them to reach their maximum weight potential without damage. Harvesting follows a unique approach, with only fully tree-ripened fruits being gathered, not plucked. These ripe mangoes are then auctioned during the summer months, with stringent criteria for colour, sugar content, and weight dictating their value. Prices per piece can range from a substantial $50 to an astounding $2500, reflecting the exceptional quality and meticulous care invested in each fruit.[32]

Mango wars carry significant economic and social ramifications. Within India's intricate mango industry, the Alphonso mango serves not only as a symbol but also as a manifestation of underlying issues such as hyper-regionalism, classism, and commodification. By reframing our understanding of the Alphonso as more than just a fruit and the mango as more than merely an ingredient, we may begin to unravel the perplexing question, at least partially: why do certain fruit varieties enjoy greater popularity than others?

About the Author
Priya Mani, a designer and food writer based in Copenhagen working to create gastronomical experiences, is particularly interested in the social interactions of making, presenting, and consuming food.

Notes
1. 'Mango', *University of Wisconsin–Stevens Point* <https://www.uwsp.edu/sbcb/tropical-conservatory/mango> [accessed 1 May 2024].
2. S.K. Mukherjee and R.E. Litz, 'Introduction: Botany and Importance', in *The Mango: Botany, Production and Uses*, ed. by R.E. Litz (CABI, 2009), pp. 1–18.
3. Garcia da Orta, *Colloquies on the Simples & Drugs of India*, trans. by Clements Markham (H. Sotheran, 1913), p. 286.
4. Niccolao Manucci, *Storia da Magor*, trans. by William Irvine (J. Murray, 1907–1908), vol. 3, p. 180.
5. Fernando do Rego, *As Mangas de Goa (The Mangoes of Goa)* (Edição de Autor, 2019), p. 83.
6. S.N. Pandey, 'Mango Cultivars', in *Mango Cultivation*, ed. by R.P. Srivastava (International Book Distributing, 1998), pp. 39–99.
7. Marshall Woodrow, *The Mango: Its Culture and Varieties* (A. Gardener, 1904), p. 25.
8. Public domain image from Hereward Carrington, *Hindu Magic: An Expose of the Tricks of the Yogis and Fakirs of India* (The Sphinx, 1913), retrieved from the Library of Congress <www.loc.gov/item/ltf91001196/> [accessed 25 February 2025].

9. Image from 'Indian Mango Tree Magic Tricks', Daily Kerala, YouTube <https://www.youtube.com/watch?v=B8qMjmNKqxo&ab_channel=DailyKeralasyllabus> [accessed 12 May 2024].
10. Wilson Popenoe, *Manual of Tropical and Subtropical Fruits: Excluding the Banana, Coconut, Pineapple, Citrus Fruits, Olive, and Fig* (Macmillan, 1920), p. 111.
11. Orta, p. 287.
12. Nandakumar Kamat and others, *Goa State Biodiversity Strategy and Action Plan* (Goa Foundation, 2002), p. 50 <https://kalpavriksh.org/wp-content/uploads/2019/05/Goa-April-2002.pdf> [accessed 2 May 2024].
13. W.S. Gruezo, '*Mangifera* L.', in *Plant Resources of South-East Asia No 2: Edible Fruits and Nuts*, ed. by E.W.M. Verheij and R.E. Coronel (Purdoc-DLO, 1992), pp. 203–06.
14. J.M. Bompard and R.J. Schnell, 'Taxonomy and Systematics', in *The Mango: Botany, Production and Uses*, ed. by R.E. Litz, pp. 21–47.
15. Sobhan Kumar Mukherjee, 'The Mango: Its Botany, Cultivation, Uses and Future Improvement, Especially as Observed in India', *Economic Botany*, 7 (1953): 130–62.
16. Mukherjee and Litz, p. 10.
17. C.P.A. Iyer and R.J. Schnell, 'Breeding and Genetics', in *The Mango: Botany, Production and Uses*, pp. 67–96 (p. 69).
18. Woodrow, p. 30.
19. Robert J. Knight and Raymond J. Schnell, 'Mango Introduction in Florida and the "Haden" Cultivar's Significance to the Modern Industry', *Economic Botany*, 48.2 (1994), pp. 139–45 <http://www.jstor.org/stable/4255600> [accessed 9 March 2025].
20. Mukherjee and Litz, p. 11, citing H.S. Wolfe, 'The Mango in Florida – 1887 to 1962'. *Proceedings of the Florida State Horticultural Society*, 75 (1962), pp. 357–91 and R.J. Schnell and others, 'Mango Genetic Diversity Analysis and Pedigree Inferences for Florida Cultivars Using Microsatellite Markers', *Journal of the American Society for Horticultural Science*, 13 (2006), pp. 214–24.
21. James Forbes, *Oriental Memoirs: Selected and Abridged from a Series of Familiar Letters Written during Seventeen Years Residence in India* (White, Cochrane, and Co., 1813). Image courtesy of the British Library.
22. Image courtesy of the LuEsther T. Mertz Library, The New York Botanical Garden.
23. Mukherjee and Litz, p. 11.
24. Abū al-Faẓl ibn Mubārak and others, *The Ā'ĪN-I Akbarī* ([Royal] Asiatic Society of Bengal, 1927).
25. Jahāngīr, *Jahāngīrnāma: Tūzuk-i Jahāngīrī*, ed. by Mohammed Hashim (Bunyad-i Farhang-i Iran, 1980), p. 239.
26. François Bernier and Irving Brock, *Travels in the Mogul Empire* (W. Pickering, 1826), p. 249.
27. Hugo Lamars, 'How to Read the Indian Mango Season,' *Mercadero*, 20 September 2018 <https://www.mercadero.nl/how-to-read-the-indian-mango-season/> [accessed 20 April 2024].
28. in Woodrow, p. 7.
29. *Handbook of Mango Fruit: Production, Postharvest Science, Processing Technology and Nutrition*, ed. by Muhammad Siddiq and others (Wiley Blackwell, 2017).
30. 'Mango', Tamil Nadu Agricultural University, 2011 <https://agritech.tnau.ac.in/govt_schemes_services/aas/mango.html> [accessed 9 March 2025].
31. Hugo A.H. Lamers and others, 'How Can Markets Contribute to the Conservation of Agricultural Diversity on Farms? From Theory to Practice', in *Tropical Fruit Tree Diversity: Good Practices for In Situ and On-Farm Conservation*, ed. by Bhuwon Sthapit and others (Routledge, 2016), pp. 263–284 (p. 270).
32. Flora Baker, 'What Is a Miyazaki Mango? Why Is It a Japanese Specialty?', *Bokksu*, 15 July 2022 <https://www.bokksu.com/blogs/news/what-is-a-miyazaki-mango> [accessed 20 April 2024].

22
Resilience of Indigenous Food Systems
A Study of Cacao Grown in *Chakras* in the Ecuadorian Amazon

Camila Marcías Álvarez

Sustainable land use has become an increasingly important subject as soil erosion, biodiversity loss, and climate change directly impact food production and farmers' livelihoods.[1] In this regard, adaptation strategies are key factors that will shape how humanity faces the impact of climate change on food production.[2] An adaptation strategy that has emerged as a resilient alternative to mitigate climate change is agroforestry. In the Ecuadorian Amazon rainforest, the Indigenous Kichwa people have cultivated in an agroforestry system known as *Chakras* for over 5300 years.[3] *Chakras* are a productive family and community space, combining staple foods, timber trees, and medicinal plants while rotating crops to manage fertility.[4]

In recent decades, Kichwas have incorporated cash crops such as cacao and vanilla as part of their rotation. This addition is a valuable opportunity but also presents challenges. The increase in cash crop monocultures in the region has displaced *Chakras*. Kichwas grow a cacao variety called *Nacional Arriba* (NA) in *Chakras*. Despite its distinctiveness, when sold in the commodity market NA has historically suffered from undervaluation, often sold as bulk cacao and mixed with the hybrid CCN-51 variety, resulting in diminished returns for farmers.[5] Low profits have led smallholders to convert to more intensively managed farms to secure short-term income.[6] These recent shifts represent a major threat to the region's biodiversity and Indigenous knowledge preservation. Globally, NA has a speciality market that pays considerably more than commodity markets. However, Ecuador fails to add value to NA.[7]

Kichwa's Food Systems
More than one million people in Ecuador identify as Indigenous, 24.1% of whom live in the Amazon.[8] Indigenous peoples have strong bonds with nature as they depend on its resources, therefore any changes in biodiversity directly affect their food systems. The Kichwas follow the *Sumak Kawsay* (SK) philosophy, an approach pursuing community living in harmony with nature, including people, animals, plants, ecosystems, natural forces, and spirits.[9] For Kichwas, nature is both wild, *'sacha'* ('forest' in Kichwa), and domesticated, *'chagra'* ('cultivated fields') (Figure 1).[10] *Chakras* involve shifting agriculture, where small areas of land are cultivated in forest gaps to fulfil food requirements, and, after years, they are abandoned, allowing the forest to recover.[11]

Figure 1. Graphic representation of a Chakra *(Atelier Pareto)*.

In a *Chakra*, 60% of crops are for self-consumption, and 40% are cash crops.[12] One cash crop is the *Theobroma cacao* tree. There is no consensus on exact origins, but researchers link cacao's origins to Ecuador.[13] Colonization brought significant changes to cacao production, including establishing international trade routes in the 1800s that favoured wealthier nations and created imbalanced power relations. Liberalized value chains (VC) have profoundly impacted producer countries.[14] In Ecuador, these impacts were felt in tariff escalations from more mature markets. In 1994, cacao management was privatized, with no state involvement in price regulation, quality control, or traceability.[15] Since then, farm-gate prices paid for NA have not been financially profitable.[16] Research conducted on the coastal region showed that without access to a differentiated VC, farmers tend to switch NA for CCN-51 because of higher yields and similar prices.[17]

The cacao industry faces several complex social, economic, and environmental issues. In Ecuador, the main issues are biodiversity loss, land ownership, ongoing poverty, and power asymmetries.[18] Despite these issues, in the last sixty years, cacao production in Ecuador has increased by 765%.[19] Since the 1990s, the cacao trade has been part of a global trend of financialization, enlarging the distance between producers and consumers.[20] In the 1970s, cacao farmers used to retain 60% of the financial value, while by 2015 their retention dropped to 4–6%.[21]

Methodology

The study area was the southernmost region of the Napo province, located in the Ecuadorian Amazonia. The focus was on *Tsatsayaku* Cacao Farmers Association. A secondary study region was Esmeraldas province, in the northwest coast of Ecuador, which was incorporated to obtain comparable information from the closest analogue of cacao production in *Chakras*. Semi-structured interviews with Kichwas and non-indigenous farmers growing cacao either in *Chakras* or agroforestry systems were conducted to validate the literature findings and obtain information that was unavailable. In total, fourteen interviews were conducted online between June and August 2023. Using interviews as a primary source enabled a better understanding of resilience and analysis of farmers' lived experiences and relevant actors in cacao production.

A VC analysis was elaborated using the Global Value Chain framework to assess the financial value retained by farmers.[22] The available literature was complemented with data obtained from interviews. This study used the farm resilience framework, which includes three resilience components: buffer capacity, self-organization, and adaptative capacity.[23] Each of these components have indicators that were measured to quantify farm resilience.

Value Chain Results

The global cacao market is divided into bulk cacao (90–95%) and Fine Flavour Cacao (FFC) (5–10%). A few major multinationals control 50–60% of the trade, with smallholders receiving minimal financial returns.[24] This can be represented as an hourglass-shaped supply chain, where smallholders produce 90% of the world's cacao but receive a minimal share of the financial value, a few companies capture most of the value, and millions of consumers are at the other end. Globally, Ecuador is the largest producer of FFC.[25] Additionally, there is also a niche market for premium-quality FFC, traded through Direct Trade (DT). In Ecuador, 100,000 households rely on the sales of cacao, of which 99% are smallholders.[26] CCN-51 is considered bulk cacao and has expanded fast, without regulatory differentiation, impacting Ecuador's reputation as an FFC producer.[27] As a result, in 1994 the Ecuadorian FFC's export status was reduced from 100% to 75%, where it continues to be.

Ecuadorian farmers have limited access to credit. Most smallholders cannot afford to hire workers, forcing them to rely on family labour. Farmers in both study regions hire daily wage workers, primarily during the high-productivity season. Cacao trees face threats from extreme weather and pests, with *'monilia'* disease (frosty pod) being a common issue reported by farmers. In Napo, farmers received training from *Tsatsayaku* to manage pests, opting not to use pesticides. Similarly, NGOs assisted farmers in pest management in Esmeraldas. While most farmers avoided synthetic pesticides, those employing a combined NA and CCN-51 system relied on them. Despite this, none of the farms were certified organic due to prohibitive costs. The Corporation of Amazonian Chakra plays a significant role in Napo. They developed a Participatory Guarantee System (PGS) ('*Sello Chakra*'). PGS are local systems that assure the quality of products, certifying producers based on collective action that recognizes the cultural, environmental, and social relevance of *Chakras*.[28]

Cacao trees produce flowers that transform into pods after pollination. Once these pods ripen, farmers collect them, but the beans, at this stage, are not yet suitable for consumption. The daily cacao referential prices are based on the futures market. This price reflects the 'freight on board', which is the price paid to farmers and intermediaries, including the transport and extra fees. Therefore, it does not reflect the price farmers receive. Information about the farm-gate price is not publicly available, but it can be calculated by subtracting intermediation costs from the referential price. Therefore, the farm-gate price negatively correlates with the number of actors in the VC.[29] According to interviews in Esmeraldas, farm-gate prices for dried and fermented NA beans ranged from $265 to $418 per 100 kg. In Napo's *Chakras*, the price of dry beans ranged between $231 and $297 per 100 kg. Farmers highlighted that there is no difference in price between CCN-51 and NA. As one farmer explained, 'The government developed policies to encourage farmers to grow CCN-51, because of higher yields. It is impossible to differentiate them because most are sold raw and mixed.'[30]

Intermediaries manage the transport and storage, connecting producers with manufacturers or exporters. Associations act as intermediaries, handling the post-harvest because smallholders do not have the infrastructure or the necessary amount of beans to dry and ferment. Therefore, in Napo, after the harvest, farmers sell their wet beans to *Tsatsayaku*. In Esmeraldas, all farmers did the post-harvest. Dry beans have a higher price because the post-harvest increases their value.[31] Farmers tend to have more than two intermediaries, and their incorporation reduces by 20% the financial value for subsistence farmers and 10% for smallholders.[32] Farmers that used DT bypass intermediaries, ensuring better traceability. Research shows that DT cacao farmers tend to perceive a higher financial value than bulk cacao farmers.[33] Results show that NA farmers, who had few or no intermediaries, received a considerably higher price compared to the local market price (+181% for Napo and +261% for Esmeraldas).

Once the beans have been dried, they are cleaned and ground, later pressed into cacao mass and butter, to transform into chocolate couverture. This stage is the most lucrative in the commodity VC, largely caused by vertical integration and VC's high levels of concentration.[34] Leading companies exert power over producers and set the price, marginalizing smaller actors.

Ecuador exports ~90% of its cacao bean production: 90% of the exports are dry beans and less than 5% is DT.[35] Approximately 70% of the exports are sold as bulk cacao due to inefficient traceability and mixing of varieties. At the retail level in the commodity market, there is a significant imbalance of value distribution. Approximately 70% of the total value is held by retailers.[36] Furthermore, regardless of farmers receiving a higher value with DT, the craft chocolate business that buys from Esmeraldas retains 85.4% of the value, and the international retailers 49.8%. Farmers selling NA from agroforestry systems via DT got a higher price than the commodity VC. Among them, farmers from Esmeraldas, selling their dry beans, retained a higher percentage (14.6%) of the value than *Chakra* farmers (11.4%). According to the literature, the financial value obtained by farmers who sell commodity beans is lower (6.6%).[37] Farmers who sold via DT mentioned that associations provided a direct link to the market, better farm-gate prices, pest management assistance, and logistics to grow agroecologically.

Farmers Resilience Results

Table 1 summarizes the results obtained from the farmers resilience analysis. The buffer capacity assessed in this study found that Napo had a higher tree diversity than cacao farms in Esmeraldas. According to the *Chakra* farmers, primary and secondary forests were present in the same area where they cultivated cacao. Members of *Tsatsayaku* had, on average, 15.5 ha of farm area, of which 2.7 ha were destined for the productive *Chakra* space. Within the domesticated *Chakra*, there were on average 27 different varieties of trees, palms, and shrubs. Esmeraldas had on average, 14 tree species. In *Chakras*, between 80 to 150 different plant species have been identified, and 62% of them are classified as food crops.[38] *Chakras* had on average 3.8 cash crops per household. The most mentioned crops were cassava and plantain. In Esmeraldas, the most frequent were corn and cassava. The average number of cash crops cultivated in Esmeraldas farms was two.

Cacao yields in Ecuador have consistently increased in time and are higher than the average global yield.[39] It is estimated that between NA and CCN-51, the yield went from 0.26 t/ha in 2012 to 0.44 t/ha in 2017.[40] Yields vary substantially according to the cacao variety: CCN-51 trees are grown in closer proximity, while NA trees are cultivated further apart.[41] Therefore, provincial figures are not comparable. However, interviewees in both regions did not know their exact yields. Therefore, data was obtained from inventories which characterize the type of farmer and variety. In 2019, the NA yield for subsistence and smallholder farmers in Napo was 0.22 t/ha and 0.19 t/ha in Esmeraldas.[42] To achieve these yields, Napo interviewees stated they did not use pesticides. In Esmeraldas, two of the interviewees used synthetic fertilizers and pesticides.

The number of income sources slightly differed between regions. In Napo, the average number of different income sources was four, while in Esmeraldas it was two. For all the interviewees in Napo, cacao sales revenue was their main income. Only the two Kichwa farmers relied solely on their *finca*'s annual income. In Esmeraldas, two of the farmers worked full-time on the *finca* with family members as day labourers. Two farmers did not live on their *fincas* and relied on family members to help. The farmers in Napo were all affiliated with *Tsatsayaku*. Among them, two were also part of other associations. In Esmeraldas, four of the five farmers interviewed were part of associations.

According to four of Napo's interviewees, in their *Chakras*, they produced almost all the necessary food and medicine to subsist. Conversely, two *mestizo* farmers in Napo relied on the outside markets to access most of their food and medicine. Overall, the calculated subsistence level for Napo was 64.2%. For Esmeraldas, only the medium-sized *finca* did not rely on the outside markets for any food. All the other interviewees relied completely on the outside markets for food and medicine (equivalent to 20% of the subsistence level).

Chakra farmers' average annual family income was $2133, while that for Esmeraldas farmers was $1440. The average annual household income in rural Ecuador in 2022 was $616.[43] The frequency of educational training is an essential part of farmers' livelihoods.[44]

In 2022, the average number of cultivation and management courses that *Tsatsayaku* farmers participated in was 4.1. Farmers in Esmeraldas participated, on average, in 1.3 courses; only the farmer that was not part of any associations did not participate in courses.

Information sources refer to a social structure in the community, such as meetings, markets, and other ways of communication.[45] *Chakra* farmers had, on average, 3.2 different sources of information. Local markets were their most important communication source, where they exchanged information about prices and management. In Esmeraldas, the mean value of information sources was four. Farmers mentioned association meetings as their most important source of information.

All the *Chakra* farmers mentioned that their main motivation to grow NA cacao was to increase their income. They also mentioned they would not switch to monoculture because *Chakras* were considered a SK way of life. Farmers associated *Chakra* cultivation with organic agriculture, soil regeneration, SK, food security and being part of an ecosystem. For most farmers in Napo and Esmeraldas, agroforestry was associated with ethical considerations, such as protecting the environment for future generations. All the interviewees mentioned that they were proud of growing NA because it helps prevent NA extinction currently threatened by the proliferation of CCN-51 monocultures and diseases. As one put it, 'We maintain the ecosystem, the flora, the fauna, and the water sources are also taken care of, it is the best way to protect the Amazon.'[46]

All associated farmers mentioned that growing NA had increased their income and access to the speciality market of FFC. Additionally, the education they had received through the associations had given them tools to improve their management. When asked about future challenges, in Napo, most farmers mentioned the increasing age of farmers, and younger generations' reluctance to preserve *Chakras*. Farmers in Napo mentioned the threat of illegal and legal gold mining in the region. Most farmers mentioned their concerns about climate change. Finally, all farmers mentioned the lack of government aid and their difficulties accessing credit.

Discussion

Results show that farmers in Esmeraldas earned higher incomes than *Chakra* farmers, especially those selling via DT due to fewer intermediaries. The presented evidence emphasizes the importance of reducing the distance between production and consumption. Additionally, this study found ethical motivations among *Chakra* farmers to grow NA cacao. Farmers cited SK as a key motivation, showing their deep connection to Amazonia through their care for *Chakras*.

The little financial value retained by farmers can be partly attributed to corporate concentration.[47] It can also be attributed to them being price takers, with no market power or alternative income sources. Regardless of different levels of poverty, the results presented show that they have other income sources. This could explain why their income is above average compared to the rural national income in Ecuador. These results align with other studies that demonstrate that the inability to access direct export markets keeps farmers at the base of an asymmetrical VC, where more work does not imply higher value.[48]

The *Sello Chakra* presents a remarkable opportunity to showcase the cultural significance and added value of NA cacao from *Chakras*. Despite farming without synthetic inputs, none of the interviewees could afford the traditional organic certifications due to cost constraints. Conventional certifications have become 'victims of their success' in the commodity trade, with corporations retaining most of the value.[49] Additionally, standard certifications are inadequate in preserving the socio-ecological values of *Chakras*. The PGSs are a new form of market governance for Indigenous peoples, with the potential to certify production that exceeds organic agriculture standards. The results show that NA cacao in *Chakras* has numerous benefits, including superior quality, better environmental management, climate change mitigation, and the preservation of Kichwas' cultural heritage. Therefore, PGS represents an opportunity to harness the strength of *Chakra* farmers to preserve SK. Although there are major challenges in the lack of market for PGS products, prioritizing food policies for cacao grown in *Chakras* and certified by PGS could foster a greater appreciation for NA cacao, ultimately benefiting farmers and associations.

This study reveals that cacao farms in Napo exhibit greater levels of resilience compared to those in Esmeraldas (Table 1). Despite minor differences in parameters, the findings strongly suggest that *Chakras* possess greater resilience due to several key indicators. The close relationship between *Chakras* and their surrounding environment, as well as the implementation of Indigenous land management practices, seem to be the primary reasons for the comparatively higher resilience levels of *Chakras*.

Regarding buffer capacity, *Chakras* are in the Amazon rainforest, a region that historically has not been exposed to the level of conventional agricultural expansion that Esmeraldas has had. In principle, *Chakras* cohabit with their environment, requiring an area of primary forest left untouched as part of the rotation system. The strong bond between Kichwas, their environment, and traditional agricultural knowledge prevents them from unnecessarily intervening in their surroundings, supporting the higher tree diversity obtained in *Chakras*. Traditional knowledge along with a higher tree and crop diversity play a crucial role in climate change mitigation.[50] The higher crop count in *Chakras* illustrates that farmers in Napo rely more on their crops for subsistence, which

Table 1. Comparison of farm resilience indicators.

	Buffer Capacity				Self-Organisation			Adaptive Capacity	
	Tree species	Crop species	Annual cacao yield (ton/ha)	Income sources	Affiliation to associations	Subsistence level (%)	Annual income (US $)	Participation in cacao courses	Information sources
Chakras in Napo	27	3.8	0.22	4	1	64.2%	$2,133	4.1	3.2
Fincas in Esmeraldas	14	2	0.19	2	0.8	20%	$1,440	1.3	4

reduces their dependency on a single crop. By cultivating a diverse range of crops, farmers were better equipped to withstand unexpected events or pests.[51] This point was largely confirmed by interviewees who mentioned that due to COVID-19, the *guayusa* market disappeared. Despite the disappearance of this market, they were still able to sell other crops, demonstrating the importance of crop diversity in building resilience.

Chakra farmers had a considerably higher subsistence level than Esmeraldas' farmers. This could be related to *Chakras'* higher tree and crop diversity. Even though most of Esmeraldas' farms were agroforestry systems, farmers relied more on the outside markets, because they did not grow enough crops for their consumption. Kichwa farmers enjoyed a more diversified diet, enhanced food security, and increased resilience compared to Esmeraldas' farmers. Family income was higher in *Chakras* than in Esmeraldas. All the interviewees had higher annual incomes than the average rural income in Ecuador. According to the VC analysis, Esmeraldas' farmers received a higher revenue from their cacao sales. However, *Chakra* farmers had more income sources and more support from *Tsatsayaku*. Regardless of how much they harvested, *Tsatsayaku* was always going to buy it if they followed the PGS guidelines. They had a more secure market and could focus on diversifying their crops, thus increasing their family income.

Throughout the interviews, a common theme emerged: farmers felt neglected by the Ecuadorian state and banks, highlighting their difficulties in obtaining credit. In addition, *Chakra* farmers expressed concern over the lack of institutional support for preserving *Chakras*. This lack is particularly concerning as younger generations do not consider this system as a sustainable means of subsistence. Without intervention, there is a risk that the *Chakras* will be abandoned, resulting in a loss of traditional knowledge and agroforestry in the region.

Conclusion

Cacao production in *Chakras* plays a key role in providing food security and well-being to Amazonian Indigenous communities. Higher resilience levels can be attributed to the close relationship between farmers and their environment. Additionally, the level of self-organization proved to be a relevant indicator of increasing resilience as it reduces power imbalances. The current expansion of commodified crops and deforestation is threatening the preservation of *Chakras*, leading to a loss of biodiversity and traditional knowledge, ultimately impacting farmers' livelihoods.

This study revealed *Chakra* farmers' motivations to cultivate NA cacao and highlighted the significance of SK in their practices. Farmers exhibit a deep commitment to their natural environment, choosing to grow NA cacao in agroforestry systems over CCN-51 in monocultures, despite potentially higher earnings from the latter. Their appreciation for biodiversity and agroecological methods underscores the need to integrate farmers' knowledge into policymaking, a discussion from which they have been excluded. The recent adoption of PGS empowers farmers and can help differentiate the market, preserving traditional knowledge and biodiversity. Future research should focus on blending indigenous and modern agricultural knowledge to enhance socio-ecological resilience on farms.

About the Author
Camila Marcías Álvarez is a pastry chef, a food writer, and a researcher, exploring the role of chefs in shaping equitable food systems. With a background in law and an MSc in Food Policy, she founded De la Raíz al Plato to advocate for Latin American food justice, earning international recognition for her work on culinary practice, migration, and Indigenous South American cacao value chains.

Notes
1 J. Jacobi and others, 'Farm Resilience in Organic and Nonorganic Cocoa Farming Systems in Alto Beni, Bolivia', *Agroecology and Sustainable Food Systems*, 39 (2015), 798–823.
2 M.A. Altieri and others, 'Agroecology and the Design of Climate Change-Resilient Farming Systems', *Agronomy for Sustainable Development*, 35 (2015), 869–90.
3 'The Amazonian Chakra, a Traditional Agroforestry System Managed by Indigenous Communities in Napo Province, Ecuador', FAO, 2023 <https://www.fao.org/giahs/giahsaroundtheworld/designated-sites/latin-america-and-the-caribbean/amazon-chakra/detailed-information/fr/> [accessed 2 May 2024].
4 O.V. Viteri-Salazar and others, 'The Challenges of a Sustainable Cocoa Value Chain: A Study of Traditional and 'Fine or Flavour' Cocoa Produced by the Kichwas in the Ecuadorian Amazon Region', *Journal of Rural Studies*, 98 (2023), 92–100.
5 T.F. Purcell, '"Hot Chocolate": Financialized Global Value Chains and Cocoa Production in Ecuador', *The Journal of Peasant Studies*, 45 (2018), 904–26, p. 916–917.
6 P. Vaast and E. Somarriba, 'Trade-offs Between Crop Intensification and Ecosystem Services: The Role of Agroforestry in Cocoa Cultivation', *Agroforestry Systems*, 88 (2014), 947–56, p 950–951.
7 Purcell, p, 916–918.
8 'Sistema Integrado de Consultas a los Censos Nacionales', INEC, 2020 <http://www.ecuadorencifras.gob.ec> [accessed 11 May 2024].
9 D. Coq-Huelva, B. Torres, and C. Bueno-Suárez, 'Indigenous Worldviews and Western Conventions: Sumak Kawsay and Cocoa Production in Ecuadorian Amazonia', *Agriculture and Human Values*, 35 (2018), 163–79, p, 171.
10 J. Garí, 'Biodiversity and Indigenous Agroecology in Amazonia: The Indigenous People of Pastaza', *Etnoecologica*, 5.7 (2001), 21–37.
11 R. Vera, H. Cota-Sánchez, and J.E. Grijalva Olmedo, 'Biodiversity, Dynamics, and Impact of Chakras on the Ecuadorian Amazon', *Journal of Plant Ecology*, 12.1 (2019), 34–44, DOI:10.1093/jpe/rtx060. [5]
12 'The Amazonian Chakra, a Traditional Agroforestry System Managed by Indigenous Communities in Napo Province – Ecuador', FAO, 2023 <https://www.fao.org/giahs/giahsaroundtheworld/ecuador-amazonian-chakra/en> [accessed 11 May 2024], p, 33.
13 R.G. Loor-Solorzano and others, 'Insight into the Wild Origin, Migration and Domestication History of the Fine Flavour Nacional *Theobroma Cacao* L. Variety from Ecuador', *PLoS One*, 7.11 (2012), e48438.
14 J. Clapp, *Food* (John Wiley & Sons, 2020), p, 76–83.
15 Purcell, p, 916–917.
16 J. Castañeda-Ccori and others, 'Unveiling Cacao Agroforestry Sustainability through the Socio-Ecological Systems Diagnostic Framework: The Case of Four Amazonian Rural Communities in Ecuador', *Sustainability*, 12.15 (2020), 5934.
17 J. Díaz-Montenegro, E. Varela, and J.M. Gil, 'Livelihood Strategies of Cacao Producers in Ecuador: Effects of National Policies to Support Cacao Farmers and Specialty Cacao Landraces', *Journal of Rural Studies*, 63 (2018), 141–56, p, 151.
18 A.C. Fountain and F. Hütz-Adams, 'Cocoa Barometer 2022', Voice Network, 2022 <https://voicenetwork.cc/cocoa-barometer/> [accessed 10 May 2024].
19 'Cocoa Bean Production from 1961 to 2022', Our World in Data, 2022 <https://ourworldindata.org/grapher/cocoa-bean-production?tab=chart> [accessed 10 May 2024].
20 A.C. Fountain and F. Hütz-Adams, 'Cocoa Barometer 2022'.
21 A. Abdulsamad and others, *Pro-Poor Development and Power Asymmetries in Global Value Chains* (Duke Center on Globalization, Governance & Competitiveness, 2015), p, 33.

22　G. Gereffi and K. Fernandez-Stark, *Global Value Chain Analysis: A Primer* (Duke Center on Globalization, Governance & Competitiveness, 2016).
23　Jacobi and others.
24　A.C. Fountain and F. Hütz-Adams, 'Cocoa Barometer 2022', p, 80–83.
25　'Fine Flavour Cocoa', ICCO <https://www.icco.org/fine-or-flavor-cocoa/> [accessed 10 May 2024].
26　'Tipos de Cacao', ANECACAO <https://anecacao.com/cacao-en-el-ecuador/tipos-de-cacao/> [accessed 10 May 2024].
27　J. Wiegel and others, *The Cacao Market in Ecuador: Opportunities for Supporting Renovation and Rehabilitation*, International Center for Tropical Agriculture (CIAT), Cali, Colombia, 2020, p, 2.
28　'PGS Guidelines: How to Develop and Manage Participatory Guarantee', IFOAM, 2019 <https://www.ifoam.bio/our-work/how/standards-certification/participatory-guarantee-systems/pgs-toolkit> [accessed 4 May 2024].
29　C.J. Melo and G.M. Hollander, 'Unsustainable Development: Alternative Food Networks and the Ecuadorian Federation of Cocoa Producers, 1995–2010', *Journal of Rural Studies*, 32 (2013), 251–63, p, 18–20.
30　Interview.
31　Díaz-Montenegro, E. Varela, and J.M. Gil.
32　A. Avadí and others, *Análisis de la cadena de valor del cacao en Ecuador* (CIRAD, 2021), p, 16.
33　'Comparative Study on the Distribution of Value in European Chocolate Chains', FAO, 2020.
34　'Cocoa Study: Industry Structures and Competition', UNCTAD, 2008 <https://unctad.org/system/files/official-document/ditccom20081_en.pdf> [accessed 20 April 2024], p, 26–28.
35　'Cocoa's Bittersweet Supply Chain in One Visualization' <https://www.weforum.org/stories/2020/11/cocoa-chocolate-supply-chain-business-bar-africa-exports/> [accessed 15 December 2024]
36　'Comparative Study on the Distribution of Value in European Chocolate Chains', FAO and BASIC, 2020 <https://lebasic.com/wp-content/uploads/2020/07/BASIC-DEVCO-FAO_Cocoa-Value-Chain-Research-report_Advance-Copy_June-2020.pdf> [accessed 1 May 2024], p, 30–32.
37　'Cocoa's Bittersweet Supply Chain in One Visualization', World Economic Forum, 2020 <https://www.weforum.org/agenda/2020/11/cocoa-chocolate-supply-chain-business-bar-africa-exports/> [accessed 10 May 2024].
38　Avadí and others, p, 114.
39　'Cocoa Bean Yields from 1961 to 2022', Our World in Data, 2023 <https://ourworldindata.org/grapher/cocoa-bean-yields?tab=chart&country=ECU~OWID_WRL> [accessed 14 May 2023].
40　'Productividad y Rendimientos Cacao', MAG, 2017 <https://www.agricultura.gob.ec/productividad-rendimientos-cacao/> [accessed 18 March 2024].
41　A. Avadí and others.
42　A. Avadí, 'Environmental Assessment of the Ecuadorian Cocoa Value Chain with Statistics-based LCA', *The International Journal of Life Cycle Assessment*, 2023, pp, 1–21.
43　*Reporte de Pobreza, Ingreso y Desigualdad*, Banco Central de Ecuador, 2022 <https://contenido.bce.fin.ec/documentos/Estadisticas/SectorReal/Previsiones/IndCoyuntura/Empleo/PobrezaJun2022.pdf> [accessed 29 March 2024].
44　Jacobi and others.
45　Jacobi and others, p, 812.
46　Interviews.
47　A.C. Fountain and F. Hütz-Adams, 'Cocoa Barometer 2015', Voice Network, 2015 <https://voicenetwork.cc/wp-content/uploads/2019/07/Cocoa-Barometer-2015.pdf> [accessed 3 September 2023].
48　J. Castañeda-Ccori and others, 'Unveiling Cacao Agroforestry Sustainability through the Socio-Ecological Systems Diagnostic Framework: The Case of Four Amazonian Rural Communities in Ecuador', *Sustainability*, 12.15 (2020), 5934.
49　'Cocoa Study: Industry Structures and Competition'.
50　M.A. Altieri and C.I. Nicholls, 'The Adaptation and Mitigation Potential of Traditional Agriculture in a Changing Climate', *Climatic Change*, 140 (2017), 33–45.
51　P. Koohafkan and M. Altieri, *Sistemas Importantes del Patrimonio Agrícola Mundial: Un Legado para el Futuro*, Rome: FAO, 2010.

23
Doomsday Plots
The High-Stakes Gardens of America's Preppers

Rebecca D. Mazumdar

In 2018, I attended Prepper Camp in rural North Carolina in the southeastern region of the United States. I was writing a dissertation on Cold War domestic spaces in fiction and cultural ephemera, and I wanted to learn how the legacies of preparedness had persisted into the twenty-first century. Prepper Camp is an annual three-day event with families arriving in their RVs (caravans) and staying on-site for the duration of camp. In addition to the socializing and vacation-like feel to the event, classes take place throughout each day with lectures and entertainment planned for the evenings. There's a marketplace that sells everything from sun ovens to surplus military gear to medical supplies. Classes at Prepper Camp cover such topics as beekeeping, avoiding abduction, fire-starting, food storage, wine-making, reloading ammunition, and a new one: 'Win Against Woke: Defeating Leftism in Daily Life'.[1] When I went, I learned about stockpiling food in buckets, making kombucha, making tents from military ponchos, and building a 'survival bidet' from a modified garden sprayer. I avoided the archery lesson zone.

What I learned is that 'prepping', or planning for a catastrophe that will substantially impact or destroy familiar rhythms of life, is a politically-polarized practice rooted in paranoia and distrust. It's also heavily militarized and weaponized. Prepper Camp class offerings change to accommodate new perceived threats; for instance, this year's camp includes a class called 'Artificial Intelligence – The Good, the Bad, and the Ugly'.[2] Past years have included classes on the basics of Islam (with a 'what-you-need-to-know' tone). The featured band at this year's camp became famous in some circles for their song 'Sad Little Man', which is about Dr Anthony Fauci, the Chief Medical Advisor to the President of the United States during the early months of the COVID-19 pandemic; one verse includes the lyrics 'Sad little man sitting deep in a lie / He's dead in his soul but he'll keep you alive / Do what he says, not what he do / 'Cause the truth is for him and the lie is for you'.[3]

Gardening is an important component of prepping, since natural or man-made disasters can make food sourcing difficult or even impossible. Maintaining a garden also extends the viability of any stockpiled shelf-stable food. Because of the political ideologies informing these garden spaces and their maintenance, I argue that so-called prepper gardens represent hyper-militarized corruptions of land-stewarding traditions like homesteading and victory gardening. I want to discuss the ways that prepper gardens cease to be merely sites of sustenance and abundance and instead become political performance spaces and potentially violent sites of conflict and secrecy. These spaces, therefore, force us to question what it is that these self-appointed survivalists are really trying to preserve.

First, a note about the terms I'll be using. The phrase 'doomsday prepper' carries a more negative connotation than, say, homesteader, gardener, or even survivalist. Without question, there are people who choose to live off-grid, living nearly entirely self-sufficient lives on their land or homesteads, whose practices contain none of the xenophobia and paranoia that inform the decisions of extreme preppers. Therefore, I'll use the terms 'prepper' and 'prepper gardens' to refer to the more polarized, extreme philosophies of self-sufficiency, which also often include stockpiles of weapons and lethal security systems. As I'll discuss below, the nearly sociopathic levels of secrecy and protection that define many prepper gardens set these spaces in direct opposition to the very traditions they claim to be continuing and the ideals they claim to be preserving.

In many ways, prepper gardens aren't much different from traditional home gardening or homesteading. Many websites offering advice to survivalists or preppers contextualize their work within the tradition of subsistence gardening and wartime victory gardens. Rick Austin, who calls himself the Survivalist Gardener and who runs the annual Prepper Camp, places his philosophy in the tradition of 'native indigenous people around the world' who 'have lived primarily on perennials'.[4] More on the irony of this statement later. Nonetheless, many resources for survivalist or prepper gardens are essentially homesteading resources, offering advice for ways to live as off-grid as possible without relying on electricity or other civic infrastructure. Other homesteaders simply choose to be as self-sufficient as possible by growing the food their family needs, including produce, livestock, and even in some cases medicinal herbs and plants. These techniques date back generations and don't necessarily carry the political connotations or paranoia that prepper motives do.

Other resources for preppers place the survival garden in the tradition of wartime victory gardens. Victory gardens in the United States freed up agricultural, transportation, and production resources to be used for the war effort during the Second World War. The programme also encouraged better health and nutrition via the physical activity of gardening and the regular access to fresh produce. In 1944, there were 18.5 million victory gardeners in the United States. The US National Parks Service reports that, by the end of the Second World War, 'American Victory Gardeners had grown between 8 and 10 million tons of food'. The programme also provided classes covering not just gardening but also the preservation of harvested foods. In especially urban or densely-populated areas, victory gardening programmes fostered community spaces where gardens and resources could be shared among neighbours. In these victory gardens, both the labour and the harvest were shared, with produce often being distributed to local schools.[5]

While most victory gardens disappeared after the Second World War, the urgency of preparedness did not. During the Cold War, preparedness took on a significantly different look and feel. Those who could afford private fallout shelters had a separate domestic space to stock and maintain, tasks that usually fell to the housewife. Building the structure itself, however, was often the husband's job. Government publications on this type of do-it-yourself preparedness evoked familiar symbols of America's patriotic mythology.[6] The 1954 comic book *The H-Bomb and You*, distributed by civil defence officials to residents of Washington, D.C., and the state of Maryland, presents civil defence as an innately American tradition.

In the early pages of this comic book, a school teacher begins a civil defence lesson with a reference to colonial America: 'Even the earliest settlers in America,' she says, 'knew they must unite if they were to survive. Then, as now, there was a job for every man, woman and child . . . [.]' The image behind her words is of a woman throwing water onto a fire inside the wooden walls of a fort, while men in Daniel-Boone-style raccoon-skin caps aim their guns at something out of frame and unseen beyond those walls. In the following frame, picturing events occurring 'some years later', two men run toward a third man who is fending off what appear to be British red-coats in the distance. The men running to Neighbor Smith's aid shout at each other: 'Hurry with your gun! Neighbor Smith is being ambushed!' and 'To the rescue! If one is in danger, all are in danger!'[7]

Today's prepper gardens are the radicalized grandchildren of that Cold War notion of civil defence: homemaking with a sheen of political performativity and a tone of existential urgency. The difference between gardening and prepper gardening is the difference between self-sufficiency and selfish-sufficiency. This sense of obligation to one's neighbour, presented in this comic book as part of the American patriotic tradition, is often completely absent from prepper garden strategies. In fact, prepper gardens corrupt the traditions of land-stewardship and self-reliance, threatening our relationships to each other, to our land, and to the ideals of American democracy.

Prepping means being ready for any number of potential cataclysms. During the Second World War, the danger in the United States was resource shortage; during the Cold War the danger became more personal and immediate as preparedness included changes to domestic structures and systems. Today's preppers are mindful of a number of possible futures, ranging from natural disasters to man-made ones. The Provident Prepper website lists the following '[p]ossible scenarios that make a survival garden critical': unemployment, hyperinflation, economic collapse or depression, crop failure or food transportation crisis, electromagnetic pulse (EMP), solar flare, cyberattack, or supply shortages caused by war, demand, or embargo.[8] Each possible catastrophe represents something outside the prepper's immediate control, another unseen enemy beyond the walls, suggesting that part of the prepper mentality is a response to feelings of profound helplessness.

Regardless of the specific disaster for which one might be prepping, the prepper garden is always-already a site of doom, an agricultural performance of agency in the face of uncertainty and instability. These gardens exist in a state of precarious and performative duplicity, being at the same time a normal, everyday garden and a potential last-chance at life in the event of cataclysm. Military language seeps into discussions of survival gardening, with terms like 'OPSEC' and 'SHTF' and even 'TEOTWAWKI'. For the uninitiated, that's Operations Security, Shit Hits the Fan, and The End Of The World As We Know It.

As mentioned above, many of the logistical considerations that go into planning a prepper garden aren't much different from less political and less radical homesteaders. However, certain potential problems need special planning. For example, making sure a survival garden has enough water means being mindful of possible futures in which water sources are inaccessible or contaminated. Therefore, water sourcing is part of the planning stages of any practical prepper garden. Recommended methods of water collection and storage include rainwater catchment, ditches, and swales.[9] The Provident Prepper also suggests considering self-watering raised beds, clay ollas, terracotta watering stakes, or reused household grey water.[10] Prepper Camp organizer Rick Austin uses grey water systems to repurpose old laundry water, as well as irrigation and swales.[11] In his self-published book *Secret Garden of Survival*, he also presents examples of above-ground and underground rainwater collection structures.[12]

Planning for unwanted pests, specifically hungry non-preppers, may be the most significant way that prepper gardens become radical sites of political performativity, losing sight of the legacies of victory gardens, homesteading, or even private gardening. I'll turn again here to the philosophy of Rick Austin, whose website 'The Survivalist Gardener' presents his philosophy and his techniques rather completely. In the foreword of his self-published book on creating what he calls 'a camouflaged food-forest', Austin invokes the legacies of 'native, indigenous peoples' to suggest that his methods and motives are both timeless and natural. However, his website reveals a more problematic tone.

As an example, one can examine two images presented side-by-side on secretgardenofsurvival.com with the heading, 'Which One of These Is a Garden?'[13] The one on the left shows a recognizable and tidy garden with plants organized in straight rows. The picture on the right appears at first glance (especially to a non-survivalist, I'll admit) to be an overgrown forest. Beneath the pictures are three questions: 'Which one of these is a garden? Which one produces more food? Which one will the zombie hordes attack?' The lesson to be learned from Austin's book is that the overgrown forest is actually a more successful and sustainable style of garden than the tidier, symmetrical one we're used to seeing. Austin's third question (the one about the zombie hordes) may at first come off as a light-hearted joke. Of all the possible modern cataclysms for which one might prepare, the zombie apocalypse seems least realistic. In fact, the phrase 'zombie apocalypse' is often used derisively to refer to unspecified total mayhem that is highly improbable (apologies to zombie enthusiasts).

However, when one clicks through the pages of Austin's site, perhaps to read the first chapter of his self-published book for free, one finds another image that casts Austin's cavalier reference to zombies in a more disturbing (and more political) light.[14] The photograph shows dozens of people wading through waist-high water, with several of them – including the man in the foreground – carrying small children. The paragraph above the picture describes the effects of natural or electronic catastrophes: 'no electricity, no water, no food, no infrastructure, no grocery stores, no gas stations, and after a short time, social anarchy will ensue, with every man for himself and the zombie hordes attacking anyone that has anything they might want, in order to feed themselves and their own family'. There they are again: 'the zombie hordes'.

Except the accompanying image, captioned with 'Zombie hordes approaching by sea' and credited as an 'AP Photo', is obviously not of zombie hordes approaching by sea. Instead, it is of human families searching for safety and help after a natural disaster.

Specifically, it's an image from 2008, captioned on Shutterstock with the following description: 'Nepalese Flood Victims Wade Through Flood Water to a Safe Zone in Sunsari District 400 Kilometer South East of Kathmandu 21 August 2008 Nearly 50 000 People Are Reported to Have Fled Their Homes and Hundreds Missing After a Koshi River Dam Collapsed in South-eastern Nepal On Monday Afternoon' [sic].[15]

In his appearance on the television show 'Doomsday Preppers', produced by National Geographic, Austin used the term 'marauders' rather than 'zombie hordes', but the effect is the same. Not only is Austin dehumanizing those who might try to (or who may accidentally) trespass in search of food, but in his use of this image he's also revealing the white supremacist foundations of his definition of 'doomsday'. He didn't, for example, use images of the attack on the United States capitol on January 6, 2021, whom someone of a different political persuasion may consider to be 'marauders' or 'zombie hordes' or even 'a basket of deplorables'.[16] In other words, Austin's zombies don't look like Austin and his wife. Data from America's Federal Emergency Management Agency, or FEMA, from 2018 reports that 81.8% of those labelled HRCs or 'Highly Resilient Citizens' are white. To be an HRC, one must be able to survive more than ninety days 'at home without utilities or outside help'.[17] That requires a great deal of socio-economic privilege, since stockpiles of food and weapons, and the land on which to homestead, cost money.

In the episode of 'Doomsday Preppers' I mentioned earlier, Austin and his wife Jane explain their reasons for wanting to live off-grid away from densely-populated areas. The circumstances they describe include their neighbourhood becoming more crowded when neighbours had to rent rooms to avoid foreclosure; according to Austin, this led to a rise in crime in the area. His wife Jane – who goes by Prepper Jane on her own website – also describes an incident when two armed men tried to carjack her as she left work one day.[18] The television episode shows the couple preparing various types of protection for their garden, ranging from thorny plants on the outermost perimeters to a homemade pepper spray dispersed by a motion-detection squirting system. They also prepare dried spices to be stored in their kitchen spice rack, which they label 'Italian herbs'. However, the herbs in that bottle were made from foxglove grown on their property. The National Capital Poison Center explains that foxglove is toxic – even deadly – because of its effects on the heart.[19] Austin explains that they have the dried leaves on hand (and in disguise) in case they have to feed 'a special meal' to 'any unsavoury guests'.[20]

Austin and Prepper Jane are extreme examples of prepper paranoia, but they aren't unique among preppers. In a post on AskaPrepper.com, author Rich M. provides various approaches to protecting one's survival garden, ranging from a privacy fence and a guard dog to the type of deceptive planting Austin promotes. Rich M. also encourages the use of poisonous plants, electric fences, solar-powered alarm systems, punji sticks, and caltrops.[21] Punji sticks are sharpened bamboo sticks used as stakes to impale intruders. They're often camouflaged and are most commonly associated with the Viet Cong during the war in Vietnam. Their combat use is banned under the Geneva Convention, and they are illegal to use on private property in the United States (as are all booby traps). Caltrops, heavy metal four-pronged spikes designed to flatten tires or injure pedestrians, are also illegal. In a stark divergence from the sense of obligation and neighbourliness emphasized by victory garden programmes and Cold War civil defence propaganda, prepper Rich M. proclaims, 'There's no sense in growing it, if we can't protect it.'[22]

A different episode of 'Doomsday Preppers', aired by National Geographic on 17 April 2020, features Lindsey and Ray, a married couple who maintain an urban homestead in Boise, Idaho, which is a heavily agricultural state in the northwestern part of the United States. In this episode, titled 'Prepper's Paradise', viewers learn not only about the hundreds of plants grown in their backyard, but also about the ways that Ray has applied his military training to the family's doomsday plans. The couple explain that they're 'preparing for the collapse of the world's agricultural system', and when that happens they'll 'bug-out' or retreat to their secluded cabin in the woods to live off-grid. At that location, they have four years' worth of food and plenty of guns and ammunition. Unlike Austin and Prepper Jane, Lindsey and Ray are open to creating prepper alliances with like-minded neighbours. Viewers of this episode watch the interview process for two potential new members of the group, a process which includes questions about their health and training and a military-style practice exercise that includes a surprise threat (an intruder at the cabin!) so that Ray can gauge the potential newcomers' ability to react quickly to neutralize the danger.[23]

In short, prepper gardens put the 'self' in self-sufficiency, often putting preppers in the contradictory position of having to disguise a large and elaborate growing operation in order not to become targets of hungry refugees. The debate about sharing one's prepper stockpile does echo back to the early years of the Cold War, when social critics and other experts argued over whether homeowners should keep guns in their fallout shelters to protect their family's chances for survival.

I don't want to paint too homogenous a picture of survivalists and preppers. There are several resources like personal blogs from survival gardeners who don't profess the same paranoia as the examples I've included above. For instance, the Seasoned Citizen Prepper suggests donating surplus produce from your homestead or garden when the proverbial S hasn't yet HTF.[24] Revival Garden suggests coordinating survival garden crops with a neighbour and trading to maximize crops grown.[25] The Provident Prepper proposes a labour bartering option, with Kylene Jones writing:

> We always produce more food than we can eat. Part of our plan includes sharing with those in need. Jon is generous, but also believes that people should work for what they get, whenever possible. He will loan them a pair of gloves and let them work to earn their food. I am more apt to just invite them to join us for dinner.[26]

While there's no way to predict how these gardeners may change their strategy after TEOTWAWKI, or even whether any of the gardens discussed here would actually survive the disasters their gardeners are prepping for, it's clear that weaponization and militarization exist in sliding scales depending on each gardener's worldview and fear.

Nonetheless, the preppers with the largest platforms, and the highest levels of visibility in media and popular culture, are quite transparent about their protection of their gardens. Their philosophies, and I'm thinking specifically here of the belief that 'There's no sense in growing it, if we can't protect it', stand in direct opposition to some of the fundamental practices and traditions of land stewardship in the United States. Environmental activist, writer, and farmer Wendell Berry has written prolifically about his life as a farmer in rural Kentucky and about the importance of preserving farming traditions and farm lands in America. In his introduction to Berry's collection of essays *The World-Ending Fire: The Essential Wendell Berry*, Paul Kingsnorth writes of the way Kentucky pulled Berry back from the life he was establishing for himself and his family in New York City in the 1960s. Kingsnorth writes,

> [Berry] left the city and went back to the land, buying a farm five miles from where he had grown up, in the area where both his mother and father had grown up before him. This is the place in which he has lived, worked and written for the last half-century. This is the place whose story he has told, and through it he has told the story of America, and through that the story of modern humanity as it turns its back on the land and lays waste to the soil.[27]

For decades, Berry and his family have lived the homesteading life touted in many prepper resources, not in response to any potential cataclysm, but in resistance to the ongoing crisis that is modern life itself.

In fact, Berry's relationship with the land is hopeful and enlightening rather than aggressively fearful. In the 1968 essay 'A Native Hill', Berry writes about what gets lost in agricultural systems that see the land as something to be conquered rather than something from which we can learn:

> Until we understand what the land is, we are at odds with everything we touch. And to come to that understanding it is necessary, even now, to leave the regions of our conquest – the cleared fields, the towns and cities, the highways – and reenter the woods. For only there can a man encounter the silence and the darkness of his own absence. Only in this silence and darkness can he recover the sense of the world's longevity, of its ability to thrive without him, of his inferiority to it and his dependence on it.[28]

Notice how Berry humbles himself to the land, finding solace in the certainty that nature will go on quite happily without him when he's gone. This humility doesn't appear in any of the prepper literature. In those texts, the land is a means for survival and little else. One doesn't get the sense that Rick Austin has given much thought to what will happen to his secret forest garden of survival once he's gone. I suppose those motion-activated pepper sprayers will run out of juice eventually, and perhaps happy families of wildlife can make their homes in the abandoned plastic structures that once collected rain.

Berry's words in the 2004 essay 'Rugged Individualism' align the prepper mentality with what Berry calls a 'tragic version of rugged individualism' that manifests itself 'in the presumptive "right" of individuals to do as they please, as if there were no God, no legitimate government, no community, no neighbors, and no posterity'. Berry bemoans the things this type of individualism has already cost us 'in lost topsoil, in destroyed forests, in the increasing toxicity of the world, and in annihilated species'.[29] And, although Berry is not speaking specifically about doomsday preppers or even homesteaders or survivalists, he nonetheless reveals the nihilism behind the prepper worldview.

Ironically, if the S ever does finally HTF, the lone prepper may – for a time – be king of his domain: government, community, neighbours, and even the promise of a predictable future will be gone. Those unseen zombie hordes beyond the spiky vines and booby traps of his property will eventually succumb to the effects of whatever catastrophe has befallen them, and our rugged individualist – that hyper-masculinized, hyper-militarized gardener of paranoia – will have finally succeeded in protecting the thing that matters most to him: his own performative sense of superiority – over the land, over his neighbours, over those less-fortunate (or of a different political persuasion) than he, and ultimately, over his own mortality. This is not the heroic settler protecting his home alongside his neighbours; this is not working the soil at the community garden to make sure others have enough of what they need. This is white American exceptionalism lying coiled up and hissing among the amber waves of grain. This is weaponized privilege and aggressive isolationism in the name of surviving only because someone else didn't. And it treats the land (in both the literal and the synecdochal senses) as tools for self-promotion rather than equal partners in the long-game of survival.

About the Author

Rebecca D. Mazumdar is Associate Professor of English at New York City College of Technology. Her research focuses on the intersections of gender, memory, empathy, and food.

Notes

1. 'Class and Event Schedules for Prepper Camp', Prepper Camp 2024 (2024) <https://www.preppercamp.com/class-and-event-schedules/> [accessed 20 May 2024].
2. 'Class and Event Schedules for Prepper Camp'.
3. Five Times August, 'Sad Little Man', *Silent War* (self-published, 2022).
4. Rick Austin, *Secret Garden of Survival: How to Grow a Camouflaged Food-Forest* (self-published, 2012), loc. 80 of 2064.
5. Megan E. Springate, 'Victory Gardens on the World War II Home Front', *National Park Service: The American Home Front and World War II* (16 November 2023) <https://www.nps.gov/articles/000/victory-gardens-on-the-world-war-ii-home-front.htm> [accessed 20 May 2024].
6. See Sarah Lichtman, 'Do-It-Yourself Security: Safety, Gender, and the Home Fallout Shelter in Cold War America', *Journal of Design History*, 19.1 (March 2006), pp. 39–55.
7. State of Maryland, *The H-Bomb and You* (1954) <https://www.ep.tc/comics/h-bomb/index.html> [accessed 20 May 2024], p. 2.
8. Kylene Jones, 'Best Strategies for Growing a Reliable Survival Garden', The Provident Prepper, 2024 <https://theprovidentprepper.org/best-strategies-for-growing-a-reliable-survival-garden/> [accessed 20 May 2024].
9. Seasoned Citizen Prepper, '6 Best Survival Garden Layouts [With Crop Lists & Square Footage]', SCP Survival, 3 March 2024 <https://seasonedcitizenprepper.com/gardening-for-survival/> [accessed 20 May 2024].

10. Kylene Jones, 'Best Strategies for Growing a Reliable Survival Garden', The Provident Prepper <https://theprovidentprepper.org/best-strategies-for-growing-a-reliable-survival-garden/> [accessed 20 May 2024].
11. Rick Austin, 'Grey Water Systems' and 'Swales, Irrigation, Micro-Climates', *Secret Garden of Survival: How to Grow a Camouflaged Food-Forest*, pp. 19–23, 24–29.
12. Rick Austin, 'Rain Water Collection', *Secret Garden of Survival: How to Grow a Camouflaged Food-Forest*, pp. 35–38.
13. Rick Austin, 'Welcome to the Secret Garden of Survival!' Secret Garden of Survival <secretgardenofsurvival.com> [accessed 19 May 2024].
14. Rick Austin, 'Secret Garden of Survival-Chapter One', Secret Garden of Survival <secretgardenofsurvival.com> [accessed 19 May 2024].
15. Somnath Bastola (image contributor), Shutterstock <https://www.shutterstock.com/editorial/image-editorial/nepalese-flood-victims-wade-through-water-safe-7836491?utm_campaign> [accessed 20 May 2024].
16. In a 2016 presidential debate with Donald Trump, Secretary of State Hilary Clinton used the phrase 'basket of deplorables' to describe some of Trump's supporters. *Time* magazine reports on her comment: 'You know, to just be grossly generalistic, you could put half of Trump's supporters into what I call the basket of deplorables. Right?' Clinton said. 'The racist, sexist, homophobic, xenophobic, Islamaphobic – you name it. And unfortunately there are people like that. And he has lifted them up.' Katie Reilly, 'Read Hillary Clinton's "Basket of Deplorables" Remarks About Donald Trump Supporters', *Time*, 10 September 2016 <https://time.com/4486502/hillary-clinton-basket-of-deplorables-transcript/> [accessed 19 May 2024].
17. John Adama, 'New Statistics on Modern Prepper Demographics from FEMA and Cornell', The Prepared, 4 August 2021 <https://theprepared.com/blog/new-statistics-on-modern-prepper-demographics-from-fema-and-cornell-university/> [accessed 20 May 2024].
18. 'Doomsday Peppers – Garden of Eden or Garden of Evil?' YouTube, 5 August 2015 <https://www.youtube.com/watch?v=y5OPgDPg8oc> [accessed 20 May 2024].
19. Serkalem Mekonnen, 'Foxglove – Toxic to the Heart', Poison Control <https://www.poison.org/articles/foxglove> [accessed 20 May 2024].
20. 'Doomsday Preppers – Garden of Eden or Garden of Evil?'
21. Rich M., 'How to Protect Your Garden from Looting Intruders', AskaPrepper.com, 18 October 2022 <https://www.askaprepper.com/how-to-protect-your-garden-from-looting-intruders/> [accessed 14 May 2024].
22. Rich M.
23. 'Doomsday Preppers – Preppers Paradise', *National Geographic*, 17 April 2020 <https://www.nationalgeographic.com/tv/episode/48454a05-746d-41a5-a458-d76e9d7e2bed> [accessed 20 May 2024].
24. Seasoned Citizen Prepper.
25. 'Food Security: How to Start Your Survival Garden', Revival Gardening, 2023 <https://www.revivalgardening.com/post/survival-garden> [accessed 20 May 2024].
26. Kylene Jones.
27. Paul Kingsnorth, 'Introduction', *The World-Ending Fire: The Essential Wendell Berry*, ed. By Paul Kingsnorth (Counterpoint, 2017), p. viii.
28. Wendell Berry, 'A Native Hill' (1968), in *The World-Ending Fire: The Essential Wendell Berry*, pp. 3–37 (p. 32).
29. Wendell Berry, 'Rugged Individualism' (2004), in *The World-Ending Fire: The Essential Wendell Berry*, pp. 265–67 (p. 265).

24
Orange Blossoms and the Holy Grail of Persian Jams

Nader Mehravari

I spent my childhood living in Tehran, the capital city of Iran, which sits at the southern foot of the towering Alborz mountains. One of our family's annual summer rituals involved packing the car for a week-long vacation on the shores of the Caspian Sea. The adventure would start with a five-hour drive on a beautiful mountainous – and at times dangerous – road over the Alborz range before arriving on its lusciously green northern base. Once we reached the coastal area, the first thing we would do – even before finding our planned lodging – was to stop at the very first roadside stand selling jars of that year's locally produced translucent pale golden coloured orange blossom jam (Figure 1). We would buy enough to last us for breakfast on every day of the week-long vacation plus a couple of extra jars to take back home as the authentic versions were almost impossible to find in Tehran.

Orange blossom jam – made from the individual white petals of orange blossoms – is the most tedious and time-consuming of jams to make, and with a tiny yield. It is

Figure 1. A jar of orange blossom jam (photo by Nader Mehravari).

the pinnacle of jam-making endeavours, demanding meticulous attention and patience throughout, hence its revered status as the 'holy grail' of jams. This paper delves into the rich history and cultural significance of orange blossoms, not only in Iran, but worldwide. It explores the time-honoured techniques of orange blossom jam-making utilized for centuries within Persianate societies, contrasting them with other flower petal jams. Additionally, the paper examines the commercial industry's preference for orange blossoms in fragrance and distillate applications, highlighting the unique allure of this precious ingredient.

A Bit of Botany and Etymology

Traditionally, the blossoms from a very specific variety of orange tree, *Citrus × aurantium*, are used for making orange blossom jam, distilling them into orange blossom water, or extracting their essential oils in perfumery applications (Figure 2). In the English-speaking world, this is commonly known as bitter orange, Seville orange, sour orange, bigarade orange, or marmalade orange. In some languages, there are distinct single-word designations for bitter orange, distinguishing it from other types of oranges. For example, in modern Greek, bitter orange is referred to as *nerantzi* and sweet orange as *portokal*. The same is true in Persian where the word for bitter orange is *nārenj* whereas the sweet orange is referred to as *porteghāl*.

Figure 2. Citrus × aurantium *tree showing previous year's unharvested fruit and current year's new blossoms (photo by Zeynel Cebeci, used under the Creative Commons Attribution-Share Alike 4.0 International license).*

The bitter orange is a round, slightly flattened fruit. Typically, it is 2.5 to 3.5 inches (5.5 to 8.5 cm) wide. Its bumpy surface features a thick, aromatic peel with a bitter taste. As the fruit matures, the peel turns a vibrant reddish-orange and develops tiny, sunken oil glands. Inside, the fruit is divided into 10 to 12 segments. These segments have bitter walls and hold a highly acidic pulp – hence one of the fruit's common names being sour orange. It is also characterized by a large number of seeds – many more than other types of citrus fruit.

The blossoms of the bitter orange tree, which are the focus of this paper, begin as tiny, tightly closed white buds on the branches, no bigger than a pea. As these buds develop, the petals tucked inside unfurl, pushing the bud's shape from round to oval. Finally, the bud bursts open, revealing a beautiful star-shaped flower. Each blossom has five white to ivory petals surrounding a cluster of yellow stamens with orange tips. The delicate, slightly waxy petals measure around 1.5 to 2.5 centimetres (Figure 3).

The bitter orange blossoms release a captivating fragrance. This aroma blends sweet citrus notes with hints of floral scents like jasmine and tuberose, along with a touch of fresh grass and nutmeg. The raw blossoms are edible, have a strong floral flavour, and some bitterness. The level of bitterness in the raw petals depends on the specific variety and origin of the tree, and the growing region's soil and climate. After being treated with boiling water (whether for distilling or making jam), the petals develop a refreshingly floral and citrusy taste. Unlike many jam fruits, the petals remain whole even after extended cooking periods.

Figure 3. Lifecyle and parts of orange blossom (photo by Nader Mehravari).

A Bit of History

Originating in Southeast Asia, the bitter orange embarked on a journey westward. By 700 CE, it had reached the Islamic world through India and Iran. Of the two kinds, bitter and sweet, it was the bitter oranges that reached Europe first, some five hundred years before the sweet oranges. Bitter oranges came to Sicily in the eleventh century and to Spain, brought by the Moors, in the twelfth century. From there, it continued its travels to Florida and the Bahamas.[1]

To the best of my knowledge, the first document citing the use of bitter orange blossoms is *The Canon of Medicine* published in 1025 CE by the Muslim Persian physician-philosopher-physicist-writer-scientist Ibn Sina commonly known in the west as Avicenna. The English translation of one of his four detailed prescriptions for dealing with symptoms of fever is:

> Bitter Orange: 4 dessert spoons of bitter orange flower is boiled in 3 glasses of water. Then its extract is strained, and it is drunk in dose of one coffee cup 4 times a day.[2]

The earliest references to orange blossom water are also associated with Ibn Sina. In the eighth century, the Iranian born Jabir ibn Hayyan, commonly known as the founder of modern pharmacy, invented a distiller called 'alembic' which he used for the chemical analysis of material (Figure 4). Ibn Sina improved the alembic by inventing and adding cooling coils to it which drastically improved the efficiency of distillation techniques of the era. Ibn Sina's improvement allowed the essence of some flowers to be captured in water, such as rosewater and orange blossom water.[3]

Rose water and orange-blossom water were initially used to make medicines taste more palatable. However, Persians, who had already been using rose petals in their cooking, started using rose water and orange blossom water in their confections. They were followed by the Arabs who adopted these practices after their invasion of Persia.

Figure 4. Drawing and description of an alembic still by Jābir ibn Ḥayyān from the eighth century (photo sourced from Wikimedia and is in public domain).

Uses of Orange Blossoms

The primary uses of orange blossoms are:

- Making jam from individual white petals.
- Making orange blossom water, a.k.a. orange blossom distillate.
- Perfumery and essential oil applications.

- Preserving by drying so they can later be rehydrated for the above three applications.
- Drying to be added to potpourris, naturally fragrant dried plant materials used to provide natural scent in residential settings.

Although bitter orange blossoms are by far the most preferred orange for all these applications, it is feasible to use other highly fragrant citrus blossoms – in particular *Citrus x sinensis* commonly known as 'sweet orange' which is a hybrid between pomelo (*Citrus maxima*) and mandarin (*Citrus reticulata*). Unless otherwise stated, however, all references to orange blossoms are assumed to be bitter orange blossoms.

While the focus of this paper is centred around the blossoms of the bitter orange tree, it should be noted that other parts of bitter orange trees are highly sought after as well. The fruit of bitter orange trees is a popular souring agent in the Persian cookery landscape. Its juice, with a pH of about 2.7, brings a highly aromatic and delicate sourish-orangish flavour to Persian cuisines that is sourer than regular orange juice (pH of about 3.5) and sweeter than lime or lemon juice (pH of about 2.2). This prized acid is especially important during Persian New Year celebrations, as it flavours the special dishes traditionally served then. The fruit's peel is preferred for making traditional orange marmalade. Petitgrain, an essential oil used in perfumery, is produced through the distillation of branches and leaves of the bitter orange tree.

Cultural Significance of Orange Blossoms

The delicate white petals and intoxicating aroma of orange blossoms have captivated people around the world for centuries. These beautiful flowers hold special meanings in many cultures, symbolizing purity, fertility, and good fortune. Their association with new beginnings is particularly strong, as seen in their use in wedding ceremonies. In fact, brides in ancient China wore orange blossoms to promote fertility, reflecting the unique nature of the orange tree – it blooms and bears fruit simultaneously. Similarly, Greek mythology links the orange tree to the goddess Hera, representing both purity and fertility.

Queen Victoria defied tradition in a big way when she married Prince Albert in 1840. Not only did she break the mould by proposing (something unheard of for a noblewoman!), but she also opted for a white wedding gown instead of the customary deep-coloured velvet dress. To top it all off, she chose a simple crown of orange blossoms over the traditional heavily jewelled tiara (Figure 5). Such traditions continue today in some Mediterranean countries as well as in Persianate societies, where orange blossoms symbolize purity and eternal love in weddings.[4]

According to the writings of French historian and diplomat Hyacinthe Louis Rabino, who had lived in Rasht, the capital of the Iranian northern province of Gilan between 1906–1912, orange blossom jam was so highly valued that the citizens of the province sent jars as gifts to the royal court of the Fat'h-Ali Shah Qajar, the second monarch of the Iranian Qajar dynasty who had reigned between 1797–1834.[5]

Orange blossoms continue to be highly embedded in the traditional rituals of citrus growing regions of Iran such as the northern Gilan province on the shores of the Caspian Sea, as well as in the central city of Shiraz, the city of flowers, poets, and wine. There is

Figure 5. Queen Victoria wearing a simple wreath of orange blossoms on her wedding day (photo courtesy of the Royal Collection Trust).

an annual festival dedicated to orange blossoms held in the Gilan province signifying the region's gratitude for the blessing of its fragrance to the region. In 2023, the festival was added to Iran's list of national intangible cultural heritage. In Shiraz, in early Spring, the whole city smells like orange blossoms as piles of freshly picked blossoms are offered for sale by street vendors.

Orange Blossom Water

Although not as popular as rose water, orange blossom water has had a key role as a liquid flavouring for baked good and sweets in Middle East and Mediterranean regions, and, in the past, even in the Moorish regions of southern Europe.[6] There are explicit references to its popularity in making sweets in ancient Persia and in the Arab world as early as thirteenth century.[7] Most recently, orange blossom water has even found its way into bartenders' modern mixology ingredients.

Today, the distillate made from the blossoms, orange blossom water [Persian: عرق بهار نارنج, Romanized: Aragh-é-Bahār-Nārenj] is a popular liquid flavouring used in traditional Persian summer sweet beverages, sharbats [Persian: شربت, Romanized: Sharbat], (not to be confused with sherbet or sorbet), and for some Persian sweets. Orange blossom water is also used in some Persian confections – sometimes as a substitute for rose water.

Perfumery and Essential Oils Applications of Orange Blossoms

The olfactory profile of orange blossoms is very rich. Along with rose and jasmine, orange blossoms form the triad that constitute the three most important floral sources for perfumery. In the perfumery industry, the essential oil produced from the blossoms of the bitter orange tree, through steam distillation, is referred to as the Neroli oil. The name 'Neroli' has been used for a fragrant essence since the seventeenth century. It originated in Italy, where Anne Marie Orsini, Duchess of Bracciano and Princess of Nerola, popularized the use of bitter orange flower extract as a perfume for her gloves and bath. Because the yields are very low, it takes about 1000 kg of flowers to obtain about 1 to 1.3 kg of Neroli oil.[8]

The great majority of the worldwide production of orange blossoms is consumed by the perfumery industry. Along with Tunisia – which is the largest producer – Morocco, Egypt, Syria, Lebanon, and Spain are among the largest producers of orange blossoms that supply the worldwide perfumery industry.

Harvesting of Orange Blossom Buds and Petals

Sour orange trees start to bloom in early spring. Depending on the variety of the tree and growing climate, harvest from a given tree lasts no more than two weeks sometime between mid-March to late-April. In a large professional orchard, the harvesting could last three or four weeks across all the trees in the orchard.

The traditional strategy for collecting orange blossoms – both in large-scale professional settings and small family settings – starts by spreading cloths or plastic tarps under the trees (Figures 6 and 7). The blossoms are either hand-plucked or naturally fall down onto the tarps along with unwanted material from the surrounding environment – falling leaves and twigs from the orange trees, wind-blown parts of other close-by spring flowering trees, bodies of dead insects, and dust. While collecting the blossoms, care must be taken not to disturb the bees – or get stung by them – as they will be busy pollinating the blossoms still on the tree.

The next step is to separate the collected buds (partially opened, fully opened, fully intact, or in pieces) from all the unwanted materials including twigs, leaves, bodies of dead insects, etc. The remaining blossoms can then directly be used in distillation processes to make orange blossom water and essential oils.

There remains, however, much more tedious and time-consuming sorting to be done to make orange blossom jam. The individual, tiny, white petals have to be painstakingly separated from all other parts of the blossoms including stamens, pistils, and sepals as well as any unmatured buds (Figure 8). For each day's harvest, the petals must then be washed with tap water to wash off any dust, soil, insect parts, and other very small unwanted debris.

Unless one has access to a large orchard of bitter oranges, there won't be enough petals to make jam from, after one day's harvest. The best strategy for conserving each day's separated and washed petals is to put them in a container, cover them with water, and put them in the least cold part of the refrigerator (closest to the refrigerator's door or in the shelves in the door). The process gets repeated each day adding more petals to

Figure 6. Professional harvesting of orange blossoms in a large sour orange orchard (photo courtesy of Robertet Group).

Figure 7. Harvesting orange blossoms in a small family setting (photo by Nader Mehravari).

Gardens, Flowers, and Fruit

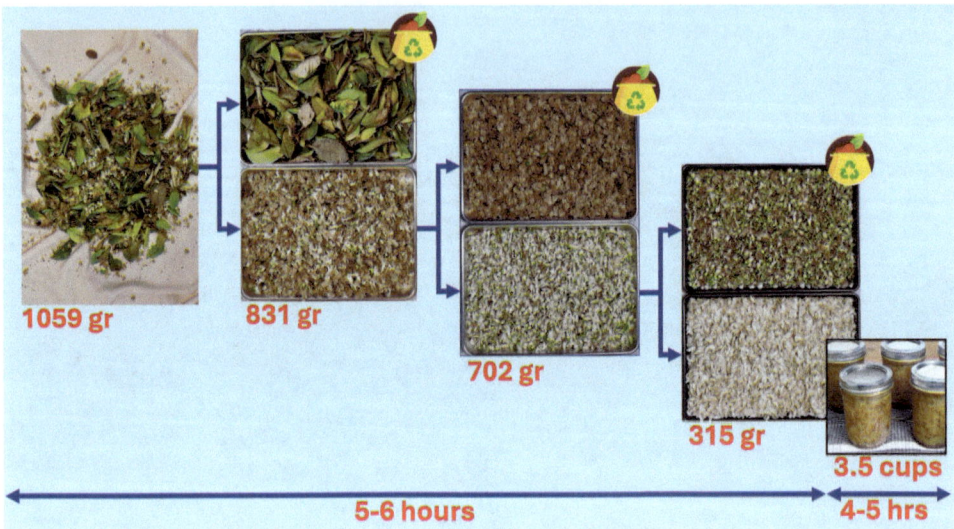

Figure 8. From the field harvest to the petals for jam making (photos by Nader Mehravari).

the same container that has been in the refrigerator. After a few days, there should be enough petals to make a good size batch of the jam. This process allows one to collect the blossoms at the height of their maturity and fragrancy (i.e. early morning before the bees arrive) while keeping them fresh over a span of a few days.

In professional settings, assuming the trees are being grown solely for their blossoms and not for their fruit, a mature tree could produce 10 kilograms of flowers (i.e. full buds – stamen, pistil, sepals, and petals) utilizing experienced pickers. An experienced picker working in a commercial large orchard can collect 3–4 kilograms per day.[9] A mature tree in a small family-run garden – typically grown both for their blossoms and their fruit – produces less than one third of that. Each kilogram of flowers results in about 400 to 450 grams of petals for jam making.

To get a better feel for the low volume and tediousness of the process, one can compare the process of orange blossom jam making with that of rose petal jam making. A typical damask rose, the preferred type of rose for Persian-style rose petal jams, provides at least fifty petals, whereas each orange blossom provides at most five petals. Moreover, it takes about a tenth the time to harvest rose petals than orange blossom petals.

Orange Blossom Jam

Persian jams are somewhat different from western-style thick jams, where the fruit is crushed and/or cooked until the fruit loses its shape or forms jellies which are firm and without fruit solids. Typically, in Persian jams, the fruit retains its shape during cooking. The surrounding syrup is translucent and has a honey-like viscosity at room temperature.

The annual supply of this jam's key ingredient, the white petals of orange blossoms, is significantly restricted due to three primary factors. First, harvesting of the petals has very low yield as each kilogram of harvested orange blossoms yields a mere 400–450 grams of cleaned petals. Second, the harvest of orange blossoms is confined to a brief window of one to two weeks during the early spring blooming season. Finally, harvesting the blossoms reduces the tree's capability for producing a significant orange harvest, impacting the small growers' financial gains.

The traditional recipe for Persian orange blossom jam calls for only cleaned fresh petals of Seville orange blossoms, sugar, water, and a bit of lemon juice. The resulting jam has a light pale golden colour where semi-translucent cooked petals swim in an aromatic pourable syrup (Figure 1).

The jam making process itself does not require any special equipment other than what is required for any typical home jam making endeavours. Once the petals are ready, the process, although time-consuming (lasting four to five hours depending on the batch size), is not any more complicated than for any other typical home jam making. Sugar is dissolved in water, the mixture is brought to boil, petals are added, and the mixture is stirred every few minutes. A bit of lemon juice is added towards the end. The jam is ready when the mixture reaches a temperature of 220°F (at which point the sugar content of the mixture is about 63°Bx using a refractometer).

Pectin plays an important role in western style jams and jellies because these are much thicker and not as pourable as traditional Persian jams. Although there is plenty of natural pectin in the cell walls of citrus fruits, there is very little natural pectin in the cell walls of the petals of the orange blossom. Properly cooked Persian jams (patiently and gently cooked long enough for the syrup to thicken naturally) do not need added pectin. If one must add pectin (maybe for the jam to be much thicker than the traditional Persian style jams or to speed up the process, which is not recommended), one can either use some form of commercial pectin products from a local grocery store, or, alternatively, go with the traditional Persian natural pectin source, seeds of quince. Soak 1 teaspoon of dried quince seeds (available in Persian markets or on the Internet) in a ½ cup of water for 15 minutes, at which point the water will slightly gelatinize. Add this to the jam at the same time that the recipe calls for adding lemon juice.

Many traditional orange blossom jam recipes start by blanching the petals in boiling water for a few minutes, then immediately rinsing them with cold tap water, and repeating the blanching process two or three more times. The blanching process is intended to remove some (not all) of the natural bitterness that comes from some varieties of orange blossom petals. There is a negative side to this blanching practice. Each time the petals are blanched, they lose some of their aroma and essence prized in the jam.

I have experimented making the jam with and without blanching the petals. My personal preference now is not to blanch the petals, preserving more of the flavour of the petals. The jam may have a delicate bitterness that, in my opinion, adds to its complexity. My recommendation for the first time that you make this jam: chew on a couple of the freshly picked petals. If you taste a lot of bitterness, then consider blanching the petals. If not, skip the blanching process. This recommendation is based on the fact that the level of bitterness in the petals highly depends on a wide range of parameters including the specific variety of Seville orange (there are many), age of the tree, the growing climate, etc.

I have seen some recipes for the Persian orange blossom jam that add other flavourings such cardamom or saffron. I do not recommend adding any flavourings. They mask the natural aroma of the orange blossoms.

Closing

Orange blossom jam stands as a crown jewel of Persian jam making. Its delicate floral fragrance and subtle citrus notes are a testament to the meticulous process involved, from the brief window of harvesting the blossoms in early spring, the time-consuming and tedious job of separating the petals, to the patient simmering that brings out their essence. While the annual yield is limited by the short blooming season and the low petal-to-blossom ratio, orange blossom jam remains a cherished cultural experience in Iran. Each spoonful offers a glimpse into the rich heritage of Persian cuisine, where time-honoured techniques and a deep appreciation for nature combine to create a truly unique and evocative flavour. Furthermore, the jam's significance extends beyond the culinary realm. Its association with weddings and national festivals underscores its role in weaving together the cultural fabric of many lands beyond Iran. As orange blossom jam continues to be passed down through generations, it serves as a reminder of the enduring connection between people, place, and the fragrant blossoms that herald spring.

About the Author

Nader Mehravari has been exploring historical, cultural, and social practices relating to the preparation and consumption of food within Persianate societies for over forty years. He is working towards a modern and innovative Persian cookery book. He volunteers as a research associate at the College of Agriculture and Environmental Science, University of California, Davis.

Notes

1. Andrew M. Watson, *Agricultural Innovation in the Early Islamic World: The Diffusion of Crops and Farming Techniques 700–1100* (Cambridge University Press, 1983), p. 81; Julia Frances Morton, *Fruits of Warm Climates* (J.F. Morton, 1987), pp. 130–33; Melitta Seiss Adamson, *Food in Medieval Times* (Greenwood Publishing, 2004), p. 23.
2. Caner Ozogul, *Healing Secrets of Avicenna – It Is Compiled from Avicenna's Work, 'The Canon of Medicine' and Then Simplified* (Lulu.com, 2017), p. 225.
3. Samir S. Amr and Abdelghani Tbakhi, 'Jabir ibn Hayyan', *Annals of Saudi Medicine*, 27.1 (2007), pp. 52–53, DOI:10.5144/0256-4947.2007.53; Iberian Coppers Lda, 'History of Alcohol Distillation', *Iberian Coppers* <https://www.copper-alembic.com/en/page/history-of-alcohol-distillation> [accessed 28 April 2024].
4. Sarah Bancroft, 'Orange Blossoms: Culture, Ceremony and Royalty – Symbolizing Purity, Fertility and Happiness for Centuries', *Fleurs de Villes* <https://www.fleursdevilles.com/post/orange-blossoms-culture-ceremony-and-royalty> [accessed 28 April 2024].
5. Mohammad Hassan Abrishami, 'Noghl, Jostari Shirin dar Goftegoye Tamadonhaye Bashari', *Ma'ārif*, XVII.1 (July 2000), p. 132.
6. Harold McGee, *Nose Dive: A Field Guide to the World's Smells* (Penguin, 2022), p. 458.
7. Melitta Seiss Adamson, *Food in Medieval Times* (Greenwood Publishing, 2004), p. 117.
8. Roberted Group, 'Orange Blossom', 3 May 2022 <https://www.linkedin.com/pulse/orange-blossom-robertet-group/> [accessed 26 April 2024].
9. Alara Dural, 'Exploring Ingredients: Orange Blossom', *Sana Jardin*, 16 April 2020 <https://sanajardin.com/blogs/news/exploring-ingredients-orange-blossom> [accessed 28 April 2024].

25
If This Field Could Talk

Considering Soil as a Site of Hospitality and Diplomacy
between Humans and Non-Humans

Jennie Moran

I have friends who live on a farm outside Ballybunnion in north Kerry on the southwest coast of Ireland. Lisa and I studied sculpture together, a while ago now. She married Rena not long after that. Rena is one of the Blakes, a well-known and deep-rooted Ballybunnion family. Rena inherited the land and farmhouse. Her ancestors were tenant farmers and, against the odds, managed to acquire the land sometime around 1928. I recall sitting in their kitchen a few years ago. It is a heavy hazy summer evening; the air sentient with midges and pollen and micro-organisms, all charged with their respective mandates. Some lazy and meandering, others more frantic. The farmhouse kitchen, with its thick walls, is cool and calm. Rena arrives in from the farm, and with her a new consignment of mischievous airborne particles. She is wiping her forehead and pouring one glass of water after another down her throat. As she adjusts to this contained, domestic setting her panting slows and after some time, there is a quiet sigh. Although I do not know her well, I glean that there is news, and I prepare to catch its softly spoken delivery.

She has come from irrigating their three potato fields. Two of the fields are situated together on one section of the land and are thriving. The other, slightly separate, has caught blight. All the crops have come from the one seed stock and have been cultivated in the same way. Her voice dips, softer still, and I find that I am leaning towards her, ears first. This evening it occurred to Rena that the field with blight is situated on the part of the land which used to be a famine village. She goes quiet now and lets the story sit with us.

Before the 'Great Famine' in 1845, Ireland was one of the most densely populated countries in Europe at over eight million people. Part of the reason for this unprecedented rise in population was the fact that poor people, previously reliant on cereal and dairy, now had a plentiful, resilient, and cheap food source.[1] Ireland was the first country in Europe to adopt the potato, initially as animal fodder, then as a winter vegetable, and, eventually, with awful consequences, we came to rely on it all year round as a primary source of food. The potato was taken from the New Word, lauded as a cheap source of food for labourers.[2] People who would previously have starved to death could now perform labour and procreate.

'Tubers are a type of enlarged structure used as storage organs for nutrients in some plants, derived from stems or roots. Tubers help plants perennate, provide energy and nutrients, and are a means of asexual reproduction.'³ I am imagining these functions transferring from plants to humans: swollen, spawning lumps of energy. They certainly helped the Irish population perennate, and I speculate that the reproductive process amongst the underfed subsistence farmers was not particularly sexy. These unsuspecting rhizomes did not sign up to be a tool for exploitation. They did not intend the precious energy they stored up to fuel the turbo expansion of empires. It would appear that the undergrowth does not appreciate extractive capitalist approaches (it's in the name, I suppose).

Is this third blight-ridden field protesting? Are these tubers marking their respect to the north Kerry cottiers that disappeared from this land during the famine? Here the soil has recognized the problematic seed more than a hundred and seventy years later. This is remarkable. It is theoretically possible for blight to remain in the soil as resting spores, but studies show that, without a host plant, these oospores are not viable after forty-eight months.[4] This North Kerry soil has been triggered by the reintroduction of potatoes and is expressing resistance, in no uncertain terms. I am imagining this underground movement, this microbial opposition to the use, over-use, and misuse of this plant. Curiously, potatoes are often used for cleaning, restoring, and reclaiming the soil, much like a memory wiping device.[5] Not this field, however. In this field there is a black box recording from the soil:

Enough now.
Enough.
I did this before. I know this plant you are pressing into me. I have held these roots in my grasp before. They were strange at first, unfamiliar. Foreign, I think. They seemed to come from very far away. Shaken from a journey undertaken unwillingly. I enveloped them in unquestioningly, like a good host. 'Say yes to who or what turns up', I was always told.[6] Now, I know them well. I reared them and plumped them up for you. Goodness. I did it again and again for you. For a long time, I loved it. It was good and right to let you have this food. You needed it. You asked me for it in prayers without words. Desperate you were. You gave offerings. We exchanged things and energy. You understood that it was a lot to ask. I was happy, and you were grateful.

What happened then, I can't be clear. My knowledge system is different to yours. Not all the information permeates down to me. Let's say we drifted apart. Well, technically I stayed put. If I drift, no one really notices. The same happens if I creep. Your drifting is more ... showy. You got distracted with tasks above ground and forgot to cast your attention downwards. I often wondered, was it because you were not burying all those hungry folk, you forgot our connection? Before, when I took in all your wasted loved ones, I felt your heart-heaviness and that salty residue dripping off you and I didn't like it. I offered you consolation. My microorganisms incorporated your kin; welcomed them unconditionally and you watched them become part of the land. This regeneration interrupted your stuffy, saline grief. You crouched down and placed your hands on me land and I think you understood: Yes, eventually we will meet as the same crumbling life matter and laugh at the absurdity of ever thinking we were different. Back then, we internalized each other. We were related, connected, mutual, commensals.

Now, you silly eejit, you forget all that. You have made a disconnection, and our channels are all withered. I have no way to send you important messages. You stopped listening out for it, and I think the language has died. Please know that I grieved you. It wasn't the same without you close by because, when you were good, you were wonderful. Your attention had value.[7] It was nourishing and invigorating. I got weary then from all the monotonous hard potato labour. Tonnes and tonnes of the same thing for years on end. I felt old. By the time the sad rot got under my skin and set into the tubers I didn't have the gumption to fight it. I decided to interpret the silence from you as a sign that the need was no longer great. I heard from my network that it was hopeless anyhow. Phytophthora infestans *is a tricky one, as blight goes. Hard to shift. It is the plant pathogen that has most greatly impacted your humanity to date.[8] It came from the same land as the potatoes and had a similar exotic feel about it, followed them over like an avenging shadow.*

Things got very quiet then. Nothing at all from you. I wondered what had happened. I used the quiet time to rest, but it was dreary enough. I know that time moves differently for you and me. You are a very recent addition, don't forget, between five and seven million years; I used to tease you by calling you a blow-in. It's all relative of course. Next to rocks, I am the Johnny-Come-Lately at only 450 million years. All this to say that I find myself thoroughly confounded by your reintroducing this same traumatic plant again here, now, after everything that has happened. I am in the dark, literally and figuratively. In the bit of time since all the potatoes went rotten and your people went hungry and died, the memory of it seems to have evaporated.

In short: you were supposed to learn something. Those rotten tubers and I were telling you the story of a plant uprooted and dragged diagonally across the Atlantic Ocean to provide a cheap source of food to labourers so that people in another place could prosper; of the exploitation of the indigenous wisdom at the plant's source.[9] Those far away farmers planted food with reverence, asking the land what it needed in return, placing offerings in with the seeds.[10] Contained within this tale is the requiem for the peaceful custodians of that distant land, most of whom were wiped out in all this extraction.[11] The spores of late blight that followed across the seas after the potatoes are telling the story too. The stinking black blight flesh was the last resort fable. We had tried countless other ways to reach you, to keep our channels open.

This has been hard for me in other ways too. Something that connects you and I is the impulse to allow in strangers, to host others and provide refuge and nourishment. You do it too. I know because you come to me when they appear at your threshold. You rush to me for food, all happy and proud when we have it to give. Your impulse to share is sometimes in competition with your desire to protect your place and the lives within it. Occasionally the latter wins out and I understand. My approach is, by nature, different. I am programmed to wait in vital anticipation for the arrival of the other, as 'the I who is uniquely chosen . . . to be host to the Other'.[12] I am here for it. That is my role – to provide unconditional welcome to other living things. It is never questioned. My belief is that you, dear pal, were once programmed like this too. I get glimpses of it now and then. Welcome issued with gusty exuberance: 'Enter quickly, for I am afraid of my happiness.'[13]

Gardens, Flowers, and Fruit

What happened with the potatoes was awful. On my watch there was the eradication of your modest sustenance and with it your prospects for survival. I failed you above ground. Below ground, the plant which I had taken into my protection was everywhere infected with the sad rot and I was powerless. Multistorey failure.

It was an undoing of everything I understood about myself – my one job. The effects have been grave. In the dark silence I have mourned all the lost life and dwelled on my role in the tragic destruction. For you to introduce this plant again now is very distressing. The familiar dull ache is back. Not long after that the sad rot returned. Perhaps it was here all along, waiting patiently. As I said before, time works differently for us. The spores are reactivated now, and we are bracing ourselves to repeat, for your benefit, the same hard story; to use the sad rot instead of words so that you will smell the foul black putridness and understand. We here in the undergrowth will not be part of this again. Every chapter of this story tells of misuse, opportunism, extractivism surrounding this plant, and we are united in our refusal to comply.

There was irreparable harm done, to you and to us. Systems were damaged. And now, because of this, my role as host has been fundamentally altered. For the first time in 450 million years, I have cause to question the new arrival in case it is the harbinger of destruction. My welcome is no longer unconditional. I have become more like you, and not in a good way. In the day to day, this is discombobulating and shameful; imagine a sort of incarceration, like an electric fence around one's fundamental impulses.

Worse, however, is the deep unsettling knowledge that this new glitch, this wariness, will result in more long spells of silence. Even for me they will feel endless. And I have a fear that they could be endless. The opportunities to tell you these stories are dwindling. Can you sense that I wonder? Can you understand on some level that an irreversible shift has occurred? The instant when hospitality from me is rationed marks the beginning of the end. If I am behaving like you, we are all doomed.

Remember that I am not one thing. I am a complex constellation. A collective. A pedosphere within a biosphere. My pronouns are 'we/our'. There is no clear distinction between me and every other organism, yourself included, by the way. I'd like to be addressing you in this same 'we', but you have dislodged yourself. You have notions of dominance, and I am beneath you now. You look down on me and use my name, 'soil' as a verb to convey visceral disgust.[14] You are busy now in your fancy new role as geological agent. Big boss of the Anthropocene. How's that going for you, by the way? Ahh curse this repugnant spite that rears its head now. I behave shamefully. See me bitter and old. It eats me up. The erosion has breached my substrate of ancient wisdom and sometimes it gets the better of me. This layer of treasure is the earth's insurance policy, buried deep in an impenetrable vault for safety. Profound in depth and in importance. That it is disturbed now is unprecedented and signals to me that I don't have long.

All the more reason to stick to the point: collectivity. Although you have embraced human exceptionalism and pride yourself as superior with all that fancy intellectual reason, you regularly crave oblivion.[15] You take actions to diminish the hard, grey loneliness of individual consciousness. Fucking. Raving. Inebriating. Praying. Tripping on Psilocybin. Singing in choirs. Running in marathons. It happens during strokes too when the left cerebral cortex is quietened, and the calculation function is cut.[16] Occasionally, during these momentary annihilations, we are we. We recognize ourselves in each other. Isolation then is inconceivable. Until the spell breaks, and you dislodge yourself again. This state that you experience sporadically is my default. Imagine that please. It is pure hospitality towards fellow earthlings. This is not the proprietorial tolerance you have come to associate with the term hospitality. Here, guest and host are interchangeable. It is the ecstatic instinct to accept the other. 'Enter quickly, for I am afraid of my happiness.'[17] This hospitality welcomes disruption and associates the new arrival with possibility, growth, and even salvation.[18] This hospitality is supposed to be limitless. To measure it and hoard it away is repugnant for it is, by nature infinite and unconditional.[19] A gushing, free-flowing renewable resource.

There is a Haudenosaunee fable, from the same land as the potato, about the creation of our biosphere, which begins when a dark watery realm is interrupted by a new bright hole in the very top of the sky. In the column of light, the creatures below spot something falling from the hole, like a speck of dust in the rays of the sun. The something is a woman called Skywoman, and she is holding a branch. While she falls, the creatures worry. The geese fly up together and catch her on their wings. This is a temporary measure because the woman is heavy for the geese. The other creatures – ducks, otters, loons, beavers, fish – form a council. A great turtle floats up and offers his back for her to rest upon. The little creatures take turns to dive to the bottom of the water to get land for her to make a home. It is a hard mission, and some of the creatures do not make it. It's the muskrat in the end who offers his life and succeeds in bringing her a handful of earth. Skywoman, overwhelmed by the bravery and generosity of the creatures, takes the earth and rubs in on the turtle's back. 'Like any good guest, she has not come empty handed.' The branch she clasps in her hand contained seeds for plants and trees and grasses and flowers and medicine. With the new stream of sunlight she opened in the sky, this new life thrived. The creatures came to live with her, and this version of earth began.[20] You see – this is the story about hospitality between you and I. This is how it should be. Here, the earth is a joint effort. A collaboration! Everyone involved needs everyone else. We are all guests and hosts of the biosphere.

This cohort of your fellow humans have held this fable and protected it through their generations. I can feel them trying to keep the channels between us open. They do not need sad rot to give them warnings because they are listening still. There will be another fable about the long quietness to come, about the reversal of this creation. At the end of the endless silence, it will be heard, but not by you, and not by them, perhaps not even by me. This story must be deliverable without words because language might have no meaning then.[21] Imagine a ceremonial unnaming, all the creatures and organisms liberated from their prescribed binomial nomenclature.[22] It will chronicle the end of our geological epoch with beauty and truth and grief. It will offer heavy-hearted insight to future stewards. It must portray the core meaning of unconditional hospitality because, if we dare to venture forth a positive legacy, something we might be proud of, it is the system we all created for welcoming others. What we need, and hope for, is that hospitality features strongly in our ending. This story must describe how we practiced unconditional welcome as a strategy to preserve our community and prolong our time; how we all gave in to it and used it to form a strong and mesmerizing mesh of connections. Let there be unexpected accounts of us spending longer and longer together in the collective consciousness mode, you included. Let our story describe in infinite detail the moments when we recognized ourselves in everything else and dismissed our suspicion; when we were wholly receptive; when we finally forgot about risk, and symmetry, and we said yes to who or what turned up.[23] Let it depict us as a complex murmuration; energy beings; all of us earthlings.

And look – now I can say we.

About the Author

Jennie Moran is an Irish visual artist and hospitality theorist. The author of a book called *How to Soften Corners*, she runs a mobile food/art project called Luncheonette. She has been awarded a research scholarship by the Getty Institute in Los Angeles, she won Best Emerging Voice at the Irish Food Writing Awards, and she is featured in the *Irish Independent*'s Ones to Watch for 2025.

Notes

1. Máirtín Mac Con Iomaire and Pádraic Óg Gallagher, 'The Potato in Irish Cuisine and Culture', *Journal of Culinary Science & Technology*, 7:2–3 (2009), pp. 152–67, (p. 154).
2. Alejandro Colás and others, *Food, Politics, and Society: Social Theory and the Modern Food Systems* (University of California Press, 2018), p. 46
3. Alan Longman, *Rooting Cuttings of Tropical Trees* (Commonwealth Science Council, 1993), p. 11.
4. Mohammad Babadoost and Carlos Pavon, 'Survival of Oospores of Phytophthora Capsici in Soil', *Plant Disease*, 97.11 (2013), p. 1478.
5. Mac Con Iomaire and Óg Gallagher citing F.S.L. Lyons, p. 155.
6. Jacques Derrida and Anne Dufourmantelle, *Of Hospitality* (Stanford University Press, 2000), p. 77.
7. Michael Cronin, *Eco-Translation: Translation and Ecology in the Age of the Anthropocene* (Routledge, 2017), p. 62
8. Erica M. Goss and others, 'The Irish Potato Famine Pathogen Phytophthora Infestans Originated in Central Mexico Rather than the Andes', *Proceedings of the National Academy of Sciences*, 111.24 (2014), pp. 8791–96.
9. Colas and others, p. 45.
10. Margaret Visser, *Much Depends on Dinner: The Extraordinary History and Mythology, Allure and Obsessions, Perils and Taboos, of an Ordinary Meal* (Penguin, 1986), p. 31.
11. Colas and others, p. 49.
12. Kim Meijer-van Wijk, 'Levinas – Hospitality and the Feminine Other', in *The Routledge Handbook of Hospitality Studies*, ed. by Conrad Lashley (Routledge, 2017), p. 47.
13. Derrida and Dufourmantelle citing Pierre Klossowski, p. 131.
14. Jay Griffiths, 'Dwelling on Earth', *Emergence Magazine*, 2019 <https://emergencemagazine.org/essay/dwelling-on-earth/> [accessed 28 February 2025].
15. Michael Cronin, 2017, p. 68
16. Jill Bolte-Taylor, 'My Stroke of Insight', TED Conference, February 2008, 18:25 <https://www.ted.com/talks/jill_bolte_taylor_my_stroke_of_insight?language=en> [accessed 28 February 2025].
17. Derrida and Dufourmantelle citing Pierre Klossowski, p. 131.
18. Luce Irigaray, 'Towards a Mutual Hospitality', in *The Conditions of Hospitality-Ethics, Politics, and Aesthetics on the Threshold of the Possible*, ed. by T. Clavietz (Fordham Press, 2013), pp. 42–56.

19 Anne Dufourmantelle, 'Hospitality – Under Compassion and Violence', in *The Conditions of Hospitality*, pp. 13–23 (p. 13).
20 Robin Wall Kimmerer, *Braiding Sweetgrass* (Penguin, 2013), pp. 3–10.
21 Michael Cronin, 'Translation, Ecology and Deep Time', YouTube <https://youtu.be/XBbaDBJP7aA> [accessed 5April 2024].
22 Ursula K. Le Guin, 'She Unnames Them', *The New Yorker*, 21 January 1985, p. 27.
23 Daniel Innerarity, *The Ethics of Hospitality* (Routledge, 2017), p. 3.

26
Common Gardens to Forbidden Forests
Food Procurement Pollution and Policy on Tokunoshima

Hanika Nakagawa

Introduction
Here, I explore the interactions of the gods of Tokunoshima, a colonized subtropical island highlighting how human presence is part of each garden's agenda.[1] From the interviews and experiences with my Elders while researching my Magistral thesis, I reflect on the many forms of garden and how the rivers, the snakes, the rocks, the gardens, and even the pests that come to consume the gardens all are considered gods on Tokunoshima.[2] Elder Susumu Machida and Elder Takefumi Tsukawa tell me they are gods even if human and non-human beings consider them dangerous pests. I examine the interactions of the various gods with reference to the forms and kinds of gardens on Tokunoshima – those that have been locked away for safe keeping, those that have been abandoned, and those that are capitalist ruins – that I have learned about through life stories told by Indigenous Elders. I will also explore the other names for gardens, the 'secret spots' of seaweed harvesters, fishers, and gatherers of other precious foods.

Tokunoshima, a part of the Amami Archipelago in Japan that has been named part of the Okinawa 'Blue Zone' for longevity and healthy lives, has many fruits and treasures.[3] One of Tokunoshima's treasures is linguistic. The word for 'papaya' has some of the most variations among islander dialects of *shimaguchi*; one can cross the street and papaya is called by a different name. In May 2024, I visited my Elder Mika Kawa to say 'hello' because she is suffering and cannot visit me as she always used to do. I interviewed Mika Kawa for my master's thesis project: she spoiled me when I was a child, and she still spoils me and my family. When we went over for tea, she then gave me and my Obaachan (grandma) a papaya to share, ripened by the winter sun, a papaya that was thought to be out of season. Mika Kawa said it would be good, yet my Obaachan and I were nonetheless both surprised at the sweetness it held, almost as though it had been raised by summer sun rays. This might be what brings me to write about papaya. My Obaachan is from Itokina, a mountainous community in the southern town of the island called Isen-cho. My Obaachan, apart from using the English word 'papaya', calls papaya '*moka*', while in the town of Tokuwase where my Ojiichan (grandfather) was stationed until his retirement from the *Nanseitouguyou* (sugarcane factory), papaya is called *manjumai*. These towns are less than twenty kilometres apart.

Once my Ojiichan retired, he grew mangoes, a hobby that turned into his work as the mangoes he grew were of a high quality. They sold for 5000 yen or more ($50 CAD) each back in 2007, and he shipped them all over Japan. I have wonderful memories of working on the tatami mats in my grandparent's house, hiding from the hot August sun, folding along the lines to make cardboard boxes to protect the mangoes. The design of those boxes is still the same; when I see mangoes for sale in stores or packaged up ready to be shipped to the mainland – mostly green with red and orange accents – I can now read the label that says Tokunoshima Mango with the individual grower's name in the corner. Tokunoshima has a mango that is recognized as a brand; therefore, who grew the mango also matters.

The Amami Archipelago, encompassing the Amami and Okinawa prefectures, exports sugar for money. In addition to sugar, Tokunoshima more recently has also exported *kabocha* (a kind of pumpkin), *negi* (green onions), and potatoes. Consequently, Tokunoshima has many gardens for production as well as for personal use. Elder Mieko Maruno gave me a tour of her garden in 2023; she and other islanders call this form of garden an *ataribate* (kitchen garden), as it is suitable for picking vegetables for cooked meals. Elder Mieko Maruno explained how she grew cabbages, staggering when she planted them, so they would be ready to harvest at different times of the year. Elder Mieko Maruno also gave me a tour of her kitchen – including showing me the insides of her freezer, with frozen *yomogi* (mugwort) for mochi making and *aosa* (seaweed) she had travelled to Agon beach to get just a few days prior. Together we cooked the gathered vegetables from the *ataribate* and her foraging. We chatted while cooking, cleaning, and eating, all while an island form of interview happened along the way. The interview is why we gathered, but food quickly took the dominant position.

Food binds us on Tokunoshima, the centre of every conversation. My grandmother watches television and says, 'we used to eat it that way.' My uncle/cousin/big brother Michio Hisaeda drops by and we talk about tapioca – listing all the ways they used to prepare it in the past, trying to retrieve the *shimaguchi* word for it, while also competing to see who could remember the most uses for it. Clearly, they think that tapioca is a potato; I don't think they're wrong. At lunch, overlooking the sea in a northern town close to Tete, we discuss the roasted chicken we are eating, and that my other cousin/uncle/big brother Takuro Nakagawa frequently buys to take to his pregnant niece and her little family so that she will be healthy and strong for her new baby. We discuss family land, and who will meet the governmental regulations for its upkeep, so that it can be passed within the Nakagawa family; everyone wants land. The land is beside our farm where we will grow enough produce to feed three or four families per year because there are many harvests; I will live there, tending to my Elders and my ancestral lands. I am charged with growing children, with or without a husband, since we are a strong matriarchy, and my grandmother approves of her headstrong, scholarly granddaughter being a single mother. I am unruly, like the weeds that overtake every untended garden, uncontrollable, unwilling to stay home and obediently tend to a partner, though I am acknowledged to be infinitely patient with children. Known as the agricultural heart of the Amami Archipelago, and acknowledging the understanding that the island has a history of abundance (just not monetary abundance according to Elder Yoko Nao), Tokunoshima has three towns, all of them ranked consistently in the top three in Japan for birth rate per capita.[4] There is pride here. That pride stems from growing fruits, tubers, leaves, flowers, animals, and humans, everything, all the time, all at once.

These islands are, however, also peppered with evidence of military occupation, both domestic and American.[5] Facing cultural, economic, military, environmental, and industrial forms of violence, here considered 'slow violence', Islanders sought some form of protection from further environmental degradation and encroachment, finally achieving it in the form of UNESCO World Heritage status in July 2021.[6] UNESCO policies dictate that 50% of a land mass must be 'untouched' to qualify for World Heritage status under natural heritage category 10, the most restrictive and protective category. Elder Takefumi Tsukasa told me that during the postwar era when people returned to the islands, the community of Tete had 700 people and records show up to 1000 people lived there, the limit of the community, resulting in the village creeping into the mountainside, expanding agricultural and urban lands. Still with all communities hosting a boom in population after the Second World War (Tete later experienced a 'bust' and has since shrunk with only 86 people in 2022), Elder Tsukasa said, Tokunoshima still maintains 50% of its forest to qualify for UNESCO status, even though UNESCO coding was based on immediate postwar records.

Figure 1. Author harvesting tsuwabuki. *Photo by Chiemi Nakagawa.*

During my data collection I was told that the former rice fields have been turned to cash-crop gardens, overlooking homes where the Elder-mother grandmas proudly show me their kitchen gardens. In other places, islanders walk or ride bike to small holdings or plots with a garden. Some islanders, like my grandmother, choose to ignore the UNESCO policies or determine for themselves the UNESCO-imposed line between urban and untouched lands when thinking about human foraging for food: they still collect *tsuwabuki*, ferns, mushrooms, and other edibles from forest gardens. Although *tsuwabuki* and *yomogi* still grow on the side of the roads and through concrete barriers, many people now grow these plants in their gardens to have some control over the pollution they consume. Since vehicular, industrial and agricultural waste cannot be adequately tracked and monitored, the only safe food is deep in the mountains where the *habu* (poisonous vipers) live – the same deep mountain forest that is to remain untouched. These practices embody the classic romantic, utopic, version of the kitchen garden, found only on the human half of the island – the urban half. On the urban half of the island, the idyllic utopia of a kitchen garden hides invisible policies of rural dispossession and 'letting die', specifically as the result of the refusal to clean up contaminants in agricultural run-off and sugarcane factory waste before this pollution enters freshwater or ocean water.[7] Pollution is one method for stealing the lands, waters, and what is grown in or on them.[8] Colonialism should not be understood merely as a series of historical harms, but rather as ongoing and evolving land relations that fail to recognize Land (including water and air) as a verb, and fail to go beyond geographic space to include contextual relationships.[9]

Forest Gardens

The places of foraging protected by the *habu* are forest gardens, natural rather than artificial, self-sustaining, dense, and robust ecosystems that have evolved over time. The forests in the mountains are a garden kept by many keepers – humans interact with root systems, insects, forest creatures, airborne seeds and particles, and winds – and together these make pathways and clearings for the Amami Black Rabbits. The Black Rabbit, now a protected species, is the face of environmentalism on Tokunoshima, even though it was once a candidate for human consumption, hunted in its forest garden. Forest gardens are the spots where acorns can be gathered in the fall, where islanders follow time-worn ancestral paths maintained for human safety from the hidden *habu* gods in the underbrush – these paths are also a place for the Black Rabbits to seek refuge from the *habu*.

Elder Tokiko Azuma remembers picking various wild fruits while gathering firewood for the family in the forest garden. Elders Tokiko Azuma and Mika Kawa recalled the questionable drinking water coming from a spring deep in the mountain's forest garden; they explained that *hatsumizu* – the first water of the year – was fetched from the spring at *shogatsu*, or New Year. The act of getting it was said to be auspicious. A number of elders remember looking at the bulls before fetching water to make sure to face the direction the bulls heads were facing when filling the vessel. Others do not remember such specifics, as is to be expected when living in different communities or *Shima* (worlds).

Elder Kamejio Morikuma (95) remembers his parents being strict about him going into the mountain as they were fearful of *habu*. The *habu* was everyone's biggest fear; the viper would come out as the sun was setting, and at the time there was little one could do if one was bitten. Therefore, he listened to his parents and stayed out of the mountains. Although during the *shinomi* (acorn) season, he did go with a group to collect the fallen *shinomi*. Elder Chiemi Nakagawa also remembers going to pick *shinomi* with friends but was warned by her mother not to go because the *habu* were out. She went anyway, and because it was unseasonably warm the group saw one under the tree. They ran home to their parents who investigated. The viper was found to indeed be a *habu*, but of the most docile type. This is how Elder Chiemi Nakagawa came to look at the weather rather than just the season to ensure the *habu* were still hibernating; she also learned to listen to the Elders. The children did not set foot off the known paths, but the mountains were still kept by the *habu* in that season; it was not yet time for humans enter that part. Forest gardens operate on a timeshare, if you will.

That is, the *habu* preys on Black Rabbits, but the *habu* also protects the garden, preventing people from going into the deep brush or from staying too long. When the *habu* hibernate (<15 degrees), other beings have their turn. The birds warn other beings that it is time to leave, thereby preventing over-picking, ensuring that sufficient food is left for the wild boars. But forest gardens are now havens for forbidden fruits, protected from humans by UNESCO policies. Consequently, forest gardens are becoming dangerous for all beings, undone by the very practices meant to protect them.

Kitchen Gardens

In the past, Elder Kamejio Morikuma explained that everyone had citrus trees in their yard, so when they were ripe people just grabbed them to eat anytime. Elder Chiemi Nakagawa even remembers using the citrus trees as a jungle gym, snacking while playing hide and seek and running from branch to branch playing tag. She wonders why no one seemed to get hurt as it would be unheard of now. The garden was a playground too. Elder Tokiko Azuma remembers play involving swings, where the ropes of the seat would be tied to a tree. Drawing on the ground and playing hopscotch was also popular. Elder Tokio Azuma recalls that *imo* (potato) was a staple food, but that the *imo* never grew very big. She remembers the *imo* being the width of her finger. She compared the *imo* to hairs or string. It was eaten as a snack as well as during meals. But such family gardens, kitchen gardens, fell out of fashion with modernization after the Second World War. Although Japan is famous too for its rice production, sweet tubers (yams) are the staple food of the peasantry.[10] For political reasons, the varieties of *imo* (yam) eaten by my Elders are sold in mainland Japan as 'Satsuma Imo' (*Ipomoea batatas*) which is recognized as a sweet potato, but my elders reject the Satsuma name as Satsuma was the colonizer. My Obaachan calls island yams 'O-imo san' giving it honorifics before and after its proper food name.

Kitchen gardens are staging a utopic comeback on Tokunoshima, after years of quiet disdain. Having been subjected to contaminants from the sugarcane factory, from industrial forms of agriculture, and from nationally sponsored modernization and development projects, kitchen gardens, organic and carefully tended, are being given pride of place just outside the entrances of people's homes. Elder Kaori Maeda (95) spent several decades in *yamatu*, the mainland of Japan, returning to Tokunoshima five years ago. A stay in Japan intended to be ten years to send her children to school became fifty years. Elder Maeda was most surprised when she found that her former rice fields and farmland where *imo* was grown were no more. In their place was a forest, bush with great trees. Elder Maeda raised her hands while saying this, displaying the circumference and the vast height of the trees, saying the land had returned to the mountain. And then she told me about how she sat on the ground and slowly gardened the forest to produce a kitchen garden, reclaiming it bit by bit.

The 'urban half' of the island as deemed by UNESCO is not a firm division. I have an auntie/cousin/big sister who lives well inside the urban centre of the island; she has a kitchen garden. My auntie/cousin/big sister Kotono was slowly bringing the garden back from the wild when she took the property over after her mother's passing. I went over there with my Obaachan to make miso one afternoon; none of us had made it before but we had heard my Obaachan's mother (Kotono's grandmother) did, so we tried (three months later it was delicious). Kotono then gave us a tour of the garden she was excavating and pointed to where she saw a *habu*. It was a taken-for-granted thing I never questioned; the garden had returned to the forest in only two years of untended-ness. The forest's keeper, the *habu*, would make a showing. Two years later I return to the same place and discover that the *habu* have frequent paths they travel along. Kotono knows her home is in the path. In short, no matter what authority asserts the garden and her home as urban land, the *habu* still see it as their forest garden to protect.

An advocate for and practitioner of traditional organic practices, my grandfather continued gardening as the island's expert in mangoes and orchids after retiring from the sugarcane factory (his entanglement with industrial agriculture). According to my grandfather, walking on the land was essential. The key principle of traditional gardens is that plants grow with the sound of people's footsteps. These words are not to be taken literally. Plants do not hear, nor do they respond to the percussive steps of humans. Rather, this saying means that plants know they are cared for. My grandfather's plants knew that he walked beside them three times per day; he could spot one pest or one small blight in time to treat it locally and immediately. These are the practices that retrieve the kitchen garden from the industrial one. These are also the practices that maintain a kitchen garden, preventing it from reverting to mountain.

Ocean Gardens

During the summers of his youth Elder Shioushi Mizomoto (87) went to Kobaru to fish with friends. Kobaru has a waterfall that is at least two stories tall, as well as freshwater springs that get heated by the blazing sun until they are too hot to touch. Elder Mizumoto says that people with chronic ailments came to these hot springs in the summer for treatment, and many came to the beautiful beach as well. He and his friends speared fish using bamboo poles about a metre long with iron plates about 30 cm long to pierce the fish. Whatever fish they caught, they sold to the people at Kobaru to earn pocket money. Elder Mizumoto and his friends would go to Kobaru bringing only the spear and a pot and build a camp on the sandy beach using the logs and leaves from the surrounding banana trees. They would camp in Kobaru for ten days, spending their days spear fishing, cooking by fire, and selling their spoils. Their catch included not just fish but also octopus and *yakogai* (a large conch shell). The small fish would hide in the *moe*, an ocean plant that grew to the same length as a person and covered the ocean floor. Elder Mizumoto said, 'you could jump in and you would be covered in *moe*.'

When reflecting on the changes to the ocean from his youth, he decries the lack of *moe*. There is so little *moe* in the sea now, he says, that you could say there is none. Elder Mizumoto believes this is why the small fish and octopi have reduced in numbers today. Although there are professionals who claim that the fertilizers used in agriculture have no effect on the growth of *moe*, Elder Mizumoto finds this hard to believe. The fertilizers used in his youth were only bull manure and plant and animal waste, while the use of synthetic fertilizers coincided with the decline of *moe* in the sea. The amount of red tide also increased due to the development of the land. Although the professionals say the chemical synthetic fertilizers are not to blame, Elder Mizumoto also finds this hard to believe and thinks it might be the result of a lack of data on their part, as adding some synthetic fertilizer to a small pool of ocean water wiped out all the living things in it.

The ocean also was a place to harvest salt. Elder Mizumoto's mother took water from the ocean and made salt every day. This salt was then traded with neighbouring communities in the mountains for rice. Those days are gone now. No one makes salt anymore. The ocean gardens are drained of life, of kin. UNESCO policies do not cover the health of the ocean, and increased agricultural pressure will result in more fertilizer use, more runoff, and more death.

Secret Spots

Boats used for fishing were made from wood from the mountains, which was carved out using a tool called a *chonou* that Elder Tokunaga Takehiko describes as like a sickle. This tool carved out the inside of the tree to form a circular boat. Later, thicker multi-material boats could be made. Using multiple materials enabled them to make a less rounded boat. This was not the boat taken to the deep sea. For deep sea fishing, sheets of wood would be formed to shape and dried for days. This created a bigger ship, but it too was still a single person boat. Boats were all only singles, but they would go out in fleets to the spot where the fish live. The place where the fish live, the *giyosou*, is called a '*sumi*' like an '*anaba*' (little known spot).

Elder Tokunaga explained that the fishers knew fish were not there on moonlit nights. Therefore, they went during *aianmu*, moonless nights. The fishers would hand paddle their boats out to the designated spot, such as a *gyuoso* (spot) for a certain type of fish that could be found there. It has a long tail, red in colour. Elder Tokunaga pointed to a red fish on a poster, saying 'this fish can only be found in that spot'. The knowledge must have been passed on how to get to the *gyusou* and where it was in the first place. Elder Tokunaga Takehiko says weather is an absolute factor; it needed to be a moonless night with calm still water. Caught in shadows of the mountain, the people of Inotabu call '*de*' meaning the tall mountain in their language. The fishers know the point where '*intan de*' and '*ineo de*' – the two mountains' shadows – overlap. This is the spot where the fish live. There are also spots for gathering roots of trees for weaving baskets, spots for seaweed like '*aosa*', and spots for gathering acorns or *kuga* (kiwi-like fruit). Part of knowing these spots is knowing the season. A spot is time, place, season, and some luck.

All the Land Is One Garden

The number one thing on Elder Isan Tougo's mind is how the mountains need to be cared for. He revealed that we humans have not been caring for the mountains, creating a need for them to be designated protected spaces by UNESCO. Elder Isan Tougo explained that the health of the mountain is connected to the health of the rivers that were once filled with shrimp and crab. Once the runoff from the mountains is healthy once again, balance should return to the rivers. He then said the rivers' health connects to the health of the ocean. If the rivers run clean, then the sea urchins and other sea creatures are given an opportunity to return. Elder Isan Tougo commented that with the return of the wildlife in the rivers and oceans, fishing can become fun and exciting once more. He emphasizes that for this change to happen everyone must take part.

Elder Isan Tougo says that, as it is now, there is a red runoff from the mountain when it rains that flows straight into the ocean. As islanders, he said, we believe that this is related to the fact that long-term plants like rice or even sugarcane have now been replaced by potatoes, ginger, pumpkin – things with a short growing season. Then the land is ploughed again and again within a short period of time; as a result, when the rain falls, there is more runoff. Sugarcane is only harvested once a year, and, after the first runoff, no matter how much rain falls, it gets integrated into the soil and root systems instead of running off into the ocean. Rice also had two harvests but still six month apart, and the land gets left unploughed.

Elder Isan Tougo says that it is clearly not an isolated thing, as other nations also are facing climate crises related to the mountain and oceans. The mountains lose their health, impacting the runoff and the ocean; the mountains themselves become unwell and the life in them withers. Elder Isan Tougo says that animals are unable to live in their natural habitats and come into the human sprawl looking for food and a place to be. Of course, this is a global issue, and global warming is happening faster than expected. He remembers being told in Heisei 15 (2003) that some ice up north only had thirty years left – and now it is gone in less time than predicted.

Sugarmakers who tap maple in Vermont suggest that they feel connected to climate change because it impacts their current way of life, that there is 'consensus among those who spoke about climate change with me was that it is a reality, and it is a threat to the long-term sustainability of sugaring in Vermont.'[11] The sugarmakers are keepers of forests that are home to the maple trees. Tokunoshima also sees the forested mountains as gardens, even though people are not necessarily the only keepers of the garden. In fact, humans are rather one of the fruits that grow from it. When I interviewed the Elders of Tokunoshima there was consensus that the lands needed to be protected from industry and military occupation. Although the Elders were careful in teaching me where food was found or grown on the island, they also told me how the garden changed throughout their lifetimes. The food landscape has changed completely. They identified many sources of change – policy, industry, national politics, and war – and climate change. The important lesson the Elders taught me was not about each individual fruit, stem, or leaf and how to cook it, but rather how to look around at the world and identify what might cause a delicious leaf to become a poison. Without ancestral lands and the autonomy to practice traditional ways, new 'modern' foods such as Spam (a 'culinary legacy' of US military presence in the Asia Pacific region) were dispensed as rations in times of scarcity.[12] What might be the reason why a plant is not growing or is growing? How can I learn to change with the changes?

The island lands, rivers, and oceans are one garden, but that does not make all of it safe. The forest never ends and enters the home; the kitchen also never ends and reaches into the crevasses of the forested mountains on Tokunoshima. I will start my adventure in the most neglected part of my garden, the pantry. I pull out the coffee cups and salt to reveal the miso I keep, one I made in Boma with my Elder mothers. I tend to my miso very well. It is still delicious, nicely aged, better than new. There ends my carefully curated pantry. Digging deeper, there is box of raisins two years expired. I open it to see they are just fine. I cook them just in case and think, does this make them overripe? I recognize that I somehow failed to tend the garden that is my pantry. We tend to forget the aspects of our lives that are not immediately visible. When my Obaachan was young, she and her friends walked through the forest daily risking encounters with *habu*, but water was gathered from the spring in the forest, rice from the fields, acorns from the mountains. The kitchen garden was walked through every day. The plants and animals grew with human footsteps and the children grew too with the steps, slithers, and brushes of other beings.

Figure 2. View of the density of trees and brush, looking from mountain path to the next mountain. Photo by author.

Therefore, this paper concludes with words of caution about protected gardens, whether they are forest gardens, kitchen gardens, ocean gardens, or secret spots. Strangers walking the designated paths of UNESCO sites conflate all gardens with the capitalist industrial one. They rush to protect 'nature' from her human enemy, not recognizing that human footsteps are necessary in caring for the non-industrial garden, and failing to recognize that humans are products of food as nourishment, and thus grown from gardens. Therefore, I conclude asking why Tokunoshima sought UNESCO World Heritage status, given that many beings are not protected by UNESCO, and that pollution and dispossession are forms of colonialism.[13]

About the Author

Hanika Nakagawa is a PhD student in Sociology and Social Anthropology at Dalhousie University in Canada, just beginning her dissertation proposal, addressing Indigenous Food Cultures in the Amami Archipelago. Hanika has interests in Indigenous methodologies, theories of empire, and the ripple effects of agricultural policies on Indigenous people's lived experiences.

Notes

1. Wendy Matsumura, *The Limits of Okinawa: Japanese Capitalism, Living Labor, and Theorizations of Community* (Duke University Press, 2015), p. 29.
2. Hanika Nakagawa, 'Indigenous Food Sovereignty: Amami Memories of a Time before Capitalist Food Systems' (unpublished masters thesis, University of Manitoba, 2023).
3. Dan Buettner and Sam Skemp, 'Blue Zones: Lessons from the World's Longest Lived', *American Journal of Lifestyle Medicine*, 10 (2016), pp. 318–21.
4. See for example 'Japanese Birth Rates Highest in Okinawa and Kyūshū Municipalities', Nippon.com, 8 May 2024 <https://www.nippon.com/en/japan-data/h01975/> [accessed 28 February 2025].
5. Jodi Kim, *Settler Garrison: Debt Imperialism, Militarism, and Transpacific Imaginaries* (Duke University Press, 2022).
6. See *Against Colonization and Rural Dispossession: Local Resistance in South and East Asia, the Pacific and Africa*, ed. by Dip Kapoor (Zed Books, 2017); on 'slow violence', see Rob Nixon, *Slow Violence and the Environmentalism of the Poor* (Harvard University Press 2011); on UNESCO World Heritage Status, see 'Amami-Oshima Island, Tokunoshima Island, Northern part of Okinawa Island, and Iriomote Island', UNESCO World Heritage Convention, 26 July 2021 <https://whc.unesco.org/en/list/1574/> [accessed 28 February 2025].
7. Tania M. Li, 'To Make Live or Let Die? Rural Dispossession and the Protection of Surplus Populations' *Antipode*, 41 (2009) pp. 66–93, DOI:10.1111/j.1467–8330.2009.00717.x.
8. Max Liboiron, *Pollution is Colonialism* (Duke University Press, 2021).
9. Liboiron, p. 43, p. 6 n. 9.
10. Matsumura, p. 82.
11. Michael A. Lange, *Meanings of Maple: An Ethnography of Sugaring* (University of Arkansas Press, 2017), p. 97.
12. Kim, p. 171.
13. Liboiron.

27
Seeding Resistance
Urban Food Forests and Indigenous Experiences

Lotta Ortheil

The plant, you need to put it in the soil. You need to give it water.
It needs sunlight. It needs wind. And it needs something everyone needs.
And that's something the planet is lacking a lot, which is love. Universal love.

—Niara do Sol[1]

Urban food forests are gaining popularity worldwide. People grow them in city gardens, parks, and industrial wastelands. Food forests (or forest gardens) differ from other forms of urban greening as they feature edible and perennial plants. As they mimic a natural forest ecosystem, they are biodiverse and very resilient. At the same time, they attract visitors as they contain a variety of nutritious plants.

In my research, I focus on Rio de Janeiro, a city marked by profound socio-economic inequalities. Within its confines, Indigenous, Black, and peasant communities often experience exclusion, where their ways of life and relations to nature are systematically denied and rendered invisible. In this challenging urban environment, food forestry emerges as a form of resistance, a means of asserting permanence amid conflicts. I argue that resistance here is a multispecies endeavour in which people together with plants seek to manage and deal with daily conflicts. Taking an ethnobotanical perspective, I explore urban food forests as plural spaces with varied ideas about nature, care, and food, involving different species – humans, animals, and plants.

This paper tells multispecies stories of urban food forestry and food preparation. It is structured in three parts. The initial section sets the context by introducing urban food forestry and the diverse edible plants from multiple perspectives. The second part delves into the culturally specific use of the chaya plant in the food forest of Niara do Sol, an Indigenous woman in Rio de Janeiro. Finally, the last section reevaluates the concept of resistance, exploring urban food forestry in Rio de Janeiro through an Indigenous viewpoint.

Framing Urban Food Forestry and Food Plants
In the mid-day sun, Niara heads to her food forest with a clear goal in mind. She goes straight to the Chaya bush and carefully picks its leaves and stems. Some milky white liquid seeps from the ends of the stems. To protect the leaves on her way back to the kitchen, she holds onto the

stems firmly. As she turns to leave, she notices the Chaya plant's leaves rustling in the breeze, almost as if they are acknowledging her presence with a gentle greeting.

When interviewing practitioners of urban food forests in European and Brazilian cities, I noticed that many of them referred to certain figures who inspired them to start planting in the city: Robert Hart, Bill Mollison, and Ernst Götsch. It raises the question: why are the references for thinking about ecological, sustainable alternatives for food systems predominantly white, Western men? Their knowledge, though systematized, often originates from regions with primarily tropical climates and diverse cultures.

Robert Hart, often regarded as one of the 'pioneers' of food forestry in Europe, drew inspiration from the agroforestry systems of the Indian state of Kerala. He gained popularity as he adapted the tropical agroforestry model to a temperate climate. An important feature of his forest garden in Shropshire is a vertical structure in seven 'storeys', with apple and pear trees constituting the 'canopy' and the other plants occupying the lower tiers.[2] In his book he mentions how he discovered agroforestry systems in tropic climates, but he does not give a single reference to a practitioner. While Hart advocated for food forestry as a means to diversify food systems into small, productive, and biodiverse units, his approach could also be interpreted as a form of colonizing traditional knowledge.[3]

During my visits to urban food forests, I observed two distinct tendencies among projects. On one hand, there are initiatives that draw inspiration from the aforementioned international figures. These projects employ similar methods and practices, such as stratification and biomass trees. On the other hand, I encountered initiatives that build upon ancestral knowledge and practices related to localized foods.

To highlight the distinction between these two tendencies more clearly, I will concentrate on plant selection. In Brazil, many urban food forests focus on a group of plants known as PANCs (*Plantas Alimentícias Não Convencionais* – Non-Conventional Food Plants), a term coined by biologist Valdely Ferreira Kinupp in 2014 to refer to edible plants that spontaneously grow in Brazil.[4] Since then, this term has been increasingly used in academic and popular contexts.

However, nutritionist Bruna Pedroso Thomaz de Oliveira raises a critical question about the concept of PANCs: 'Unconventional for whom?'.[5] She argues that the term 'PANCs' reflects a white, academic, and urban perspective which does not emphasize the traditional contributions linked to these plants. Ferreira Kinupp established 'a standard language to "facilitate" plant identification'.[6] In doing so, he pushes back the popular names of localized plants and creates a barrier of access. Similar to the international food forestry figures mentioned earlier, Ferreira Kinupp has become a national reference for a wide range of plants, overshadowing the rich traditional knowledge associated with these plants, particularly among peasant, Afro-Brazilian, and Indigenous communities.

In Rio de Janeiro, many urban food forests cite Ferreira Kinupp's work and include PANCs in their plant selection to raise awareness of these 'unconventional' plants. However, there are also community food forests that cultivate these plants without labelling them as PANCs, emphasizing the importance of preserving and honouring traditional knowledge and practices.

One of these food forests is taken care of by Niara do Sol, Indigenous daughter of Kariri-Xokó and Fulni-ô parents.[7] Her food forest is situated in the centre of Rio de Janeiro, within the Zé Keti condominium of the Minha Casa, Minha Vida programme.[8] Together with other Indigenous residents who accepted the displacement from the Aldeia Maracanã in 2013, she helped establish the Aldeia Vertical, a space of cohabitation and relation within their residential building. Behind the building, Niara tends to her food forest named Dja Guata Porã, which is a Guarani expression meaning 'walking together', 'a collective of walking well'. Guarani is one of more than 270 languages still spoken in Brazil today.

When asked about PANCS, Niara elaborates: 'Look, what happens is that I wasn't raised in the city'. She explains that she was educated by her mother, who in turn was taught by her own mother. This form of education is passed down orally: 'there was no book, nothing written down'. It becomes clear that PANCs are plants that have been known by generations of her family. Niara utilizes plants rooted in her family tradition, each with its distinct uses. One example is the jenipapo plant. Niara explains:

> Green jenipapo is used to make graphics, both on the body and on fabric. And ripe jenipapo is a vitamin, it contains a lot of iron. So if you eat it, it's very good for your health. And then you can make jelly, you can make jam, you can make liqueur, which people really like, jenipapo liqueur, and it's a fortifier.[9]

For Niara, food plants always have medicinal properties; she often emphasizes that one does not have to go to the pharmacy or visit a doctor but can use certain plants in a particular way. In addition to their nutritional and medicinal qualities, she also mentions that her plants serve as colorants and benefit the ecosystem, for instance by fixing nitrogen in the soil.

In traditional food forests, like Niara's, the aim is often to pass along family-specific knowledge. They draw from the tradition of tropical home gardens originating in the Amazon Forest, where populations developed a rich understanding of how to utilize food and medicinal plants. According to Niara, this was a long and challenging process: 'How many died before we discovered that such a plant was good for the lungs? How many died before you were sure that a certain plant was good for a child?'[10] Biodiversity increased alongside the cultural diversity of plant use. Recent studies show that the Amazon Forest developed as it did because of the presence of human beings who, in their nomadic way of life, collected and sowed seeds of various fruit trees.[11] In contrast to the Western approach of preserving forest spaces and demarcating cultivation and living spaces, in the Amerindian Amazon, family histories are closely linked to their movements through the forest.

Narrating the Chaya Plant
In Niara's kitchen, she begins by separating the Chaya stems and placing the leaves in a white bowl. Adding cold water, she gently presses the leaves down until they are completely submerged. They take on a fresh, dark green hue. One by one, she removes them from the water,

tearing them into smaller pieces before transferring them to a blender. She squeezes a lemon on top and presses the start button.

At Niara's place I was introduced to chaya. She offered me its juice the first time we met. It had a refreshing but slightly bitter taste. Later I became aware that chaya grew in almost all food forests I visited in Brazil. It is one example in Ferreira Kinupp's book about PANCs. I chose to focus on it because it became much more than a PANC through dialogues and personal stories shared by practitioners.

I will follow this plant in discussions about which parts are edible, as well as methods of cooking and preparing foods or drinks using chaya. This approach illustrates that knowledge on plants cannot simply be reduced to a scientific article about their properties and uses. It branches out, around corners, rarely in a straight line, to dissolve completely in some places.

Chaya can be easily recognized by its leaves made up of five lobes joined together about the size of a flattened hand. This remarkable plant thrives in poor soils. A small cutting placed in the soil quickly takes root and grows into a shrub, reaching heights of up to four metres in just a few months. I even saw pieces of chaya stems in the compost that started to sprout again without any contact to the soil. Practitioners are surprised by this 'incredible, impressive' plant that 'is everywhere'.[12] As it grows very rapidly, chaya provides an excellent input for soil recovery. Practitioners prune chaya shrubs, cutting branches and stems into small pieces to use as compost on depleted soil and providing a fertile base for cultivating more delicate or demanding plants.

Chaya is also called spinach tree or Andes cabbage. Some say it originated in the Andean highlands, while others say it comes from the deserts of Central America. Chaya can grow almost anywhere, but especially if there is plenty of sun and water, an average temperature of 25°C, and an altitude ranging from sea level to 1000 metres.[13] Its cultivation is rooted in countries like Mexico, Guatemala, and Colombia, among others in the Americas.[14]

Chaya was brought to other parts of the world mainly from the 1970s, when the enthusiasm for a healthy diet was strongly felt on the east coast of the United States.[15] Initially relatively unknown, the plant had already been dispersed in regions like Florida, Cuba, and the Antilles. It was introduced to other parts of the world as a food supplement, medicine, or even as a source of magic.[16] In Brazil, the spinach tree has been relatively well-known in the state of Santa Catarina since at least the 1990s, and it has gained recognition in Rio de Janeiro more recently.[17]

Chaya leaves have been consumed by humans in Mesoamerica since pre-Columbian times. However, little is known about how chaya was used before the arrival of the Spanish. Today, chaya has a wide range of culinary appearances. It can be consumed in the same ways like spinach but is more similar to cabbage.[18] Nutritionist Neide Rigo suggests boiling chaya for ten minutes, after which it can be incorporated into various dishes, including pasta.[19] In my interviews, practitioners mention that they utilize chaya leaves in various ways, such as in soups, broths, cakes, puddings, cupcakes, couscous, and pancakes (Figure 1).

Niara do Sol prepares a juice by blending the leaves together with lemon (Figure 2 and 3). In contrast, forest engineer Roberta Bicalho tells me that 'the juice is not good,

Figure 1. Niara's chaya pancakes (photo by the author).

Figure 2. Niara's chaya juice (photo by the author).

Figure 3. Niara preparing chaya leaves (photo by the author).

it has to be cooked, boiled. You can't eat it raw, it has a latex, it is from the cassava family'.[20] A report supports Bicalho's caution, indicating that uncooked chaya leaves contain cyanogenic glycosides that produce hydrogen cyanide (HCN) upon tissue damage. However, the cyanide levels are reduced by cooking or drying. According to the same article 'blending is sufficient only if the blended leaves are allowed to sit for several hours'.[21]

I remember though that at Niara's we drank the juice right after blending. Niara is familiar with this debate. She says that already her mother made chaya juice and that she was introduced to it when she was three or four years old. The important aspect to her is that 'you put lemon. A few drops of lemon' (Figure 4).[22] Nutritionist Neide Rigo elaborates: 'The Mayas make chaya water (*agua de chaya*), which is that juice with lime. It is the only way they can eat it raw [. . .]. When it is crushed, the toxic of it evaporates

Figure 4. Niara squeezing lemon into the chaya juice (photo by the author).

with lemon'.[23] Journalist Cristina Barros adds that lemon 'reduces the generation of hydrocyanic acid'.[24]

Furthermore, chaya leaves are a lot more nutritious when consumed raw. Chaya is rich in vitamin C, protein, β-carotene, calcium, phosphorus, iron, thiamine, riboflavin, and niacin.[25] Vitamin C content significantly decreases when chaya is cooked or dried.[26]

Chaya remains largely unfamiliar to many, with some younger generations perceiving it as a food of the poor.[27] However, its potential as a supplement to impoverished diets should not be overlooked. Chaya juice is a valuable alternative to the ultra-processed soft drinks that are very present in the Brazilian diet. Ultra-processed food is particularly consumed by people in lower socio-economic strata.[28] One of the reasons is that subsidies and tax incentives make industrialized products cheaper and more accessible, while providing little support for fruit and vegetables, especially from small producers.[29] Ultra-processed foods are known contributors to various diseases, including cancer, heart disease, respiratory issues, kidney problems, and hypertension, all listed among the World Health Organization's ten leading causes of death worldwide. Despite this, the ultra-processed food industry is rarely held accountable for the health problems of its consumers.

Niara already noticed the health issues related to soft drinks during her school years:

> When I went home to (my schoolmates') house, they used to drink soft drinks (. . .). When they came to our house, they only drank fruit juice. So it was quite interesting to see that many of them hadn't had fruit juice, even at that time. Almost 70 years ago.[30]

In her food forest, she aims to educate the children of the condominium, many of whom come from families in need, about the connection between food and health. She does this by offering juice from different plants in her food forest, including chaya. Initially, it was challenging to convince children to try homemade green juices: 'Some children said, "*tia* [auntie], I do not drink green juice, it's a bug thing."'[31] According to Niara, after trying it, they started to enjoy it and now always ask for her juices. Some even prefer the more unusual juices, like chaya juice, as they cannot have it at home.

One of Niara's main goals it to 'improve nutrition'. In the afternoon, around 5 pm, she prepares a lunch with fresh and healthy ingredients for the children of the condominium after spending time in her garden. In this way, the children also start questioning their parents' way of feeding them. They would say to their mothers: 'No, no, that's poison, you're poisoning me, you don't like me, you give me bread in the morning, bread is yeast, no mom we eat sweet potatoes, cassava, corn cake, at my aunt's (Niara's) house we only eat good things.'[32]

Rethinking Resistance through Traditional, Urban Agroforestry

Niara emphasizes that a life without plants is impossible for her: 'I started planting because everywhere I go, I have to plant.' In the once inhospitable environment of the Zé Keti condominium, this need became even more significant for her: 'And that wall over there was very ugly, so I opened my window, I wanted to cry, seeing how ugly it was'.[33] The large stone wall (Figure 5) around the condominium is a remnant of the Frei Caneca Prison, whose buildings were demolished in 2010.[34] Niara's flat is located next to it. There are certain plants, fruit trees, that are particularly important to her:

> Where I'm going to live, if there's space, the first plant you have to plant is annatto. The second is usually cotton. [. . .] And then there's a plant that is practically obligatory for us to have, when we live together, which is jenipapo.[35]

When Niara started to plant in the place, some residents were against it:

> The very people who live here, once came down and said, 'I'm going to pull it out. This is a leisure area for us'. I said, no. Because when you burn the meat here, the smell will reach my house and I can't stand the smell of meat. If you cooked fish, which is Indian food, that would be fine, but you want to barbecue like white people do, so I'm not going to leave any space for you to do anything.[36]

A discussion about the use of this common space quickly led to a clash of Indigenous and non-Indigenous ways of living and perceiving. Initially, non-Indigenous residents of the condominium disrespected Indigenous practices. They were against planting trees and creating spaces with woody vegetation:

> They think that trees, for example, are dirt because of the leaves that fall on the ground and then they say that they give bugs and mosquitoes and no, on the contrary, here there were a lot of bugs and there were a lot of rats when we arrived and nowadays there are almost no mosquitoes.[37]

The desire to 'clean the vegetation as if nature was something dirty' was introduced in the times of the Portuguese colonies. Until the 1930s, urban landscapes followed English or French designs and 'reflected a glorious victory of the civilization over the savage nature'.[38] This idea of a 'civilized' urban space goes against Niara's vision of a

Figure 5. Stone wall around the condominium (photo by the author).

plural space where plants, insects, birds and people with different world views coexist. Here, coexistence means they share the same space but don't necessarily agree. Niara's food forest is not an idyllic place but one marked by productive encounters and conflicts. As social anthropologist Beviláqua observes:

> The efforts also show all the creativity and hardships of an experience that is plural in many ways: Indigenous people of different ethnicities living together in a building inside a non-Indigenous condominium, and developing a garden whose fruits are destined not only for humans.[39]

Niara cares for the needs of plants just as she does for humans. When she sows, she waits for favourable lunar conditions and lets the seedling determine the place of sowing. She takes the time to listen to the '"schedules" happening with the arrangement of life cycles (involving species, climate, localized interactions, etc.) that constitute temporal niches in a particular ecology'.[40] For Niara, caring for humans and plants is driven by a universal love 'that the planet is lacking a lot'.[41]

Niara responded to the vandalism at her food forest by involving the children of the condominium: 'I started taking the children to work. Then the children became guardians. They don't let anyone touch the (plants). They look after them'. The food forest became a space for children to play and to engage in gardening, by sowing, planting and harvesting. Niara explains her approach to working with the children: 'It's like this: "Now the moon is good, what do you want to plant?", then they prepared the soil, put it in, then they said, "*tia* [auntie] do you still have peanuts?".'[42] Niara organizes seeds and small seedlings to plant with them, taking into account the moon phase, the preferences of the children, and the needs of the plants. She also reached out to schools and incorporated various celebratory days into the garden's activities, such as Mother's Day.

Over time, people began to appreciate Niara's presence and the food forest. They were even inspired to start planting themselves: 'Then other people started planting too. Asking for seedlings, plants. If you'd come here four years ago, you'd say, "I'm not in the same condominium."'[43] Niara also gained recognition for her generosity, as she would give away cuttings or seedlings of plants to visitors if they kindly ask.

However, Niara remains convinced that there will always be stigmatization of Indigenous people: 'One thing that's very sad is that it's as if Indigenous people are all idiots. Idiots, dumb, lazy, that's the reputation that exists.'[44] In a recent article, public health researchers Firpo Porto and others highlight 'radical exclusions' as a particular form of exclusion 'which ignore or despise other forms of being, living and knowing that are typical of peoples who lived in the former European colonies'.[45] Historically excluded groups, such as Indigenous people, people of African descent, and peasants, were expelled from their traditional territories throughout the nineteenth and twentieth centuries, and many of them came to the cities where they often live in degraded areas. Here, their livelihoods are once again being threatened by gentrification and real estate speculation. Niara and the other members of the Aldeia Vertical ended up in the Zé Keti condominium as their former home, the Aldeia Maracanã, had to be evacuated amidst much resistance. It was planned to be turned into a parking lot for the 2014 World Cup, adjacent to Brazil's largest soccer stadium.[46] Niara reflects on how, throughout her life in the city, she has often been treated as if she 'shouldn't be in the city, that [she] should go to the woods, that this wasn't [her] place'.[47] This reflects the common assumption that Indigenous people belong in the forest and are out of place in the city.

Niara's food forest must be seen as intricately connected to the social struggles of marginalized groups in the urban context. By planting fruit trees like annatto and introducing foods and medical treatments from her Indigenous traditions Niara engages in acts of resistance and cultivates alternatives relationships with the city. These efforts are strategic endeavours to occupy, inhabit and plant in a space whose rules are set by others. In contrast to the often exclusive and ecologically degraded urban spaces, Niara and her food forest foster relationships of plurality and liveliness.

Niara pauses the mixer and carefully pours the thin juice into a carafe. She then fills our glasses, smiling as she shares a saying from her mother: 'My mother says that when you want someone to come back to your house, they can't come to your house and leave without eating and drinking something. It could be a coffee and a slice of bread, it could be a pancake, it could be a juice. They have to eat and drink something, which means I want them to come back. And if you don't want them to come back, you don't [even] offer them a glass of water.'[48]

About the Author
Lotta Ortheil is a doctoral candidate at the Rachel Carson Center for Environment and Society and Ludwig Maximilian University of Munich. In her doctoral project she looks at agroforestry projects in Brazilian and European cities and explores agroforestry as a

practice that transforms our current, highly unjust, and unsustainable food system while addressing other social injustices in cities.

Notes

1. Niara do Sol, interview with the author, 2023.
2. Robert Hart, *Forest Gardening: Rediscovering Nature and Community in a Post-Industrial Age* (Green Books, 1996).
3. In this text, traditional knowledge refers to the wisdom of Indigenous, Black, and peasant communities. I aim to highlight these knowledge systems in contrast to dominant Western forms. According to Patrick Ngulube and Onyancha Omwoyo Bosire, traditional knowledge is generated within communities through lived experiences and is often transmitted orally. It is learned through observation, hands-on practice, and repetition. Transferring this knowledge to other contexts risks dislocating it ('What's in a Name? Using Informetric Techniques to Conceptualize the Knowledge of Traditional and Indigenous Communities', *Indilinga African Journal of Indigenous Knowledge Systems*, 10.2 (2011), pp. 129–52, DOI:10.10520/EJC61401).
4. Valdely Ferreira Kinupp and Harri Lorenzi, *Plantas Alimentícias Não Convencionais (PANC) No Brasil: Guia de Identificação, Aspectos Nutricionais e Receitas* (Instituto Plantarum de estudos da Flora, 2014).
5. Bruna Pedroso Thomaz de Oliveira, Plantas Alimentícias Não Convencionais (PANC) e Globalização: Reflexões Contra Colonialistas Sobre a Erosão da Agrobiodiversidade Brasileira, *XII COPENE (Congresso Brasileiro de Pesquisadores/as Negros/as): Democracia, Poder e Antirracismo: Avanços, Retrocessos Legais e Ações Institucionais* (2022), p. 12.
6. Pedroso Thomaz de Oliveira, 2022, p. 11.
7. The Kariri-Xocó Indigenous people, with around 2,500 people, live on an Indigenous land of around 699 hectares in the municipality of Porto Real do Colégio, in the state of Alagoas, on the banks of the São Francisco River (National Museum of Rio de Janeiro, and Federal University of Rio de Janeiro (UFRJ), *Os Primeiros Brasileiros*, 2018 <https://osprimeirosbrasileiros.mn.ufrj.br/en/presentation/> [accessed 18 April 2024]). The Fulni-ô form an Indigenous group that lives near the Ipanema River, in the municipality of Águas Belas, in the state of Pernambuco. There is no information about the year in which they were first settled. What is certain is that in the middle of the 18th century they were already known by the name of 'Carnijós' (Júlia Vasconcelos, and Tahyrine Iyalê, *'Conheça Os Fulni-ô, Povo Indígena Que Habita o Município de Águas Belas'*, Brasil de Fato, 15 February 2022 <https://www.brasildefatope.com.br/2022/02/15/conheca-os-fulni-o-povo-indigena-que-habita-o-municipio-de-aguas-belas> [accessed 20 September 2024]).
8. Minha Casa, Minha Vida is a government programme that subsidizes low-income people to purchase affordable housing built by private developers. Niara criticizes that the building is poorly constructed and that water enters her flat when it rains. She receives no compensation for this and at the same time has many costs per month, including condominium fees, light, etc. (do Sol, 2022).
9. do Sol, 2022.
10. do Sol, 2022
11. William Balee, *Cultural Forests of the Amazon: A Historical Ecology of People and Their Landscapes* (University of Alabama Press, 2013); Charles R. Clement and others, 'The Domestication of Amazonia before European Conquest', *Proceedings of the Royal Society B: Biological Sciences*, 282: 20150813 (2015), pp. 1–9, DOI:10.1098/rspb.2015.0813.
12. Vitória Gondin, interview with the author, 2022.
13. Ana Paula Santos, Rafael Cevidanes Maia, and Patrícia da Veiga Borges, '*A Trajetória Da Chaya Na Serra Da Misericórdia*', Agriculturas, 13 (2016), pp. 21–25 <https://aspta.org.br/article/a-trajetoria-da-chaya-na-serra-da-misericordia/> [accessed 14 June 2024].
14. Aurora del Carmen Orozco Andrade, '*Caracterización Farmacobotánica de Tres Poblaciones Del Género Cnidoscolus (Chaya) Con Fines de Cultivo y Comercialización*' (unpublished bachelor's thesis, Universidad de San Carlos de Guatemala, 2013) <https://catalogosiidca.csuca.org/Record/USAC.585316/Details> [accessed 14 March 2024].

15 Rodrigo Rossi Morelato, *'Eu Amo a Serra Da Misericórdia: Sobre Comunicação, Ambientalismo e Comunidade'* (unpublished master's thesis, Universidade do Estado do Rio de Janeiro, 2019) <https://www.bdtd.uerj.br:8443/handle/1/8888> [accessed 14 March 2024].
16 Jeffrey Ross-Ibarra and Álvaro Molina-Cruz, 'The Ethnobotany of Chaya (Cnidoscolus Aconitifolius SSP. Aconitifolius Breckon): A Nutritious Maya Vegetable', *Economic Botany*, 56.4 (2002), pp. 350–65, DOI:10.1663/0013-0001(2002)056[0350:TEOCCA]2.0.CO;2.
17 Ferreira Kinupp and Lorenzi.
18 Ferreira Kinupp and Lorenzi.
19 Neide Rigo, interview with the author, 2022.
20 Roberta Bicalho, interview with the author, 2022.
21 Ross-Ibarra and Molina-Cruz.
22 do Sol, 2022.
23 Neide Rigo. Chaya water is a popular drink in the Yucatan peninsula made by blending raw chaya leaves in sugar water with lemons, pineapple, and other fruits. For more information, see Ferreira Kinupp and Lorenzi.
24 Christina Barros, '*Manos a La Obra: Recetas*', *La Jornada Del Campo*, 122 (2017), p. 24 <https://www.jornada.com.mx/2017/11/18/Images/delcampo122.pdf> [accessed 15 March 2024].
25 G.S. Ranhotra and others, 'Nutritional Profile of Some Edible Plants from Mexico', *Journal of Food Composition and Analysis*, 11.4 (1998), pp. 298–304, DOI:10.1006/jfca.1998.0590.
26 Barros.
27 Ross-Ibarra and Molina-Cruz.
28 Kayná de Oliveira, '*Alimentação Do Brasileiro Ainda é Saudável*', *Jornal Da USP*, 10.13 (2020) <https://jornal.usp.br/atualidades/alimentacao-brasileira-ainda-e-saudavel/> [accessed 15 March 2024].
29 Felipe Souza, '*Salgadinho é Mais Barato Que Fruta': Subsidiados No Brasil, Ultraprocessados Causam 57 Mil Mortes No País, Diz Estudo*', *BBC News Brazil*, 3 March 2023.
30 do Sol, 2022.
31 do Sol, 2022.
32 do Sol, 2022.
33 do Sol, 2023.
34 Camila Beviláqua, '*Cada apartamento uma oca*', *Piseagrama*, Belo Horizonte, 15 (2021) <https://piseagrama.org/artigos/cada-apartamento-uma-oca/> [accessed 10 June 2024].
35 do Sol, 2023. Annatto and jenipapo play a central role in Amerindian ethnology, being used for body paintings.
36 do Sol, 2022.
37 do Sol, 2022.
38 Ricardo Cardim, *Paisagismo Sustantável Para o Brasil: Integrando Natureza e Humanidade No Século XXI* (Editora Olhares, 2022), p. 30.
39 Beviláqua, 2021.
40 María Puig de la Bellacasa, *Matters of Care: Speculative Ethics in More than Human Worlds* (University of Minnesota Press, 2017), p. 201.
41 do Sol, 2022.
42 do Sol, 2023.
43 do Sol, 2023.
44 do Sol, 2023.
45 Marcelo Firpo Porto and others, 'Emancipatory Urban Greening in the Global South: Interdisciplinary and Intercultural Dialogues and the Role of Traditional and Peasant Peoples and Communities in Brazil', *Frontiers in Sustainable Cities*, 3 (2021), pp. 1–14, DOI:10.3389/frsc.2021.686458.
46 Beviláqua.
47 do Sol, 2023.
48 do Sol, 2023.

28
The Seven Sisters and South Downs
A Culinary Exploration of Southern England's Wild Edibles

Cordula C. Peters

The East Sussex coast in the south of England, specifically the Seven Sisters chalk cliffs and the distinctive South Downs landscape that stretches beyond them, provide unique geological conditions to make this area especially interesting for nature enthusiasts as well as foragers. The chalk-rich ground, the salt marshes, and generally milder temperatures create a unique climate ideal for many edible plants, their flowers and fruits.[1]

The South Downs stretch along the eastern part of the south coast of the United Kingdom and are England's newest national park, finally receiving this designation in spring of 2009, exactly sixty years after the UK government approved the National Parks and Access to the Countryside Act in 1949.[2] While discussions had been held to declare the South Downs a national park even prior to 1949, it took another six decades to become reality and as such protect the landscape and its biodiversity, as well as re-nature lost natural habitats.[3]

The South Downs National Park stretches across 1623 km² (627 square miles) from St Catherine's Hill near Winchester in Hampshire in the west all the way past the Seven Sisters chalk cliffs to Beachy Head, near the city of Eastbourne in East Sussex in the east.[4] The park features the 160 km (100 miles) long South Downs Way, the only national hiking trail in the country that lies completely within a national park. The park also boasts a total of approximately 3300 km (just over 2000 miles) of public rights of way, routes that, even if they pass through privately owned property, are legally accessible to the public at all times.[5]

It should be mentioned here that the term 'National Park' has different definitions in different countries and therefore different laws apply around the usage of national park land.[6] This is particularly crucial to a forager. In the UK, the primary purpose of national parks was not that of conservation as was the underlying intention with the creation of the world's first national park in Yellowstone in the US. Instead, the National Parks and Access to the Countryside Act clearly states the main purpose to be for the general public to be able to enjoy the beauty of the existing countryside, be it natural or artificially altered.[7]

According to the South Downs National Park Authority, only about 25% of the South Downs National Park are currently managed for nature.[8] This also means that large areas of British national parks may be privately owned, may be lived on, and even may

Gardens, Flowers, and Fruit

be farmed. In the case of the South Downs National Park entire villages and towns and even the city of Lewes lies within the borders of the national park.[9] For a forager this is important when it comes to the legality of foraging on the land, as this means that it is indeed legal to forage on large parts of the land, in some places with and in many places even without the permission of the landowner.

Like in many countries foraging is regulated by law in the UK. There are strict laws as to what can or can't be foraged and where foraging is or isn't allowed. While there are some plants that are categorically forbidden to pick, in the UK it is allowed to forage most plants for leaves, flowers, fruits, and seeds, but it is forbidden by law to uproot wild plants.[10] It is important to notice that laws vary between countries and sometimes even counties. For example, in the UK different laws exist between England and Scotland that regulate one's right to roam and access private property, with England being much more restrictive than Scotland.[11] Therefore it is crucial to be familiar with any local laws.

The South Downs are a diverse mix of different landscapes featuring ancient forests and woodlands, river valleys, wide-open heaths, and cultivated farmed land. The heaths' acidic soil contrasts with the nutrient-rich clay found in other parts.[12] While the soil in some areas features a high concentration of chalk, other areas are distinctly sandy. Each area is home to an astonishing variety of native flora and fauna. You can find up to forty different species of wildlife within one square meter in some areas.[13] But it is the open downlands and floodplains in the east of the South Downs that can be particularly attractive to a forager.

Every year spring turns the vast meadows of the eastern South Downs into a colourful kaleidoscope of delicious flowers. Purple sweet violets (*viola odorata*), light pink mallow (*malva sylvestris*), magenta bush vetch (*vicia sepium*), and yellow primrose (*primula vulgaris*) are just some of the blossoms scattered in the grass waiting to be picked and end up in my pantry. They add great flavour and colour to any salad.

Along the chalky beaches and salt marshes of the Cuckmere River valley, the intense greens and deliciously salty aroma of young sea purslane (*sesuvium portulacastrum*) and rock samphire (*crithmum maritimum*) leaves will greet you, interspersed by lilac and yellow sea aster flowers (*tripolium pannonicum*), pink sea thrift (*armeria maritima*), and the white clouds that is the bloom of sea kale (*crambe maritime*) ready to accompany your favourite fish or seafood dish.

Throughout summer the chalky downs turn into a cornucopia of readily available nutritious food for any forager. The variety of edible flora growing in this area is breathtaking. It ranges from wild fruit trees such as crab apple (*malus sylvestris*), blackthorn (*prunus spinosa*), and wild cherry (*prunus avium*) to use for teas, jellies, and chutneys, to leafy greens ready to be picked for salads or stews such as stinging nettle (*urtica dioica*), red dead nettle (*lamium purpureum*), garlic mustard (*alliaria petiolata*), and pineapple weed (*matricaria discoidea*).

For more experienced foragers, the selection of mushrooms is a dream. Frequently popping up on restaurant menus in late April, St George's mushrooms (*calocybe gambosa*) are a common feature dotting the short grasslands covering the tops of the Seven Sister cliffs. Eventually, meaty nutty field blewits (*lepista personata*) make their appearance with their easily recognizable blue stems. And with a bit of patience and some luck bright purple amethyst deceivers (*laccaria amethystina*) will bring some extra colour to your foraging basket. Especially lucky foragers might even stumble upon a particular treasure, beautifully earthy and comparatively mild English summer truffle (*tuber aestivum*).

Autumn is the time for bramble (*rubus fruticosus*), rose hip (*rosa canina*), and hawthorn (*crataegus monogyna*) bushes to offer their fruit. Brimming with colourful berries, they dominate the landscape and provide nutrition for humans and animals alike. It's the season when a forager's kitchen transforms into a canning shed, preserving, pickling, and fermenting with abundance what nature has to offer.

Even in winter, when the trees have lost their leaves and the green grass is replaced by mud and brown leaves, the South Downs can surprise you with hidden treasures. The delicate white blossoms that graced the blackthorn trees (*prunus spinosa*) in the spring are now replaced with dark blue plump sloe berries ready to be picked after the first frost. They are now at their best to be used to infuse spirits and vinegars. Even more exciting, though, is what hides below the ground at this time of year. The chalky ground and the right trees provide the ideal environment for a very special delicacy. With the right knowledge and a bit of luck by the end of a long hike you might just find yourself in possession of some English winter truffle or Burgundy truffle (*tuber uncinatum*) with its more potent and deep flavour.

The South Downs and especially the area around the Seven Sisters, like any natural landscape, exists due to a delicate natural balance that promotes biodiversity.[14] As a forager it is important to understand the landscape you work in and to be careful as to not disturb said balance and cause irreparable damage. Not only is it important to know and understand local laws around foraging, but there are some basic rules many experienced foragers abide by to ensure the longevity of the plants they forage from, that the local wildlife have enough sustenance left to thrive, and that the plants that have been foraged are safe to consume.[15]

While there are some variations in the rules taught and followed by individual foragers or foraging groups and taught in various foraging courses, there is a general consensus on some basic foraging etiquette and safety measures:

1. Only forage plants you are 100% certain are indeed edible. If you are unsure, leave them. And while there are plenty of books and online tools such as apps available to assist a forager, learning from an experienced and knowledgeable forager is invaluable.
2. Only forage as much as you truly need and will consume, but always leave enough for what is needed by local wildlife and for the plant itself to recuperate. Depending on whom you ask the answer does vary, but most foragers and foraging groups state that you shouldn't forage more than 10%–30% of what is available.
3. Know the local laws specific to the area you are foraging in to make sure you are allowed on the land in the first place, you are allowed to forage there, and you don't forage plants or parts of plants that are illegal to forage.

There are other rules or recommendations that some foragers adhere to, such as foraging away from other people to avoid over-foraging of one area as well as avoiding potential conflicts when disagreeing on specific foraging protocols.[16] Some foragers have a more spiritual philosophy about foraging and encourage rituals to demonstrate thankfulness towards nature and giving back.[17]

What most foragers seem to agree on is that foraging, when done correctly and mindfully, is a sustainable approach to gathering food that may also promote a healthier and more nutritious diet. The act of foraging itself encourages people to spend time outside in fresh air while getting physical exercise through plenty of walking, and becoming more familiar with and knowledgeable about their local surroundings.[18]

When I moved from London to the south of the UK over two years ago to take on a chef's position, it turned out to be the beginning of an unexpected journey into foraging, recipe rediscovery, and recipe development. The sheer variety and abundance of wild edibles took my breath away and has changed how I work as a chef and how I consider seasonality and sustainability in the kitchen. I look at the South Downs as a wild and naturally grown kitchen garden with endless culinary possibilities. It has changed how I approach menu writing and ingredient sourcing with foraged produce frequently featuring on my menus.

Cooking with foraged ingredients opened the door to a lot of creativity and experimentation. However, it also allowed me to explore the past. It led me to rediscover recipes previously lost to history as natively grown and foraged ingredients disappeared from British kitchen pantries. Over time they disappeared from shops, being replaced by cultivated, farmed, and even imported foods. This process of rediscovery even steered me towards a renewed appreciation for familiar plants, whose full potential of culinary application I hadn't grasped previously.

The Elder Tree (*sambucus nigra*)

While by no means unique to this area of the UK, the elder tree thrives particularly well in the chalky, more acidic grounds in the eastern downlands of the South Downs. In late spring and early summer, the white flower umbels of the elder trees are a dominant feature all around. Many people are familiar with elderflower, mostly in the form of cordial readily available in many shops. They might be able to identify flowering elder trees and bushes. Some might even know how to make elderflower cordial or its fermented version, elderflower champagne. However, for me as a chef and lover of the sweet elderflower flavour, the flowers of the elder have many more possible culinary applications.

I spend most early mornings in May and June walking around the hills of the South Downs picking elderflowers. In the kitchen I use them to infuse honey to serve with cheese. I make elderflower vinegar to add to dressings or to finish sauces. Elderflower infused whipping cream takes any lemon cake to the next level. Dehydrated elderflowers not only make a delicious tea, but blitzed to a powder and mixed with sugar they can add that little extra finesse to many a dessert or sweet. But by far one of my favourite dishes to make with elderflower are elderflower and lemon possets. They are incredibly easy to make with very few ingredients. Yet they are so utterly scrumptious.

A posset is a traditional British dish going back to the Middle Ages, but over time what it is and what it is made of has changed quite drastically. Originally a drink, possets were used for medicinal purposes, made from curdled milk and alcohol. Eventually possets developed into a much-loved dessert still using curdled milk and alcohol, but also adding sugar and sometimes egg to enrich them, as well as bread, cream, or nuts to thicken them.[19] Nowadays possets are rather simple desserts made by combining cream and sugar and adding acidity, such as lemon juice, to curdle, and therefore thicken, the cream. An aromatic may be added for extra flavour. Possets had fallen out of favour, being seen as old-fashioned, but have recently started to reappear on restaurant menus, in recipe books, and on social media with much more enthusiasm.

This specific recipe takes advantage of the rather delicious flavour combination of elderflower and lemon with the natural sweetness of the elderflower pollen greatly complimenting the acidity of the lemon juice needed to thicken the cream.

Elderflower & Lemon Posset

Servings: 4

Ingredients:
- 300 g double cream infused with elderflowers
- 80 g sugar
- 80 ml lemon juice (1 lemon)
- lemon zest (1 lemon)

Method:
- Infuse double cream with elderflowers overnight.
- Add cream, sugar, and lemon zest in a cooking pan and bring to a boil.
 (If using whipping cream, boil for a few minutes to reduce the liquid by about 30% to increase the fat content by evaporating liquid. It will help to set the posset better.)
- Take off stove, add lemon juice, and let cool for a couple of minutes.
- Sieve and then pour into prepared dessert glasses.

Notes:
- This recipe works best with fresh elderflowers, though elderflower cordial can be used as a substitute if you do not have access to fresh flowers. You will need about 15 ml cordial.
- Instead of elderflowers, this recipe also works wonderfully with fresh lilac, lavender, or other aromatic flowers.

The fruit of the elder tree, the elderberry, is much less commonly known and for many hard to identify. Foraging for elderberries in late summer, I frequently get asked what it is I am picking, even though I am picking from the very trees that were covered in white blossoms only a few month earlier. Historically, however, in the UK, elderberry has long been used for medicinal purposes or to make cordial or wine. What few people are familiar with nowadays, though, is Pontack Sauce, an almost ketchup like elderberry condiment dating back to the seventeenth century.

As a chef, discovering old recipes such Pontack Sauce, made from berries growing wild and plentifully around us, and then reformulating those recipes for a modern professional kitchen opens up incredible opportunities to create unique and distinctly local dishes. When I first stumbled upon a recipe for Pontack Sauce I was looking for early British ketchup (or catsup) recipes, specifically those not using tomatoes, but mushrooms instead. During my research I found a recipe for an elderberry condiment with a dubious history and an even more dubious name.

Pontack Sauce may have possibly been named after a famous tavern and one of England's earliest restaurants called Pontack's Head, opened in London after the Great Fire of 1666 by the Pontac family, owners of the Haut Brion wine estate in France.[20] While a lot of the history of Pontack Sauce is based on speculation, we do know that it started to appear in recipe books as early as the 1840s, when Elisa Acton mentions a recipe for Pontack Sauce or 'pontac catsup' in her *Modern Cookery*.[21] Pontack Sauce recipes can vary greatly. While some, like Acton's, are rather liquid and more vinegar like, others are thick and more of a puree, similar to the consistency of modern tomato ketchup.

Elderberry has a very different taste than elderflowers. While elderflowers taste sweet and light, elderberries have a very earthy, deep flavour, full of umami and acidity. This makes Pontak Sauce an excellent condiment to pair with red meat, game birds, or mushrooms. While it is quite time-consuming to make, the result is worth every minute.

Pontack Sauce

Ingredients:
- 1000 g elderberries
- 1000 ml cider vinegar
- 400 g banana shallots (or red onion)
- 20 g fresh ginger
- 10 allspice berries
- 4 cloves
- 200 g demerara sugar
- salt to taste

Method:
- Thoroughly wash the elderberries and de-stalk them using a fork and combing through the elderberry umbels.
- Add elderberries, vinegar, finely sliced shallots, peeled and sliced ginger, allspice berries, and cloves into an ovenproof dish. Cover the dish with a lid (or aluminium foil), so no liquid can evaporate and cook in the oven at 120°C (250°F) for about six hours. Do check occasionally that the berries aren't drying out.

- After six hours pour the cooked berries through a tight meshed strainer and using a spoon press the berries as much as possible. Discard any skin and stones as well as the spices that get left in the strainer.
- Pour the strained berry liquid into a cooking pan, add the sugar, and bring to a boil. Simmer until the sauce has the desired consistency. I prefer the thickness of ketchup.
- While still hot pour into sterilized bottles, seal, and keep in a cool place to mature. The sauce continues to develop its taste over the next several months and will only get better.

Marsh Samphire (*Salicornia europaea*)

Often overlooked as a delicacy are the many wild sea vegetables the British coast has to offer. There are quite a variety of edible plants growing in the salt marshes around the Seven Sisters with their unique and naturally salty qualities. For example, sea purslane (*sesuvium portulacastrum*), which grows in abundance around the South Downs river system, with its strong salty flavour it introduces not only a magnificent green colour with its leaves to dishes, but it can be used very strategically as a healthier, locally grown alternative to the use of refined or unrefined salt in seasoning food.[22]

Similarly, rock samphire (*crithmum maritimum*) with its distinct carrot-like flavour, though also naturally salty, makes a great side vegetable or addition to a summer salad. As a chef, exploring healthier and more sustainable ways of cooking is especially important as many of our customers have changed their eating habits in recent years and seem to become ever more health-conscious and interested in the provenance and environmental impact of the food they consume.[23]

Some of these sea vegetables, popular once upon a time, have seemingly fallen out of favour with many chefs and home cooks alike and are hard to come by unless you know how to forage for them. Others, such as marsh samphire, are better known, but are often merely used as simple garnish for a fish or seafood dish. While marsh samphire does indeed make a beautiful green garnish, it has, like other sea vegetables, so much more culinary potential.

Marsh samphire is a small succulent that grows in the sand along intertidal areas such the South Down salt marshes. Covered in salt water for hours at a time, the stems of the marsh samphire are full of water and quite crunchy and refreshing to bite into. Raw, blanched, or sautéed they give a beautiful salty texture to fish, seafood, or vegetable dishes. Have you ever tried potato salad with marsh samphire?

With its natural ability to store liquid, the marsh samphire makes an ideal vegetable to pickle. Soaking up the pickling liquor it stays plump and keeps its bite beautifully.

Pickled Marsh Samphire

Ingredients:
- foraged marsh samphire
- equal amounts of sugar, white wine vinegar or distilled malt vinegar, water

- pink peppercorns
- sea salt

Method:
- Make a 1-1-1 pickle liquor by combining the sugar, vinegar, and water. Bring the mixture to the boil and simmer until the sugar is fully dissolved. Then leave the pickle liquor to cool down.
- Once at room temperature add the pink peppercorns and sea salt. Adding pink peppercorns when the liquor is still warm will discolor them.
- In the meantime, tightly pack a sterilized jam jar with marsh samphire. Then slowly pour the pickle liquor over the samphire.
- Once the green samphire stems are fully covered seal the jar and leave to pickle in the fridge for at least a week.

Note:
- This pickling recipe also works very well for other foraged greens, such as rock samphire or wild garlic flower buds.
- This pickle liquor is quite sweet and compliments the salty marsh samphire well. When using other greens, like wild garlic flower buds, consider salting the buds first for a couple of hours if you want a saltier pickle.

Wild Garlic / Ramsons (*Allium ursinum*)

Wild garlic is quite common throughout the UK. The South Downs are no exception. Come early spring a slight smell of garlic wafts through the air as you roam the countryside. One of the earlier plants to start growing, they are quite easy to pick out and harvest. Their unmistakable garlic smell makes them easy to identify, even for foraging novices. When wild garlic season starts around March, many weekend days are filled with foraging the leaves, flower buds, flowers, and eventually seeds. Just make sure to not uproot the plant and leave the wild garlic bulbs be, so the plant can come back the following year.

The fact that every part of the plant is usable in the kitchen and has its own culinary application is intriguing. While the leaves are mildest in flavour and great for blanching or fermenting, the young flower buds and stems are more intensely garlicky and ideal for pickling. The fully open wild garlic flowers are not only beautiful to look at and hence great for decoration, but they are also even sharper and more intense in flavour. The young green seeds are like succulent little garlic pearls exploding in your mouth.[24] I like them best just scattered on a salad.

Wild garlic is a great plant to use in the kitchen. It freezes, ferments, and pickles well and hence can be preserved easily despite its rather short season. Blitzing the leaves with some neutral vegetable oil and then freezing them best preserve the leaves. This mixture can then be used to flavour butters or sauces or to make a delicious pesto.

Wild Garlic Mayonnaise

Ingredients:
- 50 g wild garlic
- 150 g sunflower oil
- 1 egg yolk
- 5 g Dijon mustard
- 5 g cider vinegar
- salt to taste

Method:
- Blitz the wild garlic with the oil until you have a smooth puree. Strain the puree to get a gorgeous green garlicky oil.
- Whisk egg yolk, mustard, and vinegar together then slowly but steadily introduce wild garlic oil while continuously whisking until you have a nice thick mayonnaise.
- Season with salt.

Note:
- Use the wild garlic puree as the base for a pesto or mix it with soft butter to create a delicious compound butter.

There are of course many more plants to know about and many more recipes to try. However, in this paper I focused on plants that are relatively easy to identify and find, even for the most inexperienced forager. I shared recipes that are easy to follow for both practiced and novice cooks. I hope the plants discussed and recipes mentioned will inspire some readers to be more observant of and curious about their natural surroundings, be it wide open land or inner-city parks. I encourage anyone to go out and explore what food nature provides so readily. But know your laws around foraging in your area. If you are interested, find a local forager to learn from.

Yet even if you are not inclined to forage for your food, take time to explore and observe your natural surroundings. Look around you and appreciate what nature has to offer. Go to local farmers' markets or stop by local farms to see what wild edibles they might have foraged on their own lands. Look out for homemade preserves that you can't get in a supermarket. But take the time to go out and look around.

Describing the landscape of the South Downs, Virginia Woolf, who lived at Monk House, in the village of Rodmell, near Lewes, once wrote in her diary: 'As for the beauty, as I always say when I walk the terrace after breakfast, too much for one pair of eyes. Enough to float a whole population in happiness, if only they would look.'[25]

Happy foraging!

About the Author

Cordula C. Peters is a German-born artist, designer, educator, and professional chef who works as a sous chef at the Glyndebourne Opera House in East Sussex, UK, while also pursuing her interests in culinary arts, independent research, and exhibiting art.

Gardens, Flowers, and Fruit

Notes

1. The South Downs National Park Authority, '*The South Downs National Park: Special Qualities*', (2015), 3–5 <https://www.southdowns.gov.uk/wp-content/uploads/2015/03/SDNP-Special-Qualities.pdf> [accessed 12 December 2023].
2. David Adam, 'England's South Downs gain national park status', *The Guardian*, 31 March 2009, para. 2 of 12 <https://www.theguardian.com/environment/2009/mar/31/south-downs-national-park> [accessed 5 October 2023].
3. '*Key Fact*', The South Downs National Park Authority, 2024 <https://www.southdowns.gov.uk/our-history/key-facts/> [accessed October 2, 2024].
4. '"Historic day" for South Downs National Park', *BBC News*, 1 April 2011, para. 6 of 10 <https://www.bbc.co.uk/news/uk-england-12930989> [accessed June 29, 2024]; Richard Madden, 'How to Escape the Crowds in the South Downs National Park This Summer', *The Telegraph*, 29 May 29, 2024, para. 7 of 27 <https://www.telegraph.co.uk/travel/destinations/europe/united-kingdom/england/the-south-downs-national-park-guide-map-hotel-planning/> [accessed 2 October 2024].
5. *South Downs National Park: Conserving Our Unique Landscape*, National Parks UK, 2024 <https://www.nationalparks.uk/park/south-downs/> [accessed 28 June 2024].
6. C. Michael Hall and Warwick Frost, 'The Making of the National Parks Concept', in *Tourism and National Parks: International Perspectives on Development, Histories and Change*, ed. by Warwick Frost and C. Michael Hall (Routledge, 2009), pp. 3–15 (p. 6).
7. The National Parks and Access to the Countryside Act 1949, Part 2, 5.1–5.3.
8. '*Help Nature to Renature*', The South Downs National Park Authority, 2024 <https://www.southdowns.gov.uk/renature/> [accessed 2 October 2024]
9. Madden, para. 7 of 27.
10. The Wildlife and Countryside Act 1981, 13.1–13.4.
11. Fez, '*Is foraging legal in the UK?*', Totally Wild UK, 1 May 2023 <https://totallywilduk.co.uk/2023/01/05/is-foraging-legal-in-the-uk/> [accessed 28 June 2024].
12. The South Downs National Park Authority, '*South Downs Local Plan: Adopted July 2019 (2014–33)*' (2019), pp. 27–28 <https://www.southdowns.gov.uk/wp-content/uploads/2019/07/SD_LocalPlan_2019_17Wb.pdf> [accessed 28 June 2024].
13. 'Our Chalk Grassland', The South Downs National Park Authority, 2024 <https://www.southdowns.gov.uk/wildlife-habitats/habitats/chalk-grassland/> [accessed 2 October 2024].
14. Seaford Town Council, 'Your Guide to . . . Seaford Head, Local Nature Reserve' (n.d.).
15. Fern Freud, *Wild Magic: A Seasonal Guide to Foraging with Healing Recipes & Self Care Rituals* (Ebury Press, 2023), pp. 16–20; Marlow Renton and Eric Biggane, *Foraging: Pocket Guide* (Wild Food UK, 2022), pp. 10–12; Robin Hardford, *Edible and Medical Wild Plants of Britain and Ireland* (Eatweeds Press, 2022), p. vii.
16. Hardford, p. vii.
17. Freud, p. 17.
18. Freud, pp. 10–15; Renton and Biggane, pp. 7–8; Harford, p. vii.
19. Neil Buttery, 'Possets', *British Food: A History*, 28 April 2012 <https://britishfoodhistory.com/2012/04/28/possets/> [accessed 28 January 2024].
20. Stephen Harris, 'Roast Pigeon with Runner Beans and Pontack Sauce', *The Telegraph*, 25 August 2018 <https://www.telegraph.co.uk/recipes/0/stephen-harris-roast-pigeon-runner-beans-pontack-sauce/> [accessed 3 December 2023].
21. Elisa Acton, *Modern Cookery* (Philadelphia: Lea & Blanchard, 1845), p. 117.
22. A. Pires and others, 'Sea Purslane as an Emerging Food Crop: Nutritional and Biological Studies', *Applied Sciences*, 11 (2021), p. 7860, DOI:10.3390/app11177860.
23. The Behavioural Insights Team, '*A Menu for Change: Using Behavioural Science to Promote Sustainable Diets Around the World*', 2020, p. 45 <https://www.bi.team/publications/a-menu-for-change/> [accessed 5 December 2023].
24. Renton and Biggane, p. 120.
25. Virginia Woolf, *The Diary of Virginia Woolf*, ed. by Anne Olivier Bell and Andrew McNeillie, 5 vols (Hogarth Press, 1977–84), p. 269.

29
The Fall of the Acadian Belliveau
Genetics, Genealogy, and Recovery of Pomological *Matrimoine*

Karen Pinchin and Simon Thibault

Two hundred and fifty years ago, a young woman named Marie-Modeste Leblanc watched an apple tree blossom and bear fruit in her new family's orchard. In the early 1600s, her French ancestors had arrived by galleon on the verdant shores of what they called Acadie, part of a colonial project designed to extract resources from lands not yet settled by the world's major powers. With them, those new arrivals brought familiar crops – grains, vegetables, and tree fruit – and planted them on freshly cleared lands. Those lands, formerly stewarded for thousands of years by the region's Indigenous Mi'kmaq people, proved to be enormously fertile. That, among other reasons, was why, in 1755, British troops forcibly deported Marie-Modeste's family, along with thousands of other Acadians, in a massive cultural and economic trauma dubbed *le grand dérangement*. The British named the province Nova Scotia.

For deportees like Marie-Modeste, that history spread across the branches of her apple tree. After pledging fealty to the British crown, her family had eventually been allowed to return, although not to their original lands. In 1768, Marie-Modeste married a man named Frederic Belliveau, and the couple settled in Baie-Sainte-Marie, about sixty kilometres west of her parents' former home in Port Royal. On that homestead they planted an apple tree. That tree produced sweet, crisp, striated fruit unlike any other they had tasted before. Eventually, as other families asked for grafts to start trees of their own, that apple came to be known as the Belliveau.

As a cultivated plant and member of the rose family, the species *Malus domestica* holds a uniquely rarified place in the hearts and minds of human cultures more widely, functioning as symbol as well as foodstuff. In the tapestry of parallel human and pomological history, the story of the Belliveau apple and its entwined story of Acadian genealogy and the North American colonial project consists of a single thread. Yet it – and what it illuminates of the interwoven value (cultural, economic, and genetic) of a single, now nearly extinct apple species – cannot be understated.

On a landscape historically dominated by male-centric European-style capitalism and power, the Belliveau story represents a reclamation of *matrimoine*, both through the genetics of an apple, as well as through the labour of unheralded generations of Acadian women to whom one of this paper's authors owes his very existence.[1] The Belliveau apple represents a historical link to early colonial agricultural practices in Atlantic Canada, as well as the changing nature and importance of apples, as foodstuffs, cultural and agricultural markers, and community-held intellectual property.

Today's apples are no longer viewed as a seasonal, local food that is grown, stored, and used to the best of their individual qualities. Instead, they are considered evergreen, their cultivation commodified and copyrighted, with a uniformity of textures, flavours, and keeping qualities. What stems from those losses is tangible and profound, in an erasure of how cultures define themselves by the food they grow and eat. This loss of collective culinary knowledge seeds the loss of cultural and agricultural knowledge, a self-perpetuating cycle fuelled by capitalistic commodification and mass industrialization – even in the case of a single, growing tree in a sun-drenched field.

Before apple trees dotted the shorelines of Nova Scotia, the land was known as Mi'kma'ki and inhabited by the Mi'kmaq people. The French colonial conquest of those lands began in 1604, when Pierre de Gua de Mons and navigator Samuel Champlain crossed the Atlantic. They spent a difficult winter on an island between current-day New Brunswick and Maine; there, thirty-five men – more than half the crew – died from malnutrition and scurvy. Champlain later noted in a 1605 diary entry that 'the cold was so intense that the cider was divided by an axe and measured out by the pound'.[2]

In spring 1605, the remaining men sailed into the Bay of Fundy, looking for a location where winters would be milder. Upon landing, de Mons and his crew 'were welcomed by several hundred Mi'kmaw residents of the area, who permitted de Monts [sic] to construct a permanent headquarters, a "fort and habitation"'.[3] They founded Port Royal on the banks of the Dauphin River (now the Annapolis River), which is where North American apple cultivation began.

The scale of this cultivation remains amorphous, as access to original source materials on early agricultural work is scarce. Many authors and researchers have relied on oral histories and anecdotes, which makes facts and dates hard to corroborate. However, Jesuit records from 1612 note that, by that time, trees in the area were already bearing fruit.[4] By 1635, other settlements in Acadie, such as nearby LaHave, had well-established fruit trees. By 1698, a French census listed fifty-four families in Port Royal as possessing 1584 apple trees on thirty-two acres.[5] In 1699, French botanist and explorer Diereville visited Port Royal and tasted various apples there, calling it a '*petite Normandie pour les pommes*', while reporting varieties including Calville, Rambours, and Reinettes.[6]

Much of the historical record of the Acadians and their subsequent historiography starts with their deportation in 1755. The Acadians were of French ancestry but did not view themselves as French citizens – they were Acadian. Geographically, Acadie intersected both English and French trade routes and traded hands many times between the two empires in its relatively short existence. By 1713, the Treaty of Utrecht left the Acadians living in English territory. By 1755 they were told they must swear fealty to the British crown. Most did not. This marked the beginning of *le grand dérangement*, during which, between 1755 and 1762, Acadians were forced off their lands by the English. Families were separated and placed on ships to American ports, from Massachusetts to Georgia – while some hid in the province's interior, aided by the Mi'kmaq. And so the public narrative of the Acadians and their legacy largely became one of post-expulsion. Yet that expulsion, and the creation of an Acadian diaspora, established another distinct Acadian characteristic: a fierce passion for genealogy.

A common question in Atlantic Canada is, 'Who's your father?' The answer tells the recipient if they and the questioner are possibly related. For co-author Simon Thibault, the answer is '*Simon à Hector, à Ulysses, à William, à Celestin à Isadore*'. The '*à*' here is not just 'son of', but furthers the question: do we belong to one another? The interlocutor and respondent are looking for links in a chain: familial links that were broken during *le grand dérangement*.

In 1905, historian and journalist Placide Gaudet published a genealogical document entitled *Report Concerning Canadian Archives for the Year 1905*, which is the preeminent source for Acadian genealogical research. Paradoxically, the British records created during the *dérangement* enabled Gaudet's forensic reconstruction of family names and members that makes current genealogical work possible. While researching the Belliveau apple, using Gaudet's genealogical work as a primary source, Karen Pinchin – the non-Acadian co-author of this paper – came across a genetic paper trail linking Simon Thibault's family to that of the apple. And within that record resides a remnant of Marie-Modeste Leblanc.

The youngest of eight siblings, Marie-Modeste was born in Port Royal in 1744. Eleven years later, her family, alongside thousands of others who refused to swear fealty to the British crown, were forced onto one of twenty-four overcrowded ships by gun and bayonet point. For Leblanc's family, like for many other Acadian families, deportation did not mean permanent exile. In 1768, twenty-four-year-old Marie-Modeste married Frederic Belliveau. By the late 1770s, Acadians were permitted to return, but were placed in disparate parts, as the reigning government wanted to surround the French-speaking Catholics with English-speaking Protestants. Their original lush settlements had been claimed by new settlers, and the land Acadians were permitted to occupy, especially in Nova Scotia, was often of poor quality. Out of necessity, Acadian life shifted from a deeply agricultural focus towards subsistence farming, logging, and fishing. When Marie-Modeste and her husband returned to Nova Scotia, they settled in a small Acadian enclave on the Bay of Fundy called l'Anse-des-Belliveau, or Belliveau Cove. It was in this village the first Belliveau apple tree grew.

When the Acadians first settled in Baie-Sainte-Marie, many of their homesteads overlooked the ocean. Families continued to grow apples, but the land there was less arable. The land tracts Acadians owned were long and narrow concessions that went from the ocean up into the woods (Figure 1). To reduce stress on trees, Acadians cleared portions of their lands and planted most trees inland, away from their homesteads.

Yet not all fruit trees grew inland. In the case of Marie-Modeste's family, the old homestead in Belliveau Cove included one tree near the home itself: the Belliveau apple.

Gardens, Flowers, and Fruit

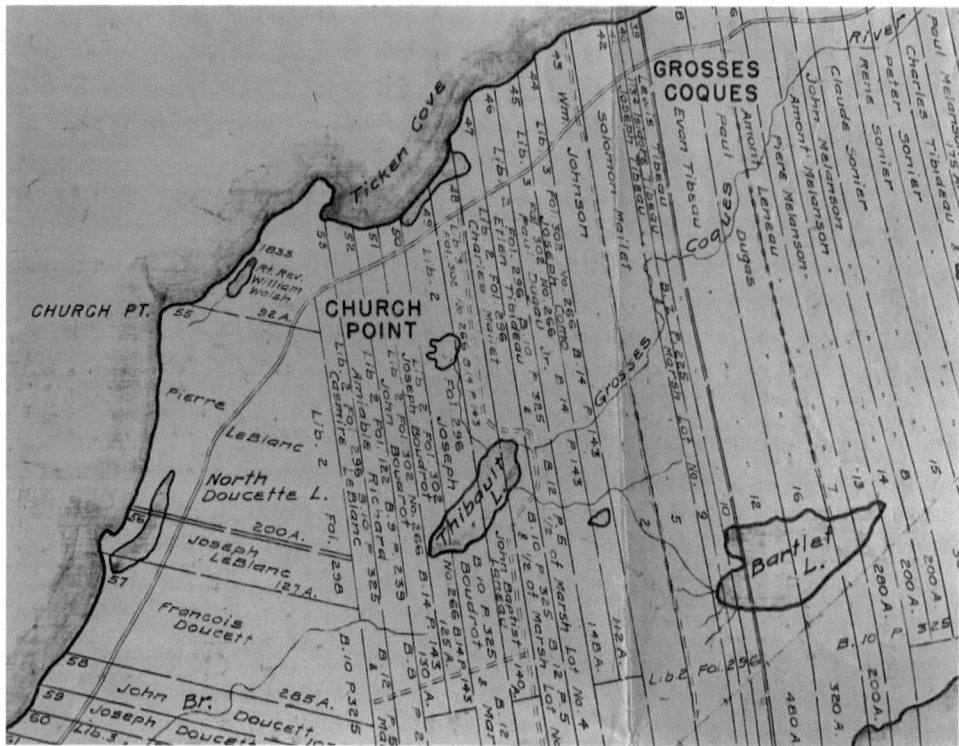

Figure 1. This early map depicts the names and boundaries of the land concessions granted to the Acadians upon their return to Nova Scotia (photo of a public domain map by Simon Thibault).

This tree was the source for scions that propagated the variety throughout the Acadian community and became a favourite for generations. Scion-sharing became part of the Acadians' (agri)cultural brain trust, as they shared apple trees commonly and freely within their community, ensuring collective food security.

Acadian women did much of this agricultural work. In 1819, Reverend Doctor Andrew Brown recounted that 'the care of the orchards fell early under the charge of the daughters of the family'.[7] In pomology, as in much of agriculture, male names tend to outlast the individual, like Cox's Orange Pippin, McIntosh, or Macoun. Though this pattern held for the Belliveau, the historical record has not completely erased Marie-Modeste. In a 1932 pamphlet for the Dominion Atlantic Railway, author François G.J. Comeau – Simon Thibault's great-great-uncle – highlighted the role of the Acadians in the development of pomiculture (Figure 2). In it, he mentioned Marie-Modeste indirectly, as the wife of Frederic Belliveau. He wrote that 'the first apples brought from Port Royal to St. Mary's Bay, Digby County, were known as the "Belliveau." These were taken [. . .] about the year 1769 or 1770 by Mrs. Frederic Belliveau from her father's orchard'.[8]

In Comeau's description of this apple, complications arise: 'She had the seeds planted the following spring, and only one tree reproduced the original apple.' Comeau is devoted to proving the theory that this apple is of Acadian origin, arguing that 'grafted apples

304

Figure 2. A woman working in a field fronts a promotional pamphlet produced by the Dominion Atlantic Railway in 1897 (photo by Simon Thibault).

will not reproduce themselves from seeds. Seedling apple (wild) trees will, however, sometimes reproduce themselves, and since the Belliveau did grow from seed, it is very clear and conclusive evidence of its Acadian origin'.[9] We believe the tree was sourced by Marie-Modeste from her father's orchard. Annapolis Royal (formerly Port Royal), where the orchard would have been, is about sixty kilometres north of Belliveau Cove. Marie-Modeste could have taken a cutting or apple from the tree when the Acadians were permitted to return to Nova Scotia, which fits Comeau's timeframe of 1769 or 1770. We may never know whether Marie-Modeste transported a scion to graft or apples to grow from seed.[10] But we know all Belliveau trees descend from Marie-Modeste's tree.

Genetically, a single apple contains about ten seeds, bound in pairs within stiff carpel chambers within the fruit. Borne from a pollinated blossom, each of those seeds carries the genetic material of the mother tree crossed with the pollen-borne mate. The plant is highly heterozygous, which means if one were to plant every seed from that single apple, ten different trees of wildly varying characteristics would result. As apple trees cannot self-pollinate and resist growing from cuttings, growers wanting to reproduce a specific variety of apple must rely on grafting, or the art of binding the flesh of two plants together.

Once natural or planned crossing results in a successful apple, each subsequent grafted tree will bear the identical genetics – and resulting fruits – of its mother tree. 'Every Granny Smith stems from the chance seedling spotted by Maria Ann Smith in her Australian compost pile in 1868,' writes Rowan Jacobsen in *Apples of Uncommon Character*.[11] In this way, given that apple trees can bear fruit for hundreds of years, the act of planting an orchard spans time as well as space. 'Because he had thought about it [. . .] with fruit on the trees and the trees in full maturity, it still contained him,' writes Helen Humphreys in *The Ghost Orchard*, of standing in an orchard planted by poet Robert Frost. 'The poet may die, but the poetry continues.'[12]

This single-track lineage makes the apple, in some respects, the ultimate time-travelling crop. It provides, when established through DNA profiling, a tangible genetic link across centuries. 'It is an intimate act, tasting an apple – having the flesh of the fruit in our mouths, the juice on our tongues,' writes Humphreys, reconstructing Quaker pomologist Ann Jessop's tasting of a White Winter Pearmain apple in 1790. 'I bite into the same kind of apple [. . .] and taste what she did,' she writes.[13] For Acadian descendants of the Belliveau family, raked across the coals of deportation, and living as a French-language minority, the value contained within this biological continuity cannot be understated. As a fruit, the Belliveau apple's skin is striated red and green and of medium thickness, while its flesh is distinctly crisp and described by Comeau as tasting redolent of oranges (Figure 3). Coline Campbell, a former member of Canadian parliament, still lives on an old Belliveau homestead, now long-renovated and modernized.[14] Alongside that home grows a huge, gnarled Belliveau tree, which she estimates as being around two hundred years old (Figure 4). She says the apple looks and tastes like a tart Gala, a popular supermarket variety brought to North America from New Zealand in the 1970s. According to some orchardists, the Belliveau takes well to grafting, something that may have fuelled its early spread within the Acadian community. Perhaps most importantly, for the purposes of pre-refrigeration apple growers, the Belliveau is a late September variety that stored well in root cellars and was equally tasty as a hand apple as in pies, sauces and cider.

Figure 3. A Belliveau apple tree bears fruit (photo by Simon Thibault).

Although the Acadians and their descendants planted and propagated apple trees on their new homesteads, their former orchards, which they had maintained and developed over a century and a half, continued to thrive. The scale of the Annapolis Valley's orchard-laden landscape

Figure 4. This Belliveau tree, located on the historic homestead of Marie-Modeste Belliveau's son, Frederic 'Tikine' Belliveau, is believed to be more than 200 years old (photo by Simon Thibault).

was so profound that even an English soldier – under orders to deport Acadians – noted it in his 1755 journals. In 1758, Captain John Knox's own record of the bounty left behind by the Acadians noted that he and his soldiers filled 'bags, haversacks, baskets and even their pockets' with fruit. 'The French have been at great pains here in clearing and planting these orchards, and indeed, finer flavoured apples, and greater varieties, cannot in any other country be produced,' he declared.[15]

Settlers from the United States, including New England Planters, farmed lands left behind by the Acadians. 'The suitability of the local environment for apple culture was demonstrated by the thickets of lusty seedlings which had sprung up in near-by fields and pastures,' wrote Charles Colby in 1925, in his 'Analysis of the Apple Industry of the Annapolis-Cornwallis Valley'.[16] They used Acadian fruit trees as root stock, grafting on scions of their own favourite New England varieties. Over the next hundred years, apple trees left behind by the Acadians formed the basis of an apple industry that fed the British Empire.

By the 1830s and 1840s, entrepreneurs were exporting Nova Scotia apples to England in large and regular numbers, with 2200 barrels shipped in 1832.[17] In 1860, they showcased their apples at the London International Exhibition. Two years later, Nova Scotian apples were put into competition at the London Horticultural Society Show, where 'no less than eight medals were awarded them, one of silver.'[18] By 1875, the 'Belleveau' [*sic*] apple appears on a report as being cultivated in Ontario, two provinces and thousands of kilometres away from its point of origin.[19]

As growers pursued efficiency and productivity with a goal of driving down costs and, by extension, price, they winnowed varieties planted in their orchards. In 1876, agriculturist R.W. Starr brought 196 varieties of Nova Scotian apples to the Philadelphia Centennial Exhibition. Eleven years later, Nova Scotia growers sent 118 apple types to the Edinburgh Apple and Pear Congress, including a mix of crisp 'dessert' apples intended for eating out-of-hand and cooking varieties intended for pies and sauces. Within fifty years, that number contracted as growers increasingly focused on popular varieties including Gravenstein, McIntosh, and Red Delicious.[20] As orchards consolidated and growers felled old varieties to graft new ones onto existing rootstock, countless apples disappeared.

For the British, the colonial project that began two centuries earlier was bearing literal fruit. In 1901, Nova Scotia farmers shipped more than one million barrels of apples to England. By 1925, more than sixty percent of apples produced in Nova Scotia were exported there, making up eighty percent of all Canadian apples sold in England.[21] Yet even by 1925, margins on apples, which were expensive to ship across the ocean into the British market, remained slim. 'It should not be thought that apple culture has proved a bonanza for these Nova Scotian farmers,' wrote Colby. 'Few men have become wealthy from their orchards.'[22] As growers continued to scrape a living from the margins of capitalism, the economic system promoted by governmental officials and marketing boards quietly oversaw the decline of varieties like the Belliveau in their quest for efficiency, uniformity, and ultimately, monetary profit, without considering what was being lost in exchange.

Like the danger posed by a single virus to a monoculture crop, the dependence of the Annapolis Valley's apple industry on one market was its downfall. By 1939, war was declared and apple shipments to England plummeted, quickly killing the demand on which the region depended. Between 1935 and 1939, Nova Scotia's annual production peaked at around six million bushels; by 1963 it was half that.[23]

After the collapse of the British export market, Canadian politicians, entranced by industrialized progress, decided to focus their attention and support on fewer agricultural growers producing more crops. By 1955, federal politicians, according to Waldo Walsh, Nova Scotia's then-Minister of Agriculture, had decided it was 'time we let the little inefficient growers be squeezed out'.[24] Many of those 'little inefficient growers,' in reality, were stewards of older trees and their genetic material that, once no longer growing, quickly faded from living memory. Through benign neglect, many apple varieties disappeared, with others preserved and protected only through the dedicated work of obsessives and hobbyists like Comeau. In 1929, Marie-Modeste's original Belliveau tree was felled (Figure 5).

Before the dawn of the twentieth century, Acadian communities depended on a collective brain trust to share agricultural knowledge. The collaborative nature of this trust

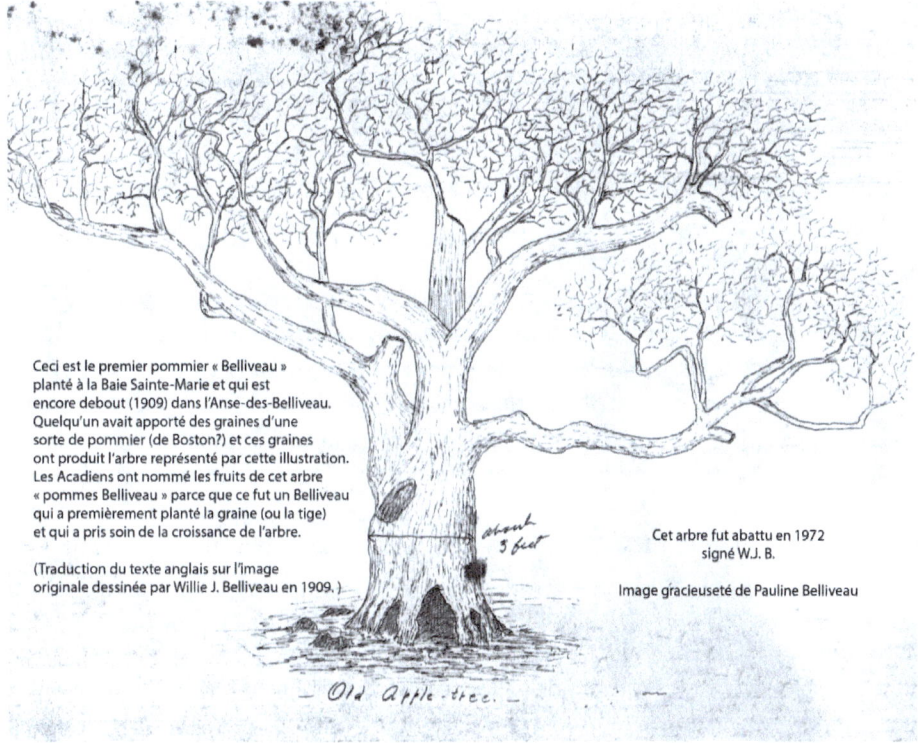

Figure 5. This 1909 drawing by Willie J. Belliveau depicts the first Belliveau apple tree, which was felled in 1929 (1972 is erroneous). (Courtesy of Pauline Belliveau.)

exemplified an inherent understanding of its value. The distribution of scions championed varieties that were well-suited to the Acadians' climate and ensured larger yields for families. The French family names of these fruits meant that agricultural propagation functioned also as cultural and familial propagation. Acadians could link directly with the food of their family's ancestors: generational sustenance.

Apples may be enthusiastically heterozygotic, but this centralization has led to a winnowing of flavours, textures, and the devaluation of non-commercially suited apples, as well as the inherent value of their less-common cousins. The most important genomes in these industries are not familial, but proprietary. Honeycrisp apples, developed in part at an agricultural research station in the Annapolis Valley, is a formerly patented and branded form of apple that has done well for farmers there. '[The] Honeycrisp can bring in five times more money for Nova Scotia growers than traditional varieties,' according to Genome Atlantic, a 'not-for-profit organization with a mission to help Atlantic Canada reap the economic and social benefits of DNA-based genomic technologies'. This shift, from community-held property to marketable commodity, is entrenched as a primary approach for considering the value of apples in our food system.[25]

Like many origin stories, the one underlying this paper has its fair share of happenstance. In 2017, Simon Thibault published *Pantry and Palate: Remembering and Rediscovering Acadian Food*, a cookbook in which he referenced his father's work growing the Belliveau and its value to his community. Soon after the book's publication, David Maxwell, a hobby orchardist and former physician living outside Lunenburg, Nova Scotia, contacted Thibault hoping to obtain scions for his own heirloom orchard. Thibault connected Maxwell with his father, who packaged and sent the other grower two small scions by mail. As far as Thibault knew, that was the end of the exchange.

Within a couple years, Maxwell's Belliveau trees, which had grafted smoothly, were thriving. For decades, Maxwell has corresponded and collaborated with Bill O'Keefe, another rare apple obsessive based in rural Ontario. On his farm, O'Keefe Grange, he and his wife Lyn have amassed one of Canada's largest collections of rare apples, numbering at over 600 varieties. He hadn't yet added the Belliveau to his collection, so Maxwell sent the Ontario farmer a handful of scions for grafting. Again, the hardy cuttings took swiftly and the young apple trees, now growing thousands of kilometres from their original home, followed their annual rhythms of budding, blossoming, fruiting, leaf loss, and hibernation.

In 2023, his last season before retirement, University of Guelph researcher Paul Kron arrived on the O'Keefe Grange for his final year of work for the Ontario Heritage and Feral Apple Project, a research programme dedicated to identifying and preserving the genetics of soon-to-be-lost and obscure varieties. He took a sample of a young leaf from one of the farm's Belliveau trees and placed it in a labelled paper envelope. Back at his lab, he pulverized the sample into a slurry to prepare it for genetic testing, and then analyzed the tree's DNA. The final result, according to Guelph plant biology professor Brian Husband, showed a slim possibility that Fameuse, an old French-Canadian variety, could be related to the Belliveau.

Every time the Annapolis Royal Historic Garden – which is located on original lands owned by the Belliveau family – posts about the Belliveau apple online, they often field a stream of questions along the same track: 'Is it possible to purchase one of these trees?' Going forward, we hope to coordinate an event that will bring together hobbyists, horticulturalists, and Nova Scotians of Acadian ancestry for an information-sharing event in Annapolis Royal that will return the results of our research to those communities. Like the ongoing work currently occurring in other communities – from Sapelo Island peas for the Gullah-Geechee, wild rice for the Anishinaabe people, and millet for India's Dongria Kondhs – we hope to support the reclamation of the Belliveau for those who seek to connect with the culinary and genealogical traditions of their ancestors.[26]

A century after F.J.G. Comeau's impassioned search for evidence of rare French apple varieties still in existence in historic Acadie, his great-great-nephew Simon Thibault found himself retracing similar steps. As a queer person raised in a traditionally Catholic community, Thibault's work attempts to make space for his Acadian identity – his *acadieneté* – by dovetailing his ancestors' agricultural and culinary traditions within an intersectional and multi-generational model. It is the work of the dispossessed seeking repossession that can help form a template for future generations.

From correspondence with growers, biologists, and passionate hobbyists, our work recalls the fleeting nature of a single apple tree, its genetic identity often known only to a single individual or as a simple X on a map of an orchard delineating a rare variety. Once that memory is erased – its map easily destroyed by water or fire or time – traditional genetic knowledge doesn't rest far from the precipice. From the living libraries of apple DNA contained on the USDA Agricultural Research Station in Geneva, New York, to the O'Keefe family's grange, these are dynamic archives that live, grow, and die alongside human societies, and they are only as healthy or protected as our own commitment to those trees and our natural environments. It is in our collective power to reclaim and protect that which we hold dear.

About the Authors

Karen Pinchin is an investigative food systems journalist, trained cook, and writing instructor at the University of King's College in Halifax, Nova Scotia. Her first book, *Kings of Their Own Ocean: Tuna, Obsession, and the Future of Our Seas*, on the global bluefin tuna trade won the Atlantic Book Awards' Evelyn Richardson Non-Fiction Award.

Simon Thibault is a writer, editor, radio producer, and journalist living in Halifax, Nova Scotia. His first book, *Pantry and Palate: Remembering and Rediscovering Acadian Food* was shortlisted for the Taste Canada Award, and his work has been featured in *The Globe and Mail*, The Southern Foodways Alliance, and many other outlets.

Acknowledgements

Karen Pinchin would like to dedicate this paper to her grandparents, Victor and Betty Pinchin, and their Riviere Fruit Farm in Streetsville, Ontario. Simon Thibault would like to dedicate it to his parents, Hector and Jeanne Thibault, his grandfather Augustin Comeau, and Sophie and Ella Halman.

Notes

1. Claire Legros, '"*Matrimoine*", ce mot ancien dont l'histoire raconte l'effacement des femmes créatrices', *Le Monde*, 13 September 2023 <https://www.lemonde.fr/idees/article/2023/09/13/matrimoine-ce-mot-ancien-dont-l-histoire-raconte-l-effacement-des-femmes-creatrices_6189119_3232.html> [accessed 25 April 2024].
2. Anne Hutten, *Valley Gold: The Story of the Apple Industry in Nova Scotia* (Petheric Press, 1981), p. 1.
3. John Mack Faragher, *A Great and Noble Scheme: The Tragic Story of the Expulsion of the French Acadians from Their American Homeland* (Norton, 2005), p. 3.
4. *The Jesuit Relations and Allied Documents*, ed. by Ruben Gold Thwaites (Burrows Brothers, 1896), vol 1 <https://www.gutenberg.org/files/44669/44669-h/44669-h.htm> [accessed 14 May 2024].
5. Hutten, p. 2.
6. Marc Lavoie, 'Les Acadiens et les "Planters" des Maritimes: une étude de deux ethnies, de 1680 à 1820' (unpublished doctoral thesis, Université Laval, 2002) <http://hdl.handle.net/20.500.11794/28592> [accessed 14 May 2024].
7. Sara Beanlands, 'Annotated Edition of Rev. Dr Andrew Brown's Manuscript: 'Removal of the French inhabitants of Nova Scotia by Lieut. Governor Lawrence & His Majesty's Council in October 1755' (unpublished master's thesis, Saint Mary's University, 2010), p. 176 <https://library2.smu.ca/handle/01/23202> [accessed 14 May 2024].

8 François G.J. Comeau, *The Introduction and Development of the Apple Industry in Nova Scotia* (Dominion Atlantic Railway, 1928), p. 7. Comeau's primary source was Placide Gaudet; it is unclear whether this information comes from Gaudet's work on genealogy or from an oral recounting, as they were contemporaries.
9 Comeau, p. 7.
10 Henri-Raymond Casgrain, *Un pélérinage au pays d'Évangéline* (L. Demers, 1887). While travelling through the former home of the Acadians in the Annapolis Valley, Casgrain wrote, 'Certain apple varieties, such as the Béliveau [*sic*], still wear the names bestowed upon them by the Acadians, and they were the first to cultivate them here' (p. 56). In correspondence between Comeau and Canadian Department of Agriculture entomologist Arthur Kelsall, the latter wrote: 'I think the weight of evidence is in favour of the Belliveau apple having originated at Belliveau's Cove as a seedling, and from that location spread, by means of grafts or buds, to Annapolis Royal and other localities' (Arthur Kelsall, letter to François GJ Comeau, January 19, 1934, fonds François GJ Comeau, Centre Acadien, Université Sainte-Anne, Church Point, Nova Scotia, Canada).
11 Rowan Jacobsen, *Apples of Uncommon Character: 123 Heirlooms, Modern Classics, and Little-Known Wonders* (Bloomsbury, 2014), p. 4.
12 Helen Humphreys, *The Ghost Orchard: The Hidden History of the Apple in North America* (HarperCollins, 2017), p. 7.
13 Humphreys, p. 4.
14 Campbell still resides in Tikine Cove, or *l'Anse a Tikine*, in the ancestral home of Frederic 'Tikine' Belliveau, one of Marie-Modeste's children.
15 Qtd. by Hutten, p. 4.
16 Charles Colby, 'Analysis of the Apple Industry of the Annapolis-Cornwallis Valley', *Economic Geography*, 1.2 (July 1925), pp. 173–97 (p. 192).
17 Julian Gwyn, *Comfort Me with Apples: The Nova Scotia Fruit Growers' Association, 1863–2013* (Lupin Press, 2014), p. 29.
18 Gwyn, p. 36.
19 D.W. Beadle (ed.), *Report of the Fruit Growers' Association of the Province of Ontario for the Year 1874*, (Ontario Department of Agriculture [Hunter, Rose & Co.], 1875).
20 James Murton, 'Subsistence Production and Commodity Production in the British Imperial Food System: The Case of Nova Scotia Apples', *Histoire sociale / Social History*, 54.III (September 2021), pp. 335–58.
21 Murton.
22 Colby, p. 175.
23 *The Current Review of Agricultural Conditions in Canada*, Economics Division, Marketing Service, Department of Agriculture, Ottawa, 16.6 (November 1955).
24 'The Blunt Truth', *Halifax Chronicle Herald*, 9 December 1955.
25 'Helping Maritime Apple Growers with Genomics', Genome Atlantic <https://genomeatlantic.ca/helping-maritime-apple-growers-with-genomics/> [accessed 10 May 2024].
26 Sonali Prasad, 'Indian Tribe Revives Heirloom Seeds for Health and Climate Security', *Mongabay*, 5 December 2018 <https://earthjournalism.net/stories/indian-tribe-revives-heirloom-seeds-for-health-and-climate-security> [accessed 10 May 2024].

30
The Rebellious Heart
Food and Flowers in the Chinese Classical Garden

Wena Poon

The Chinese classical garden has long departed from the lives of ordinary Chinese people worldwide. Nevertheless, we continue to hold it in high regard. A paved courtyard garden 庭院 is what you shoot for, if you can't marshal enough land to create the dreamy garden-forest 园林 of yore. A courtyard garden can be the size of a postage stamp, even an interstitial zone with a few potted plants squeezed in (Figure 1).

'Making a garden was among the traditional activities that a Chinese did best in the pursuit of leisure and contentment,' writes architecture professor and Suzhou native Joseph C. Wang. 'Especially when politics were corrupt, commerce was distasteful, society was disorderly, and life was harsh [. . .] some Chinese were able to seek sanctuary and solace in a garden, however small or large.'[1]

Today, the urban Chinese are forced to economize even further. Having no land at all, residents in high-rise flats would stake out a courtyard garden by curating potted plants right outside their door, claiming part of a public walkway (Figure 2).

In seeking solace in a private clump of plants, we are not unique among world cultures. What sets us apart is our passion for containment in garden design. Nature simply cannot be let alone. Her energy has to be framed and redirected. Cluttered with man-made elements, often circumscribed by high walls, Chinese gardens can sometimes be suffocating (Figure 3). Why the need for this box, I've always wondered. Could a cultural emphasis on containment hide an ungovernable inner life? John le Carré, in his memoir *The Pigeon Tunnel*, noted a similar polarity in German literary works. He was drawn to 'their classic austerity, and to their neurotic excesses. The trick, it seemed to me, was to disguise one with the other'.[2]

As a child learning Chinese ideographs, I saw wildness disguised in the word for garden 園. What is the complicated figure inside the box? 袁, the word for long robe, invokes a scholar. But it's three strokes shy of the word for ape 猿. The words for garden, scholar, and ape are homonyms: *yuan*. With these associations, the boxy word for garden 園 began to look very tense to me. If Pandora's box were ever to be opened, something deliciously wild and unexpected might leap out.

In this paper, I focus on the kind of garden found in the cities of southern China in the sixteenth and seventeenth century, examples of which can easily be visited today.[3] They resemble ornate jewel boxes hidden within resolutely blank, public-facing walls (Figure 4). Unlike Instagram users of the twenty-first century, the discerning Chinese of

Gardens, Flowers, and Fruit

Figure 1. Courtyard garden in a well-kept inn in Suzhou, China.

Figure 2. My grandmother's high-rise 'courtyard garden' in Singapore, with tied-up plants.

Figure 3. Model of a classical garden with built-in theatre in Suzhou, China.

Figure 4. Exterior of a classical residence, with blank exterior walls.

the past preferred not to publicize their private joys. Why invite thieves, jealous neighbours, and supernatural beings? Worse, what if the tax authorities came calling?

Invasion by the Mongols in the thirteenth century, and conquest by the Jurchens in the seventeenth century, made affluent Chinese homeowners more likely to hide their assets and their inner life. 'Householders, having Mongol soldiers billeted on them, tried to keep their women as much as possible confined to their apartments,' writes R.H. van Gulik in his 1961 treatise, *Sexual Life in Ancient China*.[4] 'One suspects that it was during this time that the germs of Chinese prudery came into existence, and the beginnings of their tendency to keep their sexual life a secret from all outsiders.'[5]

We see similar concealment in Edo period Japan, when sumptuary laws 'drove luxury underground', writes anthropologist Liza Dalby in *Kimono*. 'Japanese people dared not express their opinion for fear of reprisal. They had to keep their personal feelings under the cover of social convention.'[6] This led to the *iki* aesthetic, exemplified by outwardly drab kimonos with inner linings of boldly painted silk, seen only by the wearers and their intimates.[7] 'The notion that what is truly valuable is hidden has political reverberations,' notes Dalby. 'As a mode of expression, *iki* flaunted the rebellious heart even while concealing it.'[8]

The gardens of the rebellious Chinese *literati* were the most hidden of all. These men were often in exile because they offended the Emperor. Wang, the architecture professor, notes that 'their owners frequently built them as expressions of denial, self-protection, and seclusion [. . .] more often than not, the prevalent message was "the court has wronged me"'.[9] Wang gives the example of a garden in Yangzhou simply christened Ge Yuan 个园. Its original owner made clear that he and society were *à part*, for he had christened his garden 'single' 个, a branch split off from the word for bamboo 竹. We see again the scholar-ape, tightly boxed in.

Compared to the gardens of European aristocratic estates, the classical Chinese garden may not be that large. In the form they are found today in big cities, it would be difficult to put in more than half a mile for exercise. Such gardens exist for sedentary activities: looking, listening, and eating.[10]

What did people eat in this garden? We find out by visiting Li Yu 李漁 (1611–1680), a prolific, if often penniless, opera composer, novelist, book publisher, food critic, and garden designer (we Chinese do not believe in specialization). His life almost exactly overlaps with that of John Milton (1608–1674). For Li, the fruit of Eden was *fruits de mer*, especially crabs. He is known today as the Crab Sage 蟹仙. The autumnal crab season is still observed and celebrated by Chinese diners throughout the diaspora (Figure 5). During crab season, Chinese newspapers quote excerpts from Li's monograph on crabs. It appears in his compendium on the art of living, which I call *Random Jottings When In A Leisurely Mood*, or, in Chinese, *Xianqing Ouji* 閑情偶寄.[11]

Let us go back to 1671. We are in Lanxi, Zhejiang province, which is four hours' drive southwest of Shanghai. It is October,

Figure 5. Crab advertisement outside restaurant in late 2023, Singapore.

but since we are in the delightful South, it is still warm enough to sit outside. We take our place at a small, round table under a Japanese banana tree, the *Musa basjoo*.

Here comes our affable, middle-aged host. He's sixty, at the height of his powers, 'the philosopher and technician of happiness', as his literary critic Patrick Hanan calls him.[12] Everything you see in this garden, and on this table, is designed by him. He waves the little maid away so we can talk in private.[13] How he babbles. I have rendered him in English the way I, a modern Chinese writer, hear him:

> 予于饮食之美，无一物不能言之，且无一物不穷其想象，竭其幽渺而言之；独于蟹螯一物，心能嗜之，口能甘之，无论终身一日，皆不能忘之。

I admit I have the gift of the gab. On the topic of gastronomy, I'm famous for analyzing this and that, and for going on and on about any number of subjects, at great length if necessary. There is only one subject, however, where I find myself absolutely at a loss for words. I Simply. Cannot. Explain. Why. I. Am. So. Obsessed. With. Eating. Crabs.

> 至其可嗜可甘与不可忘之故，则绝口不能形容之。

If you have tasted crabs even once, you will never forget it for the rest of your life. It is that good. *Why is it that good?* Sorry. Words fail me.

> 此一事一物也者，在我则为饮食中之痴情，在彼则为天地间之怪物矣。予嗜此一生。[14]

I've been like that my whole life. Curiously and completely besotted.

Given our location, the crab Li Yu raves about would most likely be the Shanghai hairy crab, also known as the Chinese mitten crab or *Eriocheir sinensis*. Because of its life cycle and migratory pattern, it is only available for consumption in October and November of each year.[15] Its limited supply makes people even more crazy for it:

> 每岁于蟹之未出时，即储钱以待，因家人笑予以蟹为命，即自呼其钱为"买命钱"。

Each year, long before crab season comes around, I have already prepared my Crab Fund. So that the moment the crabs hit the market, I don't need to scrounge around for available cash. I can immediately dash out and buy them all. Everyone in the household says I would sell my soul for crabs. They joke about my Crab Fund. They call it my 'Blood Money'.

> 自初出之日始，至告竣之日止，未尝虚负一夕，缺陷一时。

From the first day the crab markets open, until the very last day they close, I would buy and eat crabs. Daily. I don't miss a single day. A single hour, even.

同人知予癖蟹，召者饷者，皆于此日，予因呼九月、十月为"蟹秋"。

My friends know I'm Crab Mad, so they make sure they invite me to their homes to get my fill as well. Crab season is every ninth and tenth lunar month. Also known, to us Crab Fiends, as 'Crab Autumn'.

虑其易尽而难继，又命家人涤瓮酿酒，以备糟之醉之之用。糟名"蟹糟"，酒名"蟹酿"，瓮名"蟹瓮"。向有一婢，勤于事蟹，即易其名为"蟹奴"，今亡之矣。

During Crab Autumn, there must be a continuous flow of crab for me to eat. I fear interruptions in supply. To prepare for the start of my Crab fest, I direct my family members to clean out my special Crab clay pots to make my Wines-That-Go-With-Crab, and my Pickles-That-Go-With-Crab. I used to keep a girl servant who was dedicated to All Matters Crab. I called her my Crab Butler. Sadly, she is with us no more.

蟹乎！蟹乎！汝于吾之一生，殆相终始者乎！[16]

Crab! Crab! You are my life! I never tire of worshipping at your altar.

The Crab essay is long; I present only extracts. Li Yu shines when he is at his most discerning. His complaints about the ignorant public ring true even now:

蟹之为物至美，而其味坏于食之之人。以之为羹者，鲜则鲜矣，而蟹之美质何在？以之为脍者，腻则腻矣，而蟹之真味不存。

Crabs are Nature's gift to man. A perfect and complete food. Yet there are poisonous folk who try to ruin crabs with their supposed 'recipes'. They stew the crab. They grind it up. They add oil and make it all greasy. Why? I want to cry. What has become of the Quintessence of the Crab? Completely annihilated.

更可厌者，断为两截，和以油、盐、豆粉而煎之，使蟹之色、蟹之香与蟹之真味全失。[17]

Even worse, I absolutely abhor chefs who chop the whole crab into two and fry it with some oil, salt and flour. Sacrilege! What about its colour and form, not to mention its natural taste? Utterly eviscerated.

What would he have thought of the Chef's Special crab dish that they serve in Chinese American restaurants today? They still ruin it the exact same way. Li Yu tells us how to do it properly:

和以他味者，犹之以燔火助日，掬水益河，冀其有裨也，不亦难乎？

How intolerable when chefs try to adulterate the Crab with non-Crab substances! The Crab doesn't need your aid to make it tastier. Does the sun need you to help

Gardens, Flowers, and Fruit

light it with a torch? Does the river need the pitiful bit of water in your cupped hands?

凡食蟹者，只合全其故体，蒸而熟之，贮以冰盘，列之几上，听客自取自食。

The only way to cook crab is to steam the complete, whole beauty until it is just-so. Do not overcook! Then bring it to the table on an iced platter so the crabs are cooled to the touch. Let the guests help themselves.

剖一筐，食一筐，断一螯，食一螯，则气与味纤毫不漏。

There is only one way to eat the crab. Whole. With your fingers. Focus on each crab. Break off only one piece at a time. Give your whole attention to that piece. Savor it fully, before reaching for the next. That is the only way you can get at every essential bit of goodness and not miss anything.

出于蟹之躯壳者，即入于人之口腹，饮食之三昧，再有深入于此者哉？ [18]

The sensual pleasure is unbeatable. Think about it – the sweet flesh of the crab emerges straight out of the protective shell into your mouth, down into your gullet. What immediacy. What unison. One cannot get any closer to Nature.

With the vehemence of a true *buongustaio*, Li Yu proceeds to lay down one of his many food rules:

凡治他具，皆可人任其劳，我享其逸，独蟹与瓜子、菱角三种，必须自任其劳。旋剥旋食则有味，人剥而我食之，不特味同嚼蜡，且似不成其为蟹与瓜子、菱角，而别是一物者。此与好香必须，好茶必须自斟，僮仆虽多，不能任其力者，同出一理。讲饮食清供之道者，皆不可不知也。[19]

I don't mind outsourcing labour to others. Except in only three matters – crabs, melon seeds, and water chestnuts. No matter how many pages and servants you have, *you must never let other hands peel these foods for you*. When people peel them for me, they taste awful, like chewing a candle. You must take Nature's bounty yourself, with your own fingers. There must be absolutely no intermediary between you and Nature's flesh. Then and only then can you claim to have truly tasted. Another thing – I don't care how rich you are, *you must always brew your own tea*. Only then can it be done exactly to your taste. Of course, if you are a gourmand, you already know this.

Li Yu was also a gardening fanatic. Nature is the humble scholar's friend: looking at flowers is a hobby that can take up oodles of time and potentially cost nothing at all. The world's earliest-known printed book of art was about looking at a single flower. In 1238, two centuries before the Gutenberg Bible, the Chinese used wood blocks to print Sung Po-Jen's *Guide to Capturing a Plum Blossom* 宋伯仁 · 梅花喜神譜. Sung was a Southern

Chinese poet and flower connoisseur. He captured each blossom from every possible viewing angle, on different days, in different poses. He even christened and wrote a poem to each pose of the flower. It was like one of those things people did during lockdown because they were uneasy and bored.[20]

Sung did this project during a stressful period of Chinese history. At the time, 'the northern part of China was in the hands of the nomadic Jurchens, who were, themselves, about to lose the north to the Mongols', wrote Lo Ch'ing in the introduction to the English translation of the work. 'Sung wanted to do something to encourage his countrymen to recover the north.'[21] He thought an illustrated book about a plant that blossoms in the depths of winter would spur people to stand up against invaders. To modern readers, his effort seems like a long shot. It is almost sobering to remember how much flowers could mean to people back then.

Li Yu has written reams about the art of seeing flowers. After all, he had an opinion about everything. He wrote essays about Chairs, Windows, Beds, Shoes, Women's Socks, Teacups, Porridge, Medicines, Sleeping, Sitting, Walking, and – let's not forget – How to Wash and Powder Your Face. It was a bit like ploughing through Dioscorides, Apicius, and Montaigne all at once. He wrote about twenty-three flowers, devoting an essay to each. His favourite was the peach:

Consider the Peach

凡言草木之花，矢口即称桃李，是桃李二物，领袖群芳者也。

Of all the flowering trees, only two are worthy of your connoisseurship – the peach and the plum. Together they reign supreme.

其所以领袖群芳者，以色之大都不出红白二种，桃色为红之极纯，李色为白之至洁，"桃花能红李能白"一语，足尽二物之能事。

Their flowers come in two main colours: pink and white. Peach flowers must be pink. Plum flowers must be white. Accept no other colours.

然今人所重之桃，非古人所爱之桃；今人所重者为口腹计，未尝究及观览。

Keep in mind that the peach people speak so highly of today bears no resemblance to the peach loved by our ancestors. In fact, whenever modern day ignoramuses talk about peaches, they don't care about flowers, or colours, the shape of the tree. All they care about is how the fruit tastes. That's not peach connoisseurship!

大率桃之为物，可目者未尝可口，不能执两端事人。凡欲桃实之佳者，必以他树接之，不知桃实之佳，佳于接，桃色之坏，亦坏于接。[22]

If all you care about is how it tastes, then you need to graft one peach tree to another peach tree, and hybridize them, and through endless trial and error, you might be able to breed the best tasting peach. Of course, by this point, you may have perfected the flavour, but you have completely lost the look. The flowers come out all wrong. The colours are wrong! Conclusion: you can either cultivate a tree for the eyes, or a tree for the stomach. You can't have both.

Li Yu wrote in the 1670s, during the reign of Charles II, one hundred years before the American Revolution. Even then, he thought of himself as having already missed the boat. The public of the seventeenth century, he laments, will never be able to see the trees and flowers that the Tang Dynasty poets and painters saw, seven hundred years ago. Like Milton, he speaks of a paradise from which we are forever estranged:

桃之未经接者，其色极娇，酷似美人之面，所谓"桃腮"、"桃靥"者，皆指天然未接之桃，非今时所谓碧桃、绛桃、金桃、银桃之类也。即今诗人所咏，画图所绘者，亦是此种。

The Peach of the Old Masters isn't the same as the peaches we have today. In poetry they compared the cheek and dimple of a maiden to a peach – but it's the natural peach they're talking about, not the newly-developed hybrids of peach that are common today, to which we have affixed all kinds of fancy names.

此种不得于名园，不得于胜地，惟乡村篱落之间，牧童樵叟所居之地，能富有之。欲看桃花者，必策蹇郊行，听其所至，如武陵人之偶入桃源，始能复有其乐。

The Peach Tree immortalized in classical poetry and painting is a wild tree. It can no longer be found in the city. These days, if you want to see the real thing, you have to have the time. You need at your disposal a trusty steed: horse, mule, donkey – whatever you can afford. Ride out of the city, way out into the godforsaken corners of the countryside. Ride until the only people you will pass is the odd shepherd, or the lonely woodcutter. That's when you know you're getting close. With luck, you might find *the* Peach Tree. Or even stumble upon several and find yourself in the Immortal Peach Tree Grove of Legend.

如仅载酒园亭，携姬院落，为当春行乐计者，谓赏他卉则可，谓看桃花而能得其真趣，吾不信也。[23]

In the spring, rich people saunter around with their concubines and go to flower viewing parties in one another's houses. They drink and make merry, and claim they are connoisseurs of the peach, but they're not. They're just flower-viewing. Any fool with eyes can view a flower. But it takes a rare person to truly understand what it means to look at the peach flower.

If Li complains about the public's ignorance in the 1670s, what about the 2020s? By taking pictures on our smartphones, we can now use artificial intelligence to identify flowers when we go for walks. But AI's guesses are often clumsy. Stone fruit flowers are so similar they can be maddening: for the casual observer, you have to wait till the fruit appears before you can be sure what they are. Every spring, as sidewalk trees begin to flower, Chinese netizens would post questions online about this very topic. To add to the confusion, two very different kinds of Chinese stone fruits, 李 and 梅, are both translated as 'plum' in English. With imported fruit and new hybrids, the lines continue to blur. Have we lost our way? We take pictures, but would we even know how to look at flowers as deeply as our forefathers did? Going back to Sung Po-jen and Li Yu is one way of regaining what Li Yu calls the 'Wise Eye' 慧眼. These writers did nothing casually. When they looked at a plant, or cracked open a melon seed, they instilled in the most ordinary of encounters a tremendous significance.

I would like to leave you with two simple Li Yu food philosophies that you may find useful in daily life. *Bon appetit!*

Avoid Eating When in High Passions

喜怒哀乐之始发，均非进食之时。

Avoid eating when your blood is up. When you are excessively joyous, furious, or grieving – these are *not* the times to be consuming anything.

An Excess of Delicious Things Won't Do You Harm

生平爱食之物，即可养身，不必再查《本草》。

There is really no need to second-guess yourself and look up medical encyclopedias before eating something. If you find yourself naturally drawn to a particular food, and it's nourishing for your body, go ahead and eat it.

About the Author
Wena Poon's stories have been professionally produced on the London stage, serialized on BBC Radio 4, and translated into French, Italian, and Chinese. She has published sixteen books of fiction and lives in the United States.

Notes
1. Joseph C. Wang, 'House and Garden', in *House Home Family: Living and Being Chinese*, ed. by Ronald G. Knapp and Kai-Yin Lo (University of Hawaii Press, 2005), pp. 73–98 (p. 73).
2. John le Carré, *The Pigeon Tunnel* (Viking, 2016), p 5.
3. When visiting China, most people like to see the classical gardens of Suzhou, which are on the UNESCO World Heritage list. There are also beautiful gardens in nearby Hangzhou.
4. A book so scandalous that the most sexually explicit parts were rendered in Latin. R.H. van Gulik is better known as Robert van Gulik (1910–1967), the Dutch Sinologist who also penned the popular Judge Dee detective novels, which contain a distinctly mid-century European idea of China.
5. Robert H. Van Gulik, *Sexual Life in Ancient China* (Barnes & Noble, 1996), p. 246.

6 Liza Dalby, *Kimono: Fashioning Culture* (Vintage, 2001), pp. 58–59.
7 *Iki* is popularly known as a kind of stylish chic, but the ideograph for *iki* 粋 also means essence and purity. It is a Japanese descendant of the Chinese ideograph 粹, which means quintessence, lack of adulteration.
8 Dalby, p. 60.
9 Wang, p. 94.
10 Often, the seclusion of the private classical garden allows it to serve as the perfect setting for all kinds of hanky-panky, which you can even see illustrated in the famous pornographic Ming novel *The Plum in the Golden Vase*, published in 1610. Before that, in 1598, China's most famous Kun opera *The Peony Pavilion* celebrates the garden as a setting for a cloistered young lady's first tryst with a scholar, aided by flower fairies.
11 To give you an idea of the artistic license we indulge ourselves in when translating Chinese to English, I later discovered that Harvard professor Patrick Hanan interprets the title as *Casual Expressions of Idle Feeling*. Hanan has translated many Li Yu works into highly readable, humorous English.
12 Patrick Hanan, *The Chinese Vernacular Story* (Harvard University Press, 1981), p. 165.
13 Li Yu was known for rather scandalously raising two little girls (one, a gift from his patron) and educating them in the arts. When they got a bit older, they performed his plays and doubled as his maids and mistresses.
14 Li Yu, *Xianqing Ouji* (Jiangsu Phoenix Literature & Art Publishing, 2022), pp. 164–65.
15 The Shanghai hairy crab is eaten, but not actually found, in Shanghai. It is banned in the EU as an invasive species, but if you are inspired by this paper to cook in Li Yu's style, any kind of fresh local crab will do.
16 Li, p. 165.
17 Li, p. 165.
18 Li, pp. 165–66.
19 Li, p. 166.
20 The series of stay-at-home orders that affected many countries during the critical phase of the COVID-19 pandemic in 2020–2021, during which people baked bread, ran marathons in their living rooms, and indulged in other creative, if oddball, pastimes.
21 Sung Po-Jen, *Guide to Capturing a Plum Blossom*, trans. by Red Pine (Mercury House, 1995), p. x.
22 Li, p. 178.
23 Li, p. 178.

31
Flowers and Fruit in Early Modern Mesoamerica
Chocolate as a Case Study

Kathryn E. Sampeck

Some elements of Mesoamerican cuisine are widely celebrated, but others have received less scholarly attention, even though they are central to culinary traditions in Mexico and Central America. This essay focuses on chocolate and other cacao drinks and foods to show the potential of a wider consideration of fruits and flowers in sixteenth- to eighteenth-century culinary practices. The dried, ground seeds of the fruit of the cacao tree, a key ingredient in chocolate, were often – even usually – enhanced with flowers, creating an array of flavour profiles and distinct recipes. For example, the sixteenth-century Nahuatl-Spanish document known as The Florentine Codex starkly distinguishes qualities of cacao:

> good, superior, potable cacao: the privilege, the drink of nobles, of rulers – finely ground, soft, foamy, reddish, bitter; chili water, with flowers, with uei nacaztli, with teonacaztli, with tlilxochitl, with mecaxochitl, with wild bee honey, with powdered aromatic flowers. [Inferior cacao has] maize flour and water; lime water; it is pale; the [froth] bubbles burst. [It is cacao] with water added – Chontal water [. . . fit for] flies.[1]

The poorest quality cacao drinks lacked flowers such as *uey nacaztli*, *teonacaztli*, *tlilxochitl*, and *mecaxochitl*. A singular focus on cacao impoverishes the scholarly understanding of Mesoamerican cuisines and the biodiversity of Mesoamerican agricultural and foraging systems.

The historian Marcy Norton argued that Spanish colonists cultivated the Mesoamerican flowery world by associating these new flowers and fruits with Christian notions of heaven and earthly pleasure.[2] For example, murals in the Augustinian convent of Malinalco, Mexico depict a Christian paradise or Eden of flowering cacao trees laden with fruit pods. As Norton notes, 'The Mesoamerican sensorium not only survived under colonial rule but traveled, in fragments, to Europe in chocolate, tobacco, and *xícaras*, as well as manifesting itself in an obsession with foam and floral additives.'[3] Though some, such as vanilla orchids, survived and eventually thrived far from their homeland, other highly important flowers of Mesoamerican cacao-related culinary repertoires did not.

Gardens, Flowers, and Fruit

This study focuses on cacao, but not the seed, just cacao fruit along with frequently used culinary fruit and flowers. Cacao is strongly associated with an even wider variety of plants for medicinal uses that lie beyond the scope of this study. Important sources of information for this study include early colonial vocabularies and other works in Indigenous languages, which reveal a redolent and flavourful world. This study examines the sensory qualities and uses of six culinary examples, in order of their Latin scientific names. The culinary overview of each is followed by a consideration of their ecological needs to reconstruct the biodiversity of the world of Mesoamerican cacao and chocolate.

Cacao Fruit

The almost singular focus on cacao seed products of the tree *Theobroma cacao* has directed attention away from the luscious cacao fruit itself, which people in Latin America enjoy today in a variety of ways – as a fresh fruit and in sweetened drinks and ices (Figure 1). The amount of fruit pulp surrounding the seeds varies by clone (variety) and by local growing conditions, or terroir. Flavours are generally mild, sometimes slightly tart, and gently sweet, including lychee, cherimoya, floral notes, and hints of pineapple.

Ancient and colonial evidence of enjoyment of cacao fruit perhaps dates to the initial domestication of *Theobroma cacao*, with early Ecuadorian tall-necked bottles indicating a juice or fermented wine.[4] In ancient Maya texts, fruit qualities make up almost a third of all cacao-related terms.[5] Sixteenth-century evidence includes a highly regarded recipe that was a featured drink for a feast: 'many types of cacao would promptly be brought out [for the feast], which were made very delicately, such as the following: xoxouhqui cacahuacintli, "cacao made with tender cacao cob"'.[6] The *cacahuacintli*, the cacao 'cob', refers to what people today call the pod, thus indicating the fruit, not the seed.

Figure 1. Theobroma cacao *tree, seeds, and fruit. From* O fazendeiro do Brazil . . . Tomo III. Bebidas alimentosas. Cacao. Parte III, *by José Mariano da Conceiçao Velloso (1805). Courtesy of the John Carter Brown Library.*

Flowers in Cacao Recipes

Sixteenth- to eighteenth-century Indigenous language and Spanish sources define some flowers in terms of their use in chocolate. The following six examples are not exhaustive but do include the most frequently mentioned flowers in early texts; a few also have ancient archaeological evidence of their culinary presence. Each cacao-related culinary flower has distinct sensory qualities. Each example is important for its pairings with cacao, showing that flavourings went beyond sweet honey or spicy chilli.

Chiranthodendron pentadactylon

Although the temptation might be to assume that flowers add only sweet, floral flavours, this example is dramatic in several ways. Known in English as devil's, monkey's, or Mexican hand tree and in Spanish as *flor de manitas*, *C. pentadactylon* has several Indigenous language names that usually refer to the remarkable shape of its bright red flower, such as the *Nahuatl mapilxochitl*, whose 'flowers are like the palm of a hand with its fingers. It takes its name from the palm and the fingers' (Figure 2).[7]

The flavour of the bright red flower is umami, with earthy wood smoke and tobacco notes. Observations by anthropologist Nicholas Hellmuth indicate that Maya people made honey from the nectar of this flower; he adds that '[t]oday, the tree leaves are used to wrap tamales, cheese and other foods'.[8] These leaves are described in The Florentine Codex as 'big and very dense'.[9]

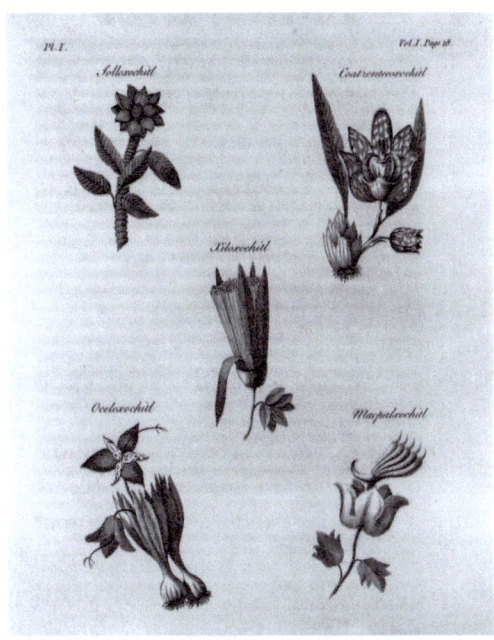

Figure 2. Five flowers, including from the tree yolloxochitl *(Magnolia mexicana, upper left) and shrub known as* mapilxochitl *or* flor de manita *(Chiranthodendron pentadactylon, lower right). From* Storia antica del Messico *by Francisco Saverio Clavigero (1787). Courtesy of the John Carter Brown Library.*

Cymbopetalum penduliflorum

Like the previous example, *C. penduliflorum* has a flower shaped like a part of the human body. The English name for this tree of the *Annonaceae* or custard-apple family, 'ear flower', nods to the common Spanish name, *orejuela* (Figure 3). The Q'eqchi Maya name is *muk*, while Nahuatl has a couple of variations emphasizing the ear shape: *uey nacaztli*, or great ear, and *teonacaztli*, divine ear.[10] Both versions imply that it is precious or highly valued 'because they are very fragrant, beautiful, and useful, for they are an aromatic spice that is commonly used in [preparing a] drink with cacao'.[11] In fact, *uey nacaztli/teonacaztli* is a hallmark for a fine cacao beverage, 'good cacao with the divine ear spice'.[12]

What was so alluring about the flowers of this tree? A further description introduces vivid colours, flavours, and effects: the Nahuatl text of The Florentine Codex describes *teonacaztli* flowers in rapturous detail, as

> of pleasing odor, fragrant. Its scent is dense; it pierces one's nose; it is strong. It is yellow; it has fuzz; it is fuzzy like cotton bolls. It is potable; the juice is extracted when it is drunk, or it is ground in cacao. Roasted first, it is drunk uncooked. A medicine, it lessens the fever. It is drunk many times and not much; for it takes

an effect on one; it makes one drunk as if it were mushrooms. I plant the teonacaztli. I transplant the uey nacaztli. I pick the teonacaztli. I put teonacaztli in the cacao. I add blossoms to it. With blossoms added, I drink the cacao. I smell the teonacaztli.[13]

This example adds important commentary on agroforest management and cultivation as well as the possible effects on human consciousness.

Orejuela is one of the most abundant flower seasonings in markets in Central America in the past and present, mentioned in The Florentine Codex as part of a trio of flowers regularly sold by merchants in the market. The flavour of *C. penduliflorum* is akin to common European and Asian spices nutmeg and cinnamon.[14]

Magnolia mexicana

This type of magnolia has a luminously white flower often associated with the Virgin Mary because of its purity and beauty.[15] The Nahuatl name, *yolloxochitl*, refers to a heart shape of the flower bud (Figure 2). Some also attribute *yolloxochitl* to *Magnolia glauca* and *Talauma mexicana*.[16] Sixteenth-century sources describe the heart-shaped flower bud of the tall tree as fragrant.[17]

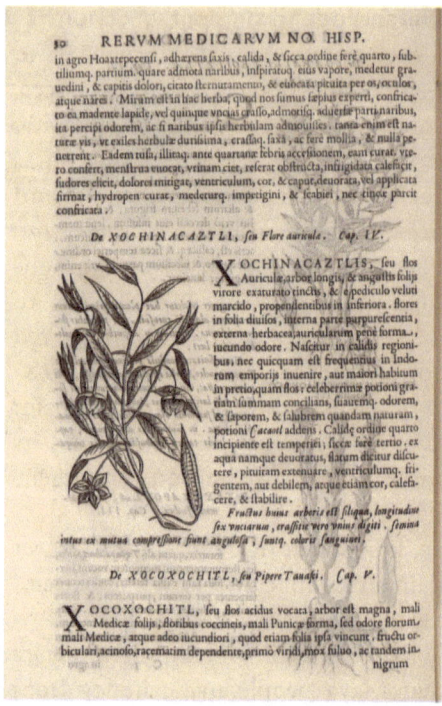

Figure 3. Branch of the xochinacaztli *or ear flower tree (*Cymbopetalum penduliflorum*). From* Nova plantarum, animalium et mineralium mexicanorum historia *by Francisco Hernández (1651). Courtesy of the John Carter Brown Library.*

The medicinal use of *yolloxochitl* calls attention to the consumption of cacao seed shells as well as seeds: 'if they [*yolloxochitl* flowers] are mixed with cacao shells or with a draft made from cacao, they strengthen the heart and the stomach'.[18] The heart meaning also may refer to its medicinal effects on human cardiac function. Scientific studies have identified compounds, a glucoside substance called talumine, that has a pharmacological effect similar to digitalis; talumine and other substances of the fruit and bark of the tree slowed down heart rate and increased and regularized the force of the heartbeat.[19] The Badianus Codex includes *yolloxochitl* in several treatments for ailments: obstructed urine, stupidity, and as a traveller's protection.[20] William Gates describes that a *yolloxochitl* infusion was 'glutinous and astringent, served in epilepsy'.[21]

Like *orejuela*, *yolloxochitl* was sold in markets seasonally. One report indicates that the tree flowers from May to July, when they are harvested and bought in great quantities in the markets in central and southeastern Mexico.[22] William Gates noted that it was a flavouring for chocolate.²³ The pleasant floral aroma complements the slightly spicy ginger/cardamom flavour, another American taste complement to cuisine of Asia and Europe.

Piper amalago

Known in English (particularly Jamaica) as 'Rough-Leaved Pepper', this shrub belongs to the same family as other well-known pepper species such as black pepper (*Piper nigrum*) and long pepper (*Piper longum*). The Nahuatl name, *mecaxochitl*, rope flower, refers to the long, upright bloom where the small, spicy fruits form (Figure 4). The Florentine Codex has the most detailed description, which calls attention to its ideal ecological setting:

> its growing place is the hot lands, at the water's edge. It is like a slender cord, a little rough. It is of pleasing odor, perfumed. Its scent is dense; one's nose is penetrated. It is potable. It cures internal [ailments].²⁴

A Nahuatl description a few pages later affirms, 'The branches are firm; they extend bristling. It is of very pleasing odor, very fragrant.'²⁵

Mature fruits are about the size of mustard seeds. The fruits are picked when full grown but not completely mature because they lose pungency and get soft once mature. Immature fruits, often harvested still attached to their stalks, are dried in the sun; once dried and ground they taste like black pepper.

Mecaxochitl has both culinary and medicinal uses, including as a treatment for traveller's protection.²⁶ It is also used to treat susto among Q'eqchi' Maya.²⁷

Sixteenth-century Indigenous language texts record *mecaxochitl* as a fruit to perfume cacao, indicating that the fruit as well as the fragrant flower was of culinary value.²⁸ Sweeter and less pungent than its black pepper relative, *Piper amalago* is still piquant. Recent

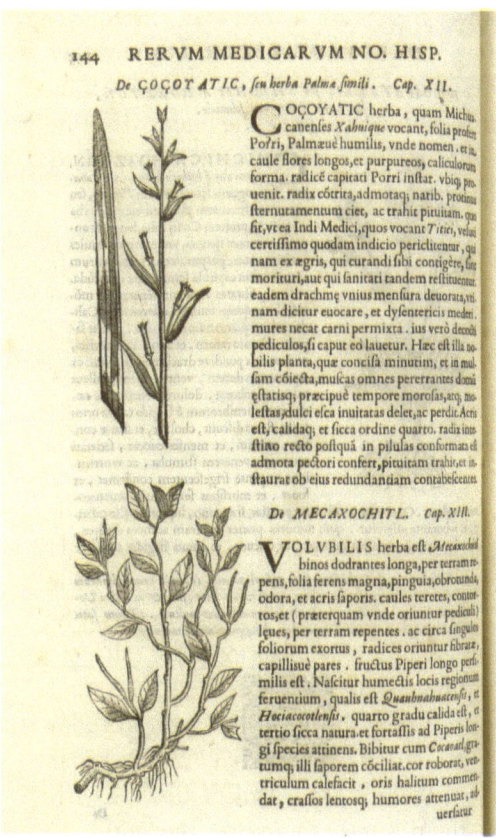

Figure 4. The bottom image is the mecaxochitl *(Piper nigrum) plant with its roots, leaves, and flower or fruit. From* Nova plantarum, animalium et mineralium mexicanorum historia *by Francisco Hernández (1651). Courtesy of the John Carter Brown Library.*

scientific studies that have identified and compared essential oils from several Piper species support the sensory experience of a pleasant, less intense sibling of black pepper.²⁹ This example adds another flavour realm – spicy – to the cacao seasoning profile that was also a close complement to European and Asian culinary practices.

Quararibea funebris

The small tree known in English as a 'Swizzle Stick tree' (because of the large, prominent stamen of the flower) or 'Funeral Tree' (because of the uses of the intensely fragrant flowers) was a culinary staple in Mesoamerica, depicted in ancient Maya ceramics. The name of the tree in Spanish is *Rosita de Cacao* or *Flor de Muerto*, the first translating the name for the tree in Nahuatl: *cacahuaxochitl* (Figure 5). The sixteenth-century description of Dr Francisco Hernández emphasized that the *cacahuaxochitl* or *cacaoatl* flower has 'heart-shaped leaves, stems a span long, purple flowers and thick, fibrous roots'.³⁰ The overwhelming, seemingly inescapable quality of *cacahuaxochitl* is its lush, enduring scent. The poetic description in The Florentine Codex gives vivid details of prized qualities:

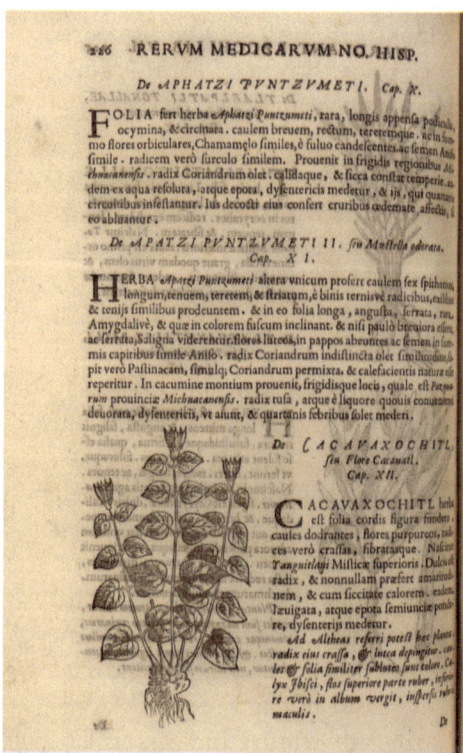

Figure 5. Cacahuaxochitl (Quararibea funebris) *tree with its roots, leaves, and flowers. From* Nova plantarum, animalium et mineralium mexicanorum historia *by Francisco Hernández (1651). Courtesy of the John Carter Brown Library.*

> It is slender, tall, like a stone column. It spreads an aroma; it is fragrant, the same as yolloxochitl. Its leaves, its foliage are slender. The name of its flower is cacauaxochitl; it is yellow, yellowish, small; the same as the acuilloxochitl. Its smell is very dense; it penetrates one's nose. It has cup-like [blossoms]; the name of its cup-like blossoms is poyomatli; a really pleasing odor is their aroma. The tree, the blossoms, its foliage, all are of pleasing odor, all perfumed, all aromatic.
>
> I cut [the blossoms], spread them out, arrange them, make a flower mat of them make a bed of flowers with them, spread them over the land.
>
> [The perfume] spreads over the whole land, swirls, constantly swirling, spreads billowing.³¹

While the Spanish text in The Florentine Codex related the intense fragrance to a well-known flower in Europe (jasmine), the Nahuatl text luxuriates in the powerful effect of such a small flower.

Besides flavouring cacao, Indigenous Mesoamericans used the flowers for preserving food and bodies. The fragrance stays in dry flowers for decades, so they were used for funeral ceremonies and were found in crypts still fragrant after many years. Dried flowers are also used at weddings and festivals. The appeal is across species: scented wood of this tree attracts fishes, so Aztecs used it as a bait for fishing.

The *poyomatli* mentioned on folio 188v of The Florentine Codex reveals another kind of potency: *poyomaxochitl* is 'the cup of the cacauaxochitl. They say that it makes one falter, that it deranges one, provokes one. It makes one falter, it deranges one, it provokes one'.[32] Francisco Hernández identified another medical use – that a mere half an ounce, powdered, can be taken to cure dysentery.[33] In addition, Mesoamericans regard *rosita de cacao* and *tejate* as a treatment for anxiety, fever, and coughs. Scientific study of *Q. funebris* flowers have detected micromorphology, volatile and bioactive compounds, and a newly discovered lactone alkaloid, named funebradiol, as well as other lactone alkaloids: funebral and funebrine.[34] These compounds possess subtle relaxing properties.

The entire flower of *Quararibea funebris* is edible. The stamen, pistil, and petals may be easily pulled out of the sepal and eaten raw as a tasty popcorn-like snack. Today, *Q. funebris* flowers are an essential ingredient in the Oaxacan cacao drink tejate. The flowers lend caramel, dried fruit, and candied fruit notes as a flavouring, including for tobacco. In addition to *Q. funebris* flowers, the root of the tree was claimed to be 'sweet, with a trace of bitterness, enough to make it hot'.[35]

Vanilla planifolia

The last example is perhaps the most familiar, even quotidian, to many, a point that we can question. In Nahuatl, the vanilla bean, described as 'fragrant pods', was referred to as *tlilxochitl* ('black flower') while the plant was *cuauhmecaexotl* (Figure 6).[36] The Nahuatl name for the bean thus invokes its appearance (black) and its flavour (flowery). The Nahuatl text of The Florentine Codex rhapsodises about vanilla:

> It is cord-like, like cords, slender. Its vine is like the tetzitzilin; it is a climber, one which sends out shoots. It has a bean. Its bean is green; when dry, it is black. It is glistening; within, it is resinous. It is of pleasing odor, perfect, superb. It is potable in cacao. It creeps, constantly creeps, travels, sends out a shoot, forms foliage, produces a bean, forms a bean.[37]

Vanilla, with a creamy, floral scent, is well known to act as a taste, like salt or umami, that reveals different aspects of foods and drinks.[38] Vanilla can change perception of 'mouthfeel', reduce perceived acidity, and for some make food or drink seem sweeter or pleasantly savoury. Vanilla can smooth over rough edges, whether sweet or savoury. The 1570s Yukatek Maya vocabulary *Bocabulario de Maya Than* defined *zizbic* as 'vanillas that one puts in chocolate, fragrant'.[39]

Figure 6. Branch, flowers, and pods of the vanilla orchid. From Voyage du tour du monde [. . .] Tome sixieme *by Giovanni Francesco Gemelli Careri (1719). Courtesy of the John Carter Brown Library.*

Combined Effects

These individual descriptions and evidence are only part of the culinary and medicinal picture. Documentary evidence suggests that these examples formed a complimentary suite, and when used together multiplied their powerful culinary and/or medicinal effects; in other words, one is not enough.[40] For example, a remedy for cough, digestive problems, and obstructions was 'That spice that they call hueinacaztli [. . .] as well as the other one that they call tlilxochitl, and the other one that they call mecaxochitl, when all these are ground and drunk with cacao'.[41]

Although medical treatments were important, combinations of these flowers also served important social and political purposes.[42] The flower-filled gourds presented to Cortes were of *yolloxochitl*, *izquixochitl*, and *cacahuaxochitl*.[43] Further evidence of the relationships among these flowers include the sculpture of *Xochipilli/Macuilxochitl* ('five flowers'). The figure and the base are covered in carvings of sacred and psychoactive plants including *Psilocybe aztecorum*, tobacco (*Nicotiana tabacum*) flowers, and a flower bud of *cacahuaxochitl* (*Quararibea funebris*).

Of the different examples, vanilla reigns supreme as the most adopted culinary plant. The close overlap of flavour effects of the rest perhaps points toward a non-American preference to swap out rather than introduce new culinary plants. It is possible, however, that the uneven global adoption of the fruit and flower examples may relate to their ecological needs as much as to the ability to transport their products. A brief overview of each plant helps assess this question as well as reconsider the overall biodiversity of the cacao culinary world.

Agroforestry Implications of Diversity of the Palate

What shows up in someone's cup or on their plate is a result of agricultural and foraging practices. Ecologically diverse agroforestry systems of cacao, vanilla, and other plants occur regularly in early colonial historical accounts in different regions of Mexico and Central America. For example, sixteenth-century accounts of orchards in today's El Salvador as well as in Chol, Mopan, and Itzá Maya areas of Mexico and Central America describe forest pluriculture that included but were not limited to cacao. As the earlier examples show, adding vanilla is only a small part of the picture. A review of plant biology provides basic information to consider whether their ecological settings complement or contrast that of cacao and the degree to which they could have been part of the same environment.

Chiranthodendron pentadactylon
This tree grows to heights of 10–27 m tall, so potentially taller than both *B. huanita* and cacao species. It could thus provide shade for cacao, and thrive in the sun it prefers; however it does best in a wet, mixed forest high in the mountains at elevations from 2000–3000 m. This tree grows also in fields from which forest has been cleared and was probably planted in some rural regions.[44]

Cymbopetalum penduliflorum
Like *B. huanita* and *C. pentadactylon*, this tree is native to southern Mexico, Guatemala, and El Salvador. It is cultivated in Guatemalan regions of Cobán and Jacaltenango, but at elevations below 800 m, in line with cacao species.[45] Generally taller than cacao, ranging from 10–23 m tall, its trunk can be 25 cm or more in diameter.[46]

Magnolia mexicana
This tree is an endangered species due to negative impacts of farming, logging, and cattle grazing.[47] Although contemporary farming has diminished its presence, it can thrive in cultivated areas, but prefers deciduous woodlands – mature humid tropical forests – and high evergreen forests at altitudes ranging between 110 m and 2900 m. This tree thus spans the habitats of low elevations tolerated by *B. huanita*, *C. penduliflorum*, and cacao and higher elevations preferred by *C. pentadactylon*. The tree can reach heights of up to 24 m, so it could serve as a shade tree for cacao.

Piper amalago
This pepper plant is a slender, often much branched, evergreen shrub or small tree, commonly growing 1.5–6 m. The slender spikes, the 'little strings', are covered by hundreds of tiny simple flowers that arise vertically from the stems. This forest shrub is 'bat-dispersed'.[48] *P. amalago* prefers moist or wet thickets or mixed forest, at elevations up to 2600 m, and thus would be compatible with ideal environments for all four of the previous trees.

Quararibea funebris
Q. funebris is in the same family as Baobab and Kapok trees. Generally a medium to large size evergreen shade tree, it can grow as tall as 25 m and has large leathery green leaves (up to 38 cm long). It grows well in moist to wet primary lowland and highland forest, at elevations up to 1600 m, and can even tolerate brief freezing temperatures, unlike cacao.

Vanilla planifolia

A 1699 account by Captain Marcelo Flores, assigned to the Presidio (fort) of Petén, in lowland central Guatemala, stated that some Chol and Mopan Maya were still living in what had been their lands, an observation made based on 'the care and tidiness of their cacao and vanilla orchards'.[49] This 'care and tidiness' is needed for the plant to fruit. *V. planifolia* is an orchid that grows as a spectacular (15–30 m long) evergreen vine. The vine can grow in full shade where it will seldom branch, but in sunlight the vine develops multiple branches. It can grow on the ground or as an epiphyte with thick, fleshy aerial roots that cling to the host tree. These aerial support roots almost never branch and indicate the younger parts of the vine; the older parts of the vine hang down through the canopy to the forest floor. The flowers, which appear in the last couple months of the dry season (April–May) are fleeting – opening in the morning and fading with rising temperatures the same afternoon. During that brief span, the flower must be pollinated to set fruit. It can self-pollinate by transfer of the pollen from the anther to the stigma, but the flower has a structure to prevent this from happening without intervention, so wild rates of pollination are only about 1%.[50] The vine takes two to four years to mature sufficiently to bear fruit, which take nine months to mature and ripen. Like cacao, the orchid does not tolerate cold, preferring temperatures ranging from 20 and 30°C, altitudes from 150 to 900 m, and moist conditions, yet with a distinct dry period to trigger flowering.[51]

Diversity of Cacao Fruit and Culinary Flowers

The examples reveal a tremendous variety, a layering of flavour, that has long been a hallmark of Mesoamerican cuisine. The examples most broadly adopted beyond the Americas – cacao and vanilla – were ones that had no easy substitute, while lychee (for cacao fruit), rose (for *B. huanita*), nutmeg or cinnamon (for *C. penduliflorum*), cardamom (for *M. mexicana*), black pepper (for *P. amalago*), or maple or dried fruit (for *Q. funebris*) could approximate some of the missing flavours that people of Mexico and Central America still enjoy and expect today.

The examples highlight the environmental impact of almost exclusive focus on cacao seeds. Complementary to the diversity of flavour, the ecological requirements for the array of comestible flowers show that Mesoamerican agroforestry systems and commodity chains depended upon variety. The evidence from the preferred habitats of common cacao flavourings is that Mesoamerican pluriculture was multi-sited, from the cool, high elevations of cloud forest (*C. pentadactylon*) to all the way down to low elevation, humid, and partial shade environment preferred by cacao tree species, vanilla vines, and *C. penduliflorum*. Human choices about local ecologies, such as the degree of canopy, affect the sustainability of insect populations, which in turn affects the productive capacity not just for cacao but also other culinary plants such as vanilla. The remaining trees and plants – *B. huanita*, *M. mexicana*, *P. amalago*, and *Q. funebris* – spanned from lower elevation, cacao-favourable environments to the cooler, higher elevation environment of *C. pentadactylon*. The cacao recipes imply movement and engagement across a breadth of Mesoamerican environments, with some species bridging from one zone to the next. The threatened status today of some of the trees calls attention to environmental degradation in these regions since the seventeenth century. Taken together, the culinary and environmental information for these examples reveal an underappreciated complexity of the world of cacao.

About the Author
Kathryn E. Sampeck writes about the social history of American foods, particularly chocolate, and culinary practices. She is a Professor of Anthropology at Illinois State University and British Academy Global Professor at the University of Reading.

Notes
1. Fr. Bernardino de Sahagún, *Florentine Codex: General History of the Things of New Spain*, ed. and trans. by Arthur J.O. Anderson and Charles E. Dibble (School of American Research and the University of Utah, 1951), Book 10, 69r.
2. Marcy Norton, 'Foreword', in *Substance and Seduction: Ingested Commodities in Early Modern Mesoamerica*, ed. by Stacey Schwartzkopf and Kathryn Sampeck (University of Texas Press, 2017), pp. vii–xiv (p. viii).
3. Norton, p. viii.
4. Sonia Zarrillo and others, 'The Use and Domestication of Theobroma Cacao during the Mid-Holocene in the Upper Amazon', *Nature Ecology & Evolution*, 2 (2018), pp. 1879–88, DOI:10.1038/s41559-018-0697-x.
5. Kathryn Sampeck, 'Chocolate and Vanilla: Seeds of Taste', in *Seeds: Proceedings of the Oxford Symposium on Food and Cookery 2018*, ed. by Mark McWilliams (Prospect Books, 2019), pp. 240–56.
6. Sahagún, Book 8, 25r.
7. 'También el árbol arriba dicho se llama macpalxóchitl, porque sus flores son como la palma de la mano con sus dedos. Toma nombre de la palma y de los dedos', Sahagún, Book 11, 190v.
8. Nicholas Hellmuth, 'Medicinal Plants of Guatemala: An Approach to Ethnobotany', *FLAAR Mesoamérica*, 2021 <https://flaar-mesoamerica.org/2021/11/30/medicinal-plants-of-guatemala-an-approach-to-ethnobotany/> [accessed 15 May 2024].
9. Sahagún, Book 11, 190v.
10. Sahagún, Book 11, 123r, 189v.
11. Sahagún, Book 11, 189v.
12. '*ioan qualli cacaoatl, teunaçaçço*', Sahagún, Book 9, 22v.
13. Sahagún, Book 11, 123r-v.
14. William Edwin Safford, *Sacred Flowers of the Aztecs, Revised and reprinted from the Volta Review, Vol. XIV, No. 2, May, 1912*, (Press of Judd & Detweiler, Inc.), p. 89.
15. Louise M. Burkhart, *Before Guadalupe: The Virgin Mary in Early Colonial Nahuatl Literature*, Institute for Mesoamerican Studies Monograph 13 (Institute for Mesoamerican Studies, University at Albany, 2001), p. 20.
16. Sahagún, Book 11, 196r.
17. Alonso de Molina, *Vocabulario en Lengua Castellana y Mexicana y Mexicana y Castellana*, part 2, Nahuatl to Spanish, (1571). Manuscript in the collection of the John Carter Brown Library at Brown University, Providence, Rhode Island, f. 41r. col. 1; Simon Varey, ed., *The Mexican Treasury: The Writings of Dr Francisco Hernández*, trans. by Rafael Chabrán, Cynthia L. Chamberlin, and Simon Varey (Stanford University Press, 2000), p. 119.
18. Varey, p. 119.
19. Forest Resources Development Branch Forest Resources Division, FAO Forestry Department, Food and Agriculture Organization of the United Nations, 'Some Medicinal Forest Plants of Africa and Latin America', FAO Forestry Paper 67 (1986), p. 212 <https://www.fao.org/4/an797e/an797e00.pdf> [accessed 12 May 2024]; Ken Fern, 'Magnolia Mexicana', Useful Tropical Plants, 2014 <https://tropical.theferns.info/viewtropical.php?id=Magnolia+mexicana> [accessed 12 May 2024].
20. de la Cruz, pp. 59–60, 98, 104.
21. de la Cruz, p. 128.

22 FAO, p. 212.
23 de la Cruz, p. 125.
24 Sahagún, Book 11, f181r.
25 Sahagún, Book 11, f195r.
26 de la Cruz, p. 104.
27 Jim Conrad, 'Excerpts from Jim Conrad's Naturalist Newsletter', Backyard Nature, 2019 <https://www.backyardnature.net/yucatan/piper-am.htm> [accessed 12 May 2024]; M. Mullallya and others, 'Anxiolytic activity and active principles of *Piper amalago* (*Piperaceae*), a Medicinal Plant Used by the Q'eqchi' Maya to Treat Susto, a Culture-Bound Illness', *Journal of Ethnopharmacology*, 185 (2016), pp. 147–54.
28 'ynn opa Techichilco nomati Mecaxochitl catqui' (Techichilco, 1568) in Nahuatl Dictionary <https://nahuatl.wired-humanities.org/content/mecaxochitl> [accessed 12 May 2024]; Gates, in de la Cruz, p. 133.
29 S.M. Cruz and others, 'Chemical Diversity of Essential Oils of 15 Piper Species from Guatemala', *Acta Horticulturae*, 964 (2012), pp. 39–46, DOI:10.17660/ActaHortic.2012.964.4
30 Varey, p. 146.
31 Sahagún, Book 11, 188v.
32 Sahagún, Book 11, 196v.
33 Varey, p. 146.
34 T.M. Zennie and J.M. Cassady, 'Funebradiol, a New Pyrrole Lactone Alkaloid from *Quararibea funebris* Flowers', *Journal of Natural Products*, 53.6 (1990), pp. 1611–14, DOI:10.1021/np50072a040; Gloria Melisa González-Anduaga and others, 'Micro-Morphology Characterization and HS-SPME-GC-MS Analysis of Floral Parts of *Quararibea funebris* (La Llave) Vischer, Traditionally Known as Rosita de Cacao', Chemistry and Biodiversity, 21.2 (2024), e202301709, DOI:10.1002/cbdv.202301709; Gloria Melisa González-Anduaga and others, 'Nutraceutical Potential, and Antioxidant and Antibacterial Properties of *Quararibea funebris* Flowers', *Food Chemistry*, 411 (2023), 135529, DOI:10.1016/j.foodchem.2023.135529.
35 Varey, p. 146.
36 Molina 1571, part 2, Nahuatl to Spanish, f. 148r. col. 1; Karttunen, p. 62.
37 Sahagún, Book 11, 195r.
38 Anne S. Bertelsen and others, 'Cross-modal Effect of Vanilla Aroma on Sweetness of Different Sweeteners among Chinese and Danish Consumers', *Food Quality and Preference*, 87 (2021), p. 104036.
39 René Acuña (ed.), *Bocabulario de Maya Than de Viena*, (México: Instituto de Investigaciones Filológicas (Universidad Nacional Autónoma de México, 1993)).
40 Sahagún, Book 10, folio 107v.
41 Sahagún, Book 11, f13r.
42 Sahagún, Book 4, 47r, Book 8, f49r.
43 Sahagún, Book 12, 24v, Nahuatl.
44 Paul Carpenter Standley and others, *Flora of Guatemala* (Chicago Natural History Museum, 1946).
45 Safford, 1912.
46 Standley and others.
47 Lilia Danae Arteaga Rios and others, '*Comparación Molecular y Morfológica entre Ejemplares de Magnolia Mexicana "Yoloxóchitl"* (*Magnoliaceae*) *del Estado de México y Veracruz, Instituto Politécnico Nacional*, 4 (2020), pp. 107–24. DOI:10.18387/polibotanica.49.7.
48 Theodore H. Fleming, 'Fecundity, Fruiting Pattern, and Seed Dispersal in *Piper amalago* (*Piperaceae*), a Bat-Dispersed Tropical Shrub', *Oecologia*, 51 (1981), pp. 42–46, DOI:10.1007/BF00344650.
49 Laura Caso Barrera and Mario Aliphat Fernández, 'Cacao, Vanilla and Annatto: Three Production and Exchange Systems in the Southern Maya Lowlands, XVI-XVII Centuries', *Journal of Latin American Geography*, 5.2 (2006), pp. 29–52 (p. 46) <http://www.jstor.org/stable/25765138> [accessed 12 May 2024].
50 Charlotte Watteyns and others, 'Vanilla Distribution Modeling for Conservation and Sustainable Cultivation in a Joint Land Sparing/Sharing Concept', *Ecosphere*, 11.3 (2020), e03056, DOI:10.1002/ecs2.3056.
51 Philipp M. Schlüter, Miguel A. Soto Arenas, and Stephen A. Harris, 'Genetic Variation in *Vanilla planifolia* (*Orchidaceae*)', *Economic Botany*, 61.4 (2007), pp. 328–36, DOI:10.1663/0013-0001(2007)61[328:GVIVPO]2.0.CO.

32
Bringing the Kitchen Garden into the Kitchen
The Research and Writing of Three Inspirational Women

Barbara Segall

Women are not centre stage in many disciplines, and even in horticulture and food history many have lacked appropriate recognition. My paper focuses on three women whose writing and research into kitchen gardens, salads and oriental vegetables, fruit, growing techniques, and cooking is important and inspirational to gardeners and cooks. These three women, all British (two now in their eighties and nineties and one deceased), have been at the fore in their chosen fields but are not as well known to a wider and younger audience as they should be. Here I look at their importance to gardening and food history in general, and in particular to my own garden and food writing and to my enjoyment of fruit, vegetables, and herbs.

All three have researched foods or techniques that were once abundant and commonplace, had declined, or were previously unknown to gardeners in the UK and through their work have helped bring them back into use or aided their reintroduction.

My trio of inspirational plot-to-plate, pre-social media influencers are, in alphabetical order, Susan Campbell, Joy Larkcom, and Joan Morgan. I know or knew them personally (Susan died on 2 January 2024). My research is based on personal interviews and material in various archives and libraries. I have read the women's published works and followed their suggestions and advice over the years as a reader, gardener, and cook.

Susan Campbell
I first encountered the work of Susan Campbell when I moved into a bedsitter in Kensington, London soon after arriving in England from South Africa. *Poor Cook*, published in 1971, by Susan Campbell and Caroline Conran was the first cookbook I purchased in my new life in the UK.[1] The illustrations, all Susan's work, are bold line drawings of cuts of meat, fish, poultry, and various kitchen tools and implements. My flatmate and I fed ourselves and our friends using Susan's frugal, imaginative recipes and found that the book certainly lived up to its subtitle: *Fabulous Food for Next to Nothing*.

My flatmate and I annotated the recipes with ticks and comments such as 'VG', 'Good', 'Will do again', or 'Delicious but too much vinegar'. I wrote 'Good' against Beef Goulash and Chilli con Carne which tells me that this was probably the first time that I had cooked them for myself. When I moved out, I took that book with me.

A Slade-trained artist who grew fruit, vegetables, and herbs, Susan wrote several other food-related books. In 1981, she suggested to her publisher that she research and write about walled kitchen gardens after her interest was piqued by visiting Thomas Pakenham's Tullynally Castle in Ireland, where she saw a working example.[2] These distinctive, productive spaces had fallen on hard times due to high maintenance costs and a lack of skilled workers. Over the next four decades Susan visited over 700 walled kitchen gardens in the UK, keeping detailed typed research notes in a series of lever-arch files. She also produced delicate and evocative line drawings to accompany her books.

Susan realized that the horticulture encompassed by these gardens was in danger of dying out and should be recorded, celebrated, preserved, and kept going. Her 1983 book *A Calendar of Gardeners' Lore*, filled with colourful anecdotes, drew the attention of the BBC and Peter Thoday, a practical horticulturist and academic, who was searching for a suitable garden for a television programme on walled kitchen gardens. Susan noted:

> I couldn't help them [the BBC] although I had already heard about the walled kitchen garden at Cottesbrooke Hall in Northampton, but it was too far east to be workable for the Bristol-based programme, which was eventually filmed at Chilton Lodge, Berkshire.[3]

The Victorian Kitchen Garden, which aired in 1987, was presented by former head gardener Harry Dodson.[4] The book to accompany the series was written by Jennifer Davies.[5] Susan's own book on life in the kitchen garden of Cottesbrooke was published in 1987.[6]

Prior to this series, the number of visitable kitchen gardens in the UK was very low, with only two working ones belonging to the National Trust. Susan told me:

> It is most gratifying to see how many gardens are now back in production and part of an international community of kitchen gardeners, so I feel my enthusiasm has not been in vain. I cannot really account for my interest in the subject. My father was a keen fruit and vegetable grower so maybe this had something to do with it.[7]

In 2001, following a meeting at Hellens Manor, Hertfordshire, Susan and the late garden historian Fiona Grant established the Walled Kitchen Gardens Network (WKGN). Grant insisted that the organization should be online, so anyone could sign up, free of charge, be involved in discussions, and ask for help with research and so on.[8] Today the WKGN numbers almost 500 members and comprises national organizations and individual experts. It holds annual symposia and forums in the UK and elsewhere. Its website (walledgardens.net) boasts a comprehensive list of walled kitchen gardens.

I included Susan's coastal garden in my 2002 book *Gardens by the Sea*.[9] Unfortunately I was not able to interview Susan myself, so a garden writer colleague, Barbara Abbs, wrote the chapter. Ever since then I longed to see the garden, a dream that was fulfilled in April 2019. Sea kale was one of the reasons I was especially drawn to visit Susan. My love of seaside gardens meant I had spent many happy hours combing the shingle beach of Aldeburgh on the Suffolk coast, where I found sea kale growing wild. However, I did not know how to use it, nor how delicious it could be. Susan grew it by forcing it in spring using a black builder's bucket on top of each plant, with a heavy stone to keep the bucket in place during winter storms, which produced beautifully coloured elongated stems and foliage.

My visit coincided with the height of the forced sea kale season, and Susan cut some for our lunch. Showing me how to prepare it and the best way to cook it, she steamed it lightly and served it with butter, and she also recommended sautéeing it in lemon and butter. I was sent away with three sea kale 'thongs' to plant at home. One has survived and will soon be joined by others given to me by Mike Kleyn, Susan's widower, when I visited in April 2024. Mike, a pharmacist by profession and a gardener by inclination, collaborated with Susan on several kitchen garden consultancies, including one at Althorp, Northamptonshire.[10]

Susan's last book, *The Garden Diary of Dr Darwin: 1838–1865* (2021), was a long time in coming.[11] Susan chanced upon the unknown diary of Charles Darwin's father Robert in 1986 and spent thirty-five years researching its background – testament to her meticulous attention to detail and accuracy. A fitting conclusion to Susan's work on kitchen gardens, the book describes the horticultural and domestic activities at The Mount, a large house with extensive gardens and pastures on the banks of the River Severn in Shrewsbury, home of the Darwin family from 1800 to 1866. Apart from revealing that Doctor Darwin made his garden available for several of his son's early horticultural experiments, it describes the plants that grew there and the animals the family raised, as well as the social interaction between the Darwins and their neighbours.

In 2023 Susan was awarded the Royal Horticultural Society (RHS) Veitch Memorial Medal for her outstanding contribution to advancing the science and practice of horticulture.

Joy Larkcom

I met Joy Larkcom, a fellow horticulturist and garden writer, in the 1980s when I moved to Suffolk. She lived forty miles away at Montrose Farm, near Diss where she created an ornamental kitchen garden that looked – and was – good enough to eat.

Joy graduated from Wye College in 1957 with a degree in horticulture, but it was sixteen years later that she combined horticulture and journalism. Frank Ward, editor of *Garden News*, offered her a regular column, in which she wrote about her kitchen garden and how she used it to feed her family.[12]

When researching her first book, *Vegetables from Small Gardens* (1976), Joy found references to intensive systems of growing vegetables that were widespread in Europe.[13] A note in her archive states that she became 'obsessed with introducing new methods of growing and refreshing old gardening techniques aimed at increasing productivity, that could also be sustainable, especially for gardeners with small spaces'.[14] She also noted that collecting the seed of old and local varieties and preserving them in a vegetable gene bank were essential: 'I wanted to find and introduce little known and undervalued edible plants [. . .] and then to persuade seedsmen to make them available commercially and to domestic gardeners.' Additionally, she wanted to persuade people that kitchen gardens could be beautiful: 'The concept that flowers, vegetables, fruit and herbs could be grown together to make the most of their intrinsic ornament, as well as their uses, was gaining ground.'

These factors fuelled her curiosity, leading her in 1976 to undertake a twelve-month 'Grand Vegetable Tour' across Europe with her husband Don Pollard and their two young children, travelling in a campervan. They visited eight countries and brought back much from which gardeners have since benefited.

Joy's main 'discoveries' were salad plants that had been forgotten or were unknown in the UK, along with the practice of intensive growing techniques, in particular 'cut-and-come-again'. This allows for individual leaf harvests and the continuation of the crop's growth for further regular harvests, which Joy 'first saw [...] in action at Jelena de Belder's arboretum in Kalmthout, Belgium'.[15] Other techniques included intercropping, planting in narrow beds, mulching, and the biological control of pests and disease.

Among the seed mixes that Joy introduced was a mixed leaf salad known as 'misticanza' in Italy or 'mezclun' in France, which she found in Italy. Having misheard the name, Joy always referred to it as 'saladini', rather than 'insalatine', which means little salad leaves. Saladini stuck. Her friends, John and Caroline Stevens, herb growers and founders of Suffolk Herbs, worked with her to produce enough to sell through their catalogue. Later, when she and the Stevens put together a similar mixed pack of oriental vegetable leaves, they called it 'Oriental Saladini'.[16] Another seed triumph was the purple-pink frilly lettuce 'Lolla Rossa' and the colourful beetroot 'Chioggia' with its pink and white interior circles.[17] At her own farm, Joy grew herb and salad leaf mixes for local restaurants, always noting how the flavours combined. Soon she was working with growers to produce the mixes at scale. This led to supermarkets following suit with the pillow packs of mixed leaves that we now all take for granted.

Joy's creativity in arranging the potager or kitchen garden attracted me to her gardening columns. The notion of 'edible landscaping' promoted by American gardeners Robert Kourik and Rosalind Creasy was gaining ground, so I was fortunate to find Joy's work in print and in person nearer to home.

In 1979 Joy met Rosemary Verey, then the *'grande dame'* of ornamental vegetable growing in the UK. They both contributed to the exhibition *The Garden: A Celebration of One Thousand Years of British Gardening* held at London's Victoria & Albert Museum. Rosemary invited Joy to use her comprehensive library, an invitation she took up often. According to Joy, it was there that she:

> found John Evelyn (*Acetaria*, 1699) and Batty Langley (*New Principles of Gardening*, 1728), as well as works of other writers of the 16, 17 and 18th centuries. I read the lengthy lists of salads that Evelyn produced at Sayes Court. I noted how Batty Langley divided salads into 'hot and biting' and 'mild and insipid' and I realised that we had only re-discovered what our ancestors had been doing centuries before.[18]

Joy's journalism was prolific and entertaining. One article in 1988 in *Practical Gardening* magazine entitled 'Slugduggery' was written from the slug's point of view. The slug could congratulate itself on living on ground that was organically run by 'that woman' – Joy – and its only fear was of her obsessive late-night, torch-lit forays to collect and deal with slugs![19]

Joy realized that there were many Asian vegetables that could be of use to western gardeners because they were fast-growing and could be sown in polytunnels and under fleece, after spring crops had been harvested in late summer, a time when there was often empty ground available. They were also nutritious and easy to use in salads and to steam or stir-fry. They had a range of spicy flavours and attractive foliage, could be grown in small areas, and were often cut-and-come-again crops.[20] In the 1970s Joy began experimental plantings in her Suffolk kitchen garden using Asian vegetable seed from Chiltern Seeds, one of the few UK companies to source it from Japan.[21]

She felt that the key to getting more gardeners to grow these plants was to widen access to the seed pool. She spent five years visiting China, Japan, and Taiwan. As a child she had spent two and a half years in China, and her grasp of the language, improved by lessons in Mandarin focusing on horticultural terms, enabled her to communicate her queries to her contacts.[22]

Joy also visited the US and Canada, where she contacted communities of Chinese, Korean, Vietnamese, and Thai people. For them, access to seed was usually through their families in their former homelands. These trips gave her detailed information about growing techniques: 'The intensity of production even in the smallest growing space was remarkable. No space was ever wasted and no part of the veg was wasted either. The produce was used at all stages: raw, stir fried and pickled.'[23]

Following her visits to China and Japan, Joy initiated links between overseas and UK seed companies, and these seeds are now catalogue staples. Among them are the packs of mixed oriental leaves sold as 'Oriental Saladini' through Suffolk Herbs and purple lab-lab beans, winter melon, and the 'Beauty Heart' radish. Joy also worked with Dobies Seeds in a joint enterprise to produce a 'Stir Fry' seed collection. Another monumental task that Joy embarked on, before the internet became a key tool for writers, was the preparation of a list of seed companies worldwide who supplied Asian seed to the amateur market.[24]

In 2002 Joy and her late husband Don (who died in 2017) moved to County Cork, Ireland, where they created an exciting, productive garden on a slope, gathering seaweed from the coast to fertilize it. Irish garden writer Jane Powers noted that 'a "normal" couple in their seventies "retiring" to Ireland might have chosen a less challenging site to make a garden from scratch, but Joy was raring to go on her sloping, stony, exposed half-acre a stone's throw from the Atlantic Ocean'. Powers asked Joy about the soil:

> I was dying to go around it with a fork, but my husband stopped me. There is good farmland around. But it's very stony near the house – may have been an old building there. Very near the sea. Mild. But very exposed on the SW which is where I believe the hurricane winds come from.

According to Powers, the goings-on in Joy and Don's garden were of great interest locally, especially the zig-zag windbreaks that Joy invented to diffuse the dreadful south westerlies and the more damaging easterly winds.[25]

Joy credits Don as the cook of the kitchen gardening life. It was Don who put the delicious produce on the table, and she acknowledges that much of her work would not have been possible without him.

I grow the lettuce 'Lolla Rosso' and mixed salad leaves, and Joy's cut-and-come-again growing method suits my space and daily needs. My small-town garden is a mix of productive and ornamental plants, not quite in the ordered, yet exuberant, style of Joy's Montrose Farm potager, nor her Irish coastal kitchen garden, but it is nonetheless a homage to her and her work. I also seek out slugs by torchlight!

Joy Larkcom was awarded the RHS Veitch Memorial Medal in 1993 and a Lifetime Achievement Award by the Garden Media Guild in 2003. In 2000 she was photographed by Tessa Traeger, for the exhibition *Escape to Eden (Five Centuries of Women Gardeners)* at London's National Portrait Gallery.

Joan Morgan

Joan Morgan worked as a research biochemist, science writer, and editor, establishing *Trends in Biochemical Sciences*, the first of what is now a family of journals. Her interest in fruit took her into the history of fruit and gardening. Now she is described as a pomologist and fruit historian and is the author of many books on fruit. She explains her interest:

> I was brought up in the countryside on a small farm in South Wales and we had orchards. Everybody had orchards, and my mother made ours (harvests) into wonderful apple tarts and produce. I started wondering about where all these apples had gone. Living in London, you could buy a reasonable range of apples, but then they gradually seemed to be getting less. So I thought I'd try and investigate this.[26]

Harry Baker, who was the Fruit Officer at the Royal Horticultural Society's garden at Wisley in the early 1980s, gave Joan free rein to sample apples, which she then wrote about: 'I looked into the dimensions of their flavour and the different culinary properties of cooking apples.'[27] Her articles were, in the main, published in the RHS magazine *The Garden*.

It was apparent to Joan that there were hundreds of apples, with an enormous diversity of flavour, that were slipping into oblivion. The situation was similar with cooking apples. Like many of her contemporaries, Joan was an enthusiast of the cookery writer Elizabeth David, but she was dismayed that David wrote in 1960 in *French Provincial Cooking* (1960):

> I have never very greatly appreciated cooked apple dishes, but from the French I learned two valuable lessons about them. First choose hard sweet apples whenever possible instead of the sour cooking variety which are used for English apple dishes.[28]

This claim was followed in 1962 in an article David wrote for *The Spectator* entitled 'Big Bad Bramleys': 'I find them too large, too sour, too collapsible [. . . .] T]here is no more chilling dish in the whole repertory of English cooking than those baked apples in their macintosh skins [. . .]. Because of the way they disintegrate, Bramleys are of very little use for the kind of apple dishes which go wonderfully well with pheasant and other game.'[29]

These statements piqued Joan's interest, and she began her experiments. Joan explained:

> In England we differentiate more than any other cuisine between cooking or culinary apples and dessert or eating apples. So I began experimenting, exploring the different properties of all the apples at Wisley. I tasted them all fresh and rated them. For the culinary ones I cooked them all and devised a way to quantify the results.[30]

She made small, sealed parcels of silver foil and placed them on a metal tray, to imitate the process of baking an apple: 'You got the idea of whether it broke down into a nice purée or whether it held its shape. And I just did them all for a standard temperature, standard time, you know, just something very basic.'

In 1980 Joan wrote an article, 'A Cooker for All Seasons', for *The Garden*, in which she described the different qualities of culinary apples, and followed that in 1982 with 'Vintage Apples', on the flavours of eating apples.[31] She observed, 'Various connoisseurs argued as passionately over their apples as their claret. Meticulously peeling them with bone-bladed knives, they reminisced about varieties and season with as much relish and in as much detail as they discussed chateaux and vintages.'[32] She noted that literature about wine had grown in scale but that most individuals had little knowledge of the flavours of apples. Joan herself used some of the language of wine-tasting, describing tastes of strawberries, aniseed, nuts, and honey.

In November 1983, Joan staged an exhibit at the Apple and Pear Exhibition and Conference in London and wrote a piece celebrating the centenary of the historic Apple and Pear Congress of 1883.[33] This work coincided with a BBC Radio 3 broadcast entitled *In Praise of Older Apples*.

Joan's interest grew to cover the market elements of food production. The main focus of this research was Victorian head gardeners, which led her to nineteenth-century gardening magazines such as *The Gardener's Chronicle* and *The Journal of Horticulture*. Most of the authors of the articles in these publications were head gardeners, who wrote with authority and knowledge.[34]

Early in 1986 a three-part programme, *A Paradise Out of a Common Field*, which Joan developed with radio producer Alison Richards, was aired on Radio 3. The series led to the publication in 1990 of a book of the same title, co-authored with Richards, looking at the importance of head gardeners in the Victorian era, and telling the story of gardening and how this changed through time.

Joan also contributed a paper entitled 'The Diversity of Flavours of the Apple' to the Oxford Food Symposium of 1987 on Taste.[35] This formed the foundation of *The Book of Apples* (1993).[36]

While studying the era of growing fruit to perfection, Joan began to focus on top fruit, particularly apples and pears, with a book on cherries now in process. In the preparation for her books on apples and pears, as well as for her cherry book, Joan has done the heavy lifting for us all, tasting and describing hundreds of varieties. As she explains:

> I ate them all, made tasting notes on them all, and then cooked all the relevant ones. It all falls into place once you dive into it because you realise it was based on what the Victorians wanted to eat. They wanted beautiful apples to serve as the finale to any dinner. At that time all head gardeners had to be good at fruit, but some were beyond excellent growing pineapples, forcing grapes out of season, and providing early strawberries for the table.[37]

According to Joan, the main emphasis was on fruit for eating fresh, as the finale to a meal, but there was also a section of culinary varieties, especially apples:

> The Victorians wanted apples that were also good for the kitchen, used to make apple sauce or to bake as hot puddings. These apples break down into a nice sharp purée, that you can make into apple sauce, so that you can cut the acidity of pork or goose or whatever you are serving them with.[38]

It is fascinating to realize that so much of today's horticulture dates from this period and references the skills of the head gardener: bedding plants, our enduring love of pot plants, and our reliance on the fruit and vegetable varieties that they tested and popularized.

In 1983 Joan moved near Brogdale, Kent, then a Ministry of Agriculture Experimental Horticultural Station called the National Fruit Trials. She was given special access to its large fruit collection to continue her research and when the Station closed in 1990 became involved in its rescue. She was a trustee of Brogdale and a tireless campaigner for its future, which was in doubt again ten years later.[39]

Brogdale opened the National Fruit Collection to the public in 1990. Joan made wonderful fruit displays for the annual Apple Days held in October: 'I used to do a big display with one half showing the dessert fruit for the dining table and the other half the culinary fruit for the kitchens.' She also traced the history of the National Fruit Collections in an article in the *Occasional Papers from the RHS Lindley Library* in March 2012.[40]

Joan applied her usual rigour in writing *The Book of Pears*, a history and guide to over 500 varieties of the fruit that is 'gold to the apple's silver'.[41] She describes a utopia where we would all appreciate the 'luscious texture, boudoir perfumes and richness of taste' of pears. Her writing also takes readers to California, Iran, Syria, and South Africa, where she talked to members of the communities who have played a role in the pear's diversity. *The Book of Pears* won the Garden Media Guild Awards Reference Book of the Year 2016 and was also the Guild of Food Writers Food Book of the Year 2016.

In the same year Joan received the BBC Food & Farming Award for Outstanding Achievement. Earlier, in 2009, she was awarded the RHS's Veitch Memorial Medal. Now a long-term member of the RHS Fruit, Vegetable, and Herb Committee, she was the first woman invited to be a committee member of the then Fruit and Vegetable Committee in 1991. She holds the Chartered Institute of Horticulture Award for excellence in horticulture (1996) and is the expert behind the Fruit Forum website (fruitforum.net) and blog (fruitforum.wordpress.com), as well as thebookofpears.fruitforum.net, which holds photographs of all the pears described in her book. She is an Honorary Freeman of the Worshipful Company of Fruiterers.

I intend to seek out samples of 'Golden Noble' noted by Edward Bunyard, one of Joan's revered apple expert predecessors, as fulfilling all the criteria for an apple pie: 'Golden before and after cooking [. . .] and every way delectable.'[42] Joan notes that the point about 'Golden Noble' is that 'not only is it beautiful – large round and golden – but it is sharp with a distinctive flavour of its own that will stand on its own. No need of cloves or lemon peel!'

Conclusion
My trio of food and garden influencers, Susan Campbell, Joy Larkcom and Joan Morgan, has led me along one of my life's most enjoyable paths: reading and writing about, growing and eating flavoursome and attractive fruit, veg and herbs. Through their detailed research, practical experience, engaging writing and total immersion in their subject matter they have made my plot-to-plate transition a culinary and horticultural delight.

About the Author
Barbara Segall is a professional horticulturist, garden writer, editor, and author of thirteen books about gardens, herbs, festive plants, and coastal gardens. In 2023 she was the recipient of the Garden Media Guild's Outstanding Contribution Award.

Acknowledgements
With thanks to the Garden Museum Archive, Susan Burton, Voltaire Cang, Mary Margaret Chappell, Naomi Clifford, Simon Edge, Tanya Hamilton, Jacqui Hurst, Mike Kleyn (and Susan Campbell), Joy Larkcom, Joan Morgan, Jane Powers, Adrian Sellars and Jane Steward.

Notes
1. Susan Campbell and Caroline Conran, *Poor Cook* (Macmillan, 1971).
2. 'Susan Campbell', *The Daily Telegraph*, 16 January 2024.
3. Susan Campbell, email interview with the author, 7 November 2023.
4. *The Victorian Kitchen Garden*, BBC2, 16 September–9 December 1987; IMDB <https://www.imdb.com/title/tt0482150/> [accessed 25 August 2024].
5. Jennifer Davies, *The Victorian Kitchen Garden* (Norton, 1988).
6. Susan Campbell, *Cottesbrooke: An English Kitchen Garden* (Ebury Press, 1987).
7. Susan Campbell, email interview with the author, 7 November 2023.
8. Lucy Pitman, co-chair of WKGN, email and visual material, 20 March 2024.
9. Barbara Segall, *Gardens by the Sea* (Frances Lincoln, 2002).
10. Mike Kleyn, conversation with the author, April 2024.
11. Susan Campbell, *The Garden Diary of Dr Darwin: 1838–1865* (Unicorn, 2021).
12. Joy Larkcom, *Just Vegetating* (Frances Lincoln, 2012), p. 9.

13 Joy Larkcom, *Vegetables from Small Gardens* (Faber, 1976).
14 Joy Larkcom, notes on *Travellers' Tales*, a screen recording held in the Joy Larkcom Archive, sent to the author by Larkcom via email, 23 November 2023. The Joy Larkcom archive is held at the Garden Museum, London.
15 Joy Larkcom, email to the author, 12 March 2024.
16 Joy Larkcom, '*Seed Files Intro* for Joy Larkcom Archive', note sent to the author by Larkcom via email, 19 February 2024.
17 Kitty Scully, interview with Joy Larkcom, Garden Masterclass <https://www.gardenmasterclass.org> [accessed 25 August 2024].
18 Larkcom, email, 12 March 2024.
19 Joy Larkcom, 'Slugduggery', *Practical Gardening*, March 1988, rpt. in Larkcom, *Just Vegetating*, pp. 144–46.
20 Joy Larkcom, notes on *A Wokful of Vegetables*, a screen recording held in the Joy Larkcom Archive, sent to the author by Joy Larkcom via email, 23 November 2023.
21 Joy Larkcom, '*Seed Files Intro* for Joy Larkcom Archive', note sent to the author by Joy Larkcom via email, 19 February 2024.
22 Larkcom, '*Seed Files Intro*'.
23 Larkcom, '*Seed Files Intro*'.
24 Larkcom, '*Seed Files Intro*'.
25 Jane Powers, email to the author, 25 March 2024.
26 Joan Morgan, telephone interview with the author, 13 December 2023.
27 Morgan, telephone interview, 13 December 2023.
28 Elizabeth David, *French Provincial Cooking* (Michael Joseph, 1960), p. 509.
29 Elizabeth David, 'Big Bad Bramleys', *The Spectator*, 26 October 1962, p. 41.
30 Joan Morgan, telephone interview with the author, 9 May 2024.
31 Joan Morgan, 'A Cooker for All Seasons', *The Garden*, 105 (1980), pp. 435–39; Joan Morgan, 'Vintage Apples', *The Garden*, 107 (1982), pp. 308–13.
32 Morgan, 'Vintage Apples', p. 308.
33 Joan Morgan, 'A Historic Centenary', *The Garden*, 108 (1983), pp. 383–88; 'A Hundred Years On', *The Garden*, 109 (1984), pp. 148–50.
34 Morgan, telephone interview, 13 December 2023.
35 Joan Morgan, 'The Diversity of Flavours of the Apple', in *Taste: Proceedings of the Oxford Symposium on Food and Cookery, 1987*, ed. by Tom Jaine (London: Prospect Books, 1988), pp. 162–64.
36 Joan Morgan, *The Book of Apples* (London: Ebury Press, 1993); *The New Book of Apples* (Ebury Press, 2002).
37 Joan Morgan, email to the author, 9 May 2024.
38 Joan Morgan, Alison Richards, *A Paradise Out of a Common Field: The Pleasures and Plenty of The Victorian Garden* (Century, 1990), p. 227.
39 Morgan, email.
40 Joan Morgan, 'Orchard Archives: The National Fruit Collection', *Occasional Papers from the RHS Lindley Library*, 7 (March 2012), pp. 3–30 <https://www.rhs.org.uk/about-us/pdfs/publications/lindley-library-occasional-papers/volume-7-march-2012.pdf> [accessed 25 August 2024].
41 Joan Morgan, *The Book of Pears* (Ebury Press, 2015).
42 Joan Morgan, 'Edward Bunyard the Epicurean Nurseryman', in *The Downright Epicure: Essays on Edward Bunyard*, ed. by Edward Wilson (Prospect Books, 2007), pp. 189–229 (p. 228).

33
Welcome to Utopia! Hungry?

Laura Shapiro

Once upon a time – but no, we can do better than that, it was right around June 1889 – a British writer named Elizabeth Corbett was leafing through magazines in her study when she fell into a deep sleep. How long did she sleep? It was hard to tell; all she could say afterwards was that when she woke up, she was very, very far from home. 'I was in a glorious garden, gay with brilliant hued flowers, the fragrance of which filled the air with a subtle and delicate perfume; around me were trees laden with luscious fruits,' she recalled in her account of the journey. The fruit looked familiar, reminding her of the apples, pears, and quinces she'd eaten all her life – 'only they were as much finer than the fruits I had hitherto been familiar with as Ribstone pippins are to crabs, and as jargonelles are to greenbacks.' (In 1889 these would have been vivid comparisons: Ribstone pippins figured among the best apples, and jargonelles among the best pears). Tasting these mysterious fruits she found them 'marvellously sweet and luscious'. In fact all the food was delicious, as she would soon learn when the locals invited her to dinner; though somewhat to her surprise, none of the 'dainty dishes' on the table contained meat. 'I resolved to discover later whether such a strange omission was of regular or only occasional occurrence.'[1]

Utopia – for this, of course, was where she had woken up – did tend to be vegetarian in those days, especially when these richly imagined communities were dreamed up by women, and very often they were. During the late nineteenth and early twentieth centuries, Utopia was one of the most popular destinations in fiction, and feminist writers found this genre irresistible. They loved devising countries like New Amazonia, the place where Corbett's narrator found herself – countries where women flourished in every aspect of public and private life, and the very notion of male superiority was laughable. In fact, the reason we can date Corbett's momentous nap so precisely is that she described in a prologue how she'd fallen asleep after reading a petition headed 'An Appeal Against Female Suffrage', which in real life did appear in the June 1889 issue of a magazine called *The Nineteenth Century*. The popular novelist Mrs Humphry Ward had circulated the petition, and dozens of socially prominent women signed their names in support. Corbett, herself a prolific novelist, went right to work. Her impassioned portrait of an ideal society ruled by feminists, a novella titled *New Amazonia: A Foretaste of the Future*, came out later that year.

Clearly more than a hint of Eden was hovering over New Amazonia, what with the 'glorious garden', and the trees, and the newcomer tempted by strange, enticing fruit. But Corbett quickly dispensed with the biblical template. Her time-traveller was greeted by an apparition so dazzling she hardly knew what she was looking at, until – '"It is a woman,"' she realized. Or possibly a goddess. At any rate, this female came from nobody's rib: it was as if the entire sex had been reinvented: 'She was close upon seven feet in height [. . .]. A magnified Venus, a glorified Hebe, a smiling Juno, were here all united in one perfect human being.'[2] Soon more of these resplendent females arrived, all with short, easy-to-manage hair and wearing graceful tunics with knee-length skirts. They were the very picture of freedom and confidence. If this was the future assigned to womankind, it was definitely an improvement over Genesis. No snake, no downfall, no lifelong sentence of pain and subservience. In Corbett's telling, anyone who took a bite of that tantalizing fruit was making a really good decision.

Corbett wasn't the only author with a fondness for deconstructing Eden. Many of the other women inventing female-centred civilizations made sure that 'luscious fruits' were on hand to give their time-travellers a taste, literally, of the world they're about to explore. In *Mizora: A World of Women*, a serial by Mary E. Bradley Lane that ran anonymously in a Cincinnati newspaper beginning in 1880, an intrepid woman set out to row across an unmapped Arctic Sea. She was caught up in a whirlpool, lost consciousness, and awakened to find herself in a 'land of enchantment', where 'birds of bright plumage' were fluttering through orchards, and 'the fragrance of tempting fruit' hung in the air. The inhabitants, lovely and courteous ladies, ushered her into a splendid dining room, where she saw 'wonderful fruit' on the table – 'clusters of grapes of a rich wine color, and clear as amethysts' as well as 'several varieties of plums, as large as hen's eggs, and transparent. They were yellow, blue and red'. She hesitated to eat this fruit, fearing it was too beautiful to be consumed – just touching it might be 'a sacrilege'. Then she saw everyone else helping themselves with 'pink tipped fingers' and decided she, too, could feast.[3]

Even male travellers were drawn to these mysterious fruits. The adventurer in the 1893 novel *Unveiling a Parallel*, by Alice Ilgenfritz Jones and Ella Merchant, climbed into his 'aeroplane' one day and flew to Mars. Soon after arriving he realized how hungry he was, and spied a tree with beautiful fruit hanging from the branches. A Martian astronomer – one of the friendly natives who have welcomed him – 'plucked a cluster of the large rich berries and gave them to me, first putting one in his own mouth to show me that it was a safe experiment'. The Martians spoke a language he couldn't understand, but the symbolism needed no translating. The berries were 'exceedingly refreshing', and he enjoyed them again at breakfast the next day.[4]

Not precisely along these lines, but charming in its own right, is the depiction of fruit in *Herland*, by Charlotte Perkins Gilman. By the time she published this novel in 1915, the omnipresence of wondrous fruit in Utopia had been well established, which was perhaps why she allowed herself to play with the idea a bit. Her travellers were three men who had heard about a possibly mythical land populated entirely by girls and women, and went off to find it. After considerable searching they came across a woodland with 'a very large and beautiful tree'. They glimpsed movement up among the leaves, they heard smothered laughter, and suddenly they could see beautiful young women perched upon the branches. '"Girls!" whispered Jeff [. . .] "Peaches!" added Terry, scarcely louder. "Peacherinos – Apricot-nectarines! Whew!"'[5]

Needless to say, these two louts did not flourish in Herland. It was a society organized around women, and for men to regard them as delicacies ripe for the picking was not only rude, it was pointless. After 2000 years on their own, nobody in Herland had the slightest need for a man: they'd been blessed with spontaneous parthenogenesis, and wouldn't know how to flirt even if they wanted to. The whole situation was confusing to the three men; they felt out of place for the first time in their lives. But – like everyone else who ended up in feminist utopia – they loved the food. Indeed, a theme throughout the genre was that when women run the world, the cuisine is universally pleasing. 'This repast with its new but delicious fruit, its dish of large rich-flavoured nuts, and its highly satisfactory little cakes, was most agreeable,' reported Van, the narrator. 'There was water to drink, and a hot beverage of a most pleasing quality, some preparation like cocoa.'[6]

The vagueness in this description is unsurprising, and not just because Van was male and couldn't be expected to have much insight into cuisine. A recurring feature of the food descriptions in feminist utopia was their lack of detail: it's rarely possible for a reader to imagine the flavours or textures of what people are eating. The fruits, as we've seen, consistently had an air of mystery about them. The cakes in *Herland* were 'highly satisfactory', the drink in *Mizora* was a 'sparkling' beverage resembling chocolate, the breakfast in *Unveiling a Parallel* featured 'some curiously prepared cereals', the tea tray in *The Day After Tomorrow* held 'delicate green leaves' and 'some creamy mixture', and the woman greatly enjoying her visit to the year 2905 in *As It May Be* simply confessed, 'Of course, I hadn't any idea what I was eating.'[7]

And yet, this indistinct food accomplished its work with stunning success. Every morsel was dense with nutritious properties – 'scientifically perfect', as the visitor to New Amazonia learned – hence all the females grew up to be impressively tall and well-proportioned.[8] They could outrun and outclimb any man, they were beautiful, they rarely got sick, and sometimes they could fly. In Mizora, they lived a century or longer but looked and felt radiantly young throughout their lives, never suffering 'decrepitude, wrinkles and imbecility'. The time-traveller learned why: it turned out that Mizoran chemists were able to extract food from the elements, hence 'earthy matter and impurities that are ever present in our food, were unknown in theirs'. It was also obvious that these highly intelligent Mizoran women made no effort to achieve hour-glass figures but rather strove to have large waists: 'Not one was less than thirty inches in circumference, and it was rare to meet with one that small.'[9] A good-sized middle was a sign of excellent lung capacity and greatly admired for that reason. Finally – and this is a feature that comes up again and again in accounts of feminist utopia – nobody ate between meals. They wouldn't dream of it. Each meal at its appointed hour was so satisfying that the temptation to nibble on snacks was non-existent. As for what we would call junk food – 'innutritive foods' was the designation in New Amazonia – there weren't any.[10] Since they supplied nothing useful to the body, they had been banished, and nobody missed them.

Remarkably, this powerhouse of a diet wasn't at all heavy or meat-centred. In fact, it was typically described as 'simple', 'light', and even 'dainty', which in that era was the term invariably applied to the sort of food nibbled at a ladies' lunch. The strong, athletic, independent women in these societies were thriving on meals that were unmistakably coded feminine – and the men were thriving, too, if there happened to be men at the table, for they put no value on machismo. Femininity, a trait usually involving frailty, submissiveness, and a dim-witted delight in second-class citizenship, had acquired radically different connotations. These were the wisest, most capable, most powerful women the world had ever seen, and they were feminine to the core. So were the men. And we can read it in the food.

What's notably absent from most accounts of dinner in Utopia is any mention of cooking. When it came to writing about the food, authors working in this genre weren't inspired by ingredients or recipes; they were inspired by infrastructure. They conjured speedy transportation systems from farm to city, and pristine public markets where standards of quality were strictly regulated and all prices were fair. Most important, nobody had to cook unless she wanted to, and, if she did, she could take it up as an honourable and well-paid profession. Food production took place in a central building that served all the families in a given neighbourhood. Meals were delivered on schedule, or perhaps a homemaker signalled for them, and the facilities bore no resemblance to the jumbled, germ-laden kitchens of the past. 'The walls of the cookery were covered with white tiles; the floor was white, the tables were immaculate, and the cooks and confectioners were spotlessly neat and clean,' reported the admiring narrator in *New Amazonia*. 'There was neither fuss, heat, nor discomfort, as is the case in England when a great deal of cooking has to be done, for the work was done systematically, and the greatest pains had been taken to make all the conditions of labour as pleasant as possible.'[11]

In short, these ideal civilizations weren't just feminist utopias – they were culinary utopias. Food had been tamed: it was no longer an instigator of guilt at the table or self-loathing in front of the mirror. And cooking, traditionally the measure of a woman's value as a wife and mother, indeed the measure of her very identity as female, had lost its iron grip on her daily life. For most homemakers at the turn of the twentieth century, even those with hired help, the question pounding at their consciousness every single day was what to feed the family and how to manage the work. But in culinary utopia there was never a hint of such stress, thanks to all the writers who recognized a good, indeed obvious, solution. They didn't bother trying to devise a slew of quick 'n' easy recipes. They just tossed the whole thing out – home cooking, that is, complete with the massive encrustation of sentiment it had acquired over the centuries. It was gone, just gone. And not a soul missed it. Take another look at those mysterious fruits, so important in this genre that they marked the very gateway to Utopia. They were a symbol of welcome and transformation, for sure, but they also symbolized exactly what they were – instantly accessible food. They grew on a tree; a moment later, they were on the table and everyone was eating. Appetizing food is important, nutritious food is important, but what's really important when 5:00 p.m. rolls around is effortless food. That's the dream, it's always been the dream, and culinary utopia zeroed right in on it.

Charlotte Perkins Gilman, an economist and socialist as well as a writer of fiction (her 1892 story *The Yellow Wallpaper* is a feminist classic), published four utopian novels, each with a somewhat different take on what would constitute an ideal world for women. But no matter how she decided to set up a futuristic society, first and foremost she got rid of the private kitchen. Homes, yes; families, yes; but no home cooking. '"One of these co-operative schemes?"' wondered John, the narrator in *Moving the Mountain*, who had just returned to America after being lost in Tibet for thirty years. During his absence, women had taken over the country and changed everything to their liking. No, his sister told him, those co-operative schemes never worked. Instead, a few women '"with a real business sense, and enough capital"' built a beautiful apartment block with nurseries, billiards, and a swimming pool, and included a '"Home Service Company,"' staffed by well-paid professionals who did the cleaning and cooking for all the apartments. The food was superb and much cheaper than when each household fended for itself. Soon the new system spread across the city, then the nation. John was dumbfounded – how could so much have been accomplished in just thirty years? Easy, said his sister: '"The women woke up."'[12]

Fresh, hot food, made available three times a day without a jot of existential stress for homemakers, remained a feature of feminist utopia for decades. '"Do you know what's in it?"' asked a sceptical dinner guest, eyeing the aluminium container that had just arrived with dinner in Gilman's 1910 novel *What Diantha Did*. '"No, thank goodness I don't,"' the hostess said cheerfully. She was delighted that a skilled cook somewhere else had planned her dinner, and that she could sit down like a guest at her own table, just as men had always done. As for the cuisine – the sceptical visitor took a cautious taste of the soup and nearly swooned with admiration: '"Why – why – it's like Paris," she said in an awed tone.'[13]

But a generation later, an even more radical change greeted visitors to Utopia. Not only was home cooking obsolete, so was food. Scientists had been able to reduce it to a vaporous state of pure nutrients, and eating was no longer necessary. In the 1930 story 'Into the 28th Century', by Lilith Lorraine (also known as Mary Maude Wright), a young man named Anthony got into a motor boat and unaccountably ended up eight centuries in the future. A wise elder called Therius welcomed him to Nirvania and was in the midst of explaining all the customs when he paused to suggest a break: '"Let us inhale, or as you would say, let's eat. Althus, bring our new comrade a flask."' Anthony started to drink from the 'jewel-encrusted flask of glittering beauty', but Therius quickly stopped him. '"Don't drink! Inhale it!"' he said. '"It's the essence of food."' And sure enough, with a single whiff 'my whole being was permeated by a delicious and seductive fragrance', and all sensations of hunger disappeared.[14]

But there was trouble in Nirvania, trouble caused by food. Despite the fact that society had achieved the ultimate in efficient nutrient-delivery systems, people were getting hungry between meals. They actually wanted to snack, possibly the first utopians to experience this urge. What's more, their favourite snack was fruit – wondrous fruit, the same fruit that had been enticing time-travellers since the start of the genre. '"We can't resist the temptation of biting into the luscious fruits that overburden our orchards,"' admitted Therius with a sigh. He was discouraged by the lack of discipline. And there was something worse afoot: people were going to parties and eagerly scarfing down '"the foamy concoctions and icy beverages that the ladies insist on serving"'. He was nervous about these foods – all of them blatantly identified as feminine – because of the dangers they posed to everyone's well-being, including a return to '"that ancient curse, the stomach ache"'.[15]

In other words, it's back to Eden, and not in a good way. The fruit, the wrong-headed ladies, and the ancient curse – for this we needed a trip to the 28th century? But Lilith Lorraine, whose politics combined Christianity with socialism, plainly had a weakness for traditional gender stereotypes and was trying her best to reinvent them as emblems of victory in the great struggle for equality between the sexes. In Nirvania, men were in charge of '"invention, mechanics, mathematics, and the more strenuous sports"'. while women were content with '"spiritual and intellectual guidance"'. No longer did they try to imitate men, she reported triumphantly; instead, women devoted themselves to protecting their '"delicacy"' and burnishing their '"femininity"'.[16]

It's the old femininity, of course, the kind wrapped up in a biblical symbol notorious across the ages. That fruit! That beautiful, tempting fruit! Back in 1889, it opened a gateway to feminism. Now it's just another apple of doom. I can see that I'm reading a moral into her story that the author never intended, and it's only fair to remember that Lilith Lorraine was writing in 1930 while living in Corpus Christi, Texas. Herbert Hoover was in the White House hoping the Depression would somehow solve itself, and, although women had been voting since 1920, little had changed in the way of rights and opportunities. Lorraine invented a land of '"true Socialism"' where disease, death, crime, and poverty have disappeared, along with '"the ancient curse of toil"'. In Nirvania, everybody spends time enjoying the arts and exploring '"the infinite"'.[17] Let's give her credit: it was certainly an improvement over Texas.

Lorraine died in 1967. I wish she'd lived long enough to eat in one of the pocket-sized culinary utopias that sprang up a few years later in the midst of second-wave feminism. Mother Courage, which opened in New York's Greenwich Village in 1972, was the first women's restaurant in the country; many more would follow. Dolores Alexander and Jill Ward, a couple of activists in the women's movement, knew nothing about running restaurants, but they decided New York needed this one. They raised money, renovated a filthy old diner, and came up with a menu true to 1970s New York – chicken curry, shrimp scampi, eggplant parmigiana, 'Chef's Cheesecake', and apple walnut cake, everything priced as affordably as they could manage.[18] Then they invited women and men to come and eat. Everybody showed up, and kept showing up for five years, at which point the founders were exhausted and had to close. But they had created a gathering place – a place where women could safely land after a long journey in search of the future. Men were always welcome, but it was women who used to say, as soon as they walked in the door, that they felt at home. I hope some of them ordered the apple walnut cake. The fruit would have been wondrous.

About the Author
Laura Shapiro is a journalist and culinary historian. Her most recent book is *What She Ate: Six Remarkable Women and the Food that Tells Their Stories.*

Notes

1. Elizabeth Burgoyne Corbett, *New Amazonia: A Foretaste of the Future* (Tower, 1889; repr. Mint Editions, 2021), pp. 13, 18.
2. Corbett, p. 14.
3. Mary E. Bradley Lane, *Mizora: A World of Women*, intro. by Joan Saberhagen (University of Nebraska Press, 1999), pp. 14, 18. First published as *Mizora: A Prophecy* (G.W. Dillingham, 1890).
4. Alice Ilgenfritz Jones and Ella Merchant, *Unveiling a Parallel* (Arena Publishing, 1893; repr. Syracuse University Press, 1991), pp. 2, 3.
5. Charlotte Perkins Gilman, *Herland*, in *Charlotte Perkins Gilman's Utopian Novels*, ed. by Minna Doskow (Fairleigh Dickinson, 1999), pp. 150–269 (p. 161). First published in *The Forerunner* (1915).
6. Gilman, *Herland*, p. 172.
7. Lane, p. 18; Jones and Merchant, p. 10; Cora Minnett, *The Day After Tomorrow* (F.V. White, 1911), p. 48; Bessie Story Rogers, *As It May Be* (Richard G. Badger, 1905), p. 23.
8. Corbett, p. 49.
9. Lane, pp. 19, 20.
10. Corbett, p. 50.
11. Corbett, p. 96.
12. Gilman, *Moving the Mountain*, in *Charlotte Perkins Gilman's Utopian Novels*, pp. 37–149 (p. 68).
13. Gilman, *What Diantha Did* (Charlton, 1910), p. 196.
14. Lilith Lorraine, 'Into the 28th Century', in *Sisters of Tomorrow: The First Women of Science Fiction*, ed. by Lisa Yaszek and Patrick B. Sharp (Wesleyan University Press, 2016), p. 116. First published in *Science Wonder Quarterly* (Winter 1930).
15. Lorraine, p. 116.
16. Lorraine, p. 122.
17. Lorraine, p. 122.
18. Alex D. Ketchum, *Ingredients for Revolution* (Concordia, 2022), p. 168.

34
Gardens, Markets, and Migrants

Jayeeta Sharma and Sarah Elton

Introduction

The springtime vegetable and herb seedlings for sale at the corner grocer opposite an east end Toronto transit station range from lettuce to tomato to mint to basil to arugula. At the start of every growing season, corner shops, garden stores, and supermarkets offer seedlings and, often, seeds. Which plants they sell depends on the neighbourhood. Customers looking for rosemary seedlings and *cicoria* seeds seek out Italian groceries. Those wanting to plant curry leaf, bitter melon, or amaranth seek out Bengali and Tamil stores. Those seeking fuzzy melon or *gai lan* head to a Chinese store. Less common plants sell out quickly. But, every summer, these ethnocultural plants become easier to buy.[1] The position of these ethnocultural seedlings in the retail foodscape is a tangible reminder of a dynamic interplay between home gardens and 'migrant marketplaces' in Toronto's translocal foodscape.[2]

Our paper explores the shifts between the past and present foodscapes of generations of Toronto's migrants.[3] We examine changes around the market gardening of ethnocultural vegetables, even as Toronto moved from a settler-colonial town of the British empire to a 'global city'.[4]

Since the 1800s, settlers from across the world moved to this imperial city carved out of the mixed forest in the Great Lakes Region of Turtle Island. They arrived with seeds and plants, with recipes, utensils, and ingredients. Plants partnered with settlers and collaborated to claim Indigenous land.[5] Cereal and vegetable plants, as well as fruit trees and vines, helped replace forests with fields, roads, villages, towns, and cities of a British colony that became Canada in 1867.

Generations of migrants have grown the city since the nineteenth century, and continue to arrive. Our paper takes a *longue-durée* perspective on Italians, Chinese, Tamil, and Bengali migrant gardeners and ethnocultural plants. Scholars have studied urban gardens as sites of emotional and sensorial experiences, market production and neoliberal values, and resistance to capitalist imperatives.[6] Such gardens do not exist in isolation. The intersectionalities of race, class, age, and gender help constitute them.[7] Our socio-ecological lens considers such gardens as sites of collaboration between human and non-human actors.[8] A food studies analysis views them as important components of culinary infrastructure, that is the 'artifacts, institutions, and media that are used to mobilise and organise food or to convey knowledge about food'.[9] These gardens are 'essential' spaces that exist in dialogue with migrant marketplaces, contributing to the city's culinary infrastructure, and to its foodscape.[10]

Gardens are integral to a food system that encompasses the material, social, and cultural assemblage that steward solar energy, water, plants, and animals into food. Bringing these gardens into the scope of culinary infrastructure helps capture the significance of seeds, soil, and fertilizer, and other inputs along with the embodied knowledge of how to save and nurture seeds and prepare plants to eat. Employing a 'culinary infrastructure' framework to explore the interconnections between home gardens and markets in creating Toronto's foodscapes, we validate actions and spaces often overlooked in food systems analysis.

Of all the foods cooked and served in the city, why focus on vegetables? Firstly, vegetables are central to the cuisines of the groups that we study. Freshly picked plant foods such as leafy greens are pillars of Italian, Chinese, Tamil, and Bengali home cooking. The home garden – the Italian *orta*, the Cantonese 菜園 or 後院菜園, the Tamil *veetu thottam*, the Bengali *bari* – are part of everyday life in those homelands. Many ethnocultural vegetables are preferred 'fresh', that is not frozen, not dried nor canned or preserved in salt.[11] To satisfy a desire for fresh produce, international supply chains connect export-focused vegetable fields in warmer climates with Toronto kitchens. The international trade in foodstuffs, with increasing amounts of fresh vegetables, has doubled between 1995 and 2018.[12]

Since the twentieth century, vegetables have been on the move, specifically in climate-controlled containers on trucks, ships, trains, and planes. Ethnocultural vegetables are big business. By 2016, the demand for ethnocultural varieties was worth over $800 million a year in Greater Toronto alone. The vegetables most in demand included bok choy, Chinese broccoli, *choy sum*, eggplant, snow pea, okra, eggplant, tomato, and bitter melon.[13] The desire for produce that satisfies ethnocultural tastes motivates both the growing of migrant gardens and the procurement and sale of certain fruits and vegetables. The simple act of selling home-grown ethnocultural vegetables and seedlings to local stores for distribution has long been part of urban culinary infrastructure that supports migrant foodways. Through a reading of that history as well as drawing on current qualitative research, we explore how four groups of migrants and their plants build culinary infrastructure.

Our Research

This paper draws on almost a decade of transdisciplinary research on urban gardens and food studies scholarship that we conducted separately, and lately, in tandem. Our mixed-method research journeys took us from archival explorations, key informant interviews, personal communications, to field-based studies, including multi-sited, digital, and multispecies ethnography and an agrobiodiversity garden survey. Our *longue-durée* telling of vegetable growing journeys connects the hyper-local food systems of backyard, market, allotment, kitchen, and home gardens with global histories of foodways, mobilities, cities, and empire. This is our first jointly-authored evolving contribution to the writing of more-than-human histories of gardens, seeds, vegetables, plants, gardens, and gardeners.

Market Gardeners and Produce Stores: Toronto's Italian and Chinese Vegetables

Italians have created homes and gardens in Toronto since the late nineteenth century. Between 1880 and 1915, millions of emigrants left the Italian peninsula after the economic impact of multi-faceted 'miseria'.[14] By 1911, 2200 men and 800 women of Italian descent lived in Toronto. A second, much larger wave arrived after the Second World War. Unlike immigrants from Britain or northern and eastern Europe, Italians were not viewed as potential Canadian farmers. In 1896, Clifford Sifton, Canada's Interior Minister, articulated a racial hierarchy that deemed Italians and Asians to be two groups that were unsuited to agricultural enterprise.[15] Italian migrants were regarded as too citified to grow food.[16] In reality, the vast majority were *contadini*, which loosely translates as peasant, or subsistence farmer.[17] They knew the challenge of producing food on limited land. When they migrated, low incomes and tradition encouraged them to continue to grow. They brought with them seeds and plants. At their request, kinfolk in Italy dispatched seeds in letters and small tree branches concealed in suitcases.[18]

For many Italian migrants, the Toronto garden acquired an essential economic function that went beyond its culinary and nutritional value. Early-twentieth-century city gardens greatly enhanced livelihoods. There was a close connection between the Italians in Toronto who tended gardens and those who sold produce. In 1907, Father Pisani of Vercelli noticed, 'Almost all the fruit and vegetable trade is exercised by Italians, mostly from Termini Imerese and Valledolmo (of the Palermo province of Sicily).'[19] The low land prices of outlying neighbourhoods such as Mount Dennis and Long Branch permitted Italians to rent or buy arable ground to cultivate.[20] Other market gardens grew up around the West Hill neighbourhood, currently part of the inner suburb of Scarborough, where, by the late twentieth century, Tamil migrants added gardens and stores.

By the mid-twentieth century, Italian migrants controlled a quarter of the greengroceries in Toronto and managed numerous market gardens.[21] Many produce vendors became well-known grocers and provisioners. Two prominent grocery chains started as small neighbourhood fruit markets run by Italian immigrants, Longos and Pusateris.[22] At the Ontario Food Terminal, Canada's largest wholesale market (North America's third largest), several wholesale leases remain with the Italian founding families. Most of those twentieth-century market gardens faded away, but numerous Italian home gardens survive, as does the desire to grow, cook, and eat ethnocultural vegetables.[23]

Michael Lo Presti, Italian-Canadian manager of a local supermarket notes that 'the old "ethnic" is now mainstream [. . .] Italian is no longer considered ethnic'.[24] He remarks: 'In produce 20 years ago we would say rapini, anise, radicchio, portobello mushrooms, Roma tomatoes, eggplant, Italian eggplant, avocados, asparagus, yellow plums, prune plums, coconuts, endive and escarole, leeks, Italian parsley, red shepherd peppers, zucchini, bok choy, Taiwanese cabbage, to name a bunch, [it] would be considered ethnic. Today none of them are.'[25] That shift of ethnocultural vegetables from 'old ethnic' to 'just items' illustrates how foodways of the migrant 'other' became mainstream, fundamentally reshaping Toronto's foodscapes.

The Chinese

Near contemporaries of those early Italians were single male migrants from southern China. Rural poverty and political upheavals led them to seek work worldwide. Racially exclusionist legislation limited their numbers in Canada to single men recruited from the Guangdong province to construct the Canadian Pacific Railway on the West Coast.[26] When rail construction ended in 1885, some moved to Toronto in search of new opportunities. That Chinese population of Toronto expanded after the Head Tax abolition in 1923, the Chinese Exclusion Act repeal in 1947, and the Kuomintang defeat in 1949.[27]

Despite the easing of immigration restrictions, systemic discrimination toward communities of colour pushed many of those migrants toward small, labour-intensive businesses that required minimal capital inputs, such as food. In 1900, Toronto had 200 Chinese residents and 95 small Chinese businesses.[28] City of Toronto assessment rolls of 1911 list one of the oldest Chinese food stores, at 16 Elizabeth Street, a wholesale grocer called Ying Chong Tai.[29] By 1924, Chinese immigrants ran 15 grocery stores and an additional 10 produce shops. As with Italian counterparts, those Chinese produce shops nurtured a symbiotic relationship with their compatriots' gardens.

Among the earliest Chinese in Toronto who grew food for a living were Jong Kee Chuck (Henry Chong) and Jong Chung Yuet (Charlie Chong). They started growing on the West Coast, among a considerable Asian community. They moved to Toronto in 1919 to start the Chong Family Farm.[30] Since the city's Chinese population was small, they offered lettuce, carrots, onions, and celery for sale around the city, hawked on streets and sold to mainstream stores. In 1950, Charlie's son Harry (Oye Suey Chong), arrived from Guangdong to take over. With a bigger Chinese customer base in the wake of increased migration, the farm could transition to a sharper focus on ethnocultural vegetables such as bok choy and bitter melon.

In the same way that Italian taste for vegetables reshaped Toronto's foodscapes, so did the cultural tastes of Chinese migrants. Multiple grocery and produce stores across Toronto neighbourhoods, and their connections to the growing, eating, and selling of ethnocultural vegetables, embody those entangled histories, languages, economies, and ecologies.[31] Across the course of modern Canadian history, the discriminatory laws and policies aimed directly at Chinese Canadians made those community stores active hubs for survival and resistance, sites of a racialized community's everyday resilience in the face of systemic oppression.[32] While the Chinese market gardens of Toronto are no more, such stores remain vital to Asian foodways. Their affordable produce is particularly crucial for low-income shoppers, irrespective of ethnicity.

Among the new supermarket chains that cater to pan-Asian foodways, one of the most prominent is T&T, founded in 1993 by a Taiwanese-Canadian entrepreneur. T&T was acquired in 2009 for $225 million by Canada's largest supermarket corporation, Loblaw.[33] This acquisition underlined the mainstreaming of foods previously ignored or relegated to the so-called 'ethnic aisle'. The influence of Chinese and Asian foodways on what is still understood to be the mainstream remains significant. It does not diminish the importance of the various migrant marketplaces.

Tamil and Bengali Home Gardens: Veetu Thottam and Bari

From the 1960s, Canada gradually overhauled its immigration policies to introduce a points system (or merit-based policy) that culminated in the Immigration Act of 1976. As the first Act to formally eliminate racial discrimination, it aimed to redress labour shortages as well as establish Canada's international image as a progressive nation in a new era of global decolonization.[34] Ostensibly, the category of ethnic group replaced that of race.[35] This set the stage for linguistically and culturally diverse migrants, such as Tamil migrants from Sri Lanka and Bengali migrants from Bangladesh. We study how their gardens and marketplaces have reshaped Toronto's foodscape in ways that echo the effects of prior migrants. The majority of the Tamils arrived after the 1980s due to political upheavals in Sri Lanka.[36] Most of the Bengalis landed as skilled or professional-class emigrants on the basis of the points system from the 1990s.[37]

The Tamils

Sri Lanka is an island country historically inhabited by Sinhalese- and Tamil-speaking communities, with the latter in a minority. Following independence from the British in 1948, the Sinhalese majority engaged in state-building measures that enhanced Sinhala-Buddhist dominance and actively discriminated against Tamils. In July 1983, anti-Tamil riots and pogroms soon plunged the island into a 26-year civil war. During those years, Tamils fled the country, many for Canada. By the 2000s, Toronto was home to a large Tamil population. Most refugees worked long hours at low-level jobs to maintain themselves, remit money, and bring kin over.

From the 1980s to the early 1990s, those migrants found it a challenge to access the ingredients vital to Tamil cuisine. The few Tamil stores that existed ran on a shoestring, with shipments of Sri Lankan produce arriving only once a month. Migrants, on the rare day off from factory labour, took buses to Chinatown to hunt for ingredients, such as long beans and okra.[38] They saved seeds from their purchases of eggplants and chillies. They tried to germinate those saved seeds in damp cloth.[39] Families rejoiced when their circumstances improved sufficiently to depart from tiny downtown apartments to suburban houses with backyards or balconies. A mother described how she planted seeds extracted from imported Jaffna vegetables. Each fall, she regaled her children with ample helpings of those tastes of home. As family members travelled between Sri Lanka and Canada, they carried over seeds and branches of cherished plants in suitcases, just as the Italians had done. When S's father visited his home district of Mullaitivu, he brought back the seeds of a particular variety of white eggplant that had been his family farm's special food crop before the civil war. Every year, that family carefully harvests those eggplants. They prize them all the more, knowing that in the war-ravaged ecology of their homeland, such heirloom varieties are in danger of extinction.

The Bangladeshis

On 25 November 2021, Sarker Foods, in Toronto's Little Bangladesh neighbourhood, staged on its Facebook page boxes of fresh vegetables – from okra and taro to jackfruit seeds, fresh turmeric, and Indian gooseberries. The store's narrow aisles bursting with fresh vegetables and fruit testified to the enhanced culinary infrastructure that ensures these global foods are available in a Canadian winter. A customer comments: 'MASHAALLAH-VERY NICE'. Another customer writes: 'Looks so fresh. For Bangladeshi fresh vegetables, I always recommend this place'.[40]

Bangladeshis migrated to Toronto in considerable numbers through the 1990s and 2000s, especially as the political and ecological conditions in the country deteriorated and Canada's points-based immigration system seemed to offer opportunity.[41] Community members tell us that Sarkar Foods was the first Bangladeshi grocery in Toronto.[42] It grew just as the nearby community of migrants did, across apartments and houses at one end of Toronto's east end transit line. Mirroring the experiences of the Italian, Chinese, and Tamil communities, culturally significant produce became increasingly available. A shopkeeper described to us the difference between then and now: 'Twenty years prior, there were no vegetables. The community people were crying for vegetables but couldn't get them anywhere. The newcomers [today] can't imagine the situation before as they are getting everything easily. They are getting the taste of home.' Over the last 25 years, numerous businesses started to import a wide range of fresh produce.[43] By the mid-2000s, global produce supply channels expanded to the extent that Sarkar Foods now imports vegetables and fruit all through the year from countries including Bangladesh, India, Sri Lanka, Thailand, Vietnam, Mexico, Dominican Republic, Brazil, and Ecuador.

Before those global supply chains were assembled, as with the other communities of migrant gardeners, many among Toronto's Bangladeshis relied on home gardens. An interviewee recollects vainly seeking ethnocultural foods in the 1990s: 'I was looking so badly for South Asian vegetables in this country. So the first year I grew red amaranth which is a very important leafy vegetable for our country, especially the women. They like this one because it has lots of iron and vitamin A.'[44] This led him to plant his own. He was not alone. We repeatedly heard from them that many Bangladeshis are serious, talented gardeners: 'In Bangladesh, the land is so fertile that you can just throw seeds on the ground and they will sprout!' They had learnt back home, growing foods such as roof gourd, a climbing squash that people train up the side of the house so heavy fruits can sit on the roof. The taste for fresh produce motivated people to find small plots of urban land, be it a backyard garden or a city allotment, to produce what they desired – and to agilely adapt plants to Toronto's soil and climate. Since so many migrants held degrees in professional fields such as agronomy they founded the Association of Bangladeshi Agriculturalists in Canada (ABACAN).[45] That knowledge and resilience produced home, community, and allotment gardens, that, as with twentieth-century migrant market gardens, are crucial to the global city of Toronto's culinary infrastructure.

The Plants and Their People, The People and Their Plants

Comparing the experiences of these four groups of migrants, we see a similarity in the interplay between market and home gardens and the retail foodscape: this interaction creates culinary infrastructure that serves and reflects ethnocultural foodways. What distinguishes the experiences of the newer arrivals from the earlier settlers is land access. Earlier Italians and Chinese migrants accessed land more easily as they arrived sooner in the settler-colonial project. By the time Tamil and Bangladeshi migrants arrived in Toronto, gentrification was already underway. Land values began a steep climb. Not only are backyard home gardens less accessible, because of the rising cost of buying a home, but the opportunity to buy peri-urban land to farm ethnocultural vegetables in the way of the Chinese and Italian market gardens has become inaccessible to anyone without significant capital.

Numerous home gardeners today would like to be market gardeners but find that access to land is prohibitive. Kabir dreams of buying a farm to grow ethnocultural produce for the market, but he continues to work a nine–to–five job and garden in his spare time. One Sri Lankan family used the pandemic's respite from in-person school and work to organically grow a wide range of ethnocultural vegetables on a peri-urban plot leased from a community member. Pandemic supply chain constraints made that produce especially welcome. But that was a one-off market garden. After the pandemic, that family reverted to the type of backyard subsistence growing of a smaller assortment that generations of migrants practiced in their vegetable gardens.

Another difference is that plants, like people, have had to adapt to place in different ways. The climate, daylight hours, growing season, and pollinating insects are different in Toronto than in Bangladesh or Sri Lanka. So techniques such as hand pollination of bottle gourds and informal seed-breeding projects that have resulted in the development of entirely new seed varieties enable both people and plants to make Toronto home.[46] Toronto's Tamil and Bengali gardeners volubly recount the challenges they faced to adapt their growing knowledge to the spatial and climatic variations they encountered in Canada. Learning what to plant in shorter periods, how to combat pests, and understanding climate change were among their major challenges he faced.

These plants and people have not only adapted, they have firmly become part of these foodscapes. On the one hand, Bangladeshi plots can be identified by red amaranth, for example, in the same way that one distinguishes a Chinese garden from an Italian one from the type of eggplant. On the other hand, many vegetables have made the same shift as those 'Italian' vegetables formerly deemed ethnocultural that became 'just items'. Tomatoes, eggplants, onions, garlic, and lettuce are common to such city gardens and grocers.

One factor currently limiting the influence of these twenty-first-century migrant gardens is land cost. While home gardens allow urban food growers to supplement their diets and incomes, and, alongside, to nurture their community, the political economy of the global city is inhospitable to small-time gardening entrepreneurs. The high cost of land and the ubiquity of cheap imported produce no longer permits the type of small-scale growing enterprise that allowed past market gardens and their migrant growers to thrive.

Conclusion

These migrant gardens helped transform a provincial settler-colonial town of the British Empire into a post-imperial metropole, a global city of North America. They continue to contribute a largely independent quality to Toronto's culinary infrastructure even if the grocery sector is dominated by five publicly-traded corporate supermarkets.[47] A continuing desire for ethnocultural produce nurtures the informal foodways of these home gardens, and urban foodscapes that still support alternatives to big chains. That urban food diversity runs counter to global supermarket trends.[48] Whether or not this will remain the case is yet to be seen.

About the Authors
Jayeeta Sharma is an Associate Professor of Food and Environmental Studies at the University of Toronto's Department of Physical and Environmental Sciences (DPES), and a founding member of the Culinaria Research Centre.

Sarah Elton is an Assistant Professor and Eakin Chair in Critical Qualitative Health Research Methodology at the University of Toronto's Dalla Lana School of Public Health.

Acknowledgements
We thank all those who shared their stories and, crucially, our student researchers: Aminah Haghighi, Jannatul Islam, Keerthana Nagaratnam, Jasleen Sohal, Geetha Sukumaran, and Emma Wan, funded by UTEA, SSHRC, and OVPR. We look forward to additional research with Afro-Caribbean, Black, and Francophone migrant gardeners and marketplaces.

Notes
1. An ethnocultural community or group is defined by shared characteristics such as cultural traditions, ancestry, language, national identity, country of origin, and/or physical traits. See Tina Chui and John Flanders, 'Immigration and Ethnocultural Diversity in Canada', *Statistics Canada*, 2011 <https://www12.statcan.gc.ca/nhs-enm/2011/as-sa/99-010-x/99-010-x2011001-eng.pdf> [accessed 13 May 2024].
2. Elizabeth Zanoni, *Migrant Marketplaces. Food and Italians in North and South America* (University of Illinois Press, 2018), pp. 3–7; Katherine Brickell and Ayona Datta, *Translocal Geographies* (Ashgate, 2011).
3. Norah MacKendrick, 'Foodscape', *Contexts*, 13.3 (2014), pp. 16–18, DOI:10.1177/1536504214545754.
4. Norah MacKendrick, 'Foodscapes', *Contexts*, 13.3 (2014), pp. 15–18, DOI:10.1177/1536504214545754.
5. Sarah Besky and Jonathan Padwe, 'Placing Plants in Territory', *Environment and Society*, 7.1 (2016), pp. 9–28 (p. 9), DOI:10.3167/ares.2016.070102.
6. On experiences see Mark Bhatti and others, '"I Love Being in the Garden": Enchanting Encounters in Everyday Life', *Social & Cultural Geography*, 10.1 (2009), pp. 61–76, DOI:10.1080/14649360802553202. On values see Nathan McClintock, 'Radical, Reformist, and Garden-Variety Neoliberal: Coming to Terms with Urban Agriculture's Contradictions', *Local Environment*, 19.2 (2014), pp. 147–71, DOI:10.1080/13549839.2012.752797. On resistance see Gerda R. Wekerle and Michael Classens, 'Food Production in the City: (Re)Negotiating Land, Food and Property', *Local Environment*, 20.10 (2015), pp. 1175–93, DOI:10.1080/13549839.2015.1007121; Jana Spilková, 'Producing Space, Cultivating Community: The Story of Prague's New Community Gardens', *Agriculture and Human Values*, 34.4 (2017), pp. 887–97, DOI:10.1007/s10460-017-9782-z; Pierpaolo Mudu and Alessia Marini, 'Radical Urban Horticulture for Food Autonomy: Beyond the Community Gardens Experience', *Antipode*, 50.2 (2018), pp. 549–73, DOI:10.1111/anti.12284; Bethaney Turner, 'Taste in the Anthropocene: The Emergence of 'Thing-Power' in Food Gardens', *M/C Journal*, 17.1 (2014), DOI:10.5204/mcj.769.
7. Elena Domene and David Saurí, 'Urbanization and Class-Produced Natures: Vegetable Gardens in the Barcelona Metropolitan Region', *Geoforum*, 38.2 (2007), pp. 287–98, DOI:10.1016/j.geoforum.2006.03.004; Pierrette Hondagneu-Sotelo, 'Cultivating Questions for a Sociology of Gardens', *Journal of Contemporary*

Ethnography, 39.5 (2010), pp. 498–516, DOI:10.1177/0891241610376069; Karen Zypchyn, 'Getting Back to the Garden: Reflections on Gendered Behaviours in Home Gardening', *Earth Common Journal*, 2.1 (2012), pp. 1–19, DOI:10.31542/j.ecj.60.

8 On socio-ecology see Fikret Berkes, Johan Colding, and Carl Folke, *Navigating Social-Ecological Systems: Building Resilience for Complexity and Change* (Cambridge University Press, 2003); Leslie Gray and others, 'Can Home Gardens Scale up into Movements for Social Change? The Role of Home Gardens in Providing Food Security and Community Change in San Jose, California', *Local Environment*, 19.2 (2014), pp. 187–203, DOI:10.1080/13549839.2013.792048; Shampa Mazumdar and Sanjoy Mazumdar, 'Immigrant Home Gardens: Places of Religion, Culture, Ecology, and Family', *Landscape and Urban Planning*, 105.3 (2012), pp. 258–65, DOI:10.1016/j.landurbplan.2011.12.020; Laura-Anne Minkoff-Zern and others, 'Role of Gardening in Mental Health, Food Security, and Economic Well-Being in Resettled Refugees: A Mixed Methods Study', *Journal on Migration and Human Security*, 12.1 (2024), pp. 3–18, DOI:10.1177/23315024231216II; Laura-Anne Minkoff-Zern, 'Pushing the Boundaries of Indigeneity and Agricultural Knowledge: Oaxacan Immigrant Gardening in California', *Agriculture and Human Values*, 29.3 (2012), pp. 381–92, DOI:10.1007/s10460-011-9348-4. On collaboration see John R. Taylor and Sarah T. Lovell, 'Exploring the Sociomaterial Dynamics of Home Food Gardening in a Black-Majority, Low-Income Neighbourhood in Chicago, IL, U.S.A', *Local Environment*, 26.11 (2021), pp. 1398–1420.

9 Jeffrey M. Pilcher, 'Culinary Infrastructure: How Facilities and Technologies Create Value and Meaning around Food', *Global Food History*, 2.2 (2016), pp. 105–31 (p. 105).

10 See Sarah Elton and Donald Cole, 'Is a Vegetable Garden Essential? Toronto Gardens as Culinary Infrastructure', *Food, Culture & Society*, 27.1 (2022), pp. 1–21, DOI:10.1080/15528014.2022.2086786; Sarah Elton and Donald Cole, 'A Prescription for Health: City Gardens Produce More Than Food', *The Conversation*, 7 July 2022 <https://theconversation.com/a-prescription-for-health-city-vegetable-gardens-produce-more-than-just-food-186019> [accessed 13 May 2024].

11 To problematize the idea of 'freshness', see Susanne Freidberg, *Fresh: A Perishable History* (Belknap Press of Harvard University Press, 2009), and Sarah Elton, 'The Relational Agency of Plants in Produce Supply Chains during COVID-19: "Mother Nature Takes Her Course"', *Journal of Rural Studies*, 98 (2023), pp. 59–67.

12 See Mengyu Li and others, 'Global Food-Miles Account for Nearly 20% of Total Food-Systems Emissions', *Nature Food*, 3.6 (2022), pp. 445–53, DOI:10.1038/s43016-022-00531-w.

13 See Glen C. Filson and Bamidale Adekunle, *Eat Local, Taste Global: How Ethnocultural Food Reaches our Tables* (Wilfrid Laurier University Press, 2017), DOI:10.51644/9781771123143; B. Adekunle, G. Filson, and S. Sethuratnam S, 'Culturally Appropriate Vegetables and Economic Development: A Contextual Analysis,' *Appetite*, 59.1 (2012), pp. 148–54; and B. Adekunle, G. Filson, and S. Sethuratnam, 'Preferences for Ethno-Cultural Foods in the Greater Toronto Area: A Market Research,' *SSRN Electronic Journal* (2011), DOI:10.2139/ssrn.1738475.

14 Filson and Adekunle, *Eat Local, Taste Global*, p. 13.

15 D.J. Hall, 'Clifford Sifton: Immigration and Settlement Policy, 1896–1905', in *Settlement of the West*, ed. by Howard Palmer (University of Calgary, 1977), p. 77.

16 See the analysis of Canadian immigration policy on Italians by Stephanie Bellissimo, 'They Were Triomphanti: The Italian Homesteading Experience in Saskatchewan, 1896–1930' (unpublished thesis, University of Saskatchewan, 2012), pp. 20–30.

17 Cristina Pietropaolo, '"His Fig Tree Came in the Mail": Growing an Italian Garden in Toronto', *Digest: A Journal of Foodways & Culture*, 2.1 (2013) <https://scholarworks.iu.edu/journals/index.php/digest/article/view/34044> [accessed 13 May 2024].

18 '"When We Come From Italy, We Take It With Us": The Italian-American Gardens of the Italian Garden Project', *The Italian Garden Project*, n.d. <https://www.theitaliangardenproject.com/blog/when-we-come-from-italy-we-take-it-with-us-the-italian-american-gardens-of-the-italian-garden-project> [accessed 13 May 2024].

19 John Zucchi, *Italians in Toronto* (McGill – Queen's University Press, 1988), pp. 29, 56.

20 See Robert Harney, 'Chiaroscuro: Italians in Toronto, 1815–1915', *Polyphony*, 6 (1984), pp. 44–49.

21 Zucchi, *Italians in Toronto*, pp. 30, 88, 90.

22. 'Our Story', Longos, n.d. <https://www.longos.com/about-us/our-story> [accessed 13 May 2024]; 'Our Story', Pusateri's, n.d. <https://www.pusateris.com/Our-Story> [accessed 13 May 2024].
23. Nicholas Keung, 'City of Gardens City Provides Fertile Ground to Grow Diversity; Immigrants Transplant their Cultures and Identities to Toronto's Backyards: [Ontario Edition]', *Toronto Star*, 2000, B01.
24. Jim Zucchero, 'The Legacy and Cultural Significance of Italian Grocers', *Accenti*, 31 August 2019 <https://accenti.ca/the-legacy-and-cultural-significance-of-italian-grocers/> [accessed 1 May 2024].
25. Zucchero.
26. Madeline Yuan-yin Hsu, *Dreaming of Gold, Dreaming of Home: Transnationalism and Migration between the United States and South China, 1882–1943* (Stanford University Press, 2000), pp. 21–22; Anthony B. Chan, 'Chinese Canadians', *The Canadian Encyclopedia*, 8 September 2016 <https://www.thecanadianencyclopedia.ca/en/article/chinese-head-tax-in-canada> [accessed 5 May 2024].
27. 'Chinese History in Toronto', City of Toronto, n.d. <https://www.toronto.ca/city-government/accountability-operations-customer-service/access-city-information-or-records/city-of-toronto-archives/using-the-archives/research-by-topic/chinese-history-in-toronto> [accessed 5 May 2024].
28. David Chuenyan Lai and Jack Leong, 'Toronto Chinatows 1878–2012', Simon Fraser University, n.d. <https://www.sfu.ca/chinese-canadian-history/toronto_chinatown_en.html> [accessed 5 May 2024].
29. City of Toronto.
30. 'Growing Our Future: Commemorating 100 Years of the Chinese Immigration Act', *Heritage Toronto*, n.d. <https://www.heritagetoronto.org/explore/explore/commemorating-chinese-immigration-act/> [accessed 5 May 2024].
31. See Camille Bégin and Jayeeta Sharma, 'A Culinary Hub in the Global City: Diasporic Asian Foodscapes across Scarborough Canada', *Food, Culture & Society*, 21.1 (2018), pp. 55–74; Michaël Bruckert, 'Diasporic Meatscapes of Tamil Community in Toronto: How Immigrants Reconfigure Food Environments & Infrastructures to Secure a Taste of Home', *Food, Culture & Society* (2023), pp. 1–22.
32. Steve Tu, 'In Toronto, Chinese Grocery Stores Defy the Whitewashed Cityscape', *Network in Canadian History and Environment*, 23 August 2023 <https://niche-canada.org/2023/08/23/in-toronto-chinese-grocery-stores-defy-the-whitewashed-cityscape/> [accessed May 10, 2024].
33. 'Our Story', T&T, n.d. <https://www.tntsupermarket.com/eng/aboutus/story> [accessed May 14, 2024].
34. Sunera Thobani, *Exalted Subjects: Studies in the Making of Race and Nation in Canada* (University of Toronto Press, 2007), p. 94.
35. Nalinie Mooten, 'Racism, Discrimination and Migrant Workers in Canada: Evidence from the Literature', *Immigration, Refugees, and Citizenship Canada*, July 2021 <https://publications.gc.ca/collections/collection_2022/ircc/Ci4-235-1-2022-eng.pdf> [accessed May 5, 2024]. Note the late-twentieth-century introduction of temporary worker schemes aimed at populations of colour.
36. Anuppiriya Sriskandarajah, 'Demonstrating Identities: Citizenship, Multiculturalism and Canadian-Tamil Identities' (unpublished master's thesis, University of Windsor, 2010) <https://scholar.uwindsor.ca/etd/57> [accessed 12 May 2024].
37. See Md Mizanur Rahman, 'Development of Bangladeshi immigrant entrepreneurship in Canada', *Asian and Pacific Migration Journal*, 27.4 (2018), pp. 404–30.
38. Vanessa Vigneswaramoorthy, 'New Places, New Palates: Tamil Cooks in Toronto's Kitchens', *Heritage Toronto*, 1 December 2022 <https://www.heritagetoronto.org/explore-learn/tamil-cooks-toronto-kitchens/> [accessed 12 May 2024].
39. Personal communications, S.
40. 'Sarker Foods', Facebook <https://www.facebook.com/p/Sarker-Foods-100076109746039/> [accessed 10 May 2024].
41. Sutama Ghosh, 'Everyday Lives in Vertical Neighbourhoods: Exploring Bangladeshi Residential Spaces in Toronto's Inner Suburbs', *International Journal of Urban and Regional Research*, 38.6 (2014), pp. 2008–24.
42. Personal communications, J. and F.
43. Valerie Imbruce, *From Farm to Canal Street: Chinatown's Alternative Food Network in the Global Marketplace* (Cornell University Press, 2015); Urban Food Security in a Global Marketplace: Tracking Produce Supply Chains to Assess Food Security & Food Sovereignty in Toronto, SSHRC-funded research project by Elton & Sharma, ongoing.

44 Personal communication, K.
45 'Association of Bangladeshi Agriculturalists in Canada (ABACAN)', Facebook <https://www.facebook.com/groups/1933529743552529/> [accessed 10 May 2024].
46 Sarah Elton, 'People-Plant Mobilities: Growing Bitter Melon and Bottle Gourd in Toronto', in *Food Mobilities: Making World Cuisines*, ed. by Daniel E. Bender and Simone Cinotto (University of Toronto Press, 2023), pp. 78–92.
47 Foreign Agricultural Service, 'Canada: Retail Foods', USDA, 18 July 2023 <https://fas.usda.gov/data/canada-retail-foods-7> [accessed 28 February 2024].
48 Sarah Elton, Jasleen Sohal, and Kyle Resendes, '"The Ontario Food Terminal: Supporting Food Access", Food Health Ecosystems Lab Working Paper #1', Toronto Metropolitan University, 2022 <https://www.torontomu.ca/content/dam/sociology/research/FHElabOFTWorkingPaper1.pdf> [accessed 10 May 2024]; Sarah Elton, Matilda Diperi, and Donald Cole, 'Public Infrastructure Supports Agriculture of the Middle: Canada's Largest Wholesale Produce Market, the Ontario Food Terminal' (under review).

35

Le Potager du Roi

The 'King's Kitchen Garden' at Versailles as Political Metaphor and Gastronomic Laboratory

Richard Warren Shepro

Gardening has long been an incisive metaphor for political and social order. In an extended allegory in Shakespeare's *Richard II*, the gardener contrasts how much care he pours into his garden with the way his king has ruled: 'O, what pity is it' that the king 'had not so trimm'd and dress'd his land / As we this garden!' (3.2:61–63). In the last sentence of Voltaire's *Candide*, after seeing hideous turmoil, Candide declares simply that we should cultivate our own gardens. Such literary references may resonate more meaningfully than many pages of political philosophy. But these literary gardens don't necessarily tell us what the garden is used for, nor whether there are culinary implications. And what of the converse: what are the implications for vegetable gardens controlled by various sorts of governments?

We can examine the history of one government-controlled, monumental (nine hectares, about twenty-five acres) vegetable and fruit kitchen garden, in France, planted in the late seventeenth-century, that has survived and changed through nearly 350 years of political upheavals and societal changes unimaginable to those who designed and planted it. This garden has had as many acts as France has had governments. The vicissitudes of this garden, the *potager du Roi*, and its successes and failures are a metaphor for – indeed, a history of – political and social change. They tell us about the actual effects of drastically different forms of government and social views on cultivation and agronomy and ecology. The efforts of this prominent edible garden have led to changes in gastronomic taste.

Nomenclature: The Idea of the *Potager*

A *potager* is a kitchen garden. The name is connected to the word *potage*, a soup profuse with vegetables (and sometimes meat). *Potagers* were designed to grow produce consumed by the adjacent household. Households might be humble, bourgeois, or noble, but the *potager* served its household and was not a commercial enterprise. In addition to annually planted vegetables, a *potager* could contain permanent crops – fruit trees, for example, though not citrus fruit trees cultivated only in specialized *orangeries* in northern France. *Potager* was also used for a precursor to the stove, made of bricks or stone with cavities filled with coals (Figure 1).[1] The range-like *potager* became common in eighteenth-century France: so the produce of the *potager* could be used to produce a

Gardens, Flowers, and Fruit

Figure 1. A seventeenth century potager *in Versailles. Source: Wikimedia Creative Commons.*

potage cooked over a *potager*. As a type of *jardin* (garden), a *potager* could also be called a *jardin potager*. A related type, originally in churchyards but later elsewhere, is the *jardin de curé* (curate's garden) that grew vegetables for a church but also flowers for church services, medicinal herbs, and sometimes grapes to make wine for the Eucharist. A *potager* that includes flowers can also be called a *potager fleuri*.

Although a sort of 'vegetable renaissance' in France has been described beginning in the fourteenth century, an influential manual about how to create and maintain a personal kitchen garden was published for the first time around 1651: *Le Jardinier françois* (*The French Gardener*) by the mysterious courtier Nicolas de Bonnefons.[2] This book joined a spate of other culinary manuals that referred to being French in their titles: 'Frenchness was in the air.'[3] The dedications of Bonnefons's book and its sequel were 'to the Ladies'. This dedication emphasized that the book was not about commercial agriculture, but about sustenance for a household, not for royalty, but sustenance at a certain serious bourgeois level. Bonnefons's sequel and equally popular book, *Les Délices de la campagne*, celebrated household cooking including the produce of the *potager*.[4]

In 1656, Jean-Baptiste de La Quintinie, a 30-year-old French lawyer, returned from a tour of Italy, visited the botanical garden at Montpellier, and abandoned his legal career, 'preferring the scent of jasmine to that of litigants'.[5] After designing a garden for the home of his previous employer in Paris, he achieved a meteoric rise, developing *potagers* for some the most prominent homes in France, including the *potager des princes* for the Prince de Condé at his chateau at Chantilly; the renovation of the old, comparatively small vegetable garden of Louis XIII at Versailles, then a small hunting lodge;

and the *potager* at Vaux-le-Vicomte, the chateau of Nicolas Fouquet, finance minister to Louis XIV. On 17 March 1670 he was officially presented to Louis XIV himself, who named La Quintinie 'Director of the Fruit Trees and Potagers of all the Royal Residences', a position created especially for him.

The *Potager du Roi*

In 1678 Louis XIV requested that La Quintinie develop a new *potager* on a sorry, nine-hectare swamp a short walk from his new palace at Versailles. Planting began after five years of preparation during which the swamp was drained; new soil brought in and fertilized; new garden drainage planned and installed; and various walls, terraces, sheds, and monuments constructed.

The famous formal gardens designed by André Le Nôtre were being developed at the same time on the other side of the palace. They cannot be seen from each other. The vegetable garden had different goals but in its own way was also grand, bold, and unprecedented; it was to be a *potager* with vast ambition for the court 'household'. La Quintinie's design created a grand central quadrangle of sixteen square plots of vegetables surrounding a dramatic circular pond. Surrounding this quadrangle were twenty-nine further enclosed fruit and vegetable gardens, surrounded by high walls to create varied micro-climates so production could be staggered. Above that was a terrace for observation by the king and his guests (Figures 2, 3, 4, and 5).[6]

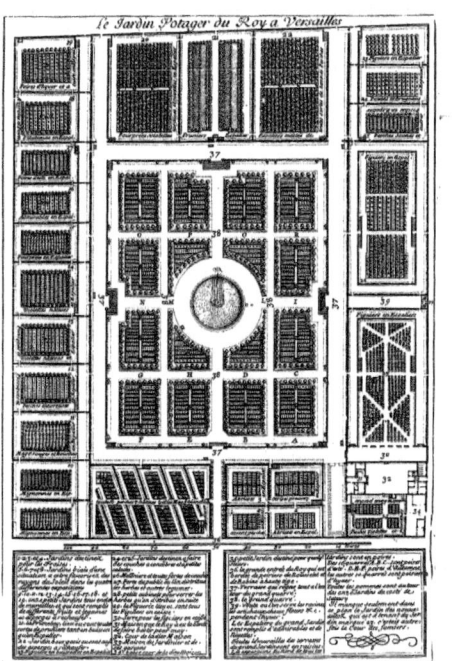

Figure 2. The original plan for the garden, which is largely in place today. Source: Wikimedia Creative Commons.

As with everything at Versailles the paramount goal for the vegetable garden was what was known as 'the pleasure of the king'. Louis XIV's meals were meticulously observed, written about, imitated, and even studied by physicians as if they were medical experiments.[7] The results from the nine hectares went far beyond this goal.

The creation of the garden and the produce it provided were part of a flourishing of artistic and scientific activities called collectively the *menus-plaisirs du Roi* that, while designed for Louis XIV's pleasure, were also enjoyed by the court and, later, a wider aristocratic and bourgeois audience. (*Menus-plaisirs*, an old phrase still in use, means 'minor pleasures'.) These *plaisirs* covered all the elite arts (gastronomy was perhaps the most minor): the *menus-plaisirs* at the palace of Versailles included many of the accomplishments of what became

Figure 3. Le potager du Roi, *by Pierre Aveline (1700). Note the many walled gardens, designed to create micro-climates. Source: Wikimedia Creative Commons.*

Figure 4. A contemporary view of the central garden. Source: Wikimedia Creative Commons.

Figure 5. A contemporary view from the edge of the garden. The vast decorative formal gardens designed by André le Nôtre are on the other side of the chateau and are not visible from the potager du Roi. *Source: Wikimedia Creative Commons.*

known as *Le Grand Siècle* (the Great Century), such as Molière's plays, Jean-Baptiste Lully's music and dances, and even the king's kitchen staff's culinary accomplishments. Varieties of pears developed by La Quintinie are still prized today. In his writings, published posthumously, he rated hundreds of different varieties of pears, placing them in categories such as Good, Mediocre, Bad, and Not To Be Planted By Anyone.[8]

Antoine Jacobsohn, in charge of the *potager du Roi* since 2007, has emphasized that the goal was to impress, and thus to emphasize the glory of France under this absolute monarch.[9] Based on the historical accounts and La Quintinie's records, many different aspects were involved in making the garden so impressive. In addition to its sheer size, its design and organization which, while not competing with the grand formal gardens designed by Le Nôtre, were nevertheless astonishing for a kitchen garden, allowing it not only to function efficiently but also to make a beautiful impression. Louis XIV often took visitors on walks around and through the *potager* and took a keen interest in the garden.

Not merely a vegetable and fruit garden, the *potager du Roi* was also a sort of botanical garden. Le Quintinie experimented with hundreds of new plants and cultivation techniques, with exotic food plants brought back by travellers and explorers, and with ways to produce out-of-season food and plants that one would not expect in northern France. These paradoxical plantings, upending the season and defying the climate, were an obsession of the age, and a corollary to the concept of taming nature that was so important in the work of French decorative gardeners. Le Quintinie trained hundreds of gardeners, personally and through his writings, and these gardeners ensured a legacy of influence of the *potager du Roi* through the centuries to come.

The garden influenced gastronomic taste considerably. However, the excitement of upending the seasons and defying the climate waned with the centuries and seems almost quaint today. The most lasting contribution may have been the *primeurs*, which were viewed as the essence of a vegetable, an example of perfection, and in many ways still are. *Primeurs* are early vegetables, not dramatically out of season but produced at the very beginning of the plausible season through careful cultivation, selection of micro-climates, sometimes early planting in greenhouses, and employing many other special techniques of master gardening. Emphasizing incredible freshness, the best *primeurs* are generally small and viewed as exceptionally tasty. Early spring vegetables such as green peas and asparagus are particularly prized as *primeurs*. La Quintinie was perhaps the first gardener to be celebrated for this endeavour. A recent edition of *Larousse Gastronomique* noted that in modern times *primeurs* have been less significant in commerce because of early or out of season imports from vegetable producers in warmer climates.[10] Nevertheless, they continue to excite a discerning public today.[11]

A period one might call pea-mania had lasting effects on French cuisine. La Quintinie has been given credit for creating 'a green-pea hybrid known as petits pois', which seems unlikely, but he did cultivate many varieties of green peas and found ways to produce them unusually early in the spring, as *primeurs*.[12] Louis XIV and other members of the court craved them, struck by their novelty, taste, and freshness, after having been brought a crate of small green peas still in their pods from Italy in 1660.[13] Thirty-six years later, the craze still continued, as recorded in a letter written in May of 1696 by Madame de Maintenon, the enigmatic secret second wife of Louis XIV, to Louis-Antoine de Noailles, the cardinal de Noailles, who became archbishop of Paris through her influence. The obsession with novelty and perfection and the endless discussion of this topic frustrated her and became all she could talk about in a letter to her old friend:

> For eight days without stopping I have been succumbing to the sadness of not hearing anything reasonable: the peas chapter is still going on: the impatience to eat them, the pleasure of having eaten them, & the joy of eating them again are the three points that our Princes have been dealing with for four days. There are ladies who, having dined with the King, and dined well, find peas at home to eat before going to bed, at the risk of indigestion: it is a fashion, a fury, and one follows the other. You have strange sheep, Monseigneur![14]

Grimod de La Reynière, the first major chronicler of French gastronomy, is commonly quoted as having said, over a century later, 'Petits-pois are without any doubt the best of all vegetables eaten in Paris'.[15]

After Louis XIV

Under Louis XIV and La Quintinie there do not appear to have been controversies about the direction of the garden. Louis XIV, at this stage in his life, was a confident monarch, and La Quintinie, who had already served the king for five years as director of the royal orchards and vegetable gardens, was unlikely to develop plans that conflicted with the monarch's goals.

When Louis XIV died in 1715, his great-grandson succeeded him as Louis XV.[16] Louis XV and his grandson, Louis XVI, were less popular kings, with less political power and less money in the treasury, and so there were changes of direction at the *potager du Roi*. King and court no longer had the goal of having Versailles be a showcase for the world. Under Louis XV, botanical science still advanced but with a smaller budget. Exoticism reigned. The head gardener, Louis Le Normand, developed a way to raise pineapples, which became a luxury fad and a lasting symbol of hospitality in France. The pineapple was introduced to France in 1730 when Louis XV was presented two buds, from which Louis Le Normand produced a pineapple by 1735 and then expanded production in his new heated greenhouses so that during the reign of Louis XVI there were 800 fruit-bearing pineapple plants at the *potager du Roi*.[17]

The experience of growing pineapples and even more with coffee plants at the *potager du Roi* illustrates the growing significance of international affairs and of diplomatic gifts as a factor in the dissemination of plants around the world.

The history of how the cultivation of coffee spread around the entire globe from its start as a wild plant in East Africa has been often revised but remains murky and increasingly controversial. However, a substantial coffee house culture had developed in Paris in the seventeenth century using imported beans, and it is well established that in 1713, a year before he died, Louis XIV was sent a live coffee plant by the burgomaster of Amsterdam to reduce reliance on imported beans. It had been propagated from a seven-foot-tall coffee plant at Amsterdam's Hortus Botanicus, possibly the first botanical garden in Europe. The plant died. A replacement set of coffee plants were sent and studied in Paris at the *Jardin des Plantes* by the French botanist Antoine de Jussieu. Enough coffee plants from that donation were eventually grown at the *potager du Roi* for Louis XV to impress foreign visitors with his own five-foot tall coffee plants and with the exotic treat of coffee made from beans grown in greenhouses on site and roasted on site.

Despite the conflicting tales of coffee transmission, it does appear that those plants formed the basis of the coffee plantations the French developed after colonizing Martinique in 1724 and later in the other French Antilles.[18] There have been assertions that '90 percent of the world's coffee sprang' indirectly from the *Arbre Mère* (mother tree) coffee plant given by the Dutch government to Louis XIV, while others maintain otherwise.[19] In any case, it is not new to have conflicting stories of the origin of colonial coffee culture. The second volume of Diderot's massive *Encyclopédie* (1752) already recognized this issue: 'As M. de Jussieu said in 1715, for the past sixty years or so that *coffee* has been known in Europe, so many people have written about it without knowing its origins, that if we were to draw up a history of those accounts, the number of errors would be too great for a single report to list them all.'[20]

The Revolution and Beyond

The fate of the nine hectares changed suddenly, and repeatedly, during the French Revolution. The best use of the garden was debated at the Estates-General but apparently not decided. Governments changed rapidly. Funds ran out: the head gardener resigned in 1793, one of the several governments of 1793 leased parcels to a new class of citizen tenant farmers, and the eight hundred pineapple plants were sold at auction. (Who purchased the pineapple plants is a mystery: these expensive plants required specialized indoor quarters and intensive care in order to thrive. The purchases seem like an unusual investment to make at a time of great economic and political instability.) Then, in 1795, a new government changed direction again, evicting the tenants and, full of Enlightenment enthusiasm, rededicated the garden to science and education, experimental again but intended for the greater glory of what was called 'modernized France' rather than the elaborate pleasures of a sovereign and his court.[21]

Napoleon, though deeply interested in gardens and botany, had little interest in food and largely ignored the *potager du Roi*.[22] Nevertheless, it was returned to the national government under Napoleon. After Waterloo, according to an 1847 commentator who was very proud of the accomplishments of La Quintinie, 'With the return of the Bourbon dynasty, this garden regained all its lustre and re-established the cultivation of the *primeurs* (early vegetables, which were highly prized).'[23] The *primeurs*, indeed, had become so crucial to the perceived mission of the garden that instead of having a single director, the garden began to be directed by two head gardeners, one in charge only of the *primeurs*, the other in charge of everything else.

Each government since has had a different approach. With the 1848 Revolution, the garden again became a school and experimental station, but when Louis Napoleon became Emperor in 1852 it became once again the personal domain of the ruler. In the early 1900s the *potager* became a showcase for intensive agriculture and became a lucrative government business.[24]

Myth, Idealization, and Legacy

In the 1920s, with France a republic and the horrors of the First World War behind it, the *potager du Roi* continued its role as a working garden. But in this period it also became, for some, a historical memory, a mythic reflection on a newly idealized past. This is reflected in a collection of essays published in 1926 by the chef Éduard Nignon. Nignon included an essay called 'Les Légumes' (Vegetables) in which he observed that all kitchen gardens are more than just a vehicle for production: 'The atmosphere and the view of a vegetable garden have a unique charm that is difficult to express.'[25] He singled out the work of La Quintinie at Versailles and marvelled at the enthusiasm the *potager du Roi* had engendered in Louis XIV.

Today, few of us who have studied records of the extravagant banquets held at the court of Louis XIV would think of the period's dishes as exemplars of finesse, but Nignon saw something different in the early work at the *potager du Roi* under the direction of Louis XIV and La Quintinie:

It was at this time that the cultivation of *primeurs*, which had previously been completely ignored, was introduced. It was this cultivation that was to prove so enjoyable for gourmets. It is clear that the pleasures of the palate that we owe to the vegetable garden are of a different order and quite different from the other 'pleasures of the table', which seem to be due to heavier and less delicious foods.[26]

It is no wonder that Nignon's nearly forgotten books made him the favourite food and recipe writer of many of the chefs who lightened French cuisine in the Nouvelle Cuisine movement of the last century – especially Michel Guérard, whose own kitchen gardens, called the *jardin de curé* and *Potager de la Ferme aux Grives*, create a central focus for his celebrated restaurant in the southwest of France today (Figures 6 and 7).

Since 1995, the *potager du Roi* has been under the direction of the l'École nationale supérieure de paysage (ENSP) (the National School of Landscape), the premier landscape architecture school in France, quite separate from the Domaine du Château de Versailles that oversees all the rest of the buildings and grounds. This legal separation from the chateau fits in with the director Antoine Jacobsohn's referring to it as 'an ideal urban garden'.[27] This old garden has always been urban but as it has become more hemmed in by the town of Versailles, which is now really a suburb of Paris, the *potager du Roi* has begun to seem more like an (unusually grand) urban garden and less a part of the

Figures 6 and 7. The *jardin de curé of the great chef Michel Guérard (1931–2024), one of the modern* Roi des Potagers.

chateau. A recent North American book on urban agriculture devotes a chapter to it as the precursor of the modern urban garden.[28]

But in keeping with many controversies of the current age, debates about the condition, goals, and methods of gardening are carried out within the government, in the newspapers *Le Monde* and *Le Figaro*, in public petitions, and in the scientific gardening community.[29] A group calling itself *les Amis du Potager du Roi* who want the nine hectares planted in a historical manner as part of the enormous palace of Versailles complex (in essence saying, 'Let's restore the *potager du Roi* without endangering its historical and pedagogical values') question such things as the styles of pruning and whether the pear trees are properly espaliered.[30]

Some conservationists and many experimentalists ardently believe (not always in the same way, so agreement is difficult) that heritage preservation is not the proper goal to be pursued today, and that the nine hectares should be used to advance science and promote caring for the land, emphasizing zero phyto (no pesticide), biodynamics, natural agronomy, and responses to climate change. Some aspects of this argument actually fit in with questions in the debate over historical preservation, since, in a frugal solution to the complex problem of improving the soil after draining the swamp, La Quintinie fertilized the new soil with manure from the royal stables. Vincent Piveteau, director of the ENSP, has said in a somewhat grand but aspirational statement that the real 'question is whether the *potager du Roi* is the best place to prepare for the "third agricultural revolution, that of the 21st century," which will have to feed nine billion people', and urges entry into the debate between permaculture and agroecology.[31] It appears the only current point of agreement is that all groups applaud the sale of whatever produce is grown at a public market.

Yet the prominence of the garden remains. In 2008, eighty-two years after Nignon's essay on vegetables, Christian Millau, the chief chronicler of the Nouvelle Cuisine, quipped that when he telephoned famous chefs, for years he 'inevitably' was told, 'the chef can't talk now, he's being filmed'. But now, '*plus chic*', was the answer, 'I'm sorry but the chef is in his *potager*.' More seriously, though, he discussed the grand twenty-first-century revival of *potagers* and the elevation of vegetables and especially *primeurs* grown by or at the direction of modern chefs such as Michel Guérard, Jean-André Charial, and Alain Passard. His advice was to cultivate the practical within the spirit of what was perceived as the ideal: 'Unable to afford the *potager du Roi*, one makes oneself a *roi du potager*.'[32] This is the mythic spirit of celebration that led La Quintinie to become a gardener, that led Nignon to identify the special qualities of the vegetable garden, and that can lead all who seek it to find inspiration in the *potager du Roi*.

About the Author
Richard Warren Shepro is both an international lawyer and a food scholar. He teaches at the University of Chicago and is the author of nine Oxford Symposium papers. He is the président of the Alliance Française de Chicago and a former editor of the *Harvard Law Review*.

Notes

1. Jean-Robert Pitte, *Gastronomie française: Histoire et géographie d'une passion* (Fayard, 2005) pp. 97–98.
2. See Susan Pinkard, *A Revolution in Taste: The Rise of French Cuisine* (Cambridge University Press, 2009), pp. 35–43. Nicolas de Bonnefons, *Le Jardinier françois que enseigne à cultivar les arbres & herbes potagères avec la manière de conserver les fruits et faire toutes sortes de confitures, conserves & massepains* (Pierre des Hayes, 1651). Although Bonnefons emphasizes household produce, he does give advice about selling excess produce.
3. Priscilla Rarkhurst Ferguson, *Accounting for Taste: The Triumph of French Cuisine* (University of Chicago Press, 2004) p. 36. Other examples include La Varenne's *Le Cuisinier françois, Le Pâtissier françois*, etc., the influence of which are described so thoroughly by Barbara Ketcham Wheaton in *Savoring the Past: The French Kitchen and Table from 1300 to 1789* (University of Pennsylvania Press, 1983), pp. 113–38, that French historians frequently cite Wheaton's work.
4. Nicolas de Bonnefons, *Les Délices de la campagne* (Pierre des Hayes, 1656). Pierre Gagnaire and Hervé This wrote a detailed modern commentary surrounding their own edition of this book: *Alchimistes aux fourneaux* (Flammarion, 2007). See also Wheaton pp. 123–26.
5. Qtd. by Michèle Barrière in the historical appendix to her evocative novel set in 1683, an early year for the *potager du Roi* (*Meurtres au Potager du Roy: Un roman noir et gastronomique sous Louis XIV* (Livre de poche, 2008), p. 267).
6. Stéphanie de Courtois, *Le Potager du Roi: The King's Kitchen Garden* (Actes Sud, 2023), p. 11.
7. For a detailed summary, see Stanis Perez, *La Santé de Louis XIV: Une biohistoire du Roi-Soleil* (Champ Vallon, 2007).
8. See Antoine Jacobsohn, *Le Potager du Roi: Dialogues avec La Quintinie* (Éditions Artlys, 2017), pp. 109–11, taken from details in Jean-Baptiste de La Quintinie, *Instruction pour les jardins fruitiers et potagers* (Claude Barbin, 1690; facsimile repr. Actes Sud, 2016).
9. See Jacobsohn, and an interview by the Fondation Louis Bonduelle, 'Interview with Antoine Jacobsohn – Potager du Roi', 11 July 2018 <https://www.youtube.com/watch?v=-QFwhutkjKg> [accessed 9 April 2025].
10. Joël Robuchon, ed., *La Grande Larousse Gastronomique* (Larousse, 2012), p. 712.
11. For example, Elaine Sciolino, 'Spring Brings Caviar in a Pod', *The New York Times*, 13 April 2012.
12. Sciolino.
13. Marie-Hélene Baylac, *Dictionnaire gourmand du canard d'Apicius à la purée de Joël Robuchon* (Omnibus, 2014).
14. Letter from Mme de Maintenon to the cardinal of Noailles, 18 May 1696 (Françoise d'Aubigné and Marquise de Maintenon, *Lettres de Madame de Maintenon* (1757), trans. by the author). Somewhat surprisingly, this is sometimes inaccurately attributed to Mme de Sevigny, also famous as a letter writer, including in de Courtois, pp. 21–22.
15. As one example, see the web site of the village of Clamart, between Versailles and Paris, which became known for petit-pois of high quality after the initial successes at the *potager du Roi* <https://www.clamart.fr/fr/actualites/sur-les-traces-des-petits-pois> [accessed 9 April 2025]. Grimod's writings are extensive, and this quotation may not be authentic. See, for example, the 1106-page recent edition, Alexandre Balthazar Laurent Grimod de la Reynière, *L'Almanach des Gourmands: Servant de guide dans les moyens de faire excellent chère (1803–1812)* (Éditeurs de Gastronomie, 2014), or Chikako Hashimoto, *La naissance du gourmand: Grimod de la Reynière et la Révolution française* (Presse universitaires François-Rabelais, 2019).
16. Louis XV's great-uncle, Philippe II, Duc d'Orléans, ruled France as regent until Louis XV reached the age of maturity (13 years) in 1723.
17. Garth Sanewski, Duane P. Bartholomew, and Robert E. Pauli, *The Pineapple: Botany, Production and Uses*, 2nd edn (CAB International, 2018,) p. 3; de Courtois, pp. 30–31.
18. de Courtois pp. 26–31. It is not clear, however, whether this grant led to coffee production also in French Guyana or whether those coffee plants were stolen from previously established plants in Dutch Guyana; see Reay Tannahill, *Food in History* (Paladin, 1975) pp. 252–53.

19 Marie Nadine Antol, *Coffee Through the Ages* (Square One Publishers, 2002) pp. 13–17. For opposing views, see for example Reay Tannahill's claim that coffee spread world-wide from Dutch plantations in Java (pp. 252–53).
20 Denis Diderot, *l'Encyclopédie*, vol. 2 (1752), p. 527, Collaborative Translation Project, University of Michigan <https://quod.lib.umich.edu/d/did/did2222.0000.065/--coffee?rgn=main;view=fulltext> [accessed 9 April 2025].
21 Joseph Adrien Le Roi, *Notice historique sur le Potager du Roi à Versailles* (Imprimerie de Montalant-Bougleux, 1847); de Courtois, pp. 42–45.
22 See Ruth Scurr, *Napoleon: A Life in Gardens and Shadows* (Chatto & Windus, 2021) and Eric Glatre, *Histoire(s) de la gastronomie* (Éditions du Félin, 2022), pp. 7–17.
23 Le Roi, p. 15.
24 de Courtois, pp. 47–58.
25 Édouard Nignon, *Les Plaisirs de la table* (1926, rpt Éditions de gastronomie, 2016), p. 251, trans. by the author.
26 Nignon, p. 251, trans. by the author. Though Nignon was highly attuned to vegetable pleasures he is also articulate in his discussion of the pleasures of meat, fish, and fowl.
27 Fondation Louis Bonduelle, 'Interview with Antoine Jacobsohn – *Potager du Roi*', 17 January 2020 <https://www.youtube.com/watch?v=-QFwhutkjKg> [accessed 9 April 2025].
28 Jennifer Cockrall-King, *Food and the City: Urban Agriculture and the New Food Revolution* (Prometheus Books, 2012).
29 For example, see Marc Mennessier, '*Versailles: le Potager du roi est bel et bien en péril*', *Le Figaro*, 30 October 2019 <https://www.lefigaro.fr/jardin/versailles-le-potager-du-roi-est-bel-et-bienen-peril-20191030> [accessed 9 April 2025].
30 '*Peut-on encore réussir à conserver un métier d'art unique?*', Les Amis du Potager du Roi <https://www.amisdupotagerduroi.org/wp-content/uploads/2017/11/interview-JB.pdf> [accessed 9 April 2025].
31 Mennessier, trans. by the author.
32 '*Faute de pouvoir s'offrir le Potager du Roy, on se fait roi du potager*' (Christian Millau, 'Potager (Les Rois du)', in *Dictionnaire amoureux de la Gastronomie* (Plon 2008), p. 640).

36
The Garden
Keynote Address

Carolyn Steel

> *'Il faut cultiver notre jardin'*
> —Voltaire

What is a garden? The idea is so deeply embedded in our culture and imagination that the question seems faintly absurd. We all have images of gardens in our minds: perhaps those we played in as children; those we might enjoy relaxing or working in now; ones we've had parties or barbeques in; famous ones we've visited or seen in films. For Christians, Muslims and Jews (whether or not they believe in an afterlife) there is also arguably the most powerful image of all, the Garden of Eden. The significance of Eden, with its mythical depiction of an ideal habitat from which humans have been expelled (and to which, according to certain beliefs, we might return if we lead good lives) would be hard to overstate. For roughly half the global population, it holds within it something fundamental about our understanding of ourselves and of our place in the universe. It is both a race memory of a real past life (when our ancestors lived as hunter gatherers) and an acknowledgement that, in our rush to embrace modernity, we've left something essential behind.

Most of us in the modern world see gardens as a kind of refuge from the stresses of everyday life: as spaces of contemplation, peace, and healing, as well as of fun, relaxation, and renewal. In many ways, such feelings echo the way Indigenous peoples have reportedly felt about their environments: the sense that the landscapes they inhabit are benign, comforting, and magically life-giving seems quasi-universal.[1]

Our modern response to gardens and those of Indigenous peoples to their environs are in essence the same. Just as gardens are modified versions of nature, so are Indigenous habitats. Despite the assumptions made by early modern European explorers that the landscapes they encountered in the 'New World' were wild and untouched by human hand, they were often something else entirely. Although the original inhabitants of North America and Australia were predominantly hunter-gatherers, they nevertheless used highly sophisticated methods to modify their territory to suit their needs, notably by using controlled burning to create clearings and clear bush in order to improve their hunting. In Australia, such modifications produced landscapes that were so neatly groomed and naturally pleasing to British travellers that many described them in terms such as 'very picturesque' and 'like an English Park'.[2]

The irony of this misreading was that the British who so admired the Australian landscape assumed that it was natural, despite the English parks of which it reminded them being profoundly artificial. English landscape gardens (whose mid-eighteenth century popularity coincided with the British colonization of Australia) purported to be natural, yet were in fact created through a variety of contrivances such as the planting of trees in scenic clumps in the midst of grazed lawns, the use of ha-has (disguised ditches) to exclude farm animals, the damming of rivers to create scenic lakes, and even the removal of entire villages so as not to spoil the view from the great houses whose command over nature they were designed to express.

This clash of cultures reveals the power of the garden as an idea. Had they but realized it, the Europeans who gazed upon the inhabited vastness of the New World were encountering real-life versions of Eden. Like the mythical Eden, the landscapes they admired had been adapted to suit human needs; not by God, but by their human inhabitants. Indigenous Americans, Australians, and Africans might not have called their habitats 'gardens', yet neither did they think of them as wild. To them, the landscape was a living entity full of meaning and possibility, as familiar as fields and streets were to Europeans, yet without any such artificial divisions. The categories by which Europeans described and valued their land – wilderness, forest, farm, orchard, garden, and so on – were unknown to them; yet of all such categorizations, the one that perhaps comes closest to capturing their sense of intimate connection with the living world is 'garden'.

What then, returning to our original question, is a garden? The word itself derives from the Old English *geard*, meaning fence, from the Old French *gardin*, of Germanic origin.[3] Etymologically speaking, then, a garden is an enclosed space, in common with related words such as yard, *garten*, *jardin*, *giardino*, *hortus*, paradise, park, *parc*, *parquet*, court, *hof*, and town.[4] In essence, a garden is an area of land distinct from nature in its wild, unmodified state. Erecting some kind of barrier such as a hedge or fence effectively creates a garden, since it interrupts natural flows across the site, which in time will lead to a different ecosystem within the enclosed area. Such barriers are of course usually erected with some greater purpose in mind, such as the provision of amenities for the owner, whether for sustenance, pleasure, or show. Despite the etymology, however, the creation of a garden need not involve any enclosure: many Indigenous gardens, for example, consist of 'curated' sections of forest, where clearings have been made and certain plants selected for cultivation.

Whatever defines its boundaries, a garden is simply distinguished as an identifiable patch of land that has been modified in some deliberate way. Quite where this leaves the Garden of Eden, which supposedly embodied the entire world in its pristine state (albeit made by God to suit mankind) is moot. In many ways, this ambiguity lies at the heart of the garden's power, both as a physical space and metaphor. Consciously or not, every garden poses the same two questions: what is our human relationship with nature, and what, from our point of view, might an ideal form of nature look like?

Great gardens of the world have invariably addressed these questions. Among other things, they have represented perfect worlds that have been lost: idealized versions of nature in the present, and future paradises we might regain, either here on earth or in some heavenly realm. As such, gardens through time have represented no less than a curated philosophical and spiritual enquiry into the nature of human existence.

Simultaneity

The capacity to be simultaneously a living space and a metaphorical device is a quality shared by all gardens. This is due to the fact that the metaphysical substance of gardens (the fusion of nature and culture) is inseparable from their mode of production. Placing a potted bay tree on one's doorstep may not seem a grand gesture, yet the ontological space it creates is no less meaningful than that which animates the great horticultural masterpieces of the Italian Renaissance. Since all gardens are born of our relationship with nature, they cannot help but express it. While the gardens of the Villas d'Este or Lante are based on complex allegorical programmes detailing our human journey towards civilization or the contrasted paths of Hercules or Apollo, the bay tree on one's doorstep creates a cultural space that speaks of its owner, suggesting that he or she cares about nature and probably enjoys cooking. Odd though it may seem, the well-spring of these seemingly disparate gestures is one and the same.

A key question concerning gardens is the extent to which their purpose is to feed us. Once again, the paradox of Eden is invoked here, since the Garden is unambiguously geared towards feeding humans, albeit in their pre-Fall, fruitarian state. This is the sense in which the curated landscapes of the New World were closer to Eden than any then remaining in early modern Europe, which had already been sub-divided into categories that designated, *inter alia*, their relative productiveness and consequently their perceived value. In societies that contrasted city and country, field and street, wild and tame, the garden's place was always ambiguous. While the primary purpose of farms was clear (producing food), that of gardens was less so; while kitchen gardens had the same clear purpose, others were purely ornamental, and many were somewhere in between.

The vast variety of forms and meanings that gardens have assumed makes them a unique reflector of cultural ideas and values over time. Indeed, to trace the history of the garden is to track, not only the evolution of our species as it entered modernity, but that of civilization itself. This evolution is most evident in Europe and Asia, where gardens emerged in tandem with agriculture and cities, with which they were in constant dialogue. Indeed, gardens of the global north can only be understood in the context of the urban civilizations to which they have belonged and of which they provide a unique mirror.

Ancient Origins

The origins of gardens stretch back much further than the orderly enclosures we might think of today, to our ancestors' efforts to modify the landscape with fire, which recent evidence suggests date back at least 90,000 years.[5] Whenever such labours began, it seems likely that they would have coincided with an emergent imagination among our forebears that led some 45,000 years ago to the first works of art: evidence that our ancestors were becoming aware of themselves as independent actors in the world.

This moment of self-consciousness and the vast, unanswerable questions that it raised – who we are, what the world is, and what our place is in it – reverberate throughout garden history. At their core are two threads that have at times coincided and at others diverged: our dual quests as humans for sustenance and meaning. As well as being living entities, gardens bear the imprint of their makers' science, philosophy, spirituality, and power. Gardens both embody the cultures from which they spring and reinforce them.

The failure of eighteenth-century British travellers to realize that the beautiful vistas they encountered in Australia and North America were to some extent curated, just as theirs were back home, sprang partly from the contrasting world views that gave rise to such landscapes. Whereas Lancelot 'Capability' Brown famously saw the natural world as a canvas upon which to impose a heightened, idealized version of itself, Indigenous Australians saw their land as a territory created by God during the Dreaming, a time that would last for all eternity. Their duty, according to the Law by which they lived, was to look after the land and to leave it for future generations just as they had found it. As Bill Gammage has noted, aboriginal Australians did not see their land as 'nature' as Westerners did; for them it was, quite literally, heaven on earth.[6]

In many ways, the arrival of Europeans in the New World represented civilization coming full circle, revealing not only what had been gained during roughly ten millennia of technological advancement, but also what had been lost. While Australians had found a stable way of dwelling in their land that provided them with both sustenance and meaning, the sculpted parks of Capability Brown spoke only of dominance over nature and over those whose stories didn't fit the narrative. The fact that English gentlemen's parks often occupied land confiscated through enclosure and that the farms that fed the nation (including slave plantations overseas that provided much of the cash) were excluded from view speaks volumes. English landscape gardens were remarkable, not just for their beauty, but also for what was left out of the frame.

How humans went from dwelling in nature to declaring command over it is the civilizational story that gardens both reflect and refract. Gardens have much to reveal about our changing relationship with nature and to suggest how we might regain the spiritual symbiosis that once characterized our sense of our place in the world.

Agriculture: Egypt

The fact that gardens first emerged alongside agriculture is no accident. The domestication of plants and animals that began around 12,000 BCE in the ancient Near East and subsequent planting of fields, orchards, and kitchen gardens in order to feed emergent cities established conditions under which other kinds of enclosed outdoor space would eventually emerge. In both Egypt and Mesopotamia, planting food crops necessitated the building of levees and channels in order to control the annual floodwaters upon which both civilizations depended, as well as protective walls to keep out marauding animals. By around 3500 BCE, the riverine banks of the Nile and Euphrates featured dense patchworks of farms and gardens. Cereals such as wheat and barley grew on the floodplains, while date palms flourished on the levees alongside apple trees, figs, and pomegranates. Meanwhile kitchen gardens grew enough produce to please any chef, including chickpeas, lentils, beans, onions, garlic and leeks, cucumbers, cress, mustard, lettuce, and grapes.

Whether or not such gardens were the first to be created (evidence suggests they go back as far as 6000 BCE), what seems likely is that farms and gardens emerged in tandem in the ancient Near East, from the same essential motive of producing food. The fact that the region just east of the Mediterranean known as the Fertile Crescent was the real-life Garden of Eden emphasizes this connection. The story told in both the Bible and Quran of man's expulsion from Paradise speaks of a real place where our ancestors' long transition from life as hunter-gatherers to farmers began: a transformation that would set the course of human history.

If sustenance was the motive behind the world's earliest gardens, meaning was the second. In both Mesopotamia and Egypt, the annual floodwaters that brought fertility to the land gave rise to creation myths whose cyclical themes of death and resurrection mirrored the agricultural seasons. While the Tigris and Euphrates frequently changed course and caused catastrophic flooding, however, the Nile was steady, laying the same fertile strip of black silt (albeit of differing heights) in the red desert every year. This reliability created an urban society of unmatched longevity and a belief in eternity to match that of Indigenous Australians. Much attention in Egypt was given to making preparations for the afterlife, especially for the pharaohs whose divine task it was to preserve *Maat* (cosmic order), who were expected to live on in the splendour to which they had become accustomed in their earthly life.[7]

Temple and palace gardens (which were sometimes merged, since pharaohs needed gardens in both their earthly and heavenly lives) were surrounded by high walls and combined date palms, fruit trees, vegetable patches, and fish ponds with pavilions and vines trained over pergolas to give shade. The most famous image of such a garden – from Sennufer's tomb in Thebes, dating from 1400 BCE – combines all these elements in a grand design that feels remarkably similar to those one can still find in the Middle East today. Tellingly, the entrance gate opened directly onto water, on the symbolic boundary between life and death.[8] Temple sanctuaries were sacred spaces where priests performed rituals reenacting the moment of creation: at Karnak, for example, a pyramid-like mound rose from a sacred lake, representing the first emergence of land from the primordial flood, where geese were released, as the god Amun was said to have done at the dawn of time.

Geography gave rise in Egypt to a unique society in which the Nile brought ceaseless fertility and the desert afforded ideal protection. Once the Egyptians had learned to harness the river's ebb and flow (which took little effort, since the annual floods brought fresh topsoil and water every year), there was little reason for them to doubt such a way of life could last forever. Indeed, rural stretches of the Nile still bear a remarkable resemblance to depictions of farming to be found in 3000-year-old tombs. What is missing from this picture, of course, is the vast state machinery required to build the monumental structures that housed the pharaohs through eternity. Monumentalism requires hierarchy and workers who must be fed, as the vast grain silos that formed part of every major temple complex testify.[9]

In Egypt, clear distinctions thus emerged between sacred representations of nature in temple sanctuaries, agricultural landscapes, and a range of domestic and palace gardens that combined elements of both. Egypt thus helped to establish the ambiguous space in which gardens in urban societies would henceforth sit.

Cities: Mesopotamia

Far more is known about Egyptian gardens than those of Mesopotamia, due to the fact that much more evidence of the former is preserved. Yet, as the real Garden of Eden and cradle of agriculture and urbanity, Mesopotamia also nurtured gardens as we know them today.

Early evidence of gardens in Mesopotamia comes from the *Epic of Gilgamesh*, a 4000-year-old text that describes the Sumerian city of Uruk (often cited as the world's oldest) as being roughly one-third date-groves.[10] Uruk is also described as a 'sheepfold', suggesting walled enclosures to keep the animals safe from predators. In the *Epic*, King Gilgamesh and his wild friend Enkidu visit a sacred cedar grove outside the city, a wondrous place that enchants them: 'They stood there marvelling at the forest, gazing at the lofty cedars'.[11] Against Enkidu's advice, however, Gilgamesh kills the keeper of the forest Humbaba, with dire consequences for them both (Enkidu dies, leaving Gilgamesh heartbroken).

The *Epic of Gilgamesh* is remarkable in the degree to which it can be read as a philosophical treatise on the need to balance urban life with respect for the natural world. Gilgamesh's failure to understand this lesson proves his downfall, one from which he spends the rest of the *Epic* trying to redeem himself. The fact that such thinking was embedded in Sumerian culture suggests a powerful love of nature that continued to flourish in the region. That the world's second most famous garden (after Eden), the so-called Hanging Gardens of Babylon, were created here suggests a deep continuity among the Sumerians, Akkadians, Babylonians, and Assyrians who lived here from 5000 to 700 BCE, whose shared love of gardens would lay the foundations for garden culture across Europe and Asia.

The Assyrian empire was a vast territory teeming with wild game including lions, wolves, boars, jackals, and gazelles. As farming gradually replaced hunting as the principle means of feeding people, the latter became a favourite royal sport: by the first millennium BCE, Assyrian kings had established hunting parks close to their cities and boasted of their prowess in the chase.[12] The spread of empire also created new kinds of gardens, as conquering kings brought home plants and animals from distant lands and established them in royal parks that were effectively the world's first zoological and botanical gardens. But perhaps the most influential gardens of all were the famous 'hanging' ones, now thought to have been not in Babylon, but Ninevah. A 700 BCE bas-relief of King Sennacherib's garden in that city shows a magnificent series of terraces planted with date palms and fruit trees connected by water channels fed from the river. On top of the hill is a pavilion, in an ideal position to catch the breeze and enjoy the view. Whether or not Sennacherib's terraces were the mythical hanging gardens, they certainly displayed a scale of ambition and level of sophistication to suggest that the era of pleasure gardens had fully arrived.

Paradise on Earth: Persia

One more element was needed before such gardens evolved into arguably the most popular and widespread type of all, the paradise garden. Around 550 BCE, the Achaemenid King Cyrus the Great built a walled garden at Pasargadae in Persia, divided into four sections separated by stone water courses. The first known example of a *chahar bagh* (from Persian *chahar*, four + *bagh*, garden), its quadripartite design echoed pottery images of the world dating back as far as 4000 BCE. Geometrical representations of the earth were therefore ancient, yet their incorporation into garden design would bring new meaning for a religion in which gardens would reflect paradise on earth, Islam.

As Islam spread across Asia, India, North Africa, and Europe from the seventh century onwards, the making of gardens became a way of embodying, celebrating, and taking pleasure in the perfection of god's creation. Since representations of the human form were considered idolatrous by Muslims, geometry became the dominant element in Islamic art, with elaborate tiled patterns and ornate designs echoing the beauty and complexity of nature. Paradise gardens (from the Persian *pairi*, around + *daeza*, wall) were mostly walled *chahar baghs* whose perfect geometry, fragrant fruit trees, and flowing water mirrored the perfection of heaven. Their four water channels, which often met in a central fountain, invoked the four rivers of water, wine, milk, and honey which were said in the Quran to flow out of the Garden of Eden. As well as representing the four corners of the world, therefore, such gardens represented heaven, reinforcing the connection between the two.

Such ideas gave rise to some of the most famous and beloved gardens in the world, from the Alhambra in Spain to the Taj Mahal in India. Brilliantly adapted to the hot climates in which they were typically built, such gardens reflected not only places of delight and contemplation, but a way of life that sought to mirror the perfection of god's creation on earth.

Contemplating Nature: China and Japan

India marked the eastern extent of paradise gardens, since in China another religion, Taoism, already flourished, and with it, a very different view of nature. According to the fourth-century BCE *Tao Te Ching*, the main goal in life for both humans and animals was to live in harmony with the *Tao* (universe), which their spirits would all join after death. A complex concept whose meanings include the eternal process of transformation that creates reality, *Tao* describes not only the world, but also the lifelong task of learning to live in harmony with it, often translated as the 'way'.

Taoism's deeply spiritual view of nature gave rise to a powerful garden tradition. Chinese gardens were effectively miniaturized landscapes, featuring craggy rocks to represent mountains (revered as a symbol of nature's creative force) and stretches of water to represent the lakes and rivers that were the earth's life-giving arteries. Designed to stimulate the imagination and the senses, such landscapes were intended to draw people closer to nature and invite contemplation of the universe in all its vastness. Winding paths often linked a series of pavilions and bridges where one might stroll and talk, write poetry, or gather in order to admire the blossoming of a cherry tree or rising of the full moon. Natural phenomena such as the movement of planets and the changing of seasons were revered, since they echoed the creative power that gave rise to nature's ever-shifting forms.

Never meant to be seen all at once, Chinese gardens revealed themselves slowly in a series of carefully composed tableaux, as one strolled along their meandering paths, perhaps pausing on a bridge to admire a weeping willow or gently flowing stream, whose sinuous curves ensured one could never see its end. William Chambers, whose 1757 *Dissertation on Oriental Gardening* was pivotal in bringing knowledge of Chinese gardens to Europe, observed that water could occupy one third or more of a Chinese garden. 'They compare a clear lake to a rich piece of painting,' he wrote, 'upon which the circumambient objects are represented in the highest perfection; and say, it is like an aperture in the world, through which you see another world, another sun and other skies.'[13]

Garden design and landscape painting were closely aligned in China, to the extent that gardens might imitate paintings, and great gardens were often designed by famous landscape painters. The two arts were inseparable, since their aim was effectively the same: to observe nature and, through such observation, to gain a sense of wonder and peace: in short, to learn the *Tao*. As Ji Cheng wrote in his influential garden treatise of 1631 *Yuanye*, 'Let your feelings dwell among hills and valleys; there you may feel removed from all the unrest of the world. In your fancy you enter a painting.'[14]

Surrounded by high walls, Chinese gardens were meant to bring a sense of tranquillity, even in the heart of the city. The intimate relationship between house and garden – and more broadly between nature and architecture – was essential to everyday life in China (and later in Japan, where Chinese garden culture had great influence). Life was lived as much outdoors as indoors, and house and garden were reflective parts of one continuous whole, where one might gather on a terrace in order to eat, read, meditate, or perform, or in a pavilion in order to watch as the full moon cast bamboo shadows across the garden wall or the sliding translucent screens of the house.

As Osvald Sirén observed, the Chinese love of gardens was 'doubtless [. . .] due to the uncommonly intimate kinship with Nature that is part of the Chinese temperament. They have listened to the thoughts of Nature and felt the beat of her pulse in quite a different way from ourselves. Consequently, the garden has meant something more for them than for us'.[15]

Natural Science – Ancient Greece

The Chinese reverence for nature was far from unique in the ancient world. In ancient Greece, the gods were said to reside on the wild slopes of Mount Olympus, and immersing oneself in nature was similarly seen as the key to spiritual awakening. The Greek landscape was dotted with shrines and sanctuaries dedicated to various deities, with features such as springs and caves associated with particular gods, muses, or sybils, who, it was believed, could predict the future. In short, the natural landscape was as bound up with sacred myth and symbolism as it was for Indigenous Australians.

The essential difference was that the Greeks lived in an urban-agrarian society, and therefore had as much reverence for Demeter, the goddess of earth and agriculture, as they did for Artemis, the goddess of hunting. As was the case in Egypt and Mesopotamia, the Greek calendar mirrored the agricultural seasons: at harvest time, Athenians decamped to Eleusis (the mythical fields where Demeter was said to have first taught man to cultivate) to celebrate the Eleusinia, a nine-day festival of ritual, fasting, and feasting.

Leaving the city in order to find oneself in nature was an idea endemic to ancient Greece. Young men went hunting in the woods, not only to develop their physical strength and agility, but also to find their spiritual selves and thus progress to manhood. This emphasis on developing mind and body was also central to the practice of philosophy. Plato's Academy and Aristotle's Lyceum were both situated in sacred groves just outside the city, where teachers and pupils took part in naked gymnastic exercises, before putting on robes in order to study natural science or debate the nature of existence while strolling through the groves. Aristotle's Lyceum also included a physic garden for the study of plants, which his pupil Theophrastus took over after his death, publishing an *Enquiry into Plants* that became an important influence on Renaissance science, earning its author the title 'Father of Botany'.

The fact that these twin crucibles of Western thought were situated in garden-like spaces somewhere between the city and wilderness feels significant. Like the Mesopotamians before them, the Greeks were acutely conscious of the need to balance urban life with nature; indeed, the Athenian *agora* – the democratic city's food market and early political space – resembled a garden more than an urban territory: irregular in shape and dotted about with ancient shrines, with an earthen floor and clumps of trees planted for shade, the *agora* probably felt more like a sacred grove to Athenians than a city square.

The *agora*'s attraction as a bucolic space in the heart of the city was no doubt intensified by the fact that Greek cities were generally compact and fortified, with little space for gardens. Although larger houses from the fourth century BCE developed peristyle garden courts, the Greeks' chief contribution to garden history is arguably their input to the narrative within which gardens sit, namely that of the relationship between nature and culture, and the question at its core: that of how to eat.[16]

Aristotle argued in his *Politics* that acquiring food from nature was 'natural' for humans: since we need to eat plants and animals in order to survive, he reasoned, they must exist 'for the sake of man', since 'nature makes nothing without some end in view'.[17] This somewhat utilitarian outlook (which mirrored that of the Old Testament) placed humanity firmly at the top of the food chain, in command of the natural world, but without the constraints placed upon Adam and Eve to avoid the acquisition of knowledge. Indeed, Aristotle's thirst for knowledge led him to become the first to try to explain natural phenomena through observation, effectively making him the father of modern science.

Aristotle took a similarly pragmatic approach to the question of how cities should feed themselves, concluding that the ideal arrangement would be for each citizen to have a house in the city and a farm in the nearby countryside from which to feed it. Such a system of *oikonomia* (household management) would render the *polis* self-sufficient, provided it remained small enough for the arrangement to work.[18] This model – which effectively outlined the economic principles of the city-state for the first time – would prove remarkably influential, not least in the utopian movement of which it effectively formed the blueprint. Cities and farms, the Greeks suggested, were locked together in a symbiotic dance in which gardens would henceforth have to find their place.

Otium and *Negotium* – Rome

With a capital city of one million citizens by the first century BCE, the Roman reverence for farming is perhaps unsurprising. The inhabitants of the world's first metropolis were acutely aware of their dependence on distant lands to feed them: much of the empire's expansion was driven by it. In place of *Oikonomia* came *pastio villatica* (villa farming), in which villas and farms close to the city produced foods such as fruit, vegetables, poultry, and game for the luxury market. *Pastio villatica* made farmers a fortune, yet not everyone approved: as Martial noted, Rome had once grown its own grain and imported roses from Egypt; now it merely grew roses and relied on Egypt for its bread.[19]

The Romans considered *ager*, the cultivated land around a city, to be sacred alongside *civitas*, the city itself. Together, city and country formed civilization's essential partnership. As Cicero wrote, 'We sow corn, we plant trees, we fertilise the soil by irrigation, we dam the rivers and direct them where we want. In short, by means of our hands we try to create as it were a second nature in the natural world.'[20] Unlike the Greeks, however, the Romans had little time for wilderness. For them, *foris* (forests and woods) were merely uncivilized tracts of land waiting to be tamed, while *saltus* (unproductive land, such as mountains) was far beyond the pale.[21]

Their view of nature gave Romans an especial love of gardens. Here was nature as it was meant to be: orderly, neat, and productive. Town and country villas were typically arranged around gardens, entered through a garden-like *atrium*, with a central pool open to the sky to collect rainwater, often with a peristyle garden beyond. Like those in China, Roman gardens were distinctly theatrical, yet in an almost opposite sense: while Chinese gardens aimed to inspire wonder in nature, Roman ones mostly expressed dominance over the natural world. This did not, however, lessen the Romans' love of nature; on the contrary, villas took on special meaning for them, as a welcome respite from the noise and hassle of the metropolis. After a hard day of *negotium* (business) in the city, Pliny noted with satisfaction, he could retire to his villa for some much-needed *otium* (rest). The idea of the country villa as a refuge from the city set up a dynamic that we still recognize today.

The most spectacular such villa is that of Hadrian at Tivoli: a vast complex of gardens, buildings, and water features whose various elements memorialized the emperor's travels through his empire. Alongside such personal spaces as the Maritime Theatre (a private house, art gallery, and library set within a small lake) were spectacular set-pieces such as the *Serapheum*, a great hemispherical dining hall through which water coursed from a *nymphaeum* to the rear into the *Canopus*, a large reflecting pool representing the Nile. Feasting in his great hall, the emperor could survey his *ecumene* (inhabited world), while immersed in its elemental power, enjoying its earthly fruits and blessed by divine spirits.

Rarely have the arts combined to create such a complete expression of temporal power, animal enjoyment, and divine inspiration all at once.

With its monumental scale and combination of theatrical bravado with intimate reflection, Hadrian's Villa took garden design to new limits, signalling a new sensibility that would reverberate through Europe for centuries to come. Implicitly and occasionally explicitly, Hadrian's Villa formed the blueprint for gardens of the powerful in which two contradictory themes – global conquest and inner reflection – would sit in uneasy juxtaposition.

Hortus Conclusus – Medieval Europe

The centuries of instability that followed the fall of Rome meant that gardens in Europe once again retreated behind fortified walls. In the limited space afforded by castles and monasteries, gardens took on special resonance, both as places of production and contemplation. Like the paradise gardens of which they were in effect the Christian counterparts, cloister gardens used geometric forms to convey sacred meaning. Typically square in shape with central fountains and quadripartite beds planted with herbs or fruit trees, they symbolized the four-quartered earth as well as Heavenly Jerusalem.

Their dual function as spaces of production and contemplation meant that gardens played a central role in the Christian search for a virtuous life. The founding fathers of Western monasticism, St Augustine and St Benedict, both advocated a life of work, service, and prayer, a programme that early monasteries evolved to accommodate. The ninth-century Plan of St Gall, which shows an idealized scheme for a monastery, gives a rare glimpse into the role of gardens in early monastic life. A rectangular precinct is shown containing a church and cloister, library, school, infirmary, kitchens and dining halls, guesthouses, and cemetery, surrounded by a range of gardens, farms, and outbuildings. Cowsheds, stables, and sheds for goats, sheep, and pigs are sited close to the brewery and bakery, giving the animals ready access to the leftover bread and grain, while geese and chicken houses are sited next to the vegetable garden, presumably so the birds could run about gathering garden scraps, fertilizing the soil as they went. A herb garden is beside the apothecary, while the cemetery doubles as an orchard, so the brothers' decomposing bodies could bring forth new life to hazel, almond, pear, and chestnut trees.

The ecological circularity of such arrangements, which today would be recognized as highly sustainable, was typical of medieval cities, which, like the castles and monasteries they resembled, were compact and fortified. Gardens played a crucial role in all such communities, not only as precious sources of sustenance, but also for much-needed respite from the cramped environs. Herb and vegetable gardens, orchards, and vineyards were places to sit and reflect in the welcome embrace of living greenness.

The introspective nature of medieval gardens was epitomized by the tautological *hortus conclusus* (enclosed garden), a first mentioned in the Hebrew *Song of Solomon* in allegorical reference to a spotless maiden later taken by Christians to symbolize Mary's immaculate conception.[22] A small square of perfect lawn surrounded by neatly trimmed bushes of fragrant herbs and flowers or a raised grassy seat, with perhaps a white rose to signify Mary's purity or a red one for Christ's suffering, the *hortus conclusus* was a contemplative space heavy with symbolism, intensified by the tradition of painters such as Fra Angelico of representing the annunciation as having taken place in just such a garden.

Beyond the monastery walls, however, a very different view of nature was developing in northern Europe. Clearances for agriculture, combined with the spread of Frankish and Gothic tribes for whom hunting was essential to social prestige, brought new pressures to bear. Disputes over territory became common as various powers – including the monasteries – vied for exclusive rights over the forest. When William the Conqueror defeated Harold in 1066, he lost no time in annexing one quarter of his new kingdom as 'royal forest', territory in which only he was allowed to hunt. The fact that this 'forest' included tracts of pasture, meadow, farmland, and even towns left locals used to snaring the odd rabbit for the pot excluded from the woods upon which they had always depended.[23] With the Norman Conquest, forest transitioned from being seen as a threat to civilization to a vital resource which the rich, whose need for it was the least, nevertheless claimed for themselves.

Third Nature – Renaissance Italy

The tumultuous years out of which the Renaissance would spring gave gardens a somewhat polarized role. Confined within protective walls, the *hortus conclusus* retreated into a world of symbolism and hope for a better life hereafter, while a wilder, more pagan version of a good life raged outside. The contrast between these two worldviews – between spiritual and earthly life, or the Christian and classical worlds – recombined to make Renaissance gardens among the richest ever seen.

As war gave way to trade in fourteenth-century Italy and city-states began to flourish, villas proliferated in the countryside once again, just as they had in ancient times. Like their Roman forebears, villa owners combined farming for profit with a sense of retreat and reflection: balancing the *vita contemplativa* with the *vita attiva* became the new *otium* and *negotium*. As Cicero had noted, country life was not so much about rest, as study. When Cosimo de Medici welcomed some distinguished guests to his villa at Careggi in 1462, he told them, 'I come to the villa of Careggi, not to cultivate my fields, but my soul.'[24]

The rediscovery of classical texts during the Renaissance led to renewed interest in man's place within the natural world, complicated in 1415 by the discovery of perspective, which created the so-called 'subject-object problem' in which human subjects were divided from their chief object of contemplation, nature. Instead of being part of the natural world, humans now saw themselves as somehow outside it, studying it and perhaps controlling it too.

Humanist philosophers distinguished between *natura naturans*, nature's creative force, and *natura naturata*, nature as modified by man. Such distinctions made the place of gardens deeply ambiguous. If, as Cicero had suggested, agriculture represented a kind of 'second nature' distinct from wilderness, what, then, were gardens? In 1541, the Italian humanist Jacopo Bonfadio suggested that garden design effectively created a kind of 'third nature', the culmination of humanity's progress from living in wilderness, through farming, to the application of art and culture to nature.[25] The view from Bonfadio's garden, in which a tidy parterre gave way to green pastures with mountains looming in the distance, reflected not only man's place within nature, but the story of human development itself.

Such themes fed directly into the great gardens of the *cinquecento*, such as those of the Villa d'Este at Tivoli (1560–) and Villa Lante at Bagnaia (1568–). With plentiful references to classical mythology, such gardens were effectively commentaries on the story of civilization.[26] At the Villa Lante, for example, one entered through the *bosco* (wild forest), symbolizing the Golden Age, then climbed up to the water source at the top of the hill, which referenced both to nature's creative force *natura naturans* and the Flood, following the water down through a *catena d'aqua* (a 'chain of water' flowing down the carved handrail of a stair), past a fountain of the river gods to a terrace with a vast stone dining table, also with water flowing through it, in the manner of the ancient Romans. Below the dining terrace, two pavilions overlooked a formal parterre, representing *natura naturata*, beyond which was a view over the town, representing urban civilization.

In the villas of the Italian Renaissance, the philosophical quest for knowledge, the desire to express power through mythical narrative, supreme theatricality, and even humour combined to create what many see as the apotheosis of Western garden art. Yet it was arguably another development in gardening taking place at the same time that would have a greater influence over future generations, and that was not to do with art, but science.

From Alchemy to Botany

Renewed interest in natural sciences during the Renaissance was greatly enhanced by the sea exploration that was then changing the face of the globe. As mariners brought back exotic foods, spices, and plants from distant lands, Europe experienced a botanical Renaissance, as physic gardens, where herbalists had long studied plants for their healing properties, received a sudden influx of unknown species, leading to the foundation of greatly expanded gardens designed to house and study the newly global community of plants.

Padua University's *Orto Botanico*, founded in 1545 by the Venetian Republic conveniently close to the port where all the plants were arriving, was among the first. The oldest botanical garden still on its original site, it was laid out as an 84-metre circle within a pond (representing the world surrounded by the sea) divided into four parts aligned to the cardinal points to represent the four corners of the globe. Today, the *Orto Botanico* houses a library of 50,000 botanical volumes, including some of great historical significance. Similarly august was Leiden University's *Hortus Botanicus*, founded in 1593 close to the ports of Amsterdam and Rotterdam, where the Dutch East India Company (VOC) brought live plants and dried specimens at the behest of Leiden's prefect Carolus Clusius, who during his lifetime cultivated some 1000 different plants and was responsible for bringing a number of species to Europe, including the tulip, potato, and horse chestnut.

Enthusiasm for growing exotic plants led to renewed efforts in greenhouse design in Europe. Although greenhouses had existed since Roman times (Pliny the Elder mentions *specularia*, wheeled beds used to grow crops such as melons that could be moved into the sun or covered with transparent stone to keep in the warmth), the use of glass and heat were now essential if tropical fruits such as pineapples were to be cultivated in northern climes.[27] Perhaps unsurprisingly, the first to achieve such a feat, Pieter de la Court, lived near Leiden. After more than a century of botanical endeavour (de la Court's first pineapple ripened in 1675), the Dutch were now the world's leading horticulturalists, a position they maintain to this day.

Pineapple Fever

In many ways, pineapple cultivation in Europe represented the birth of modern horticulture. The sheer difficulty of growing such a tropical plant in the north (which only fruits after three years and must be kept above 16°C with a high degree of humidity throughout) tested horticulturalists to their limits and required extraordinary ingenuity and resources. The key was to bury the pineapples in spent tanners' bark, or 'tan': wetted oak bark that gave off enough heat while fermenting to keep the beds heated to between 24–30°C for a remarkable three to six months.[28] Combined with glass houses heated by wood-fired stoves (known for this reason simply as 'stoves'), pineapples began to thrive, not only in botanical establishments such as the Chelsea Physic Garden, which built its own stoves for the purpose in 1723, but also the gardens of the rich and powerful.

Such challenges, combined with the fruit's exotic appearance and deliciousness (described by one Spanish traveller in 1535 as a mixture of quinces, peaches, and melons, 'but surpassing in excellence all those fruits'), made the pineapple an object of the highest prestige.[29] King Charles II was so thrilled to be presented with one by his gardener in 1675 (almost certainly imported rather than cultivated by him), that he commissioned a portrait to commemorate the event. By 1733, techniques had advanced sufficiently for King Louis XV to be presented with a pineapple that actually had been grown in his vast kitchen gardens at Versailles.

By the mid-eighteenth century, possession of a pinery was not only an essential accoutrement among the aristocracy, but a matter of intense rivalry. John Murray, the Fourth Earl of Dunmore, was so proud of his pinery that he surmounted it with a fourteen-metre-high likeness of the fruit, known as the 'Dunmore Pineapple'. So great was demand in fashionable society and so prohibitive the cost of growing the fruit (the equivalent of around £11,000 today), that it was often served as the decorative centrepiece of a display, surrounded by lesser fruit that was actually meant to be eaten. This way, pineapples could be served multiple times, until they started to rot – a practice that gave rise to a thriving trade in pineapple rental.

Power over Nature – Enlightenment France

Pineapple Fever gives some sense of the passions evoked by gardening in early modern Europe. The scientific and philosophical breakthroughs that began with the Renaissance and accelerated through the age of sea exploration and the Enlightenment was one of astonishing growth, discovery and confidence, in which the relationship between man and nature (always at the core of gardens and gardening) was constantly evolving.

This evolution reached its apotheosis when two mathematical geniuses – René Descartes and Isaac Newton – effectively laid the foundations of modern science, forever altering humanity's relationship with nature. In 1637, Descartes invented a way of mathematically mapping every point in the universe (through the x, y, and z coordinates that constitute so-called Cartesian space), while simultaneously declaring '*cogito ergo sum*', thus effectively placing himself and all thinking humans outside the grid.[30] In his *Principia Mathematica* of 1687, Newton gave Descartes's cogitating human the tools with which to manipulate his newly abstract world, by outlining the principles of classical mechanics such as the laws of motion and gravity. Three centuries later, these two men's calculations were enough to take humans to the moon.

The growing sense of mankind's power over nature, together with a related feeling of detachment, combined to produce arguably the greatest expression of horticultural grandeur and mastery ever seen, at Louis XIV's Palace at Versailles. Having encountered the work of the celebrated gardener André Le Nôtre at Vaux le Vicomte in 1661, Louis lost no time in poaching him in order to start laying out his own monumental garden.[31] Louis' self-appointed title *Le Roi Soleil* expresses the scale of his ambition, not seen since the days of the pharaohs.

Until his death in 1700, Le Nôtre used his extraordinary skill and agility to create a garden to match Louis's escalating power and personal myth-making. Having established an overall sense of order with a grand central axis, the *allée royale*, he ran various radiating and intersecting pathways off it, creating a series of squares and circles that could be landscaped into formal *bosquets* (groves), elaborate fountains, or theatrical set-pieces. In 1669, in response to the newly gathered court at Versailles, Le Nôtre extended the *allée royale* with a mile-long canal along the same axis, with two transverse sections to allow distant parts of the garden to be reached by boat. The rigidity of the garden's geometry was mirrored by the tall trees and hedges that defined its various outdoor rooms, clipped to angular perfection by a retinue of 7000 gardeners. The completed garden covered 800 acres, with eight miles of walkways and an astonishing 1400 fountains, which together consumed more water than the whole of Paris.[32]

Versailles's dominant mythological theme was the sun god Apollo, in whose reflected glory Louis basked. A vast orangery was built to house 2000 trees, whose golden fruit embodied the solar royal theme. The garden reflected both the human triumph over nature and Louis' brilliance in concentrating power around himself. The vast theatre of subjugated nature that radiated outward reflected the ranks of emasculated nobles who bowed before the King. At the core of the whole display was the *allée royale*, stretching westwards from the King's apartment like a ray of light towards infinity. It was a device designed, not to inspire awe in some unseen god, but in the all-conquering 'gaze of the King himself.

Genius Loci – The English Landscape Garden

Given the vast concentration of wealth, power, and will needed to create and maintain a garden such as Versailles, it is perhaps unsurprising that, in a continent as turbulent as Europe, the garden didn't last long. Even before Louis's death in 1715, it was falling into disrepair, as a series of disastrous wars and growing national debt forced the King to retrench.[33] When Antoine Watteau arrived in Paris in 1712, he found a 'picturesque neglect' in its parks and gardens that his paintings helped to frame as the dawn of a new era. Partly inspired by the works of Jean-Jacques Rousseau, a new romanticism was taking hold in France, which compelled people to 'walk in nature and solitude where previously they had [. . .] only walked between tortured yews'.[34]

Long before Versailles's downfall, however, there had been a rebellion against its principles across the Channel. Members of the Royal Society objected to its Cartesian abstraction, arguing that our human response to nature should be based on our understanding of specific places, species, and conditions; one should aim to create natural 'histories', not 'systems'.[35] The new sensibility was summed up by Alexander Pope in his famous exhortation to garden designers to 'consult the genius of the place in all'.[36]

This appeal to *genius loci* lay at the core of the English Landscape Garden. Rather than command nature with geometry, the idea was to tease out the essence of the natural landscape and enhance it, with suitable mythological allusions along the way. For Joseph Addison, such gardens not only revealed the true beauty of nature, but were, in effect, representative of freedom. As Addison noted in 1711, works of art could never compare to the 'vastness and immensity' of nature, whose 'rough careless strokes [. . .] afford so great an entertainment on the mind'.[37]

The new appreciation of 'natural' forms required artful framing: as Shaftesbury noted, a sweeping landscape was best observed from a regular terrace, and a 'haha' could keep cows and sheep in their correct 'second nature' positions while giving the illusion of gardens stretching to the horizon. As Horace Walpole noted of William Kent (the first English gardener to adopt a haha), 'He leaped the fence, and saw that all nature was a garden.'[38]

The idea of gazing at 'first nature' via a pleasingly theatrical 'second nature' was of course directly drawn from classical precedent. As well as sources such as Cicero and Pliny, the inspiration for this vision came from painters working in Italy including Claude Lorrain and Nicolas Poussin, whose mythological scenes set in dreamy landscapes perfectly channelled the recaptured classical spirit.[39] Just as in China, garden design and landscape painting became inextricable; as Pope remarked, 'all gardening is landscape painting.'[40]

The idea of nature as an object of contemplation reached its apotheosis in Europe with the works of Lancelot 'Capability' Brown, whose name remains synonymous with English landscape gardening. Universally beloved today, Brown was far from unanimously admired in his own time. His chief sin, according to his critics, was to have removed all classical allusions from his landscapes and to let nature speak for herself. In place of nymphs, temples, and grottoes, Brown gave his clients clumps of trees, serpentine lakes, and rolling swathes of lawn that swept right up to their houses.

The affrontery of dispensing with classical references was too much for such luminaries as Sir Joshua Reynolds and Sir William Chambers, who lined up to condemn Brown's naturalistic approach. 'Gardening, as far as gardening is an art,' harrumphed Reynolds, '[. . .] is a deviation from nature; for if the true taste consists [. . .] in banishing every appearance of art, or any traces of the footsteps of man, it would then be no longer a garden.'[41] Sir William Chambers, whose fondness for Chinese gardens lent him an especial disdain for Brown's naturalism, declared that Brown's landscapes 'differ very little from common fields'.[42]

The irony of such condemnation was that it was only through supreme artistry that Brown managed to create the illusion of an ideal, untouched English landscape. His contemporaries' blindness to this mirrored their failure to spot the ways in which Indigenous Australians used similar skills to curate their environment, albeit with very different aims in mind.

The Unseen

What was not controversial in Brown's day but is glaringly obvious now, is the problematic nature of the social, political, and cultural conditions that enabled his work. As has been true of every great garden in history (particularly those featuring idealized vistas stretching to the horizon), these included plenty of land and wealth under singular ownership: in other words, extreme inequality. In the case of English landscape gardens, these means were principally founded on land enclosure, aristocracy, and colonialism.

Land enclosures had occurred in England since Elizabethan times, but it was only with the onset of the Agricultural Revolution that they really took hold, as 'improvers' such as Charles 'Turnip' Townshend urged ever-greater efficiencies, which meant, among other things, turning peasants off the land so their fields could be amalgamated to create more productive farms. This new 'second nature' (the most radical transformation of the countryside ever seen) played little part in the framing of landscape gardens, whose only nod to farming was to allow the odd flock of sheep or herd of cows to wander into view. Entire hamlets were removed if their presence interrupted the vista, occasionally to be replaced by idealized farmer's cottages (*fermes ornées*) designed to animate the landscape, much as Poussin or Lorrain might have placed a group of allegorical figures in their paintings.

What was being constructed was an idyllic fantasy of rural life that has, in many ways, remained deeply rooted in the British psyche. Also excluded were the colonial landscapes (particularly Caribbean sugar plantations) that often funded the entire enterprise. The colonial era saw a new breed of landowner emerge, keen to build great estates to emulate those of noble families with historic claims to land. One such was the First Baron Harewood Edwin Lascelles, a Barbadian-born planter and slave-owner who commissioned a vast mansion on his family estate in Yorkshire, Harewood House, later turning to Capability Brown to remodel the surrounding landscape.

Completed in 1771, the house looks out across a formal terrace to a partially wooded valley, at the base of which is a gently curving lake, created by damming the local river. A 1798 painting by J.M.W. Turner shows the far slope grazed by a flock of sheep, thus fulfilling the perfect 'three natures' view from the house. Missing from both from real and painted views, however, was any reference to the magnificent kitchen garden, whose hotbeds and greenhouses were essential to the life of the household, but whose angular, fifteen-foot-high fruit walls would have disturbed the scenic tranquillity of the landscape, and were therefore concealed on an island one third of a mile from the house, shrouded from sight by trees.[43]

The Cavendish Banana

Kitchen gardens were not, however, destined to be hidden from view for long. The expansion of botany during the nineteenth century enabled by sea exploration and evolving science gave new impetus to commercial plant breeding in Britain, whose gardeners and botanists were ideally placed to gather specimens from around the world and to seek new ways to study and exploit them.

One such botanist was Joseph Banks, who travelled with James Cook on his first voyage to Australia in 1768, collecting so many specimens that Botany Bay was later named in his honour. Fateful for the Indigenous Australians whose lives they devastated, such voyages were vastly enriching for those who made it back home: Banks returned to instant fame, becoming President of the Royal Society and advising King George III on the setting up of the Royal Botanic Gardens at Kew. Through the commissioning of botanic expeditions around the world, Banks is credited with bringing 30,000 plant species to Britain, of which eighty now bear his name.

This botanical bonanza was in many ways the precursor of the global food system we know today. Tainted with all the disruption and exploitation of colonialism, it combined the widespread disregard of Indigenous food cultures with the discovery, extraction, and cultivation of certain foods that would later come to dominate global cuisines, first through the magnifying lens of imperialism, and later through the global trade systems that followed in its wake.[44]

One fruit that exemplifies these complex origins is the Cavendish banana. While working as head gardener for the 6th Duke of Devonshire William Cavendish at Chatsworth, the brilliant polymath Joseph Paxton (whose innovations in greenhouse design led to the Crystal Palace) succeeded in cultivating some bananas from Mauritius, which he named *Musa cavendishii* after the Duke.[45] Subsequently shipped off to the Canary Islands, the bananas were reimported to Britain by Thomas Fyffe in 1888, the significance of which only became apparent in the 1950s, when the devastating Panama disease began wiping out Gros Michel bananas, then the dominant commercial variety. A frantic search to find a banana resistant to the disease led to the Cavendish, which today makes up 47 percent of all bananas traded globally.[46] This happy ending may prove short-lived, however, since it now appears that Cavendish bananas are susceptible to a new strain of the disease.

Industrialization and Romanticism

The Cavendish banana makes a perfect metaphor for our modern relationship with food – and the role of gardeners in creating it. The botanical fervour occasioned by sea exploration morphed seamlessly into a global search for plants that could be cultivated at scale to feed increasing populations – a search that was vastly accelerated in the nineteenth century by the arrival of railways, which opened up vast tracts of the 'New World' to European settlers, who lost little time in converting native forests and grasslands into grazing and grainfields. Within a few decades, the basic tenets of modern farming were established, as local flora and fauna were replaced by imported ones, Indigenous people were displaced by settlers, complexity was lost to standardization, and traditional methods dismissed in favour of mechanization.

The folly of all this was not lost on some who sensed its coming; indeed the roots of environmentalism lie in the Romanticism of writers such as Jean-Jacques Rousseau, who stated in his 1755 *Second Discourse* that it was 'iron and corn which first civilised men, and ruined humanity'.[47] Rousseau's antipathy towards modernity and yearning for a simpler life resonated strongly across the Atlantic, where writers such as Ralph Waldo Emerson wrote of the glory of nature and its capacity to heal mankind, while his acolyte Henry Thoreau declared that 'In wilderness is the preservation of the world'.[48]

The new-found reverence for nature in the US had unfortunate consequences for Native Americans, however, since wilderness evangelists such as the Scottish geologist John Muir now demanded that 'special temples of Nature' such as Yosemite Valley should be free of all human inhabitation and designated a national park, conveniently forgetting that that his beloved landscape was partly crafted by the Ahwahneechee, whose home it had long been.[49]

The mutual antagonism of industrialism and Romanticism left gardens in something of a vacuum. The reverence for 'pristine' nature that led to the creation of national parks such as Yosemite effectively robbed gardens of their primary function: to bring humans and nature into intimate contact. National parks were effectively the opposite of gardens, creating a fictional 'nature' from which humans were excluded, yet which they were still supposed to venerate. In this schema, first nature reigned supreme, second nature was invisible, and third nature (the artifice without which no such 'pure' condition could have existed) was denied. Meanwhile, the landscapes that 'second nature' was actually producing – the vast monocultural grainfields of the Great West and the Chicago stockyards where millions of animals were slaughtered every year – were firmly out of sight and mind.

The nineteenth century ushered in a polarized era in which nature was either to be unsentimentally exploited to the hilt, or worshipped as sacred from afar. As the heavenly counterpart to hellish factory farms, national parks represented a new form of exclusion, akin to a second Eden from which humanity had voluntarily imposed a second Fall.

Garden City

The Romantic veneration of nature was in stark contrast to the experience of most people in the nineteenth century, the vast majority of whom still lived on the land. In 1800, just three percent of the global population lived in towns with 5000 inhabitants or more, and in the mid-century sixty-four percent of North Americans were still farmers, the vast majority working on small to medium-sized family farms.[50]

The great transformation wrought by industrialization created a schism between people and land that is still playing out today. The rupture of this ancient bond represents a trauma that modern societies have struggled to heal, giving gardens new meaning. The English Arts and Crafts movement, for example, was a conscious reaction to industrial horrors and an attempt to reconnect humans with the beauty of nature. In his utopian fantasy of 1890 *News from Nowhere*, the socialist and craftsman William Morris imagined an England transformed by revolution into a bucolic idyll whose cities had been rewilded so their streets were once more 'pleasant lanes' and meadows, culverts reverted to 'bubbling brooks', and Trafalgar Square was an orchard full of apricot trees.[51]

One man for whom Morris' vision resonated was Ebenezer Howard, whose answer to the agricultural depression of the 1870s was to construct a network of 'garden cities' in the countryside: small city-states with protected 'green belts' of farmland, that would reinvigorate the rural economy and thus relieve the migratory pressure on cities. Backed by a group of major industrialists, construction began in 1903 on a garden city in Letchworth, Hertfordshire, designed by leading Arts and Crafts architects Parker and Unwin. All went well at first, but the project soon struggled to attract enough investors, leading Howard's backers to renege on their promise to cede most land rents to a community trust, prompting him to resign from the board. Although Letchworth eventually prospered, it fell far short of Howard's original vision to unite city and country through progressive land reform.[52]

Growing Your Own

Despite this failure, the garden city's spirit found fertile soil in post-war Britain, where its ideals of social independence and closeness to nature resonated with a population ravaged by war. In fact, the British had already embraced suburbia in the previous century, as low-density 'commuter belts' followed the spread of railways to create residential districts that supposedly combined the best of urban and rural life. Dubbed 'Metroland', suburbs along the London Metropolitan line were marketed with alluring posters advertising a rural tranquillity available to city-workers for the first time. The British penchant for suburbia was formally enshrined in the 1919 Housing Act, which provided for generous back gardens of 400 square yards in all new public housing, with grants of up to seventy-five percent to help councils build this new vision of the good life. Between the wars, one million such homes were built.[53]

Numerous gardening magazines catered to the new British obsession, advising suburbanites on how to manage their precious patches of grass, crazy paving, and flower-beds; 'mowing the lawn' became a classic Sunday activity for men, while their wives remained indoors cooking the family roast. Today, seventy-eight percent of Britons own a garden of some sort, and gardens represent a kind of freedom for many who cherish their own small patch of nature to look after in a nation where land is at a premium.

The British suburban garden, where millions of sausages are charred to a crisp each year, is one solution to the ancient dilemma of how to reconcile our human needs for society and nature. Another almost equal and opposite answer is the Russian dacha. Originally pieces of land granted by the Tsar to favoured nobles, seventeenth-century dachas were elegant summer residences which, much like Italian Renaissance villas before them, became cultural outposts for the aristocracy. During the Soviet era, however, ongoing agricultural crises led the government to allot small plots of land close to the city for ordinary residents to grow their own food, with simple dwellings no bigger than 25 square metres to allow them to stay overnight. Today, sixty percent of Russians own a dacha, growing a remarkable 40 percent of all the food produced in Russia. Despite the lack of piped water, plumbing, and other conveniences, many Russians spend weekends and much of the summer at their dacha; the Mayday traffic jams out of Moscow, when most people plant their crops for the summer, are notorious.

The mixture of pleasure and necessity represented by dachas has its British counterpart in the allotment. Common across Northern Europe (more people in the south retain links to a family farm), allotments spring from the same dispossession that gave rise to dachas; namely, the progressive exclusion of people from the land. In England, such allotments date back to Elizabethan times, accelerating during the years of parliamentary enclosure to culminate in the 1908 Smallholdings and Allotments Act, which requires councils to maintain 'adequate provision of land' for those who wish to grow their own food. Today, an estimated 330,000 Britons have an allotment, and waiting lists are longer than at any time since the Second World War.[54] For many, the appeal is not just the satisfaction or security of growing one's own food, but the sense of freedom and tranquillity that such activity brings.

Back to the Garden?

The industrialization of farming and gardening raises urgent questions about our relationship with nature. Yet the wonder and pleasure that gardens have evoked for millennia remains intact; indeed, private gardens and family farms arguably represent a last bastion against the industrialization and commodification of nature itself. The right to grow one's own food remains a key component of the idea of a good life for many in the global north as well as the south, where it remains a daily reality for some two billion people. No matter who one is or where one lives, access to a patch of land upon which to work, rest, and play remains a prerequisite for the kind of sovereignty and freedom our distant ancestors took for granted.

How to live well in balance with nature is a question as old as humanity. Today, as more of us live in cities and our lives become ever more disembodied, the role of gardens will be crucial. In our urbanized, digital world, the need to rebalance our relationship with nature has become existential. Reuniting city and country – a task one might describe as inventing *Oikomonia 2.0* – is in fact already underway across the world, as projects ranging from food co-ops and organic box schemes to community gardens and urban farms recognize the benefits of bringing society and nature closer together.

Permaculture, first outlined by Australians Bill Mollison and David Holmgren in the 1970s, embodies such ideals, proposing a recalibration of human habits and habitats to live more in harmony with nature, harnessing natural synergies to create productive ecosystems that will mature over time to provide year-round nourishment with little effort.[55] The idea is the chief inspiration behind forest gardens, which mimic forests to create curated productive ecosystems that sustain a balance between inputs and outputs, just as wild ecosystems do.

In many ways, we've come full circle. Recognition that nature knows best – so clear to our hunter-gatherer ancestors yet so obscure to us – is returning, and with it a call to live differently. Whether this call will be heard in the defining roar of super-modernity remains to be seen; yet if we are to come to our senses, gardens and gardeners will play a key part. For five millennia and more, gardens have been spaces of creativity, contemplation, freedom, and sustenance. Gardens, in short, epitomize much of what we humans need in order to flourish. As Voltaire wrote at the end of his madcap adventure *Candide*, 'we must cultivate our garden'. Today, as we seek meaning and stability in a topsy-turvy world, the sentiment has never been more potent. Gardens are our natural home, and our renewed sense of duty to nurture them is what will ultimately remind us what it means to be human.

About the Author
Carolyn Steel is a leading thinker on food and cities. She is the author of the award-winning *Hungry City: How Food Shapes Our Lives* and *Sitopia: How Food Can Save the World*.

Notes
1. There are numerous examples of this. For instance, a Mbuti elder interviewed in the Congo in the 1950s stated that 'The forest is a father and mother to us, and like a father or mother it gives us everything we need: food, clothing, shelter, warmth and affection', while the Australian anthropologist Mervyn Meggitt noted how the western Australian Walbiri viewed the landscape as a living embodiment of their ancestors, whom they believed to have created it in the past time known as the Dreaming. See Colin Turnbull, *The Forest People* (Simon and Schuster, 1962), p. 14 and Tim Ingold, *The Perception of the Environment* (Routledge, 2011), p. 21.
2. Quoted in Bill Gammage, *The Biggest Estate on Earth: How Aborigines Made Australia* (Allen & Unwin, 2012), p. 16.
3. OED.
4. See Tom Turner, *Garden History: Philosophy and Design 2000 BC–2000 AD* (Spon Press 2005), p. 1.
5. Mike Cummings, Study Offers Earliest Evidence of Humans Changing Ecosystems with Fire, Yale-News, 5 May 2021 <https://news.yale.edu/2021/05/05/study-offers-earliest-evidence-humans-changing-ecosystems-fire> [accessed 17 April 2025].
6. Gammage, p. 123.
7. Egyptians, like their Australian counterparts, believed that heaven was here on earth.
8. For further discussion of Egyptian gardens see Turner, pp. 23–48, from which much of this section is taken.
9. The Ramesseum at Thebes could hold enough grain to feed 3400 families, the population of a medium-sized city (Barry Kemp, *Ancient Egypt: Anatomy of a Civilization* (Routledge, 1989), p. 195).
10. *The Epic of Gilgamesh*, trans. by Andrew George (Penguin, 1999), p. 2.
11. *The Epic of Gilgamesh*, p. 39.
12. As King Ashurbanipal boasted on the walls of Nineveh, 'I killed the lion' (Turner, p. 83).
13. The book was based on Chamber's trips to China during the 1740s. Qtd. in Osvald Sirén, *Gardens of China* (The Ronald Press Company, 1949), p. 17.
14. Sirén, p. 10.
15. Sirén, pp. 4–5.
16. Peristyle courts are courts surrounded by colonnaded walkways, similar to the stoas that surrounded the agora.
17. Aristotle, *The Politics*, trans. by T.A. Sinclair (Penguin, 1981), p. 79.
18. From the Greek *oikos*, house, + *nomos*, management.
19. Neville Morley, *Metropolis and Hinterland* (Cambridge University Press, 1996), p. 88.
20. Cicero, *De Natura Deorum*, qtd. in John Dixon Hunt, *Greater Perfections: The Practice of Garden Theory* (Thames and Hudson, 2000), p. 33.

21 This idea is literal: when Romans founded a city, they ploughed a sacred furrow – the *pomoerium* – inside which was civilization, and outside of which was wildness and barbarity. See Joseph Rykwert, *The Idea of a Town* (Faber and Faber, 1976), pp. 91–93.
22 '*Hortus conclusus soror mea, sponsa, hortus conclusus, fons signatus*' ('A garden enclosed is my sister, my spouse; a garden enclosed, a fountain sealed up'), Song of Solomon, 4:12.
23 Simon Schama, *Landscape and Memory* (Fontana Press, 1996), p. 144.
24 The guests included Marsilio Ficino, founder of the Neoplatonist movement. Qtd. in David Coffin, *The Villa in the Life of Renaissance Rome* (Princeton University Press, 1979), p. 9.
25 Dixon Hunt, p. 34.
26 The Villa d'Este was designed by Pierre Ligorio for the Cardinal Ippolito II d'Este, mostly constructed between 1560–1569. The Villa Lante was designed by Vignola for Cardinal Gambara, commencing in 1568.
27 See Pliny the Elder, *Historia Naturalis*, Book 19, 23:64.
28 For a fascinating and comprehensive account of pineapple growing in Europe, see Susan Campbell, *Charleston Kedding: A History of Kitchen Gardening* (Ebury Press, 1996), pp. 151–65.
29 Qtd. in Campbell, p. 152.
30 René Descartes, *Discourse on Method and Meditations on First Philosophy*, trans. by Donald A. Cress (Hackett, 1998), pp. 18–19.
31 This pivotal moment in garden history was accompanied by Fouquet's being imprisoned (on possibly trumped-up charges of embezzlement) Louis also poached the architect Louis Le Vau and the painter Charles Le Brun from Vaux's owner Nicolas Fouquet, shortly after which the unfortunate Fouquet was imprisoned by the King. See William Howard Adams, *The French Garden 1500–1800* (George Braziller, 1979), pp. 75–84.
32 See Adams, p. 88.
33 The annual cost of maintaining the royal gardens at Marly fell from 100,000 *livres* in 1698 to less than 5,000 by 1712 (Adams, p. 104).
34 Adams, p. 103.
35 Quoted in John Dixon Hunt and Peter Willis, *The Genius of the Place* (MIT Press, 1988), p. 8.
36 Alexander Pope, *An Epistle to Lord Burlington*, 1731, quoted in Dixon Hunt and Willis, p. 212.
37 Joseph Addison, *The Spectator*, No.414, 25 June 1712, qtd. in Dixon Hunt and Willis, p. 141.
38 Horace Walpole, *Essay on Modern Gardening* (Kirgate, 1904), p. 55.
39 The celebrated gardens at Stourhead, created from 1741 onwards by Henry Hoare, were directly inspired by Lorrain's 1672 painting *Aeneas at Delos*, which Hoare owned.
40 Qtd. in Joseph Spence, *Observations, Anecdotes and Characters of Famous Men*, ed. by James Osborn (Clarenden, 1966), vol I, p. 252.
41 Sir Joshua Reynolds, '13th discourse to students at the Royal Academy' (1786), qtd. in Dixon-Hunt, p. 80.
42 Sir William Chambers, *A Dissertation on Oriental Gardening*, 1772, quoted in Dixon-Hunt, p. 80.
43 Kitchen gardens at this time were reaching something of an apotheosis. Through the use of hotbeds, heated greenhouses, advanced breeding, planting and pruning techniques, and tall fruit walls up which a variety of fruit trees were trained – orientated in varying directions so as to extend the seasons – such gardens were capable of feeding large households all year round with a remarkable variety of produce. For a fascinating discussion of this see Campbell.
44 One early example of this commodification of food came with breadfruit, discovered by Banks in Tahiti during his voyage with Cook. It occurred to Banks that this cheap, high-energy food would be ideal for feeding slaves working on sugar plantations in the Caribbean, so he persuaded King George III to offer a cash bounty for anyone who succeeded in bringing the fruit to British colonies there, a task famously accomplished, after a mutiny and many further mishaps, by Captain William Bligh.
45 Paxton's work on how to maximize light through the use of lightweight structures, ridge-and-furrow roof construction, and outsize panes of glass led to the 1841 construction of the Chatworth Great Stove, a vast glasshouse which so impressed Queen Victoria that it gave her the idea for the Crystal Palace, which Paxton co-designed.
46 This happy ending may be short-lived, however, since it now appears that Cavendish bananas are susceptible to a new strain of the disease.

47 Jean-Jacques Rousseau, *The Social Contract* (1762), *The Social Contract and The First and Second Discourses*, ed. by Susan Dunn (Yale University Press, 2002), p. 120.
48 Henry David Thoreau, *Walden, or Life in the Woods* (1854, repr. Oxford University Press, 1997), p. 84.
49 Muir's advocacy led to Yosemite's designation as a national park and the removal of its Native inhabitants, who in the words of Simon Schama were 'carefully and forcibly edited out of the idyll' (p. 7).
50 'The Seeds of Change', Growing a Nation <https://growinganation.org/content/show-content/the_seeds_of_change/> [accessed 17 April 2025].
51 William Morris, *News From Nowhere and Other Writings*, ed. by Clive Wilmer (Penguin Classics, 1993), pp. 61, 77.
52 For a detailed discussion of the building of Letchworth, see Peter Hall, *Cities of Tomorrow* (Blackwell, 2002), pp. 97–101.
53 See Michael Gilson, *Behind the Privet Hedge* (Reaktion Books, 2024), p. 21.
54 Sarah Marsh, 'Losing the Plot: Fears Huge Rent Rises Will Price Many Out of UK Allotments', *The Guardian*, 21 April 2023 <https://www.theguardian.com/business/2023/apr/21/uk-allotments-rent-hikes-400-per-cent-councils> [accessed 17 April 2025].
55 Bill Mollison and David Holmgren described permaculture as 'an integrated, evolving system of perennial or self-perpetuating plant and animal species useful to man' (*Permaculture One: A Perennial Agriculture for Human Settlements* (1978, repr. Tagari, 1990), p. 1).

37

Ayurvedic Renaissance

Exploring Fruits, Flowers, and Well-Being in Colonial Western India

Maithili Tagare

The advent of print technology in Western India, along with the expansion of educational opportunities in both vernacular languages and English and the material transformations brought about by colonial rule, catalyzed significant shifts in lifestyle practices from the late nineteenth century onward.[1] Among these changes, the realms of food and well-being experienced notable developments. By the early twentieth century, evidence of evolving gastronomic practices and health trends emerged through periodicals and cookbooks, which increasingly featured 'modern' recipes that incorporated Western, particularly English, ingredients and introduced new cooking techniques such as baking. Simultaneously, magazines began to include advertorials and advertisements promoting ready-made products, reflecting the growing influence of commercialized convenience. This paper examines the shifting trends in lifestyle and well-being during the late nineteenth and twentieth centuries, focusing on the intersections between colonial influence, culinary practices, and emerging consumerism. In particular, it analyzes the (re)emergence of Ayurveda and the renewed use of flora in the domain of health and well-being, highlighting how traditional indigenous knowledge systems were adapted and promoted during the broader nationalistic transformations of this period.

From the nineteenth century onwards, coinciding with the consolidation of colonial rule in India, the appeal of aligning oneself with the perceived 'civilized' cultures of the West gained significant prominence. The introduction and adoption of Western practices became emblematic of modernity and progress. In the realm of health and well-bring, such a cultural shift is exemplified, for instance, by an advertisement for iodized sarsaparilla syrup in the magazine *Manorama* (Figure 1).[2] Dr Vaman Gopal, who produced the syrup, marketed it as a tonic to purify blood. Interestingly, sarsaparilla, a perennial vine native to Central America and Mexico, had long been valued for its medicinal properties by Native American communities. Its transfer to Europe in the sixteenth century, as part of the broader Columbian exchange, further integrated it into European medicinal practices.

By the nineteenth century, such remedies, once indigenous to the Americas, were incorporated into the expanding colonial medical frameworks and introduced into India. The dissemination of so-called European health practices was facilitated by the broader mechanisms of colonial governance, including trade, print, and education, which sought

Figure 1. Advertisement for Dr Vaman Gopal's Iodised Sarsaparilla. Ārogya Mandir, *November 1942, p. 77.*

to reconstitute local understandings of well-being in alignment with Western medical and scientific paradigms. The spread of such practices underscores the broader colonial strategy of asserting the superiority of Western knowledge systems over indigenous practices, marking a significant shift in the conceptualization of health and modernity in India.

However, the rise of colonial modernity, especially in the realm of health and lifestyle, was not without resistance. Indigenous and alternative narratives began to challenge the dominance of European ideas, offering counter-perspectives that sought to reclaim and valorize local practices and knowledge systems. These contestations were part of a broader cultural negotiation, in which Western modernity was selectively appropriated, adapted, and, at times, rejected in favour of indigenous epistemologies. Thus, while European practices of health and well-being gained prominence, they were simultaneously met with critiques and adaptations, reflecting the complex dynamics of cultural exchange during colonial rule.

In response to the increasing influence of Western colonial practices and the deliberate appropriation of 'modern' habits, an alternative movement emerged that sought to revive and promote ancient Indian systems of health and well-being, with a particular emphasis on Ayurveda. This resurgence of Ayurveda was employed to construct an alternative narrative, underscoring India's 'rich' scientific, culinary, and medicinal heritage – one that not only predated colonial rule but also stood in contrast to the paradigms introduced by the West, with an emphasis on locally available flora, herbs, and vegetables, among others. This movement gained particular momentum during the early to mid-twentieth century, a period defined by the burgeoning nationalist sentiment in India.

In this context, Ayurveda was mobilized as a symbol of India's cultural and intellectual sovereignty, offering a framework that countered the hegemonic imposition of Western medical and health practices. The revival of indigenous practices was not merely a cultural gesture but a political act of reclamation, positioning India's traditional knowledge systems as markers of national identity. Jayanta Sengupta's work explores this phenomenon, particularly in Bengal, where food and cuisine became critical sites for contesting colonialism and articulating nationalist ideologies.[3] Similarly, Rachel Berger's research on food, indigenous medicine, and Ayurveda in early twentieth-century North India reveals how discussions about food and health were intricately tied to larger historical, political, and economic discourses concerning the Indian nation during this period.[4]

However, while such focused studies exist for regions like Bengal and North India, analyses of similar trends in Western India are relatively underexplored. This paper seeks to address this gap by examining how the revival of indigenous health practices, particularly the use of regional fruits, flowers, herbs, and shrubs, intersected with the socio-political developments in Western India, offering a nuanced understanding of how regional dynamics contributed to the broader nationalist narrative.

In Western India, the revival of ancient traditions formed a crucial part of a broader nationalist endeavour aimed at resisting the cultural hegemony of Western colonialism. Counter-narratives to colonial modernity, particularly in the domains of lifestyle and well-being, are evident in numerous articles published in wellness periodicals such as *Pathyabodh*. These periodicals frequently featured detailed recipes for *kādhās* (herbal concoctions) and offered recommendations for organizing domestic spaces in accordance with ayurvedic principles, alongside a variety of other guidelines rooted in indigenous practices.

By the late nineteenth century, the preparation of these elaborate concoctions and the arrangement of household spaces as prescribed by ayurvedic texts required a diverse array of ingredients such as shrubs, herbs, flowers, oils, and *ghee* (clarified butter), all of which had to be manually procured and processed. However, the early twentieth century witnessed significant shifts with the proliferation of advertisements for ready-to-consume medicines, tonics, and syrups that began to feature in Marathi-language periodicals. These products were often marketed as modern solutions to health, yet inherently aligned with traditional herbal principles. The onset of such hybridization, which utilized local flora, herbs, and shrubs, but which were packaged in modern bottles, signalled the increasing commodification of wellness practices as the local economy became progressively integrated into emergent capitalist markets.

This paper is structured in two sections. The first section examines the monthly magazine *Pathyabodh*, established by ayurvedic *vaidyas* (doctors) in the late nineteenth century with the aim of popularizing an ayurvedic lifestyle to its readers. Of particular interest here is the inclusion of recipes and propagation of lifestyle habits as delineated in ayurvedic texts. Very often, these guidelines utilized locally available fruits and flowers to promote internal and external well-being. For instance, recipes for *kādhās* were prescribed for overall health. Suggestions to organize spaces, such as the sleeping area during summer, which were to be enhanced by utilizing regional flora, were advocated. Both these initiatives were aimed at showcasing tangible manifestations of a 'glorious' and 'advanced' Indian heritage against the backdrop of emerging colonial modernity. The first section will thus argue that by prescribing tangible lifestyle practices that affected the body both internally and externally, *Pathyabodh* marked the beginnings of visible symbolic resistance to foreign influence while fostering a sense of cultural identity.

The second section explores the nationalistic response to colonial modernity, highlighting how the burgeoning nationalism in Western India in the early and mid-twentieth century influenced and changed the prescriptive trends of wellness manuals like *Pathyabodh*. Instead of furnishing elaborate recipes and wellness guidelines, which included a host of herbs, shrubs, fruits, and flowers, among others, the early twentieth century saw a shift in the presentation of ancient prescriptions and lifestyle suggestions, as industries began to encapsulate ayurvedic principles into marketable products like ready-to-drink tonics which can be witnessed through advertisements in Marathi periodicals. Thus, ancient principles of well-being were now propagated through modern mediums. A particularly notable feature of this period was the resistance to tea, which symbolized British dominance. Small-scale manufacturing units actively chose to boycott the leafy brew, and instead promoted *Swadeshi* alternatives – products made within the country. These herbal beverages were positioned not only as health-conscious choices but also as a patriotic rejection of colonial influence. This section will thus demonstrate that the transformations underway during the twentieth century reflected the evolving relationship between ancient practices and emerging capitalist markets, signalling a more commercial approach to Ayurveda in response to both colonial modernity and growing nationalistic sentiments.

Fruits, Flowers, and Well-Being in *Pathyabodh*

In Western India, the call to return to traditional practices reflected cultural resilience and a conscious reclamation of India's ancient heritage. In this connection, ayurvedic *vaidyas* were responsible for disseminating ancient knowledge, with many publishing magazines focused on the ayurvedic way of life. A notable example is the magazine *Pathyabodh*, published in the Marathi language from 1888 to 1901. The term *Pathyabodh* combines two words, '*pathya*' (diet or dietetics) and '*bodh*' (knowledge). Wellness manuals like *Pathyabodh* meticulously expounded the properties of various fruits, flowers, herbs, and shrubs, and provided nuanced guidance for their incorporation into everyday life. Remarkably, flora came to play a pivotal role in sustaining a healthy body, serving not only as an internal consumable but also through its external use.

Notably, articles on *Rutucharya*, which can be loosely translated as 'conduct appropriate to the season', reflected the belief that health-related issues during specific seasons arose from a lack of ancient knowledge about appropriate dietetics. The commencement of the new year in Maharashtra, starting from the month of *Chaitra* and concluding in *Phalgun*, establishes a chronological sequence of six seasons, each spanning two months.[5] These seasons are delineated as *vasanta* (spring), *grīṣma* (summer), *varṣa* (monsoon), *śarada* (autumn), *hemanta* (pre-winter), and *śiśira* (winter). With regard to the seasons, the *vaidyas* note:

> Personal habits, food, drinks, and even behaviour must be modified to suit the season. Even a healthy person has the natural tendency to change his diet as the seasons and weather change. And such a tendency is not without logic.[6]

The logic cited by the *vaidyas* was frequently drawn from the wisdom contained in ancient Indian texts concerning health and medicinal practices. Notably, within the framework of *Pathyabodh*, the *vaidyas* relied on Vāgabhata's *Aṣṭāṅgahridayam* as a primary source of guidance and authority.[7] This reliance on classical texts served as the bedrock for ayurvedic principles and methodologies, which shaped their therapeutic approaches, as can be seen in one *vaidya*'s advice for the *griśma* season, where he advocated a return to the ancient ways of conducting life:

> The afternoons during the summer season are harsher than in the spring. Even in such conditions, the sun's rays do not touch the ground in thick forests with dense, sky-high trees. In such thickets, vines of grapes are grown. And so that the area may smell aromatic, creepers of *mogrā* [Jasminum sambac, or Arabian jasmine] are grown between the vines. In such a forest, one must build a house made of *velu* [a kind of bamboo]. Cloth curtains must be hung around this house, soaked in cold, aromatic water. The tender branches of mango trees should be hung alongside these curtains. A bed should be prepared from the branches of banana stems, the *kalhāra* flowers [Nymphaea, commonly known as water lily], and lotuses with their stems, along with various other flowers and tender leaves. One must lie on this bed and even sleep on it.[8]

This recommendation to lie on a bed made of banana branches, *kalhāra* flowers, lotuses, and other flowers during the hot months underscores the therapeutic significance of specific flora in ayurvedic practices. In the face of evolving lifestyle norms and shifting gastronomic trends, this nuanced approach to healthcare and lifestyle indicated a complex interplay between the historical context of British colonial modernity and the beginnings of Indian nationalism.

As the *Swadeshi* sentiment gained momentum, *Pathyabodh* emerged as a harbinger of indigenous wellness practices in Maharashtra.[9] Consider, for instance, a *vaidya*'s recipe for a *kādhā*, crafted from a blend of regionally available fruits, flowers, vegetables, and aromatic ingredients, which was prescribed to treat a diverse range of ailments, including different types of leprosy, inflammations, thirst, delusions, anaemia, skin irritations, and haemorrhoids:

> Take four-*tolās* [approx. 46 grams] each of *padval* [Trichosanthes dioica], *nimba* [Azadirachta indica], *kutki* [Picrorhiza kurrooa], *daruhalad* [berberis aristata], *dhamāsā* [fagonia indica], *pittapāpadā* [flumaria indica], and *trayamān* [Gentian kurrooa]. Pound them roughly before adding them to 512-*tolās* [approx. 6 litres] of water. Reduce the water to 768 millilitres [about 25.97 oz] [by boiling all the ingredients]. Later, add one *tolā* [approx. 11 grams] each of *mustā* [Cyperus rotundus], *kirāit* [Andrographis paniculate], and *ćandan* [Santalum album], along with 48-*tolās* [approx. 576 grams] of pure, clean cow *ghee*. Mix all the ingredients to prepare a *ghrita* [medicated ghee].[10]

This versatile elixir, along with a myriad of other ayurvedic remedies, began to circulate within the Marathi public sphere from the late nineteenth century, primarily through printed publications such as *Pathyabodh*. These publications introduced a structured framework for implementing tangible lifestyle modifications rooted in ayurvedic principles, which claimed to produce observable effects on the body. The practices and recipes advocated in these texts extended beyond internal well-being to encompass visible external comportment, thus encouraging the adoption of mannerisms informed by indigenous health systems. This phenomenon signalled the emergence of a symbolic resistance to external cultural influences, particularly those introduced under colonial rule, while simultaneously fostering a heightened sense of cultural identity anchored in India's ancient past. The notion of a glorious precolonial heritage – encompassing advancements in culinary, gastronomic, medical, and scientific domains that predated Western civilization – gained increasing traction at the turn of the century.

The twentieth century, however, witnessed a significant transformation in the dissemination and adaptation of these ancient principles, spurred by the processes of industrialization and the integration of the native economy into emergent capitalist markets. Urban centres in Western India, such as Bombay, became hubs of industrial growth with the rise of textile mills, iron and steel factories, and other large-scale enterprises.[11] This industrial expansion facilitated the growth of a capitalist economy, yet its benefits were unevenly distributed, exacerbating existing socio-economic disparities among the Indian population. Prominent nationalist thinkers such as Dadabhai Naoroji and Romesh Chandra Dutt articulated influential critiques of the economic exploitation underpinning British colonial rule, highlighting the phenomenon of the 'drain of wealth' from India to Britain.[12]

In this context, the *Swadeshi* movement emerged as a critical response to growing awareness of colonialism's detrimental impact on native industries, embodying a broader nationalist struggle for economic self-reliance. The movement not only sought to promote indigenous industries but also played a pivotal role in the commercialization of Ayurveda in Western India. By linking ayurvedic remedies to the ideals of economic nationalism and self-sufficiency, the *Swadeshi* movement contributed to the revitalization of traditional knowledge systems as part of a broader effort to resist colonial economic dominance. The following section explores how the *Swadeshi* movement's emphasis on indigenous production intersected with the promotion of Ayurveda in Maharashtra, advancing both the cause of economic nationalism and the reassertion of indigenous health practices.

Reclaiming Ancient Wisdom: Nationalistic Response to Colonial Modernity

Prominent political figures of Maharashtra, particularly Bal Gangadhar Tilak (popularly known as Lokmanya), had called for the boycott of foreign goods and the patronage of indigenous industries as a means of challenging British economic dominance. The nationalistic call for self-reliance resonated well with the masses, especially manufacturers, which is evidenced by many advertisements in Marathi-language magazines endorsing *Swadeshi* alternatives to imported products. By the mid-twentieth century, there was a noticeable transition from home-prepared Ayurvedic medications as prescribed by the *vaidyas* to commercially available bottled tonics and concoctions.

In December 1930, at the peak of nationalism in India, the magazine *Manorama* featured an advertisement for Dr Gowade's *Jvarabindu*, a concoction against fever (Figure 2). The endorsement of *Jvarabindu* by the influential leader Lokmanya Tilak undoubtedly lent considerable credibility to the product. Tilak's support professed the efficacy of traditional ayurvedic principles in enhancing health and well-being. Such a recommendation from a well-known public figure not only bolstered confidence in the product's effectiveness but also underscored the broader cultural significance of embracing indigenous healing traditions amidst the backdrop of nationalist fervour. Indeed, the convergence of the *Swadeshi* movement and nationalist enthusiasm moved beyond the realm of influential figures advocating traditional practices: small-scale manufacturing units, too, capitalized on this collective spirit, eager to align themselves with the burgeoning movement. In their endeavour to counter colonial practices and promote indigenous alternatives, these units turned to ayurvedic herbal supplements as a means to overcome the widespread cravings for tea, a symbol of British influence.

Figure 2. Advertisement for Jvarabindu, *from* Manorama, *December 1930, p. 23.*

The history of tea in India is intricately linked to British colonial endeavours. Introduced by the British East India Company in the seventeenth century and motivated by economic interests, the British sought to establish a domestic source of tea production to reduce reliance on imports from China and strengthen their trade position.[13] Accordingly, Lizzie Collingham notes, the British attempted to cultivate tea in various areas of the subcontinent, including Darjeeling, Ootacamund, the Nilgiris, the lower Himalayan foothills, and Munnar in Kerela; by the end of the eighteenth century, tea had become 'the' British drink.[14] Despite it being strongly associated with India in British and Australian minds, Indians themselves did not drink tea.[15] In this context, Collingham and Rachel Berger demonstrate how Indians were gradually taught to incorporate it into their everyday activities, through extensive tea campaigns undertaken by the British.[16] Berger further notes that the tea-drinking campaigns focused on upper-caste and upper-class families with the aim that the practice would trickle down.[17] By the end of the First World War, the consumption of this leafy brew became ingrained in Indian culture and society, especially due to the setting up of tea stalls around factories, mines, and mills where labourers provided a ready market for the consumption of the sugary drink. Collingham notes:

> By 1919, the tea canteen was firmly established as 'an important element in industrial concern' [and] thus, tea entered Indian life as an integral part of the modern industrial world that began to encroach on India in the twentieth century.[18]

It was no wonder that tea posed a unique opportunity for marketing and advertising. In North India, *Bharat Chai* (Indian Tea) entered the picture 'to sway the Indian masses away from their affinity for tea as a British commodity and to appreciate it as an Indian one instead'.[19]

In Western India, the role of tea transcended mere economic implications and assumed symbolic significance as a manifestation of colonial domination over Indian resources and markets. Within this context and amidst the burgeoning *Swadeshi* movement, the negotiation of tea consumption presented an avenue for Indians to reassert their agency and assert both cultural and economic sovereignty. Small manufacturing units, inspired by the call for self-reliance and indigenous production, sought to promote ayurvedic herbal supplements as alternatives to tea, thus challenging British hegemony and reconnecting with India's 'rich' culinary glory. The *Śrī Siddheśwar Kāryālaya* in Satara especially evidenced this ideology through their advertisement that provocatively questioned the necessity of consuming tea in the presence of superior *Swadeshi* alternatives (Figure 3). Notably, the *Kāryālaya* (a manufacturing unit) promoted its *swadeshi śantipeya* as a cooling and refreshing drink made from home-grown ingredients and thereby provided a counter-narrative to Dr Gopal's sarsaparilla syrup.

Positioned as a wholesome alternative to tea, the *śantipeya* was advertised as being composed of pure *vanaspati*, reminding one of the *ghritas* (medicated *ghee*) prescribed by the *vaidyas*. At the same time, however, it was marketed as offering health benefits reminiscent of Ovaltine, thus underscoring the prevailing perception that foreign brands and products were considered superior, reflecting the deep-seated influence of colonial hierarchies in shaping consumer preferences.[20] The *śantipeya* was deliberately advertised as an indigenous means to satisfy cravings for tea while simultaneously aiding digestion, soothing the mind, and serving as a holistic health tonic. Although the *śantipeya*, like other products that came to stock the markets in Western India, harped on the trend of going back to India's 'glorious past' and reclaiming ancient heritage using culinary principles as a site to counter Westernization, nevertheless the very concept of ready-made foods signalled the adoption of Western techniques. During the earlier period, publications prescribed detailed recipes of *kādhās* for home preparation. Instead, by the mid-twentieth century, there

Figure 3. Advertisement for Śantipeya, *from* Manoramā, *December 1930, p. 23.*

emerged a trend favouring pre-packaged health and well-being products. This transition, thus, marked the advent of an emerging hybrid modernity, where traditional Ayurvedic recipes were encapsulated in modern bottles. These pre-packed products not only heralded convenience but also embodied the integration of indigenous knowledge into a burgeoning capitalist economy.

Another notable example of this trend was the Pearl Company, which introduced the Pearl *Kādhā*. The company named its factory the *Āryoshadi Kārkhānā*. While *kārkhānā* means a manufacturing unit in Marathi, the word *āryoshadi* is made up of two words, *ārya* and *aośadhi*. Molesworth's Marathi-to-English dictionary defines *ārya* as proper or suitable, and *aośadhi* is a tree or shrub of a medicinal nature.[21] Given the context of this advertisement, it would not be wrong to say that Pearl Company's *Āryoshadi Kārkhānā* promoted its medicinal products prepared following the principles of the Āryans or Āryan knowledge, tracing the legacy of their supplements to Ayurveda. Thus, through its naming and branding, the company sought to instil a sense of assurance among its consumers regarding the unwavering sense of commitment to upholding Ayurvedic principles and practices. The tendency to regard ancient principles as infallible appears to have gained traction with the progress of the twentieth century.

An advertisement, published in the magazine *Āhār*, attests to this trend, intelligently utilizing the domestic space to evoke nostalgia for homemade remedies and employing graphic elements to captivate its readers and consumers (Figure 4). Depicted within the advertisement are a man and a woman, likely a married couple, in what appears to be a contemporary home. The man, wearing a shirt, is depicted seated on a chair positioned before a table by the window. Notably, he sits in a manner divergent from the traditional Indian custom of sitting on the floor for a meal, which hints at his modern sensibilities. Before him lies a plate with two bowls and a small pitcher of water. Meanwhile, his wife, draped in a saree, stands beside him, serving food from another utensil. Noteworthy here is the gesture of the man's raised right hand, appearing to convey a gesture of refusal. Below this quintessential image of the homely, dutiful Indian wife (*ārya mahila*) serving her husband in a rather modern house setup is a description of issues caused by body heat alongside its solution, a picture of the packaged *kādhā*. Towards the end, the advertisement invokes the notion of *anubhavik* (tried and tested) remedies reminiscent of homemade concoctions, as the company aimed to assure its consumers further of the authenticity and efficacy of its product.

Thus, the strategic marketing of Ayurvedic products as credible, authentic remedies highlighted the broader trend of returning to traditional knowledge systems while adapting them to the demands of

Figure 4. Advertisement for Pearl *Kādhā*, from Āhār, September 1947, p. 8.

modern consumerism. Advertisements in Marathi-language magazines like *Manoramā* and *Āhār* played a pivotal role in reinforcing these trends by leveraging the nostalgic appeal of traditional remedies while simultaneously presenting them in modern, convenient formats. By invoking the concept of *'anubhavik'*, these advertisements connected with consumers' desire for authenticity and reliability, thereby reinforcing the legitimacy of ayurvedic products in a rapidly changing socio-economic landscape.

Conclusion

The exploration of ayurvedic science in nineteenth- and twentieth-century Western India provides a nuanced understanding of the complex dynamics between colonial encounters, indigenous knowledge systems, and nationalist aspirations. Through a detailed analysis of writings and advertisements in journals such as *Pathyabodh*, *Manoramā*, *Āhār*, and *Ārogya Mandir*, this paper has demonstrated how ancient knowledge was both disseminated and appropriated in everyday life. These practices served not only to promote physical and spiritual well-being but also as a symbolic form of resistance against the encroaching forces of colonial modernity. While Western influences permeated culinary practices and lifestyle norms, the conscious re-articulation of indigenous systems like Ayurveda – framed within the larger context of rising nationalism – underscored a deliberate effort to preserve and assert cultural identity.

The recommendations found in *Pathyabodh*, such as those advocating for seasonal conduct, the use of seasonal flora, and the re-adoption of long-forgotten pathways to well-being, promoted a holistic approach that simultaneously contested the Eurocentric ideals of progress and modernity. Conversely, the early twentieth century witnessed a critical transition from home-prepared remedies to the growing consumption of packaged ayurvedic products, reflecting a hybrid form of modernity emerging in the context of Western India. This synthesis of traditional and modern forms represents a dynamic adaptation to shifting material conditions and consumer preferences, signifying not merely an acceptance of modernity but an active reconfiguration of it through indigenous frameworks.

Thus, this study moves beyond the conventional discourse that contrasts colonial and native bodies – often typified by the characterization of the British Raj as masculine and the native body as effeminate. Instead, it redirects focus on the examination of fruits, flowers, and regional ecology, which, at the intersection of food, lifestyle, and well-being, are contextualized within the framework of indigenous knowledge traditions. By considering evolving material conditions and the simultaneous rise of nationalism in Western India, this analysis argues that the strategic revival and adaptation of traditional practices represented a conscious effort to reconnect with India's cultural heritage. This initiative, grounded in ayurvedic science, functioned as a symbolic act of resistance and as a means of cultural preservation amid the transformative forces of colonial modernity.

About the Author

Maithili Tagare is a PhD Scholar in Food History at the Indian Institute of Technology, Gandhinagar. Her research explores the intersection of caste, gender, and food against the backdrop of colonial modernity in late nineteenth and early twentieth centuries in Western India (Maharashtra).

Notes

1 For the purpose of this research, and ease of convenience, I refer to Western India synonymously with present-day Maharashtra, although such a political, geographical, and linguistic entity did not exist officially. In other words, Western India can be understood as the Marathi-speaking regions of the Bombay Presidency.
2 Sarsaparilla is a perennial trailing vine native to Mexico and South America. The vine was introduced from the New World to Europe, where it remained a popular cure for syphilis. Furthermore, iodized sarsaparilla was recommended to be diluted with water and consumed to purify blood. See Figure 1 and 'Bottle of Blood Purifying Mixture', Science Museum Group <https://collection.sciencemuseumgroup.org.uk/objects/co195968/bottle-of-blood-purifying-mixture-bottle> [accessed 9 April 2025].
3 Jayanta Sengupta, 'Nation on a Platter: The Culture and Politics of Food and Cuisine in Colonial Bengal', *Modern Asian Studies*, 44 (2009), pp. 81–98 (p. 81), DOI:10.1017/S0026749X09990072.
4 Rachel Berger, 'Between Digestion and Desire: Genealogies of Food in Nationalist North India', *Modern Asian Studies*, 47 (2013), pp. 1622–43 (p. 1622), DOI:10.1017/S0026749X11000850.
5 The Marathi month of *Chaitra* begins towards the end of March or early April. *Chaitra*, together with the following month of *Vaiśākh*, forms the spring season which lasts approximately through the end of May or early June.
6 Anonymous, 'Rutucharya' [Conduct Appropriate to the Season], *Pathyabodh*, Kārtik 1810 (November-December 1888), p. 10, trans. by the author.
7 *Aṣṭāṅgahridayam* is a text on diagnosis, medicine, and surgery which includes citations from many ancient texts on Ayurveda such as *Charaka Samhita*, *Sushrut Samhita* and others.
8 Anonymous, 'Rutucharya' [Conduct Appropriate to the Season], *Pathyabodh*, Vaiśākh 1811 (April-May 1889), p. 161, trans. by the author.
9 The *Swadeshi* Movement emerged as a part of the broader Indian nationalist movement. It primarily aimed at making India a self-sufficient country, especially by strengthening village industries. One of the most significant symbols of the *Swadeshi* movement was '*khadi*'. Khadi is a natural-fibre cloth which is hand-spun and woven. Mahatma Gandhi had called for every Indian to spin and weave his own cloth and reduce dependence on expensive, imported cloth. Overall, the movement sought to make India a self-reliant country by promoting indigenous industries.
10 Anonymous, 'Rutucharya' [Conduct Appropriate to the Season], *Pathyabodh*, Kārtik 1810 (November-December 1888), p. 10, trans. by the author.
11 See Thomas Blom Hansen, *Wages of Violence: Naming and Identity in Postcolonial Bombay* (Princeton University Press, 2001); Frank F. Conlon, 'Dining Out in Bombay', in *Consuming Modernity: Public Culture in a South Asian World*, ed. by Carol A. Breckenridge (University of Minnesota Press, 1995), pp. 90–127; Ira Klein, 'Urban Development and Death: Bombay City, 1870–1914', *Modern Asian Studies*, 20.4 (1986), pp. 725–54, DOI:10.1017/S0026749X00013706.
12 See Bipan Chandra, *History of Modern India* (Orient Blackswan, 2009).
13 Lizzie Collingham, *Curry: A Tale of Cooks and Conquerors* (Oxford University Press, 2006), pp. 187–208.
14 Collingham, p. 191.
15 Collingham, p. 194.
16 Berger, p. 1641.
17 Collingham, pp. 187–208; Berger.
18 Collingham, p. 195.
19 Berger, p. 1641.
20 Ovaltine is a Western brand of ready-to-drink milk flavouring product, typically made with malt, sugar, and whey. It was often marketed as an energy boosting drink in colonial India. See Berger, p. 1642.
21 Baba Padmanji, *A Compendium of Molesworth's Marathi-English Dictionary* (Education Society Press, 1863), pp. 51, 74.

38
Opening the Garden Gate
Wild Gardens and Indigenous Culinary Knowledge

Rachel Thomas Tharmabalan and Jeremy Morell

Community Gardens as Hedges Against Precarity

Even before the supply chain disruptions due to the COVID-19 pandemic, large scale nutritional precarity has been a shadow hanging over policymakers and academicians, formed by the dual rise of urbanization and globalization. This has raised serious questions regarding sustainability, stability, food security, and economic opportunity (Lindner 2021). Some researchers have been eager to identify more locally sourced foods that can not only resolve malnutrition but also conserve heritage food as a socially integrated answer to these problems.

Historically, community gardens have often emerged in response to catastrophe and have been central to community recovery (Camps-Calvet and others 2015). A community garden is a 'small plot of land developed and managed by a neighborhood or non-profit association in which agricultural activities take place' (Villas-Boas 2006: 19). During the late nineteenth century, many communities provided disadvantaged residents with the chance to produce food on city-owned vacant lands. During and following both World Wars, community gardens were employed to enhance the availability of food with minimum transportation and boost the morale of the people (McKelvey 2009).

In anticipation of the increasing catastrophic impacts of anthropogenic activities, wild gardens have again started gaining traction. This movement is not something new, as research has shown that the Indigenous people in British Columbia intentionally cultivated 'forest gardens' over 150 years ago (Fox 2021). These gardens, comprised of fruit trees and berry bushes, were deliberately planted around ancestral settlements and sustained a diverse mix of plant species. The Ts'msyen and Coast Salish people's woodland gardens on Canada's northwest coast are still flourishing despite over 150 years of neglect (Brehaut 2021). They used to plant and care for native fruit and nut trees, bushes, and medicinal plants along the Pacific coast. These gardens show the sustainable methods of Indigenous tribes. Despite initial scepticism from ecologists, research shows that these forest gardens persisted and thrived, providing food for both humans and wildlife through prioritizing biodiversity, ecological balance, and support for native wildlife.

Recently, there have been a few notable research initiatives that aim to intentionally combine and connect community to local native species (Mumaw and Mata 2022). While there has been an increase in wild gardens, there are still gaps in how wild gardens can be leveraged in the social, ecological, and social-ecological domains. As such, this research aims to connect the social dynamics and implications of wild gardens not only for Indigenous communities like the Orang Asli in Malaysia, but also for any community invested in cultural vitality and variety. We show empirical evidence of the beneficial effects of wildlife gardening on self-reported well-being using the Orang Asli case study. We present a summary of research on the social dynamics and implications of wild gardens for community well-being. We then discuss the ecological benefits of wildlife gardening, including its ability to preserve biodiversity. We wrap up by discussing the implications at the nexus of social and ecological ties, including how to foster stewardship practice and ethics across temporal, spatial, and political boundaries.

Orang Asli

Recent estimates place the number of Indigenous people living in 90 different countries around the world at approximately 476 million (United Nations 2023). Every Indigenous community has distinctive customs and characteristics that distinguish them from the wider population in the ways that they live, think, relate to one another, and structure their society. The Orang Asli, the Indigenous people of Malaysia, comprises three ethnic groups, which are the Semang, Senoi, and Proto Malay. Their low economic standing, marginalization, lack of representation, and subpar living conditions are notable in contrast to the emerging standards of living in Southeast Asia's fifth largest economy.

We concentrate on the Semai in this essay – a subdivision of the larger Senoi ethnic group. The majority of Orang Asli in West Malaysia are Semai. The Semai are semi-sedentary tribes that cultivate horticulture; they have given up shifting agriculture and are now engaged in a variety of economic activities including agroforestry, subsistence hunting and gathering, and paid employment (casual, part-time, or full-time) in nearby towns. It is debatable where exactly they originated. Geographically speaking, the Semai are widely dispersed throughout Malaysia; while approximately 25% of them still live in the highlands, the majority have settled on lower terrain.

The Overgrowth of Monoculture

Malaysia is the twelfth most megadiverse country in the world according to the Forestry Department of Peninsular Malaysia (2016). With an annual deforestation rate standing at 86% from 1900 to 2005 (FAO 2006), Malaysia has been losing its diverse flora and fauna due to rapid modernization and urbanization. Of note, out of the 15,000 species of vegetable plants available, only 300 species Indigenous to the country have been used as food (Ministry of Agriculture 1996).

In Malaysia, the terms forest plantation and planted forest have been regularly used by the government since the 1990s. The definition provided by the Department of Statistics for forest plantation is:

> an area planted with trees or forest plants, whether from local or foreign species, the method of cultivation as wide open no less than 50 hectares. Forest plantations

can include areas that are located within or outside the permanent reserved forest (qtd. in Sahabat Alam Malaysia 2020).

This definition has found use together with the definition provided by the FAO for planted forests, albeit with criteria emphasizing intensive management and monoculture characteristics. This breadth of definitions has allowed for monoculture plantations to be considered forested areas, blurring lines for conservation efforts. Sahabat Alam Malaysia has written that monoculture plantations cannot be considered forests, as they lack the ecological complexity and biodiversity essential for forest ecosystems. Plantations, characterized by uniformity and intensive management, are seen as ecologically inferior to natural forests, which perform vital ecological functions such as water regulation and habitat provision. The impacts of monoculture plantations are severely felt by local communities, from disruption of food chains to the destruction of habitats for fauna, soil erosion to the depletion of soil nutrients.

Cultural Erosion and Indigenous Knowledge Runoff

Among many Indigenous groups around the world, cultural erasure is a common theme. In Malaysia, the government has played both a passive and active role in the erasure of Indigenous culture through various strategies that include assimilation, Islamization, and resettlement schemes, which can all be included within the ethnocide notion (Endicott and Dentan 2004). As reported by Amar-Singh (2019), despite efforts from NGOs and the government to help improve the economic status of the Orang Asli, the poverty rate still stood at 89.4% in 2020 as compared to the national poverty rate, which stands at 6.2% (Statistica Research Department 2023). The primary reason for this high rate of poverty is land dispossession, cutting off their main source for livelihoods. The impacts of deforestation have compounded this dispossession for communities still resistant to assimilation, threatening them with increased marginalization and worsening living conditions. It's important to note that the Orang Asli are not opposed to change; rather, they seek to benefit from development while retaining their ethnic and cultural identities.

The Orang Asli are said to be uniquely positioned as gatekeepers of knowledge because of their historic and cultural connections with their ancestral land (Kardoni and others 2014). They have been using plants from the rainforest for their food, fodder, nourishment, and medicine. Indigenous knowledge held by the Orang Asli people includes information about their culture, customs, and surrounding flora and fauna. However, because of widespread deforestation and regulations that affect land use and foraging, and the many developmental programmes aimed at modernizing Malaysia, there has been a rapid decline in biodiversity which has had a spill-over effect on the well-being of the Orang Asli. Consequently, the Indigenous knowledge associated with the environment is slowly being lost due to the poor intergenerational transfer of knowledge which has brought about changes in their socio-cultural relationships, health, and economic problems. Additionally, the younger generation of Orang Asli is gradually being affected by cultural assimilation due to a disconnection between nature and culture, which results in a lack of curiosity about both.

Wild Edible Plants as Community Root Anchorage

As defined by the Food and Agriculture Organization, 'wild edible plants (WEPs) are plants that grow spontaneously in self-maintaining populations in natural or semi-natural ecosystems and can exist independent of direct human action' (Heywood 1999). Hundreds of millions of people worldwide rely on WEPs for nutrition, subsistence, and income generation, particularly benefiting women and children in traditional communities. These population groups also act as agents of preservation for these crops, and often the knowledge and classification of these plants are locked up in Indigenous knowledge systems. During droughts, famines, and conflicts, WEPs act as a safeguard against malnutrition and starvation and offer dietary diversity. Although not typically part of the cash economy, they can contribute to income generation and livelihood improvement when gathered for sale. WEPs thrive in marginal lands, offer low input and higher yield, and are often unaffected by the international community of plant diseases and pests that plague major cash crops which then allows for sustainable production systems. Many are resistant to extreme environmental conditions or are suited to adverse anthropogenic conditions, such as salinization, global warming, and desertification.

WEPs are known for not only their high nutritional profile but also their antioxidant activity. These plants' strong antioxidant content has made them valuable as traditional medicines. Their use has also been associated with a lower risk of diabetes, cancer, heart disease, high cholesterol, and other disorders (Keatinge and others 2010; Ngo and others 2011).

As global food demands increase and supply chains become brittle through centralization and monoculture, many researchers recognize the importance of integrating WEPs into food production alongside staple crops. This shift can decrease reliance on cultivated crops while enhancing sustainability, preserving biodiversity and building resilience in light of climate extremities and the significance they play as crop wild relatives (Khan and others 2023). WEPs boast genetic diversity and require minimal care until harvest, offering potential resistance to climate change effects. Elevating the status of WEPs is crucial for documenting their gene pool, usage, and associated knowledge before it is lost. Domesticating WEPs can contribute to crop improvement and the development of new, climate-resilient vegetable varieties (Raghuvanshi 2001).

Furthermore, the significance of traditional food knowledge presents academics and nutritionists with an understanding of the social significance that food serves for an individual, thereby enabling them to make associations to the cultural significance of food while advising individuals about the most suitable course of action for bettering their health (Douglas 1984; Wahlqvist 2004).

With that said, wild gardens are a potential solution to help build resilience in the environment. This is because cultivating a diverse array of plant species, including native plants, community and wild gardens provides essential habitat and resources for a variety of wildlife, and also has the potential to support biodiversity and enhance ecosystem resilience. Traditional and Indigenous food systems can offer a sustainable solution for balancing food production with human and environmental health as it is rooted in a strong sense of place (Ahmed and others 2022). This is because wild food is part of the cultural identity of the people and the surroundings. This is particularly true for Indigenous populations as many of their food traditions are rooted in the land where they live. As such, WEPs represent an opportunity to test the prevalent agricultural system in a more socially representative manner than accomplished by organic and other alternative food movements.

Methodology

With the manyfold challenges and the failure of Jabatan Kemajuan Orang Asli (JAKOA, also known as the Department of Orang Asli Development) to improve the overall well-being of the Orang Asli community, several non-governmental organizations (NGOs) have formed partnerships with Orang Asli villages, working together to start community gardens in which the sales of a portion of the produce would be sold to local supermarkets. Two sites, located in Pahang, utilizing aid from these NGOs were selected for this study. Semi-structured interviews were conducted among five informants for each case study.

In the Kampung Ulu Gumun site, the NGOs provided seeds that are more resilient and adapted to the local environment, and they ran classes to teach the Orang Asli to cultivate and care for some of the commercialized crops that would be included in the project. Similar projects conducted by other NGOs have garnered international awards as they not only helped with alleviation of malnutrition among these villages but also provided job opportunities and fostered a sense of community and environmental stewardship among the younger generation. A major criterion for the success of these projects was the extent of collaboration in working with the Orang Asli. Previous projects, similar in intent, have failed by limiting the participation of the Orang Asli in their planning and execution.

In consultation with an NGO, the Orang Asli in the Telimau site have also initiated their own wild gardens to have access to wild plants for their consumption, as they have very limited access to the forest. However, not all the wild plants customary to their diet can be domesticated due to the environmental conditions associated with growing these plants. These are communal wild gardens in which members of the community share the responsibility of tending to these gardens.

Nutritional Abundance

Nine wild edible plants commonly consumed by the Orang Asli were analyzed by the researcher for their nutritional content. Another comparative study was done with popular locally available produce. The analysis showed that these wild plants are nutritionally superior to commonly consumed commercial vegetables in Malaysia. Despite their lower average protein content and incomplete amino acids, wild edible plants (WEPs) exhibit higher protein levels when compared to vegetables such as 'Kang Kung' (*Ipomoea aquatica*), 'Bok Choy' (*Brassica campestris* L.), and 'Gai Lan' (*Brassica oleracea*) (Tharmabalan 2021). These traditional plants have conspicuously higher fibre, mineral, and protein content in comparison. Furthermore, wild edibles typically possess very high moisture content, ranging from 72.3% to 89.7% per 100 grams, resulting in a significantly higher energy density. This attribute is advantageous as it reduces calorie intake while promoting satiety.

Conceptual Proliferation

However, even with a high nutritional content and points of market availability, the urban populace has been resistant to the introduction of wild edible plants into their everyday diet. As such, to reintroduce these wild plants onto the plates of the urban population, these plants need to be made more appealing to the imagination of the Malaysian urban population. Dewakan Restaurant, located in Malaysia, is a two-star Michelin restaurant which utilizes Indigenous plants and herbs to showcase the rich biodiversity of Malaysia's forests. Michelin-star chefs are no strangers to using wild plants as ingredients in their dishes not only to showcase their creativity, but also to connect diners with nature and highlight the diversity of flavours available in local ecosystems. These Indigenous plants are often known as a 'poor man's food' and are often perceived as a staple of impoverished diets, so showcasing these plants being as prestigious ingredients in fine dining may elevate their status for some diners but may erect even more perceptual barriers to the majority.

In essence, the dualism of wild plants as both a staple of impoverished diets and a coveted delicacy in haute cuisine highlights broader societal disparities and ethical considerations within the culinary landscape. Addressing these disparities requires a nuanced approach that prioritizes food equity, cultural sensitivity, and environmental stewardship in the sourcing, preparation, and consumption of wild plants. For example, researchers in Cambodia have collaborated with the local Hmong community to create a pop-up restaurant called Forest (Feuer 2019). This initiative seeks to creatively integrate sustainably sourced wild edible plants and mushrooms into modern Khmer cuisine while remaining affordable and accessible for working-class people. Their stated aim is to create an ethical space for discovery, experimentation, and biodiversity.

Harvesting Success

The community garden project in Kampung Ulu Gumun proved to be successful in several key metrics. Not only did average community nutritional status improve, but they also created a surplus in income which allowed them to buy motorbikes, expand houses, support education, store seeds, and purchase other equipment. As these projects became more successful, to date they have been replicated by four other settlements who have managed to harvest 300–2500 kg on three acres of land with a cumulative sale of RM27000 in July 2019 (Kon 2019).

The elders in Telimau also initiated a programme (implemented before the pandemic but successfully continued into 2024) in which these wild plants would be utilized in the school lunch programme to address disparities in hunger and to shore up cultural education, as they recognize the disconnect happening between the younger and older generations.

As these projects are beginning to be replicated by other communities, traditional knowledge networks are starting to emerge among successful Orang Asli farmers that not only act as a knowledge-exchange hub, but also cultivate a culture of learning and teaching. These hubs provide a foundation for community cohesion and intergenerational learning, as well as contributing to the preservation of cultural heritage.

The farming community exemplifies solidarity through shared facilities and equipment, fostering a collective effort towards achieving economies of scale and securing higher prices for their produce. Mutual respect is ingrained in the community's ethos, with leadership responsibilities shared and decisions made on a consensus basis. This democratic approach is facilitated by representation from five leaders, ensuring that farmers' voices are heard and respected. Moreover, equity and justice prevail within the social enterprise, which is open to all who wish to farm, with farmers representing eighteen of the fifty families involved. Inclusivity and diversity are celebrated, as individuals of all ages, genders, and religious backgrounds, including animists, Christians, and Muslims, actively participate in the farming activities, highlighting the community's commitment to embracing differences and fostering unity.

In both case studies, the younger generation has been actively participating alongside the elders of the community to create and maintain wild gardens: they develop a shared sense of purpose, deeper connection to their surroundings, greater appreciation for the ecological heritage of their community, and ownership over these spaces. By working together with the older groups of Indigenous people to plant native vegetation, maintain habitats for local wildlife, and promote biodiversity, younger individuals collaborate towards a common goal of enhancing the natural environment in their urban surroundings.

Wild gardens can become spaces for shared purpose and environmental stewardship by serving as communal areas where not only young Indigenous people but also the elders of the community come together to engage in activities related to gardening, conservation, and environmental education. This allows the Orang Asli to demonstrate a commitment to sustainability, conservation, and ecological resilience.

Gardens from the Ground Up

For this model to be successful elsewhere, self-determination is key. Although there has been strong advocacy to allow the Orang Asli the right to self-determine their way of life, nothing has been implemented simultaneously with a development model that emphasizes the relationship between the political, economic, environmental, and socio-cultural aspects and their well-being (Nicholas 2000; Nordin and Witbrodt 2012). However, as both case studies have highlighted, allowing the Orang Asli to self-determine their food systems would allow them to react and adapt to their own necessities, by applying culturally based Indigenous food and deciding on resource allocation decisions.

Food serves as an important link between political, economic, environmental, and socio-cultural factors. It is crucial to recognize the shared connections which can then be used to explore further into the self-determination rights of the Orang Asli in regard to their ancestral land, customary way of life, and their social relationships.

The Orang Asli must be enabled to develop their unique solutions to mitigate the negative effects of modernization on their environment. The right to self-determination in economic, environmental, and sociocultural dimensions is essential for people to achieve cultural, ecological, and physical well-being. These NGOs have found success with a polycentric model, which allows for a bottom-up approach that includes the

Orang Asli, as opposed to a top-down approach when dealing with the Orang Asli. This strategy helps to acknowledge the knowledge that the Orang Asli possess while enhancing the local capacities of NGOs, which in turn helps the government and NGOs create endogenous resources that prioritize the interests of the Orang Asli. But these initiatives could also benefit from the knowledge of other government agencies, such the Economic Planning Unit and JAKOA.

Conclusion

The findings from this research study highlight the long-lasting positive effects of human impact on the environment and emphasize the importance of incorporating Indigenous knowledge into conservation efforts. Additionally, the study sheds light on the social complexity of Indigenous societies in the absence of traditional Western agriculture practices.

With the increasing fragility of the global food system manifesting during crises, we must look for creative, resilient alternatives to dominant, industry-controlled monocultures. To address these challenges, innovative solutions are necessary to promote food security, diversity, and sustainability. To transform to agroecological methods of farming, a transitional pathway via the cultivation and consumption of wild plants could challenge the dominance of industrial food systems on path dependency while ensuring food security, environmental protection, nutritional adequacy, and social equity. The growing interest in the advancement of so-called neglected and underutilized crops presents a timely opportunity to investigate a viable competitor to a narrow regime in agriculture.

Wild gardens are intricately woven into the fabric of communities where wild vegetables play a vital role in daily life. These gardens, adapted to the local environment, harbour a rich tapestry of Indigenous knowledge and practices. Unlike conventional community gardens, which often lack cultural specificity, wild gardens thrive on the knowledge and communal identity passed down through generations. They not only celebrate cultural diversity but also serve as reservoirs of genetic richness and a repository of shared knowledge. When envisioning successful community gardens, especially those tailored for Indigenous populations and multicultural societies, we must embrace this blend of tradition, biodiversity, and community building.

About the Authors

Rachel Thomas Tharmabalan is an assistant professor at the University of Wisconsin Systems. Her research interests are transdisciplinary in nature and revolve around nutrition and well-being, and the revitalisation of traditional food and forgotten crops. She is part of the ulam school project, a transnational interactive food education that intends to alleviate the impact of non-communicable diseases on public health in Malaysia, Vietnam and Cambodia.

Jeremy Morell is an independent researcher and documentary filmmaker whose interests include indigenous rights, sustainable farming practices, immigration issues, politics, and arts.

References

Ahmed, S. and others. 2022. 'Role of Wild Food Environments for Cultural Identity, Food Security, and Dietary Quality in a Rural American State', *Frontiers in Sustainable Food Systems*, 6, DOI:10.3389/fsufs.2022.774701

Amar-Singh, H.S.S. 2019. 'Malnutrition and Poverty among the Orang Asli (Indigenous) Children of Malaysia (Submission for UN Special Rapporteur on Extreme Poverty)', Office of the United Nations High Commissioner for Human Rights <https://www.ohchr.org/sites/default/files/IndigenousChildren.pdf> [accessed 8 May 2024]

Brehaut, Laura. 2021. 'Ancient Indigenous Forest Gardens Still Yield Bounty 150 Years Later: Study', *National Post* <https://nationalpost.com/news/canada/ancient-Indigenous-forest-gardens-still-yield-bounty-150-years-later-study> [accessed 1 May 2024]

Camps-Calvet, Marta and others. 2015. 'Sowing Resilience and Contestation in Times of Crises: The Case of Urban Gardening Movements in Barcelona', *PARTECIPAZIONE E CONFLITTO*. 8, pp. 417–42. DOI:10.1285/i20356609v8i2p417.

Douglas, M. 1984. *Food in the Social Order* (Russell Sage Foundation)

Endicott, Kirk, and Robert Knox Dentan. 2004. 'Into the Mainstream or Into the Backwater? Malaysian assimilation of Orang Asli', in *Civilizing the Margins: Southeast Asian Government Policies for the Development of Minorities*. ed. by Christopher R Duncan. (Cornell University Press), pp. 24–55

FAO. 2006. 'Forestry Profile World Resources Institute', *World Rainforests* <https://rainforests.mongabay.com/20malaysia.htm> [accessed 30 April 2024]

Feuer, Hart Nadav. 2019. 'Forest', Hart Nadav Feuer <https://www.hartfeuer.net/forestrestaurant/> [accessed 18 April 2024]

Heywood, Vernon.1999. 'Use and Potential of Wild Plants in Farm Households. FAO Farm Systems Management Series', Food and Agriculture Organization <http://www.fao.org/docrep/003/w8801e/w8801e00.htm> [accessed 18 April 2024]

Kardooni, R., F. Kari, S. Yahaya, and S. Yusup (2014). 'Traditional Knowledge of Orang Asli on Forests in Peninsular Malaysia', *Indian Journal of Traditional Knowledge*, 13, pp. 283–91

Keatinge, J.D.H. and others. 2011. 'The Importance of Vegetables in Ensuring Both Food and Nutritional Security in Attainment of the Millennium Development Goals', *Food Science*, 3, pp. 491–501

Khan, M.K. and others. 2023. 'Crop Wild Relatives: The Road to Climate Change Adaptation', *Crop & Pasture Science*, 74.11, pp. i–iii, DOI:10.1071/cp23253

Kon, Onn Sein. 2019. 'Community Farm Enterprise, KG OA ULU GUMUN', The Social Solidarity Economy Resource Website <https://base.socioeco.org/docs/asec_6_community_farm_malaysia.pdf> [accessed 18 April 2024]

Lindner, Chloe. 2021. '"Rooted in Community": The Importance of Community Gardens', *Liberated Arts: A Journal for Undergraduate Research*, 8.1, article 3 <https://ojs.lib.uwo.ca/index.php/lajur/article/view/13648/11234> [accessed 8 May 2024]

McKelvey, Bill. 2009. 'Community Garden Tool Kit', University of Missouri Extension <http://extension.missouri.edu/explorepdf/miscpubs/mp0906.pdf> [accessed 8 May 2024]

Ministry of Agriculture. 1996. 'Malaysia: Country Report to the FAO International Technical Conference on Plant Genetic Resources', Food and Agriculture Organization <http://www.fao.org/fileadmin/templates/agphome/documents/PGR/SoW1/asia/MALAYSIA.pdf> [accessed 12 April 2024]

Mumaw, L., and L. Mata 2022. 'The Socio-Ecological Benefits of Wildlife Gardening. Report Prepared for Gardens for Wildlife Victoria,' <https://gardensforwildlifevictoria.com/wp-content/uploads/2022/04/The-socio-ecological-benefits-of-wildlife-gardening-Mumaw-and-Mata-v1-31Mar22-lowres.pdf> [accessed 8 May 2024]

Ngo, S.N., D.B. Williams, and R.J. Head. 2011. 'Rosemary and Cancer Prevention: Preclinical Perspectives', *Critical Reviews in Food Science and Nutrition*, 51.10, pp. 946–54

Nicholas, Colin. 2000. *The Orang Asli and the Contest for Resources. Indigenous Politics, Development and Identity in Peninsular Malaysia* (Centre for Orang Asli Concerns)

Nordin, R., and M.A. Witbrodt. 2012. 'Self-Determination of Indigenous Peoples: The Case of Orang Asli', *Asia Pacific Law Review*, 20.2, pp. 189–210

Raghuvanshi, R.S., and R. Singh. 2001. 'Nutritional Composition of Uncommon Foods and Their Role in Meeting Micronutrient Needs', *International Journal of Food Science and Nutrition*, 52, pp. 331–35

Sahabat Alam Malaysia. 2020. 'Plantations Are Not Forests', FOE Malaysia <https://foe-malaysia.org/articles/plantations-are-not-forests/> [accessed 15 April 2024]

Statistica Research Department. 2023. 'Poverty Rate of Rural and Urban areas in Malaysia from 2007 to 2022', Statistica <https://www.statista.com/statistics/795371/poverty-rate-of-rural-and-urban-areas-malaysia/> [accessed 8 May 2024]

Tharmabalan, R.T. 2021. 'Nutritional Analysis of Five Wild Edible Vegetables Traditionally Consumed by the Orang Asli in Perak', *International Journal of Food Science*, 1–7, DOI:10.1155/2021/8823565

United Nations. 2024. 'International day of the World's Indigenous People', United Nations <https://www.un.org/en/observances/indigenous-day> [accessed 1 April 2024]

Villas-Boas M.L.S. 2006. 'How Community Gardens Function: A Case Study of "Complexo Aeroporto", Ribeirao Preto' (unpublished masters thesis, University of Ohio) <http://rave.ohiolink.edu/etdc/view?acc_num=ohiou1149463363> [accessed 18 April 2024]

Wahlqvist, M.L. 2004. 'Requirements for Healthy Nutrition: Integrating Food Sustainability, Food Variety, Health', *Journal of Food Science*, 69.1, pp. CRH16–CRH18

39

English Commercial and Private Garden Production from the Sixteenth Century until the Coming of the Railways

Malcolm Thick

Thomas Hill, who wrote the earliest printed book in English on gardening in the 1560s, quoted with approval the recommendation of the Roman authors Varro and Palladius that gardens 'be placest neere to a Citie', because 'gardens placed far from the city do rather hinder the apte bringing of all kinds of Hearbes and flowers unto the market to be solde'.[1]

This geography of market gardening fits neatly into the model of agricultural land use propounded by German economist Johann Heinrich von Thünen in the early nineteenth century. Imagine a flat, round country with uniform soil and climate with one large city in its centre. Agriculture will increase in intensity as one gets closer to the city because transport costs are less and delicate produce – garden vegetables – arrive at the market in the best condition. Very close to the city, the innermost ring, gardeners might be using extra heat and light to raise crops speedily or out of season: delicate crops demanded by richer citizens. In contrast, in the area furthest away, the outermost ring, one would find animals grazing on rough pastures.[2]

This symmetrical system of rings of intensity will be distorted if a navigable river is introduced running through the countryside and past the city. This makes transport much easier for produce grown near it: the inner rings of intensive commercial gardening would extend along it. Soils favourable to gardening some miles from the city may induce pockets of intense production at a distance from the built-up area. A pre-existent body of expert gardeners may have a similar effect, as will a population of small peasant farms with an abundance of family labour to grow vegetables intensively. The result is a complex map of horticulture near large population centres, but one which explains the geography of intensive gardening.

The theory is particularly applicable to London in the eighteenth century. The 1798 fold-out map of agriculture in the *General View of the Agriculture of Middlesex* shows two areas of commercial gardening. A small area adjacent to the city, to the northeast, was largely composed of market gardens and nursery grounds. This area had favourable soils and was very close to the city. The main area, to the west of the capital, also had good gardening soil but additionally, easy access to London down the Thames. The river distorts the ring of gardens, dragging it towards the west. This map only covers Middlesex, to the north of the Thames; in Surrey, south of the river there is a similar pattern.[3]

This is a static model, but we can add changes over time: the main feature in our period being urbanization. Most towns increased in size in early modern England, none more so than London whose estimated population was:

5,000 in 1550,
200,000 in 1600,
400,000 in 1650,
More than 500,000 in 1700
Nearly a million in 1800
1,400,000 by 1821[4]

The eighteenth century also saw the beginnings of the expansion of northern industrial towns. Manchester, a parish of some 5000 souls in 1650, by 1821 housed 108,000 people.[5] By 1750, when many towns had been mapped, all towns of any size in England had a rim of market gardens.

In the countryside many families had some land to raise crops or feed a few animals, but most inhabitants of large towns had no connection with the land and were totally dependent on the market for food. As the population of towns grew, the landless town-dwellers became ever more dependent on produce brought to market, including garden crops. There was a corresponding increase in market gardening, with gardens nearest towns becoming ever more intensive in their production. Expanding towns meant that gardens close to towns and cities were often built upon, pushing the intensive garden rim ever wider. The historian of London, John Stow, observed this process at the end of the sixteenth century: 'a fair Field' near Houndsditch to the east of London was partially occupied by a gardener, 'one that served the markets with herbs and roots', but his garden was superseded by 'many fair houses'.[6] C.W. Shaw made exactly the same point much later in 1879: 'In the more immediate neighbourhood of London market gardening is considerably on the decrease, owing to the land being required for building and other purposes'.[7]

The development of market gardening in England was not, however, a smooth upward curve following population growth. It grew in 'fits and starts', and it is this uneven history that I wish to outline.

The earliest shock to the system of vegetable gardening for the market is difficult to establish due to thin evidence. The shock was the dissolution of the monasteries by Henry VIII in the 1530s. At the time religious orders in England between them owned about a quarter of English farmed land, and most had gardens near or within their precincts. Monasteries invariably had infirmary gardens where medicinal herbs were grown. Kitchen gardens produced fruit and vegetables to supplement the diet of monks and nuns.[8]

What happened to these gardens when the monasteries were closed down? We have very little direct evidence of monastic gardens passing directly into lay hands, but there is indirect evidence that such gardens became commercial enterprises after dissolution. Take Norwich Priory as an example. In the early fifteenth century, as well as supplying the monks with vegetables, onions, colewarts, apples, pears, leeks, beans, and other foods, surplus produce was sold to the city. By 1527, under a decade before the dissolution, no produce was sold. The gardens were not cultivated by the Priory: instead rent was received from laymen for using the 'Great Garden', 'the Small garden', and the garden next to the Chamberlain's house. Robert Castyr, who was employed in the gardens when they were producing food for the Priory in 1480–1481, in 1483–1484 paid £1 6s 8d to the Priory 'for farm of the Great garden leased to him for the term of ten years'. Thus Robert and maybe other gardeners were already carrying on their trade using these rented monastic gardens before the dissolution. The dissolution simply meant, to them, a change of landlord.[9]

The next jolt to commercial gardening in parts of southern England and the London suburbs also came from the supply side of the business, the arrival, in the second half of the sixteenth century, of Protestant refugees fleeing religious persecution in the Low Countries. These Protestants (usually called 'Dutch') initially established themselves in towns on the east coast of southern England, most prominently at Sandwich in Kent, and Colchester, Yarmouth, and Norwich in East Anglia. Many of the refugees were skilled workers, and quite a few were market gardeners. Their capacity to produce large quantities of vegetables from intensively worked gardens gave a new impetus to English market gardening. The importance of these immigrants became apparent in the middle of the 1590s when the European grain harvest was bad for three years in a row. The worst year was 1596, when 'never ceasing raine' left English corn 'utterlie rotted and corrupted'. London was particularly hit by the dearth of bread-grains. Londoners' hunger was partially relieved by the import from East Anglia of large quantities of roots, mainly carrots. Most were shipped through Yarmouth, outport for Norwich. Between October and March 1593–1594, 281 tons and 600 bushels of roots were sent to London; in 1597–1598, 600 tons and 600 bushels were sent, and in 1598–1599 639 tons and 1 'last'. The roots were sent by Dutchmen, gardeners or their agents, who had been busy for some years supplying the poor of Norwich, England's second city. In 1575 they were said to 'digge and delve a grete quantitie of grounde for rootes which is a greate succor for the pore'.[10]

The acceptance of roots by the poor during the near-famine years of the 1590s may indicate that many were already used to eating them. William Harrison claimed the poor were eating 'melons, pompions, gourds, cucumbers, radishes, skirrets, parsnips, carrots, cabbages navews, turnips, and all kinds of salad herbs' in 1575. The physician Thomas Cogan in 1596 said of carrots and parsnips, 'The rootes are used to be eaten of both, first sodden, then buttered, but especially Parseneppes: for they are common meate among the common people, all the time of Autumne, and chiefly upon fish daies.' Certainly, there is evidence that Londoners continued to consume roots *after* the dearth years. In 1629 the apothecary and gardener John Parkinson found that some of the poor were eating so many roots that they developed 'moist and loose flesh' – waterlogged tissues and swollen limbs caused by excessive consumption of turnips.[11]

Ben Jonson humorously warned of the consequences of eating summer vegetables when he wrote of a voyage up the river Fleet, a tributary of the Thames rapidly becoming an open sewer, observing:

> how dare,
> Your daintie nostrills (in so hot a season,
> When every clerke eates artichokes and peason,
> Laxative lettus and such windie meate)
> Tempt such a passage? When each privies seate
> Is filled with buttock?[12]

Soon after the Restoration, in 1662, Thomas Fuller thought it 'incredible how many *poor people* in *London* live thereon [i.e. roots], so that in some seasons, *Gardens* feed more *poor people* than the *Field*.'[13] This is a significant observation: the London authorities also recognized that roots were an important element in diet of the poor in the first half of the seventeenth century. A long-running court case over the Gardeners' Company's attempt to control the gardeners of Fulham, Chelsea, and Kensington, to the west of the City, was settled in favour of the gardeners, not so much on the strength of their case against control by the Company, but on the contribution they made to London's food supply:

> And we finde that by this manner of husbandry and ymployment of their grounds the Cittys of London Westminster and places adjacent are furnished with above fower and twenty Thousand loads yearly of Rootes as is credibly affirmed unto us as wee believe whereby as well the poore as the ritch have plenty of that victuall at reasonable prices.[14]

The high demand for garden wares encouraged Dutch gardeners to move from Sandwich and other places where they had first settled to the Surrey bank of the Thames, across the river from London. Robert Child, in the 1650s, had talked to old men who remembered their arrival:

> Some old men in Surrey, where it flourisheth very much at present; report, that they knew the first gardiners that came into these parts, to plant Cabages, Colley-flowers, and to sow turneps, carrets, and Parsnips, to sow Raith (or Early ripe) Pease, Rape, all which at that time were great rarities, we having few, or none in England, but what came from Holland and Flanders. These Gardiners with much ado procured a plot of good ground, and gave no less than 8 pound per Acre: yet the Gentleman was not content, fearing they would spoile his ground, because they did use to dig it. So ignorant were we of Gardening in those days.[15]

Commentators on diet in the eighteenth century continued to emphasize the importance of roots, greens, and other common vegetables in the diet of the poor and middle classes. The agricultural author John Mortimer in 1716 commented that carrots 'are the most universal and necessary Root that this Country affords', and in 1760 the gardener and botanist Philip Miller thought that carrots 'provide great comfort to the poor'.[16] Joan Thirsk detected an increase in the use of field peas and beans in the early eighteenth century. She cites social commentators such as William Ellis noting their use in poorer households.[17] A French visitor in 1719 said: 'The common People of *England* run away with a Notion, that the *French* live upon nothing but Herbs and Roots [but] tis very certain, that Herbs, Pulse, and Roots, are more used in *England* than in *France*'.[18]

The most ubiquitous vegetable on the table of the poorer sort in the eighteenth century was boiled cabbage, 'greens'. The author of a 1744 book on vegetable growing and cookery comments, 'The various Kinds of this Plant are endless to describe'.[19] Because of the many varieties, greens of some sort could be available the year round, which may help to explain their ubiquity. Certainly, they were sold throughout the year, by greens-sellers in London streets. Bacon, or pork, with greens, turnips, or beans, was a very common meal for the common man in London at this time.[20]

The middle classes dining at home also ate 'homely dishes' with liberal amounts of cheap vegetables. A French visitor in the 1690s describes dining with the 'middling sort' of people. If they dined on a piece of boiled beef, they 'besiege it with five or six Heaps of Cabbage, Carrots, Turnips, or some other Herbs or Roots, well pepper'd and salted, and swimming in Butter'.[21] In 1748, a Swedish visitor also noted that butchers' meat formed the centrepiece of dinners in London (ones probably experienced at inns), but 'they take turnips, potatoes, carrots, &c from the dish and lay them in abundance on their plates'.[22]

In parallel with a general acceptance of vegetables, by the early eighteenth century another 'jolt' to market gardening was increasing consumption of certain types of garden produce by the gentry and aristocracy. They consumed 'delicate' garden fare, influenced by French cuisine. One notable feature of their tables was dishes of costly out of season produce. This was ridiculed by some commentators: 'And verily the vanity of some deserves our wonder, who are of the Heliogabalian Stomach, to which nothing doth relish which is not dear . . . only loving Pease, when they are scarce to be had.' John Evelyn was contemptuous of forced vegetables like 'impatiently longed after [. . .] Early Asparagus'. Richard Steele in 1710 summed up the dictates of fashion by observing, 'They are to eat everything before it comes in Season, and to leave it off as soon as it is good to be eaten'. Richard Bradley observed 'the pride of Gardeners about London chiefly consists in the production of Melons and Cucumbers at times either before or after the natural Season.'[23] Bradley, in one of his publications, records prices for garden produce month-by-month in London markets thus: 'Colleyflowers, of the right sort [could be bought for] 5s each' in May, and 'Kidney Beans raised in Hot-Beds were about 3s or 4s per Hundred'.[24] The most dramatic annual rise and fall in prices involved peas. In May 1723 he found 'Forward Peas were sold this month for Half a Guinea per pottle-basket'. This is equivalent to 7 guineas or 1764 pence per half sieve: in July they sold for 6 pence for the same measure.[25] An inordinate liking for out of season peas was the reason why a prisoner was, in 1741, in Newgate Prison facing the death sentence for robbery. He described how he got unto 'bad Company' and enjoyed luxury which:

> plunged me into such Extravagance, that I soon acquired the Name of Mr. EPI-CURE [. . .]. I used to frequent the Play-house every Night, and have often given a Guinea for a Quart of PEASE, 7s. for a Gill of Strawberries, and 5s. apiece for

Cucumbers; in short, nothing would suit my voluptuous Taste, except in its greatest Bloom and Glory, and when it bore the highest Price.[26]

Such prices for rarities makes Richard Bradley's observation plausible: 'It is not very rare to see Bills from Fruiterers and Herb-shops, of one Winter's standing to amount to Sixty, Eighty, an Hundred and Fifty Pounds, where Families are large'.[27] An analysis of average garden produce prices at Covent Garden over a year, published in 1824 shows big seasonal variations.[28]

The quest for novelty was aided by seed importers such as Stephen Switzer. In 1728 he published a pamphlet advertising his latest imports subtitled: *A compendious method for raising Italian broccoli, Spanish cardoon, celeriac, fenochi, and other foreign seeds*. In the pamphlet he mentions English enthusiasts for new foreign vegetables, men like 'the Right Honourable the Earl of Peterborough' who, said Switzer, was almost the only one who had previously imported and understood the use of Florence Fennel.[29]

The influence of this steady demand for novelty was therefore another jolt to the market gardening industry, spreading gardening skills. One cluster of gardens which over several centuries became famous for developing new technology to meet this demand were the Neat House gardens in Westminster. They quickly achieved a reputation for raising delicate or out of season vegetables. In 1632 they were noted for muskmelons and asparagus. The Neat Houses was sited in a bend of the Thames near the present Tate Britain Gallery. The first gardens, started in about 1610, were near the riverbank but later ones were created inland as far north as the present Victoria Station. A few Neat House gardeners in the mid-seventeenth century had Low Countries surnames and may have been descendants of Dutch immigrants who first saw the potential for gardening here.[30] The gardens were ideally situated, just to the west of the London conurbation. They were relatively small, almost all under 10 acres (4 hectares). To some extent sheltered from the prevailing cold north-west wind, their soil was alluvium, deposited by the Thames in former times. Low lying, they often had drainage problems in the winter. Narrow lanes and paths gave each gardener access to the Thames.[31]

No plans of commercial gardens dating from the seventeenth century exist, but we can reconstruct them from illustrations in gardening books and inventories taken by appraisers methodically going round the gardens noting equipment and growing crops. Crops were grown in beds – usually quite narrow so crops could be tended from each side. The mud walls round gardens often had sheltered narrow beds beneath them. Sometimes the beds were boarded up to form raised beds which might also be 'hot-beds'.[32]

The boats which were navigated down the Thames for the London markets returned full of dung to contribute to the high output achieved by the gardeners. One observer thought so much dung was dug into the Neat House soil that it was totally renewed every three or four years. Dung and glass were the two major reasons for the Neat House gardens' success. Dung was carefully managed – formed into long mounds and regularly turned, it was ready for general use when it was cold and largely odourless. Much glass was used by the gardeners, either bell-glasses (bell-shaped glasses for covering individual plants) or glass-lights (rectangular frames with flat panes of glass set in them which were placed on top of rectangles made of boards). With this technology plants were grown out of season, and exotic crops could be grown in this way.[33]

Hot-beds were the foundation of the Neat Houses' fame as market-gardens. They were made by placing a layer of rotting (i.e. fermenting and hot) dung on the ground, sifting several inches of fine soil on top of this, and planting crops in the topsoil. The dung warmed and fertilized the soil above it. If a layer of glass – bell-glasses or glass lights – was added above the plants, one had, in effect a 'mini-greenhouse'. In the words of Richard Bradley, 'A Hot Bed is the common Help made use of by gardeners to forward the Growth of a plant [. . .] when the Season itself is not warm enough.'[34] Hot-beds could last up to two months, and the gardeners' skill was in not allowing them to get too hot or, if they cooled, carefully renewing the layer of rotting dung.

The rise in the use of glass in the seventeenth century thus gave another 'jolt' to market gardening. It was possible now to protect plants from the weather while allowing sunlight to enter. Although patents for glass manufacture had been issued in 1552, the price of glass made it uneconomic for large-scale use in gardens until the end of the seventeenth century. Amounts used were initially small. At the Neat Houses, the following valuations of glassware were made when gardeners died: William Pearce, 1679 £18; Philip Luke, 1684 £16; John Lee, 1684 £3 2s; Kathleen Weston 1687 £2 10s. Weston's glass consisted of '7 doz Old Glasses, One Dozen of Bell Glasses & two ould Glasse fframes'.[35]

By 1695 technical advances in glassware production halved the cost of bottle-glass compared with a few years before, and this change is reflected in the increased use of glass. Robert Gascoine of the Neat Houses, who died in 1718, had three gardens full of glass. Each had boxes and lights, and he had a total of 1240 whole bell-glasses.[36] Evidence of widespread commercial use of glassware by mid-century comes from details of damage from the market-gardeners south of the Thames who were caught by a freak hailstorm in July 1750. Forty-three gardeners' losses were quantified in inventories, and in three-quarters of them damage to glassware figures prominently (their crop losses were mostly cucumbers grown under glass). Total damage to glassware was over £2800; eleven individuals lost over £100 worth of glass.[37]

Richard Weston, in 1773, gives details of the use of glassware to grow garden crops in market gardens the year-round. He remarks that 'the large expences attending them require some extraordinary profit; nor are there many gardeners who [. . .] can afford to sink so much money as the glasses cost'. After describing growing several crops in succession on the same ground, Weston estimates the cost per acre at £30, and the produce at £120 5s 10d.[38] Earlier, in 1721, Richard Bradley found that the Neat Houses 'abound in Salads, early Cucumbers, Colliflowers, Melons, Winter Asparagus, and almost every Herb fitting the Table'.[39] He singled out Mr Jewel, a Neat House gardener who sent the 'fine sort [of asparagus] To market first, about the 14th of May' [1723] and was the 'first gardener in England that raised the young Sallad Herbs for the Winter Markets, and Kidney Beans in Hot beds'. Bradley thought 'there is no where so good a school for a kitchen gardener than this place'.[40] Just a few miles away in Fulham, in the mid nineteenth century a piece of ground might have cabbages planted 25th October. After this crop was cleared the ground was trenched for celery, with lettuces in rows in between. In March it was dunged and sown with onions and lettuces. When the onions came off it was trenched and sown with cauliflowers.[41] With intensive cropping the Neat Houses paid rents in excess of what builders could afford until the 1830s. Their fame was enhanced by the number of people who visited them: several gardens were open in the summer, selling wine and fresh melons to customers who arrived by boat. In 1666 Samuel Pepys bought a melon there on a boating trip, and two years later he went with business companions 'to one of the Neathouses, where walked in the garden'.[42]

Another sort of jolt shook the market for tree fruit. Orchards were an important part of the rural economy of both Kent and the West Country, and royal patronage created demand for new types of tree fruit. Henry VIII employed agents to search abroad for such novelties. One agent, Richard Harris, was said to have been the first to introduce French grafts of cherry, pear, and pippin apple trees, setting up an orchard at Teynham in Kent. The royal gardens were some of the first in England to grow gooseberries. Wolf, another royal agent, probably introduced the apricot into England in 1524. In the next century, John Tradescant the elder, royal gardener to Charles I 'laboured to obtain all the rarest fruits he can hear of'.[43] Two areas of market gardening which were some distance from the main markets were operating from at least the eighteenth century: the Vale of Evesham and Sandy in Bedfordshire. These areas do not seem to owe their origins to any of the stimuli that have been mentioned. At Evesham small farmers with a surplus of family labour was an important factor, plus complaisant landlords who did not look closely at their production methods and did not object to deep digging. In the words of J.M. Martin, 'For humble men the attraction of gardening was that [. . .] it utilized what lay to hand: access to garden property and commons, family labour, and time.'[44]

Throughout this period, surplus produce from private gardens was sold in markets. Some monasteries sold their surpluses until they were dissolved in the 1530s. Two centuries previously there was a thriving garden produce market in London near St Paul's. For many years the gardeners of the 'Earls, Barons, Bishops, and citizens of London' were accustomed to sell their 'pulse, cherries, vegetables, and other wares to their trade pertaining', there. By 1345, however, this fruit and vegetable market had grown to such an extent, and had become so crowded, as to hinder 'persons passing both on foot and horseback', and the 'scurrility, clamour, and nuisance' of the gardeners and their servants had become obnoxious 'to the people dwelling in the houses of reputable persons there'. Casual traders, who only came to market when they had surpluses to sell may have swelled markets: twenty-five percent of the tolls from gardeners at Covent Garden in the mid-eighteenth century were from casual traders and many did not pay the toll. A commercial trade paper in 1843 claimed that unfair competition from private gardens was harming commercial growers.[45]

From the 1840s onwards, gardeners were faced with the biggest jolt of all to their businesses – the coming of the railways. There is no space to discuss this advent – suffice to say that distance to market and the state of roads were no longer barriers to exploiting favourable soils and climate. New market gardening areas arose: for instance, new potatoes and daffodils came to London from Cornwall, and strawberries from Warsash near Portsmouth.[46] Existing areas reached their full potential: as F. Beavington wrote of Sandy in Bedfordshire, the railways 'provided the essential link between the cities and the rural market-gardening districts which by lavish inputs of fertility sustained high outputs of produce'.[47]

About the Author
Malcolm Thick is a longtime symposiast and the author, most recently, of *William Ellis: Eighteenth-Century Farmer, Journalist, and Entrepreneur*.

Notes
1. Thomas Hill, *The Arte of Gardening* (Edward Allde, 1608), pp. 1–2.
2. Johann Heinrich von Thünen, *Der isolirte Staat in Beziehung auf Landwirtschaft und Nationalökonomie* (Wirtschaft & Finan, 1826).
3. John Middleton, *General View of the Agriculture of Middlesex* (B. Macmillan, 1798), frontispiece fold-out map; William Stevenson, *General View of the Agriculture of Surrey* (R. Phillips, 1809), frontispiece map.
4. Roy Porter, *London: A Social History* (Penguin, 1994), pp. 42, 97–98.
5. 'Public Intelligence: Manchester's Population Over Time', Manchester City Council <https://www.manchester.gov.uk/download/downloads/id/25393/a20_1086-2016_manchester_population.pdf> [accessed 21 March 2024].
6. John Stow, *The Survey of London* (Elizabeth Purslovv, 1633), p. 122.
7. C.W. Shaw, *The London Market Gardens* (n.p., 1879), p. 1.
8. Jane Whitaker, *Raised from the Ruins: Monastic Houses after the Dissolution* (Unicorn, 2021), pp. 7–8.
9. *Norwich Cathedral Priory Gardens Accounts 1329–1530*, ed. by Claire Noble (Norfolk Record Society, 1997), vol. LXI, pp. 70–75.
10. Qtd. in Malcolm Thick, 'Roots and Other Garden Vegetables in the Diet of Londoners, c.1550–1650, and Some Responses to Harvest Failures in the 1590s', in *Staple Foods: Proceedings of the Oxford Symposium on Food and Cookery 1989*, ed. by Harlan Walker (Prospect Books, 1990), pp. 228–35 (pp. 232–33). A 'last' is 1976 kg.
11. William Harrison, *The Description of England*, ed. by Georges Edelen (Folger Shakespeare Library, 1968), p. 264; Thomas Cogan, *The Haven of Health*, 1596, p. 3; John Parkinson, *Paradisi in Sole* (Humphrey Lownes and Robert Young, 1629), p. 509.
12. Ben Jonson, 'CXXXIII: On the Famovs Voyage', *Epigrammes* (W. Stansby, 1616), ll. 166–69
13. Thomas Fuller, *The Worthies of England* (J.G.W.L. and W.G. for Thomas Williams, 1662), p. 7.
14. Qtd. in Malcolm Thick, 'Root Crops and the Feeding of London's Poor in the Late Sixteenth and Early Seventeenth Centuries', in *English Rural Society: 1500–1800: Essays in Honour of Joan Thirsk*, ed. by John Chartres and David Hey (Cambridge University Press, 1990), pp. 279–96 (p. 294).
15. Samuel Hartlib, *Samuel Hartlib his Legacie* (J.M. for Richard Wodenothe, 1655), p. 9.
16. John Mortimer, *The Whole Art of Husbandry* (J.H. for R Robinson, 1716), vol. II, p. 134; qtd. in Malcolm Thick, 'Superior Vegetables' in *Food Culture & History*, ed. by Gerald and Valerie Mars (London Food Seminar, 1993), p. 140.
17. Joan Thirsk, *Food in Early Modern England* (Bloomsbury, 2007), pp. 171–77.
18. Henri Misson, *M. Misson's Memoirs and Observations in His Travels over England* (D. Brown and others, 1719), p. 125.
19. *Adam's Luxury, and Eve's Cookery, or, The Kitchen Garden Displayed* (R. Dodsley, 1744), p. 14.
20. Information gleaned from The Proceedings of the Old Bailey Online <https://www.oldbaileyonline.org> [accessed 12 April 2025].
21. Misson, p. 314.
22. Qtd. in Thick, 'Superior Vegetables' p. 135.
23. Qtd. in Thick, 'Superior Vegetables' p. 143.

24 Qtd. in Thick, 'Superior Vegetables', p. 143 n. 27; Richard Bradley, *New Improvements of Husbandry and Gardening* (J. Peele, 1721), p. 143.
25 Richard Bradley, *A General Treatise of Husbandry and Gardening* (T. Woodward and J. Peele, 1726), pp. 41–43, 108, 148. A sieve equals a bushel.
26 Old Bailey Proceedings Online, *Ordinary of Newgate's Account*, 18 March 1741 (OA17410318) <https://www.oldbaileyonline.org/record/OA17410318> [accessed 12 April 2025].
27 Bradley, *General Treatise*, p. 150.
28 J.C. Loudon, *An Encyclopaedia of Gardening* (Longman and others, 1824), para. 7514.
29 Stephen Switzer, *The Country Gentleman's Companion* (T. Astley, 1732), pp. 10–13.
30 Malcolm Thick, *The Neat House Gardens* (Prospect Books, 1998), pp. 79–88.
31 Thick, *The Neat House Gardens*, pp. 90–93.
32 Thick, *The Neat House Gardens*, pp. 98–101.
33 Thick, *The Neat House Gardens*, pp. 101–03.
34 Bradley, *New Improvements*, p. 103.
35 Thick, *The Neat House Gardens*, pp. 104.
36 Thick, *The Neat House Gardens*, pp. 105–06.
37 Thick, *The Neat House Gardens*, pp. 106–07.
38 Richard Weston, *Tracts on Practical Agriculture and Gardening* (S. Hooper, 1773), p. 54.
39 Thick, *The Neat House Gardens*, p. 107.
40 Bradley, *New Improvements*, pp. 108, 116, 127, 138, 156, 160.
41 James Cuthill, *Market Gardening Round London*, 185, p. 1.
42 Thick, *The Neat House Gardens*, pp. 127–30, 147; Samuel Pepys, *The Diary of Samuel Pepys*, ed. by H.B. Wheatley (G. Bell, 1920), vol. V, p. 365.
43 Joan Thirsk (ed.), *Chapters from The Agrarian History of England and Wales* (Cambridge University Press, 1990), vol II, pp. 49–51; Malcolm Thick, *William Ellis: Eighteenth-Century Farmer, Journalist, and Entrepreneur* (Hatfield, 2022), pp. 50–51.
44 J.M. Martin, 'The Social and Economic Origins of the Vale of Evesham Market Gardening Industry', *Agricultural History Review*, 33.1 (1985), pp. 41–51.
45 Alicia Amherst, *A History of Gardening* (Bernard Quaritch, 1895), qtd. in Malcolm Thick, 'The Sale of Produce from Non-Commercial Gardens in Late Medieval and Early Modern England', *Agricultural History Review*, 66.1 (2018), pp. 1–17.
46 Indy Almroth-Wright, 'The Strawberry Coast', BBC News: Hampshire and Isle of Wight, 27 June 2008 <https://www.bbc.co.uk/hampshire/content/articles/2008/05/03/history_strawberry_feature.shtml> [accessed 21 March 2024].
47 F. Beavington, 'The Development of Market Gardening in Bedfordshire, 1799–1939', *Agricultural History Review*, 23.1 (1975), p. 4.

40
The Emperor, the Major, and the Jackfruit Tree Disaster

Marcia Zoladz

Figure 1. A view of Rio de Janeiro from the Tijuca Forest. To the left, the Corcovado Mountain, with the Sugar Loaf in the middle, across the Guanabara Bay. Photo: Marion Strecker.

O pomo que da Patria Persia veio,
Melhor tornado no terreno alheio.

The fruit arrived from the land of Persia
turns out better in alien ground
—Luís de Camões in 'Os Lusíadas'[1]

As Luís de Camões's poem implies, the effort to control newly conquered South American regions by Portugal included the domination of local nature – animals, trees, grasses, plants, flowers, parasites – or even its substitution, with the removal or conversion of local populations. One of these substitutions, an exotic plant introduced in a new land, did not turn out well.

In May 1500, Pero Vaz de Caminha, a member of the first Portuguese expedition to Brazil, sent a letter to the King describing the voyage from the moment they left Lisbon until they moored in the south of today's state of Bahia.[2] In the text he describes local inhabitants and writes about the landscape and its variety of plants. And, surprisingly given his lack of precise knowledge about the quality of the soil, he wrote, perhaps impromptu, 'em se plantando tudo dâ', which freely translated means everything planted in this soil will thrive. To the initial goals of colonialism – taking possession of the lands and converting the inhabitants – he added growing an abundant harvest.

However, looking back after five centuries, it is clear that not every attempt to relocate plant species was successful. The first Portuguese colonizers brought with them European and exotic seedlings, some of them, such as rice, already adapted in Europe.[3] Other plants arrived later from their possessions in Asia.

The jackfruit tree – which triggers ongoing discussions in Rio de Janeiro – was among several species imported, but it turned out not to be central to the colonial enterprise. However, the trees are a good example of the terrible conundrum in the city: on one hand, they must be contained because of their extensive expansion; on the other hand, five hundred years after their arrival, should they be considered an exotic or a local species? As someone born in Rio, I can attest that this dilemma remains present in the life of its citizens.

The seeds or seedlings arrived in Brazil at two different times. The first ones arrived from Goa, an important commercial hub of the Portuguese Empire in Asia, around 1680.[4] This importation was part of a plan to introduce new crops in the South American colony since its production of sugar, then its main crop, was suffering from competition with French and British colonies in the Caribbean. It seems the tree was also brought to Rio de Janeiro in 1809, along with other specimens from the *Habitation Royale des Épiceries*, known as La Gabrielle, in Cayenne, French Guiana.[5] At the time Rio de Janeiro was controlled by the Portuguese government as part of a political alliance with the British during the Napoleonic wars. The seedlings and seeds were sent to the botanical garden of Belém, in Pará, with a recommendation from the crown to send them forward to newly established botanical gardens in other cities. The tree was probably introduced in Rio de Janeiro itself around this time.

Nowadays, in Rio, jackfruit trees are in parks, in sidewalks, in the surrounding mountains, in parking lots, in squares, in backyards, in college campuses, and in the Atlantic Forest surrounding the city.

Why Was the Jackfruit Tree Sent to Brazil?

Given the gigantic spread of the species over five hundred years in Brazil, one is bound to ask why the tree was sent from Goa to Brazil in the first place. The answer lies in the plant itself, but also in understanding the workings of a colonial empire. Colonial business in the Luso-Brazilian colony at the start of the sixteenth century consisted mostly of exploiting wood. The main tree species exported were rosewoods – *Pau-Brasil* (*Paubrasilia echinata* (Lam.) Gagnon, H.C. Lima & G.P. Lewis) – a type of timber used for furniture and to make a red/orange dye in the textile industry. Only later, at the end of the sixteenth century, did the Portuguese Crown start to invest in the commercial viability of its colony.

But for Portugal and Spain, the largest spice merchants of the early modern age, the lands in South America, especially in the Amazon region with its tropical climate, were understood as an opportunity to expand spice production. The new arrivals list included samples of cinnamon (*Cinnamomum verum*), black pepper (*Piper nigrum*), cloves (*Sygyzium aromaticum*), and the jackfruit tree (*Artocarpus heterophyllus* L.).

For a Portuguese bureaucrat in India, the jackfruit tree seemed to have a good possibility to thrive in another colony with similar hot and humid weather. In Goa, and in many other cities in Asia, jackfruit is a native species. Its wood was widely used in construction, for making furniture, and for carving decorative objects. However, five hundred years later one can argue whether its wood was really valued, especially compared with the huge variety of native trees in Brazil.

Whyever they were brought to the colony, jackfruit has had a long interaction with the city's population, especially in the more populated region where the city meets the surrounding forest. There, as described in an article about the jackfruit tree's spread, geographers Victória B.S. Ferreira and Isabel Ávila found vestiges of thirteen old charcoal ovens from the nineteenth and twentieth centuries.[6]

What About the Tree?

The adaptation of botanic species was understood as an important part of colonial business, and it started very early within the commercial relations among the Portuguese colonies.[7] There was a constant search for botanical treasures that could thrive in other lands besides their original environments. The first experiment started in the Madeira islands in the fifteenth century, when the business model of combining sugarcane farms with sugar mills producing molasses and sugar in the same place was perfected. The key was the intensive use of enslaved African workers.

By the middle of the sixteenth century, the possibilities of business were expanding, and there were already inquiries and lists of plants circulating inside the Portuguese Empire. One of the first was a 1563 book by Garcia Orta, *Coloquio dos Simples e das Drogas da Índia* (*Colloquies on the Simples & Drugs of India*).

This mix of searching for botanical knowledge with imperialistic expansion was well underway at the end of the seventeenth century. Until the start of the eighteenth century, however, Brazil was visited by single travellers, sometimes monks or priests like Jean de Léry who arrived with the French in Rio in 1555.[8] There were also adventurers and soldiers like Hans Staden, who described in detail the everyday habits, including cannibalism rituals, of the *Tupinamba*, who lived in Rio de Janeiro and along the coast south of the city.[9]

At the end of the eighteenth century these early travellers were joined by official scientific expeditions. Botanists, entomologists, astronomers, doctors, ichthyologists, and other specialists moved around the country, accompanied at first by artists and later by photographers, to document and collect specimens. Every landscape, fauna, and flora, was described, published in books, transformed into atlases, and made into etchings to decorate walls all over Europe. The time when the New World became known in Europe through a new literary genre – the travel book – was also the moment when Carl Linnaeus (1707–1778), using specimens received from professional and amateur botanists around the world, created his system for the classification of the species. Not long after, Alexander von Humboldt (1769–1859), while traveling in South America, observed that altitude was an important factor in the adaption of species.

Importing plants had two goals: commercial exploitation and enlarging local food production. Okra, cowpeas, bananas, coconut, rice, oranges, lemons, and oranges were added to the list of native plants, such as cassava, corn, guava, pineapple, cashew, passion fruit, and various beans. By the second half of the eighteenth century, the country started to produce coffee.[10]

Jackfruit must have seemed an obvious addition. It was an important protein source in its original region in Asia, where meat is often a religious taboo. Its leaf has anti-inflammatory qualities, and the latex extracted from unripe fruit is similar to rubber sap (studies of its qualities are ongoing). The raw seeds are not edible, but when cooked taste like chestnuts, and they can be cooked and salted, or cooked in a sugar syrup, like *marrons glacés*.

The jackfruit tree proved to be a tropical plant that was well adapted in the Amazon region, and it grew well along the coast of Brazil, and later inland. Most of today's population does not know that it is an exotic plant – just like nobody thinks mangoes came from India, or okra from Africa, as they are thoroughly integrated in Brazilian foodways.

What Is a Jackfruit Tree from a Brazilian Point of View?

The jackfruit tree (*Artocarpus heterophyllus*) is an exotic tropical plant in Brazil. It can reach twenty metres high, and its trunk easily achieves a diameter of one metre. The tree starts bearing fruit four years after being planted, and in the right environment it is quite prolific as it fructifies during all its life, which can stretch one hundred years or more. The largest producing countries are India and Bangladesh, with an average 1.4 million metric tons of fruit annually. In Brazil, according to the Tropical Fruits division of EMBRAPA, the Brazilian government's enterprise for developing agriculture, the market is not developed enough to record production or sales numbers.[11] Instead, production is distributed in backyards and by small farms in rural regions, and in cities is quite common to see jackfruit sold in the streets and squares.

Each tree can produce an average of eighty fruits per year, and some produce up to one hundred and fifty fruits a year. It is a summer fruit: the harvest starts in November in some states in the North and Northeast, and it goes until April or May in the Southeast and South. The fruit is quite unique, and the *Encyclopaedia Britannica* describes it in a very clear and simple way:

> Jackfruit is the largest tree-borne fruit in the world, reaching up to 60 cm (about 2 feet) long and weighing up to 18 kg (about 40 pounds). It is ellipsoidal and aggregate, composed of multiple 'bulbs' of seed-containing flesh around a stringy core, all of which is enclosed by a bumpy rind.[12]

Jackfruit is difficult to handle. Aside from its huge size, it is not very easy to cut it open as the peel is quite thick, more like a bark when not fully ripe. The fruit is not only difficult to open, but when cut it also releases a sticky glue (the latex) making it tricky to handle. Videos showing the best way to cut a jackfruit open without getting one's fingers stuck in it have become common on YouTube.

The bees love it, and monkeys and opossums are main reasons for its prolific propagation in cities and forests alike. When a fruit falls, each one containing more than five hundred seeds, each one of them is like a seed bomb, as a friend of mine put it.

The Jackfruit Is Not an Easy Fruit

Nina Horta (1939–2019), a well-known food chronicler for the Brazilian newspaper *Folha de S. Paulo*, has one of the best descriptions of jackfruit in one of her books – moreover, her text explains our incapacity of absorbing and profiting from its many qualities. Horta questions her own difficulty to define the fruit: 'is it a rhino or an elephant turned into some fruit?'[13] This sort of perplexity is quite a common feeling in Brazil whenever some exotic plants are mentioned. There are other species with similar controversial opinions, such as okra (*Abelmoschus esculentus*), gherkins (*Cucumis anguria*), and scarlet eggplant (*Solanum aethiopicum*).

The flavour of the jackfruit flesh around the seeds has a unique smell, quite strong, somewhat close to a mix of sugar and ammonia when ripe. The levels of love or revulsion for the fruit in Brazil varies, as some people experience the ammoniac like taste stronger than others. The flesh around the seed is very soft in the mouth when the fruit is ripe: it sort of slides on the tongue. Beloved by some, hated by others, but one rarely hears an 'I don't know . . .' response to its flavour and texture.

Here the fruit is less often used while unripe in savoury dishes, which is the way it is commonly consumed in its native countries. In Brazil, the green jackfruit meat, as the pulp is called in Portuguese, is sold as an animal protein substitute in vegan supermarkets. According to the newspaper *O Estado de São Paulo*, in an article of August 2023[14], about seven million persons declare themselves vegan or vegetarians in the country – a small number in a population of over 203 million, according to the 2022 Brazilian Government census.

The country has been since the start of the sixteenth century the largest producer of sugarcane of the world – according to the National Council of Food Provision (CONAB), over 610 million tons of sugarcane harvested in 2022/2023 were used for making sugar and ethanol.[15] This historically large stock defined the palate of the inhabitants, leading towards an excessive use of sugar in preserving both native and exotic fruits, but also in most of the traditional desserts.

Cakes, puddings, ice-creams, fruit pastes, and fruits comfits use an inordinate amount of sugar when compared with similar recipes prepared in Europe. Brazilian sweet recipes are strongly influenced by the conventual confections made in Portuguese Catholic convents, which in Portugal consist of sweets prepared with egg yolks and sugar. In Brazil, coconut and cassava were added to the original recipes, creating new ones.

Even avocados, part of the savoury food repertoire in its original region of North America, is used in ice-creams, or in a typical smoothie with milk, sugar, and a little lime juice. As to the jackfruit, one can buy them as a compote, a marmalade, or comfit in syrup. However, it is quite difficult to buy the raw fruit in supermarkets, after all they are available just around the corner, or in quite a few backyards.

How the Jackfruit Disaster Began

At the end of the nineteenth century, the main fresh water source for the central part of Rio de Janeiro dried out, due to intensive deforestation in its surrounding mountains. The city was founded in 1565, between Guanabara Bay, a large arm of the Atlantic Ocean surrounded by a narrow strip of land, and the foot of a central chain of mountains. The water supply in Rio de Janeiro was always problematic, as the mountains, where most of the water came from, were exploited at first for their valuable wood, then used for sugar-cane farms, and after 1760 by coffee plantations.

The main water supply of the city, the Carioca River, was used since before 1500. The river even gave its name to the Indigenous population in Rio – Cariocas. It starts in the Tijuca Mountains, where there is a Park today, and meanders down until it reaches the bay of Guanabara. Just before reaching the plains, part of its water was captured and sent towards an aqueduct built in the centre of the city in 1744. As the water reached the city, it was distributed to several downtown neighbourhoods by a system of fountains.

The history of the city of Rio de Janeiro records several projects to solve its constant water supply problem. The first one, after the aqueduct was built, was undertaken in 1861. It was quite a novel proposal – to reforest the region – from the Emperor Pedro II (1825–1891). The plan consisted of buying the devastated lands in the mountains from its owners and declaring them of public interest. The reforestation project of the city slopes around its water sources started in the region where the Tijuca Forest is today – at that moment completely devastated. In a rather symbolic way, the region was named a 'protective forest', as its mission was to save the water resources of the city.

The Emperor contracted his friend Major Archer (1821–1907), who was well known for his interest in forestry activities. For the next twelve years his workers planted one hundred thousand trees. They included native and exotic species, although in his final report the jackfruit is not mentioned.[16] It is possible that, by the second half of the nineteenth century, the tree was already established in the plantations and gardens in the mountains around the city.

The Major and the Emperor are popular characters in the imagination of the population, even today, after the end of the empire in 1889, as their project offered to its inhabitants a strong identity as nature lovers. It is as much a symbol of the city as the beaches along the coast.

Later, at the end of the nineteenth century, there was a renovation of the forest area in Art Nouveau style, including resting places to enjoy the view with such suggestive names as the Emperor Table or the Chinese View.

The plan worked quite well, and today Rio de Janeiro has one of the largest urban forests in the world. It is integrated with the life of the city as a recreation area, but, more importantly, the forest grew, and today the National Park of Rio de Janeiro, which unites other conservation areas, includes 39.58 km² (3972 hectares).[17]

The Carioca River does not provide water to the city inhabitants anymore, but it supplies the forest with water, and it plays an important role in containing the mountain slope. As it meanders down from the mountains it is polluted by illegal sewage, so much so that in 2016 it was classified as sewage by the city officials.

Combating the Spread of the Jackfruit Trees

According to Anderson Araujo, the spread of the jackfruit tree is being controlled by different actions: 'In the last five years, 55,662 seedlings were uprooted, 1921 small jackfruit trees were cut, and 881 ones were girdled.' Such measures are necessary to curtail such an invasive spread, notably because the seeds of the fruits attract monkeys and coatis (*Nasua nasua*). Besides the increase of the growth of the area populated by the trees, there is a secondary but equally important effect: the animals who eat its fruits also eat the birds' eggs. Araujo notes that the forest is losing its birds and becoming silent.[18]

As an exotic plant in Brazil, the jackfruit tree does not have natural predators, such as parasites, and therefore they live longer and grow larger, casting shadows which allow neither undergrowth nor the fauna and flora that usually thrive close to large trees. Their size dislocates native trees: with so many seeds, the large number of seedlings does not allow the development of local species. Their roots are a problem because they have allelopathic qualities – they do not allow undergrowth around a tree's perimeter. The roots also impact the lives of citizens, lifting sidewalks and undermining house walls. Because of the weight of the fruits (some can achieve twenty kilograms or more), they simply fall when ripe, damaging cars parked underneath the trees. Their number is excessive: a census of the trees indicated that in some areas of the forest there are two hundred jackfruit trees per hectare, a reasonable number would be up to one hundred trees.[19]

The spectacular spread of the trees in Rio de Janeiro is such that TripAdvisor, the tourism site, even offers a visit to a wood of jackfruit trees.[20] However, in a city classified as a World Heritage site by UNESCO, with an important part of its revenue acquired from tourists' visits, that does not seem like a good idea. Commentary on the site proposes eliminating them to allow space for native species.

To limit jackfruit trees' growth, since 2000, the Tijuca National Park has established a management programme. They systematically girdle a certain number of them. The method consists of cutting a ring in the bark around the trunk, thus preventing the circulation of the sap, and after a few years the tree dies. The environmental engineers of the park have tried other methods too, such as increasing the local fauna: they reintroduced the Brazilian agoutis (*Dasyprocta leporine*) a seed eater, in the hope they would spread other trees besides the jackfruit tree.

The Difficulties around the Jackfruit Trees

The environment of the Tijuca Forrest has another menace besides the spread of the jackfruit trees. The constant growth of legal and illegal buildings in the mountains around its perimeter is narrowing the limits of the forest, forcing animals, such as monkeys, to move to the periphery of the woods and into urban areas. These monkeys include the native robust capuchin (*Cebus Apella*) and the common marmoset (*C. Jacchus* L.), originally from the northeast of the country. The common marmoset is sweet looking, with white tufted ears, but it spreads toxoplasmosis, rabies, and yellow fever, according to biologist Helena Bargallo, in a 2008 interview in the literature and politics magazine *Piauí*.[21]

Figure 2. Jackfruits form clusters in the tree trunk at the farm of Nilda de Oliveira Almeida, in Cruz das Almas, Bahia, 2024. Photo: Nilda de Oliveira Almeida.

It took some time for the city administration to understand the full extent of the problems, the destruction of environmental diversity, caused by spread of jackfruit trees. The expansive growth of the trees is still happening in 2024.

Is It an Exotic Specimen for the Population?

The geographers and environmentalists Victória B.S. Ferreira and Isabel Ávila have proposed an interesting question: they think it is time to change how the city understands the jackfruit. They question whether the large numbers of trees should not be understood as part of environmental change, whether the larger number of jackfruit trees in the Tijuca Forrest should not be embraced. After all, the trees have been around hundreds of years, and they have a cultural value for the population. Instead of considering the trees as invaders of the land, maybe today they should be considered as a part of the environment in the city.[22]

The fact remains, though, that for the moment the growth of the jackfruit forest is understood – by specialists and the general population alike – as a menace to the forest and to the city's identity.

About the Author

Marcia Zoladz is food-writer and historian born in Rio de Janeiro, and is based in São Paulo. She writes about Brazilian foodways, its ingredients, and the several influences that makes it so delicious.

Notes

1   Luís de Camões, *Os Lusíadas* (*The Lusiads*), canto X, verse 58 <https://www.gutenberg.org/cache/epub/3333/pg3333-images.html> [accessed 15 April 2024]. Camões was a soldier who travelled extensively in

Asia and was a strong advocate of colonial conquests in his epic, the most important poem of the Portuguese language. The text cited here refers to the fights of the Portuguese against the Arabs during the conquest of Ceuta, in Africa.

2. Pero Vaz de Caminha, 'Letter to the King D. Manuel of Portugal', 1 May 1500 <http://www.dominio publico.gov.br/download/texto/bv000292.pdf> [accessed 15 April 2024].

3. Warren Dean, *A Botânica e a Política Imperial: Introdução e Adaptação de Plantas no Brasil Colonial e Imperial* (IEA – USP, 1989) <https://www.iea.usp.br/publicacoes/textos/deanbotanicaimperial.pdf> [accessed 15 April 2025].

4. The first mention of the jackfruit tree in the Portuguese Empire was by Garcia Orta (1501–1568), a physician who lived in Goa and in Bombay. *Coloquio dos Simples e das Drogas da Índia* (*Colloquies on the Simples & Drugs of India*) was published in 1563. Organized in alphabetical order, the *Coloquio* was written as a dialogue between two fictional characters, questioning each other about plants and their uses.

5. Nelson Sanjad, *Os Jardins Botânicos Luso-Brasileiros* (*The Luso-Brazilian Botanical Gardens*), Ciência e Cultura, 12.1 (2010) <http://cienciaecultura.bvs.br/scielo.php?script=sci_arttext&pid=S0009 -67252010000100009> [accessed 15 April 2024].

6. Victória B.S. Ferreira e Isabel Ávila, *História Ambiental da Jaqueira (Artocarpus Heterophyllus Lam.) no Maciço da Tijuca: Introdução, Distribuição e Produção de Novos Ecossistemas* (Departamento de Geografia, Pontifícia Universidade Católica – PUC, s.d) <https://www.puc-rio.br/ensinopesq/ccpg/pibic/ relatorio_resumo2015/resumos_pdf/ccs/GEO/GEO-3433_Victoria%20Ferreira%20e%20Isabel%20Avila .pdf>[accessed 15 April 2024].

7. Nelson Sanjad, *Os Jardins Botânicos Luso-Brasileiros* (*The Luso-Brazilian Botanical Gardens*), Ciência e Cultura, 12.1 (2010) <http://cienciaecultura.bvs.br/scielo.php?script=sci_arttext&pid=S0009 -67252010000100009> [accessed 15 April 2024].

8. Jean de Léry, *Histoire d'un voyage fait en la terre du Brésil, dite Amérique* (La Rochelle, 1578) <https://fr .wikipedia.org/wiki/Histoire_d%27un_voyage> [accessed 15 April 2024].

9. Hans Staden was a German soldier who travelled twice to Brazil, in 1547 and in 1549. He described his life with the Tupinamba in a bestselling book: *Die Wahrhaftige Historia und Beschreibubg eyner Landschafft . . .* (*The True History and Description of a Landscape . . .*). In 1593, Theodor de Bry, a German engraver and editor, published the book with a collection of etchings that included a series of descriptions of cannibalism. <https://digital.bbm.usp.br/handle/bbm/4570> [accessed 15 April 2024].

10. Marcia Zoladz, 'Cacao in Brazil or the History of a Crime', in *Food and Morality: Proceedings of the Oxford Symposium on Food and Cookery 2007*, ed. by Susan Friedlander (Prospect Books, 2008), pp. 309–20.

11. Jackfruit production numbers are not consolidated, but the following site is quite helpful with the values of their export and import in Brazil: 'Jackfruit', Tridge <https://www.tridge.com/intelligences/jackfruit/ BR> [accessed 15 April 2025].

12. 'Jackfruit', *Encyclopaedia Brittanica Online*, 24 March 2025 <https://www.britannica.com/plant/ jackfruit> [accessed 15 April 2024].

13. Nina Horta, *O Frango Ensopado da Minha Mãe, Crônicas de Comida* (*The Braised Chicken of My Mother: Food Chronicles*) (Companhia Das Letras, 2015), p. 71.

14. Redação Paladar, 'Veganismo no Brasil: Vegetarianos e Veganos se Multiplicaram nos últimos 10 anos', Paladar <https://www.estadao.com.br/paladar/radar/veganismo-no-brasil-vegetarianos-e-veganos-se -multiplicaram-nos-ultimos-10-anos/> [accessed 15 April 2025].

15. Mariana Grilli, '*Cana de Açúcar: produção de 2022/2023 cresce 5,4% aponta Conab*', Exame <https://exame .com/agro/cana-de-acucar-producao-22-23-cresce-54-aponta-conab/> [accessed 15 April 2025].

16. Major Archer's report about the accomplishments of his forestry efforts with a list of the trees he planted is at Fundação Biblioteca Nacional, Rio de Janeiro, Manuscript section #1427958.

17. The Parque Nacional da Floresta da Tijuca (National Park Floresta da Tijuca) is a good source of information: <https://parquenacionaldatijuca.rio/historia-do-parque-nacional-da-tijuca/> [accessed 25 April 2025].

18 Anderson Araújo, '*O Perigo das Jaqueiras*' ('The Danger of the Jackfruit Trees'), *Tijuca-RJ – O seu Bairro na Internet*, 3 May 2009 <https://tijucarj.wordpress.com/2009/03/05/o-perigo-das-jaqueiras/> [accessed 15 April 2024].
19 Douglas Duarte, '*Assassinas! A Floresta da Tijuca caiu na mão de alienígenas*' (Murderers! The Tijuca Forest is in the hands of aliens), *Piauí*, 21 June 2021 <https://piaui.folha.uol.com.br/materia/assassinas/> [accessed 13 April 2025].
20 For a while TripAdvisor listed a tour of an area called the Jackfruit Trees Wood, but someone posted a comment that it is not an interesting place to visit: <https://www.tripadvisor.com.br/ShowUserReviews-g303506-d311246-r194208803-Parque_Nacional_da_Tijuca-Rio_de_Janeiro_State_of_Rio_de_Janeiro.html> [accessed 15 April 2025].
21 Duarte.
22 Ferreira and Ávila.